One Lucky Canuck
An Autobiography

David A. Barr

© Copyright 2006 David A. Barr.
All rights reserved. No part of this publication may be reproduced, stored in a retrieval system, or transmitted, in any form or by any means, electronic, mechanical, photocopying, recording, or otherwise, without the written prior permission of the author.

Note for Librarians: A cataloguing record for this book is available from Library and Archives Canada at www.collectionscanada.ca/amicus/index-e.html
ISBN 1-4122-0139-x

Printed on paper with minimum 30% recycled fibre. Trafford's print shop runs on "green energy" from solar, wind and other environmentally-friendly power sources.

Offices in Canada, USA, Ireland and UK

Book sales for North America and international:
Trafford Publishing, 6E–2333 Government St.,
Victoria, BC V8T 4P4 CANADA
phone 250 383 6864 (toll-free 1 888 232 4444)
fax 250 383 6804; email to orders@trafford.com
Book sales in Europe:
Trafford Publishing (UK) Limited, 9 Park End Street, 2nd Floor
Oxford, UK OX1 1HH UNITED KINGDOM
phone 44 (0)1865 722 113 (local rate 0845 230 9601)
facsimile 44 (0)1865 722 868; info.uk@trafford.com
Order online at:
trafford.com/04-0591

10 9 8 7 6 5 4 3 2 1

Table of Contents

Early Days	5
Upper Canada College	14
Uncloistered Halls	34
Horses & Packers	48
Introduction to Babies, Helicopters, Perrennial Snow & Ice, BC, 1954	52
Win Some - Lose Some	63
Exploration in Northwestern British Columbia – 1959	78
Exploration in Northwestern British Columbia and Northwest Territories - 1960	93
Galore Creek, B.C. 1961-65	104
The Berg and Lost Claims	136
African Safari	157
Labrador - Kiglapait, Harp Lake and Voisey Bay	179
DuPont of Canada Exploration	186
Early Characters	255
Resource Land Conflicts	279
The Windy Craggy Saga	295
Resource Land Conflicts Continued	312
The Sherwood Mine Saga, Strathcona Park, BC	338
Survival Adventures & Searches	358
Travel Trekking and Climbing	378
Astounding China - 1990	398
Blood Donations – Friends for Life	420
Sporting Events	425

In Retrospect	460
Acknowledgements	462
Selected Bibliography	463
Index of Proper Names	465
End Notes	477

Early Days

I am reasonably confident that I was born in Cali, Colombia on January 27, 1928. In fact, there is overwhelming evidence to this effect based on my birth certificate which duly notes my birth date and place of birth as stated; sex as male; name and surname of father as David Anderson Barr - monotonously the same as mine; name and maiden surname of mother as Edyth Maude Barr, formerly Willis. There are sundry other declarations, not the least of which was that my 'entry of birth' was the first in the Register of Births kept at this Vice Consulate and so witnessed and ascribed on August 1, 1928. Note the difference between the date of my birth and its official registration! Has anything changed?

Before proceeding further I would like to state at this very early stage of my so-called being and well before any recollections, that I am deeply appreciative of the events that must have transpired, presumably on or about the end of April, 1927 that led to my birth.

In addition, the idea of recording many of my subsequent experiences and those of mutual acquaintances, mostly involved in the mining profession, stemmed from a very limited edition written by my mother. It was called 'Those were the Days,' and was completed by Nina in 1973 when she was 79 years old and four years after the death of my father. It was printed for her as six small hard-covered booklets by Michael Anti-Rose of Duncan. Nina dedicated the book to her 'three very dear grandchildren'.

At the time of writing, which has spanned almost a decade in 1995, it would be highly presumptuous for me to anticipate dedicating a book on my recollections to grandchildren, which will explain my more guarded but enthusiastic dedication.

To my knowledge, both my parents were truly individualists who elected, as many others of British descent did, to sever their roots to explore and possibly settle in more distant lands, following World War I.

My father, born in 1900 in Steveston, near Kilmarnock, Ayrshire in Scotland, was the eldest of five children born to William Muirhead (1859 - 1909) and Mary Muirhead (1873 - 1956). William was one of 16 children born to James Barr (1802 - 1879) and Ann Montgomerie (1815 - 1895).

My father left a farming household at Steveston at the age of 19 to seek a future in radio-telegraphy in the New World. Although slightly built and of boyish appearance based on photos of the time, I recall him mentioning his being slurred by Canadian troops returning on his ship to Canada following World War I in 1919. The implication was that he should have been in uniform in spite of his youthful appearance. I really felt for him, considering the years passed, even though I had never been through a similar experience at age 17 following the end of World War II. I only recall being chided by an older classmate, when I expressed some regret at not having had an opportunity to be directly involved. Such was the patriotic fervour in Canada in 1945.

Dad spent a short time in Vancouver in 1919-20 before being relegated to telegraphic duties at Bamfield on the West Coast of Vancouver Island. I have several albums of his, which include photos of this period showing his obvious enjoyment of the West Coast. Included are beach scenes at English Bay and at least one photo of him on top of Goat

Mountain on the North Shore. He must have hiked there from the top of the old tramway on Lonsdale in North Vancouver, returning to Vancouver, all in a single day!

By late 1921 he had accepted a position in South America as a radio-telegraphist with All American Cables, a subsidiary of ITT. He was to spend the next 30 years in Central and South America with that company, eventually occupying positions as general manager in various Latin American branches in the latter part of his career.

My mother was born in 1894 in Swindon, Wiltshire, England, a date she never admitted to. She had two brothers and three sisters born to Harry Willis (1864 - 1933) a railway worker in that prominent railway junction, and Annie Peart (maiden name Goldsmith, 1869 - 1928). The Willis daughters were to be blessed with longevity, three of them living for 89 - 95 years. Pat, the youngest, born in 1907, was still very much alive in 2002.

My mother, although named Edith at birth was to have several nicknames including Midge, Nino and Nina. She generally preferred Nina. Her favourite sister was Pat, whom we have visited on numerous occasions and who still lives in Swindon in a senior apartment complex.

Nina's background is by no means as well documented as my father's who has remarkable photos supporting his many travels throughout western Canada in 1919 - 21, mostly in the lower mainland but reaching as far north as Whitehorse and north-easterly to Banff in that period. It's quite evident that Nina felt that she had no significant early recollections to pass on to her grandchildren – or anyone else.

A 30-year period is covered in her booklet in the following paragraph:

Born in England and loving that beautiful country, and never thinking to leave it, I took two years training in hospital as a nurse. Then suddenly – Would I like to go the Argentine, and so it came about.

My parents met in Argentina at Villa Mercedes Province of San Luis. While in Buenos Aires, following her arrival in Argentina and debating her future, she received a telegram from friends made on her trip from England, both of whom had been recently injured, asking for her help. She accepted and made her way 700 kilometres inland to Villa Mercedes.

My father was transferred to the general area at this time and in May 1925 they were married on the estencia where they had met a couple of years earlier. They were to spend an eventful 28 years in various countries in South and Central America before retiring to Duncan on Vancouver Island in 1952.

I have somewhat galloped through the years in the last paragraph. While in Villa Mercedes in 1926 with what were to be long-time friends, Becky and Herbert Cuttrell, my parents' first-born, John, arrived. Three months later they were transferred to Cali, Colombia where shortly after, John died.

My childhood memories of life in Cali are quite restricted. My parents' friends and acquaintances were virtually all foreign representatives of business concerns operating in Colombia. My mother recalls giving Christmas parties which would be attended by children of ten or twelve different nationalities.

My greatest friends were Gunar Lindahl of Swedish parents and Sheila Johnston, an American. One of my earliest and more vivid memories was cutting Sheila's hair, by mutual consent, when we were both about five. Although we were both delighted with the results, I must have had certain misgivings.

As my mother recalled, the episode occurred one afternoon while she was having a siesta - a Latin American custom, gratefully adopted by the 'European and North American' non-nationals. She was awakened by Sheila and I holding hands and Sheila saying,

"Nina, look what David has done", as we both smiled happily. Nina asked, "Where is the hair?", to which Sheila replied, "Oh, under the wardrobe."

My mother was non-committal in her recollections as to the consequences of this event, although I recall it as my big first and only "spanking" by my father. Sheila and I were to share much time together - including sailing with our parents on the same ship to England on one of my father's three-month leaves, which occurred every three years.

At one time Sheila announced to my mother, "Nina, I'm going to marry David." Before she could comment I apparently said, "You are not – I wouldn't marry a mummy with freckles!" When we left Cali in 1934 for Panama, I was never to see her again. However, by a strange coincidence I belatedly learned that she had been in a convent in Toronto while I was attending Upper Canada College in the 1940s.

My father was a prolific writer - although I was not to be aware of this until long after his death in 1969 when I was required to take possession of virtually all of my mother's effects prior to her death in 1985. Much of his life during the 1934-8 period following his transfer from Cali to Panama City as a Travelling Auditor for All American Cables, Inc. is described in letters he wrote to my mother at weekly to bi-weekly intervals. These were retained by my mother and I have only read them in some detail quite recently.

A letter to my father from the Assistant Comptroller for AAC in New York dated January 2, 1935 describes the position of Travelling Auditor. The first letter of 41 written to my mother between July 2, 1936 and April 8, 1938 was from Guayaquil, Ecuador and suggests that he had been away from Panama City for some time and was looking forward to returning home.

My mother describes little of our life in Panama City in her booklet. It must have been a lonely existence for her. She again describes this two to three-year period quite cryptically:

> Then finally we were transferred to Panama for two or three years. But David and I were alone and Grandpa saw very little of David after that as he was inspecting all the cable stations in the nearby and far away places. Then it was time to take David home to school.

In Panama, we lived in an apartment in a three-storey building near the Canal Zone overlooking a large undeveloped area with tall grass and shrubs. Beyond I recall a 'beer garden,' which was a quite lively dining place with loud music playing in the evenings.

My friends were again of mixed nationalities with a preponderance of Panamanians and Americans. I attended first and second grades of the 'Instituto Pan-American'. The school was the only one in Panama City providing a partial English education, influenced in the 1920s by the desire of many Panamanians to have their children learn English as well as Spanish. Classes were offered from grades 1 - 6 in the Grade School Section followed by a four year Commercial course.

In 1936 I was only one of two children in a class of 24 born of English-speaking parents. Attendance at the school must have influenced my preferred language as I recall being instructed by my father to remember to communicate with my mother in English!

We walked to school through part of Panama City and I can still remember a corner store which I frequently passed which exuded strong spicy odours. Although the smells were memorable from both markets and some stores, this particular store offered for sale shrunken human heads which left a lasting impression. Derived from Panamanian natives following local combats, these were actually displayed at the store entrance hanging from nails.

I can recall little of the school curriculum. However, two incidents provide an indication of the control or lack thereof exerted by teachers over students. One day a very large woman arrived with a cane and entered the class unannounced. The teacher and the students all watched in astonishment as she rapidly advanced down an aisle and ran after one of the students, a young boy, who tried to elude her. She finally caught him and marched him to the head of the class, where she held him up by one his arms and beat him several times on his behind with the cane. Finally, satisfied and without a word, she left the class, leaving her sobbing son behind.

On another occasion, we arrived at the class for an examination. Unfortunately, the teacher had caught someone cheating on a previous exam. The whole class had to line up and pass by her desk where she ripped off a piece of adhesive tape and applied it across each students' mouth. It would have made an unusual photograph. I can still recall the problem in peeling off the tape at the end of the class.

A more painful event also occurred to me while in Panama. Several friends and I became aware of grapefruit growing on several trees on private property in our residential area. The property was directly accessible through a gate attached to a wrought-iron picket fence. We thought it adventuresome to jump over the picket fence from several boxes piled on top of each other to try to pick off a grapefruit, exiting via the gate. Whether it was my first jump I cannot be sure, but it was certainly my last. In jumping, the top box dislodged and I was impaled through the back of my left knee on the fence. Apparently I yelled loud enough to attract two neighbourhood residents who pulled me off the fence and carried me home, where I was eventually attended by a doctor.

While we were in Panama, we experienced several earthquakes, but none as severe as some survived by my parents. The worst of them occurred at Guayaquil in Ecuador in the 1940s while I was at school in Toronto. About 400 people were killed including several friends of my parents.

Our apartment in Panama had several large cracks resulting from earthquakes. One night my father reached for a light switch on the wall in my parents' bedroom and was surprised to have the dark object move. He finally turned on a light and discovered a large tarantula on the floor which he killed with a slipper. My mother wanted to show it to me the next morning, but when we went to view it, all that was left was a leg. It had been taken apart by large ants which we could see disappearing into one of the cracks in the wall with the remains of the spider.

In May, 1937 we left Panama for England on one of my father's leaves. My parents had prepared over a two-year period for having me attend a boarding school in England. I doubt that I was aware of their plans until after our arrival there. We spent almost two months visiting my parents' families in Wiltshire and Ayrshire prior to my father leaving England at the end of July to return to Panama and later Bogota, Colombia.

They had planned to have me share the common plight of children of ex-nationals living abroad where English education was not available. They selected Milstead Manor, formerly Deal School and founded in 1820 as Sandwich School in Deal, Kent – a rural community about 60 miles northeast of Brighton, Sussex. Their selection was based on a recommendation from friends in South America whose two sons had attended the school.

Unfortunately the school had degenerated in the ensuing period and during my brief and unpleasant tenure which ended in early November, 1937, only seven students remained.

My parents had planned that Nina would remain in England during my first year. She settled into a flat in Hove, which adjoins Brighton where I visited her briefly on one occasion during a two-month period. My departure from Milstead Manor was apparently provoked by abuse from the head master and, I suspect, combined with homesickness, although the details are not specifically discussed in available letters and were not traumatic enough to be recalled.

I was fortunate in being relocated at the Brighton, Hove & East Sussex Grammar School, a well-established institution in northern Brighton near the edge of The Downs. There were about 30 - 40 boarders at the school and I made many friends there and had no problem accepting the boarding school environment.

My mother left Brighton to rejoin my father in Guayaquil, Ecuador in April, 1938. During her stay in Brighton she had developed a strong friendship with Mabel Pope, a woman of her own age who shared my mother's dedication to Christian Science. Miss Pope was to act as my guardian during my stay in England and I still remember her kindly nature most fondly.

As at most boarding schools of that period, our daily routine was quite regimented. We arose to the clanging of a bell at the same time each morning, attended meals in the dining hall at precise times, were encouraged to participate in late afternoon sports and attended supervised home-work period prior to bed time.

We were accommodated in dormitories with capacities for about one dozen individuals, each dormitory containing boys of equivalent ages. The dormitories were similar to old hospital wards, with beds arranged cheek-by-jowl on opposite sides of long, narrow rooms. I can still recall the bare radiators interspersed along the walls as they provided not only heat during the winter, but a warming spot for bread that we occasionally secreted carefully inside our jackets at mealtime to be eaten after lights-out. The bread was referred to as 'scrape' as it was pre-buttered, a barely visible streak on one surface, presumably after being scraped onto the bread from a source pat.

Food was always a welcome commodity as we never seemed to get enough. This was understandable during the latter part of my stay through July 1940 as a result of rationing following the outbreak of World War II. Our weekly menu as at most institutions of this type was predictable except for Fridays as the war progressed. We were instructed to touch nothing on our main plate courses, unless we could eat it all! Fridays' meal included a generous quantity of leftovers.

I was only caned on one occasion while at the grammar school and this was provoked by a meal-time incident. Another boy sitting next to me knocked over a glass of milk on the table, the contents of which rapidly advanced across the table in my direction. Unfortunately I was observed by a senior who noted that I was directing the flow away from myself to cascade over the edge of the table. The caning was not severe as it was applied after lunch on my rear end while still garbed in pants. Later canings at Upper Canada College were virtually all applied to boarders and were conducted on bottoms protected only by pyjamas. As I will undoubtedly recall later, some were particularly sadistic affairs applied by a housemaster named I.K. Shearer.

I was to live through several particularly dated periods during my early school years as events will unfold. One of them was the recognition even at the prospectus stage for boarders, that they were entitled to have 'tuck boxes'. What in the world were tuck boxes one might ask? They were wooden boxes equipped with locks designed principally for the storage of edible items. Again, their presence at boarding schools, generally in a basement area, suggested that the ravenous boarders were simply not getting enough to eat! Many of the schools also had 'tuck shops' where confectionery items could be purchased.

My only recollection of food items stored in my tuck box while at Brighton were pancakes baked by my Aunt Isa in Scotland and mailed to me at irregular intervals!

Among Nina's letters, report cards and informative documents forwarded to her from the grammar school was the following item in a 'Memoranda for Parents of Boarders - April 1938':

"All tuck brought back should be packed in a boy's tuck box, not among his clothes. The amount should be as small as possible, and preferably limited to jam or fruit. NO TINNED FOODS ARE ALLOWED; any that are found will be confiscated at once."

Although I had only my limited experience as a school boarder to draw on prior to arriving at Brighton, I never felt that the regulations at the grammar school were particularly onerous. Several of the general regulations perhaps revealed a personal bias by a person or persons in authority, as shown by the following examples excerpted from the above-referenced memoranda:

"Brown suits are not allowed for Sunday wear. Parents are asked to note that the permissible colours are black (coat and waist-coat), <u>dark</u> grey, or navy blue for senior boys only. LIGHT GREY FLANNEL SUITS ARE NOT ALLOWED FOR SUNDAYS."

"BROWN SHOES are NOT ALLOWED for School wear."

"Parents are asked in future to supply only black shoes for gymnasium."

"<u>Third Sunday</u>. Boys with parents or friends in Brighton may visit them on the "Third Sundays" without special leave asked except in the case of special invitations. Leave cannot be given at other times, except for special reasons, such as a visit of parents from a distance to Brighton, and then only on a written application to the Headmaster. Verbal applications through the boy cannot be considered."

The restricted visitation policy, although not applicable to me after Nina returned to South America in May 1938, appears particularly authoritarian, especially since only about 20 percent of the pupils were boarders. At Upper Canada College, with a similar percentage of boarders, many of whom actually had parents living in Toronto (which I could never fathom), boarders were free to spend portions of weekends away from the school and visits by parents were not nearly so restricted.

As an 11-year old in mid-1939, I was well entrenched in the school routine and like others my age, I was not overly concerned or necessarily knowledgeable about the deteriorating political conditions in Europe, attributable to Herr Hitler and the Nazi party's hopes of world-wide domination. However, following the outbreak of World War II, my father wrote Mr. Barron, the principal, expressing his concerns, and the possibility of my being moved to another school in Wales or the south-western part of England. He received reassurance from Mr. Barron in a letter in late October, that Brighton was relatively safe.

Anticipating war with Germany, the British government had already issued gas masks to the public. I happened to be with Miss Pope on September 3 when war was declared. That afternoon an air-raid siren sounded and we went into a garden outside her apartment with our gas masks, and soon heard an all-clear siren.

The school routine was seriously disrupted as early as October 1939 by the necessity of sharing the school buildings with another evacuated school. This led to a decision to have our school's classes in the morning and those for the evacuated school in the afternoon.

Although there were air-raid warnings in the early 1940s in the Brighton area, they were rare. Eventually anti-aircraft guns were established in the district, several on the edges of our playing field. However in this period I had little personal contact with others directly affected by the war. One new boarder I met was there as a ship evacuating him from England had been torpedoed just off the coast. Unbelievably, this had occurred to him on a previous occasion!

Arrangements to evacuate me from England were made by my parents in mid-1940. I was not to return to Brighton until May, 1977 when I arrived there from Canada during a vacation with Rene. We had no problem locating the school and I entered Marshall House - my old abode where a vice-principal of the school spent considerable time with us explaining what had transpired in my absence. Shortly after I left in 1940, Marshall House was occupied by personnel from the Canadian army who manned the anti-aircraft guns in the playing-fields. Following the end of the war, the school reverted to its pre-war curriculum and by 1950 there were about 60 boarders - similar to the number in 1940. However, by 1977 there were only 28 and the school had become co-educational with expanded facilities. Plans were in progress to form the school into a college, i.e. university in 1978.

In mid-June 1940, I left Brighton by train for Scotland where I stayed for several weeks with my grandparents at their home in Steveston, Ayrshire. This was shortly after the British troops and others were miraculously evacuated with such minor loss of life from Dunkirk.

A cousin was among the returning troops who arrived in Scotland during my brief visit. Like many others of my age, I had started a collection of assorted paraphernalia issued to the armed forces, which mostly included uniform buttons and badges along with defused and live ammunition. My collection was relatively small and I still have it. Several prized items were badges from the Royal Tank Corps, Royal Air Force, Royal Army Medical Corps, Royal Scottish Fusiliers, The Queen's Royal Regiment, 48th Highlanders and The Royal Artillery. Bullets included several 8 mm tracers and even an ancient bullet that had survived the Boer War.

In retrospect, I marvel at the generosity of the many individuals who were delighted to provide a small boy with a part of their uniforms that obviously had to be replaced. My cousin, like other survivors of Dunkirk came back with a minimum of equipment, having had to either wade out or swim to one of the incredible variety of vessels in the flotilla that rescued them. However, he had a clip of ammunition left, and insisted on defusing the bullet he provided to me as a souvenir.

I recall the spirit of the Scots being very high. An invasion by Germany was quite anticipated by the general public. I overheard an elderly Scots farmer on a bus saying that he would be ready for any German paratroops with a pitchfork in the backside. By then the sign-posts on all the country roads were being turned about or taken down to confuse the anticipated arrivals.

I spent the night before my departure from England in a hotel in Liverpool. That night there was an air raid - but I slept right through it, apparently much to the consternation of the management who were unable to waken me (and who presumably had no spare key?). I boarded the "S.S. Cythia" of the Cunard White Star Line along with hundreds of other evacuees - mostly children bound for North America.

We were escorted by a naval ship and partly by an aircraft for a couple of days, spending a total of eleven days crossing the Atlantic before reaching landfall in New York. The trip took far longer than the normal five to six days because of the zig-zagging course elected by the Captain. Friendships were rapidly made, even though they were to be of such restricted length. Midway through our voyage there was an abrupt commotion aboard ship which drew most of the passengers excitedly on deck, including me. A submarine had surfaced on that lonely sea not far from the ship - perhaps to identify itself to a friendly ship. It was British and the yells and cheers directed its way must have been heartening.

On arrival in New York on July 7th, I was met by pre-arrangement by Herbert and Becky Cuttrell, long-time family friends of my parents who had first met them at Villa Mercedes in Argentina at the time of their marriage. I spent several days with them at their home in Hempstead, Long Island, enjoying the company of Jerry and Patricia, the Cutrell's two youngest children who, fortunately, were on holiday. By a strange coincidence we all went to a movie one day in Hempstead and were startled to see me and other children aboard the S.S. Cythia, waving to the nearby camera on our arrival in New York.

I had a relatively uneventful trip on the Santa Maria, a Grace Line ship, through the Panama Canal to Guayaquil, Ecuador to meet my parents. However, our arrival on July 28th was far from uneventful and almost disastrous. There being no docking facilities for large ships, passengers and visitors were required to travel from ship to shore via small vessels. My parents arrived from shore alongside the ship in one of these, while I watched with other passengers from a deck above. It was necessary for those boarding to step from the launch on to a small platform at the base of a gangway suspended alongside the ship. Just as my father attempted to step over to the platform their launch was struck broadside by a large fire launch, obviously out of control. It tipped the other vessel over, throwing my father into the water by the platform. Pandemonium broke loose as other boarding passengers tried to reach the platform. Dad reappeared - apparently with a cigarette still dangling from his mouth. My mother was assisted on to the platform by another person. Such was my reception to Guayaquil!

Considering that I spent almost two months in Guayaquil, I have very few memories of that time. My father was an ardent golfer and he belonged to a golf club which had its course and club house about 10 kilometres inland and at a significantly higher elevation. Father had a strange but somewhat practical ritual which involved at least a weekly outing at the golf club to which he was driven by the company chauffeur - generally on a weekend. Following his outing on the course, he would walk back to Guayaquil. This was not for reasons of penance - it was to keep in shape.

The course was literally cut out of the jungle as I was to witness on at least one occasion. The golf course was frequented by residents other than golfers. I remember seeing caddies 'chasing' tarantulas which had inadvertently located an easy access through the jungle and were clubbed as though they were a significant threat, which was far from the case.

One hole on the course was particularly challenging although only a par three. Access was by a steep 'stairwell' cut through the jungle to the tee area, with a relatively narrow slot cut out to exhibit the green, which was surrounded by a moat filled with jungle debris. Your golf ball either landed on the green or you were obliged to have another shot - there was no other alternative except admitting to being a wimp who could not finish the game. Terminate your membership? In Guayaquil such conduct would have been socially unacceptable!

My father had considered various alternatives for my continued education, which included an English-speaking school in Chile and, mercifully, Upper Canada College in Toronto. The latter was a recommendation from a British Consul in Guayaquil who actually had a prospectus of the College in his office. Hey - UCC was founded in 1829!

Dad wrote a letter dated August 2, 1940 to the Principal of UCC, which described my background and included an application form. Time for a response to provide for my arrival in Toronto, if accepted for the fall term was a concern. In anticipation he requested a cable, charged to his company account in Guayaquil with the following comment, if I was accepted:

"(Deferred Rate) BARR GUAYAQUIL ACCEPTED"

A favourable response was received by cable on August 9 from Mr. McDermot, who had also referred his letter to Alan Stephen, Headmaster of the Preparatory School. In his letter Dad had made reference to his being "a Scot". Stephen, in a short response to Dad's letter included the following reference to clothing, which reflected his perceptive nature:

"If you are in any difficulty about clothing or equipment, our Scottish nurse will gladly make the necessary purchases for you in Toronto. I can promise that anything bought, will be well bought."

It was to take some time for me to become aware of my parents' financial circumstances. Basic annual fees for boarders in 1945-6, the year I graduated, were $850, compared with $300 for day boys. Clothing, medical and dental fees, and vacation requirements, probably added an additional $500 or more, depending on individual situations. In 1952, when I visited my parents in El Salvador, just prior to my father's early retirement, he confided that he was earning $7,000 U.S. per year. My education at UCC must have absorbed as much as 20 - 30% of my parents' annual net income.

In any event, I certainly sensed the need to help out financially at an early stage and beyond, managing to obtain summer employment from 1943 onward until graduation from UCC in 1946 on two different farms in Ontario under the War Time Farm Assistance Program, at a lumber mill and even some employment in the relatively short Christmas and Easter vacations on several occasions.

Arrangements were hurriedly made for my departure to Canada from Guayaquil via the "Santa Lucia", which sailed on August 30th and arrived in New York on September 9th. I stayed overnight with the Cuttrells who saw me off from New York by train for Toronto on the next day.

Dad had written Mr. Stephen concerning my travel arrangements which included a request to have me met at Union Station in Toronto. He wrote, *"To assist you in identifying him, he is quite a small chap, weighs about 4½ stone, wears glasses, but is otherwise normal"*. He followed this up with the following paragraph:

"I have noted in the College prospectus that you lay stress on the use of correct English, a point which I appreciate. David is inclined to be lazy in this respect, as, like many youngsters of his age, he has copied many Americanisms of which we do not altogether approve, and which we do not wish encouraged. He can speak with what my wife is pleased to call a "Cultivated English accent" – when he likes, and we should like his masters to insist on this point. He appears to be weak in maths as you will no doubt find out for yourself, but in other respects appears to be normally intelligent. As I had not seen him for three years, I may be inclined to be rather critical of his progress in education up to the present, but as you will have gathered from my original letter, he has had rather a mixed schooling, so perhaps one should not be too critical."

In my own defence, I cannot imagine having cultivated American slang, following three years in England and preceded by two years in Panama where I mostly spoke Spanish. However, I was not immune to this perversion, for years later in 1949 after having worked for a summer with a seismic crew in central Alberta composed almost entirely of Americans from Texas, Oklahoma and Arkansas, another Canadian and I had both developed noticeable drawls. However, I soon lost this affliction, which is more than I can say for either my mother or father who after 50 - 60 years in Latin America and Canada both maintained their native accents throughout their lives.

Upper Canada College

I was excited at the prospect of exposure to Canada with its vastness, wild Indians, Royal Mounted Police, mountains, ice and snow. It has surpassed my expectations until recently, as the aspirations of certain of its inhabitants have led to unrealistic land claims by the Aboriginals, excessive alienation of other lands by environmental activists and their supportive governments, and of course, the growing demand by Quebec for special status - including the right to veto.

This current Canada is not the Canada that I was exposed to in the early 1940s - but that was over 50 years ago and conditions change. The Canada I arrived in was not directly threatened by war and yet it was obviously almost totally dedicated and involved in supporting the war effort in distant Europe. I was to appreciate this only with time.

My earliest friend at UCC was Harry Beatty, youngest son of an admiral in the Royal Navy. Through this friendship, I was invited to spend part of the Christmas holiday with Harry and his brother, in addition to two other British evacuees, at a summer cottage north of Toronto. The two others were Milo Cripps, son of Sir Stafford Cripps who served as Minister of Aircraft Production in Britain from November 1942 to 1945 and later as Minister of Economic Affairs, and John Cooper, son of the head of the Ministry of Information in Britain during World War II. All four were at UCC and under the care of Fara Bartlett, their guardian while in Canada.

I wrote a letter imploring my parents to allow me to accept the invitation - which they did. The cottage was actually a spacious, beautifully furnished log cabin set among birch and evergreen trees in a hilly area, quite secluded from nearby buildings. Fara had assured that we were well equipped for our one-week stay and had generously arranged for presents for all of us. Although I was to spend a considerable amount of time in later years in the 'northern Canadian bush', this was my first introduction to its more accessible parts.

I had been skiing on one or two occasions at Norval, near Toronto and we all had skis with us as we planned spending the final part of the Christmas vacation at a ski camp at Dwight in the Lake of Bays area with about 50 other students under the supervision of Charles Carlson, a master at UCC assisted by two other masters and a nurse. While at the cabin, I decided to go out on my own one particularly cold day. After several runs down a fairly modest hill near the cabin, I fell awkwardly near the bottom of the slope. After picking myself up, I felt a very minor weakness in my right leg, but I had no problem hiking back up the hill toward the cabin. Still feeling somewhat uncomfortable, I checked my leg and saw that I had torn my ski pants on the upper part of my thigh. Lowering my pants, I saw a bloody red hole almost an inch in diameter and almost passed out. Apparently I had managed to fall directly on top of one of the upturned ski poles, which in that era had a pointed protrusion extending from the base of the pole, with a diameter of about 3/4 inch. The point of the pole had not hit my femur, but had almost penetrated the back of my thigh.

I explained my predicament to Miss Bartlett on reaching the cabin and she immediately phoned for a doctor who did his best to close the puncture with several stitches.

My first year in Canada passed very quickly. Apart from learning how to ski and skate, I was fortunate enough to attend a camp during part of the summer vacation where I learned the rudiments of sailing and completed an intermediate life-saving course. On the scholastic side, I had no real problems adjusting to minor differences between British and Canadian curricula at the Grade B level - called Lower Remove and Upper Remove at UCC. My principal weakness was in Science, and to a lesser extent, Mathematics. My Spring 1941 report indicated that at 13 years, 2 months old I weighed 68 pounds and was 4'-8" high - still a shrimp!

As indicated above, the few female contacts many boarders had included rare encounters with the House Nurse and probably just as infrequent contacts with Miss Barbara Barrow, Head Nurse and a legend in her own time. Relatively young when I was quarantined with measles in 1942, she was a mother and inspiration to all her dependants and admirers throughout her lengthy career lasting over 50 years. Graduates from UCC visited her and corresponded with her throughout her career and she maintained a regular column in the 'Old Times' indicating the whereabouts and recent activities of 'Old Boys'.

In addition, I remember attending dance classes at UCC in the Upper School. I believe that the female participants were from nearby Bishop Strachan School (BSS), which also had a boarding contingent. I don't believe that my dancing capability advanced beyond an acceptable generative step, however I did make a couple of contacts, which led to others.

As will probably become fairly evident in my future ramblings, I was brought up in a mainly male world because of this boarding-school environment and my prolonged absence from my parents. In retrospect I have been most fortunate to have experienced a relatively close family life made possible by Rene and our children, none of whom were ever separated from more than one parent at a time.

In September 1941, I entered the Upper School to face a similar transition to that encountered by others graduating to high school and from there to university. However, at UCC there were several significant differences in the early 1940s. As a boarder I entered Seaton's House, the other domicile for boarders being Wedd's, each of which provided accommodation for about 60 boys ranging in age from 13 to 19. Supervision was provided by a Senior and Junior Housemaster and a House Nurse, all of whom were residents in three buildings forming part of a cloister dominated by a statue of Sir John Colborne, later Lord Seaton, the founder of UCC in 1829.

Instead of being one of numerous occupants of a dormitory, as had been the case in England and the Lower School, I now shared a room with two others. That appeared to be the only benefit. On the negative side, UCC remained a male bastion with many of its boarders deprived at this early stage of their development from any frequent association during their daily lives with members of the 'fairer' sex. For boarders such as myself, any such association was at best restricted to weekends. My situation was also compounded by years of absence from my family life, in addition to being an only child.

In 1942 UCC was probably the only private school in Canada which retained the British public school practice of 'fagging', bestowed on first-year entrants to upper schools. In today's vernacular, the term 'fagging' suggests some sexual connotation, which was certainly not the case. As in the English system, a fag acted as a servant to an older schoolmate; however at UCC the older schoolmate was generally in his last year and was expected to provide a reciprocal benefit by looking after his fag's interests.

I'm not at all sure what this was expected to entail. However, there were no written guidelines. I have few distinct memories of my own experience. I was assigned as a fag to a prefect in Seaton's House and I can only recall one incident in which I had apparently neglected to polish his 'regimental' buckles and belt. When he complained that he would have to do it himself, unless I completed it as assigned, I stupidly agreed that it was apparently my job. I cannot recall a single contribution he made on my behalf - nor did I expect any.

As indicated above, I neither rebelled against the fagging system, nor was I adversely impacted by it. In retrospect, after reading James Fitzgerald's publication 'Old Boys' on Upper Canada in 1995, which contains diversified recollections by former students of the pros and cons of a UCC education, many of which I consider totally unreliable, there are only a few references to the fagging system. Fagging was discontinued in 1945 and although there is a reference in 'Old Boys' to the anger and frustration felt by some who had fagged but 'lost out' on having a fag, it was good riddance as far as I was concerned.

UCC placed a major emphasis on athletics to the extent that all boys were expected to not only participate in some sport, but to engage in the manly art of boxing, unless exempted from this requirement. In general the most popular boys were the best athletes. Although this facet of the UCC system was a problem for many students, it wasn't to me, despite my lack of expertise or physique, as I cherished any kind of participation in sports.

Because of Canada's heavy involvement in World War II and UCC's unusually long-standing recognition of its graduates' participation in various conflicts dating back for over a century, a heavy emphasis was placed on participation in the UCC Rifle Regiment which had actually seen service during the Fenian Raids (1866-7) in addition to exemplary service by graduates in later wars. This could not be regarded as a negative factor, for there was a war on and discipline and training were obviously advantageous for those who would be directly involved.

I made many wonderful friends in my early years as a boarder in the Upper School at Seaton's House. Included were very early room-mates Mickey McFarland Jr. and Jimmy Biddell, both of whom I had to fight in final boxing matches, losing in my first two years in finals of my weight class to Mickey, and winning in my last three years. (My later boxing 'career' is described elsewhere.) Mickey came from a totally different background to Jimmy Biddell and myself. Like his father, he cut short his education to become involved in construction - a field that his father dominated in the Port Hope area during World War II and for many years thereafter. Jimmy was at UCC as a boarder, for similar reasons to myself. His father was an executive with Brazilian Light & Traction, stationed in Sao Paulo, Brazil. Jimmy had been sent to Canada rather than Britain because of World War II, to continue his education in English rather than Portuguese.

During early 1942, I had spent part of an Easter vacation with a very restricted group from UCC supervised by Eric Wiseman, then Junior Housemaster at Seaton's. A most congenial individual, he had in the past supervised several summer hiking expeditions involving UCC students into the Canadian Rockies, in addition to Easter excursions into ski areas in Quebec.

During the Easter vacation I thoroughly enjoyed both downhill and cross-country skiing with Eric's small party between Sainte Adèle and Saint Jérôme in Quebec. Consequently, Eric wrote to my parents suggesting that I could be accommodated during part of the summer vacation on a bicycle trip that he had planned in the Canadian Rockies. In his letter, dated April 19, 1942 he mentioned that in 1941 "I took a group of 15 boys on a 7-week camping tour in the Rockies. We travelled light, carrying sleeping bags and tuck snacks and did quite a bit of hiking." He then mentioned that he planned a similar trip but that he had decided to focus on cycling in order to eliminate / reduce the payloads. My not owning a bicycle was no problem, as Eric could obtain the loan of bikes as required.

My parents supported Eric's proposal and I was one of about 10 individuals aged 13 - 18, mostly from UCC who were involved. We followed Eric's revised itinerary as he had presented it, to a certain degree, although there were several modifications in practice.

We could not have had a more amenable person than Eric to lead our group. At the outset, he drew me aside and said "Dave, for the next six weeks we will be sharing many responsibilities. I would appreciate your calling me Eric and not 'Sir' as you have been accustomed in the past!"

That was not really a problem as events transpired. For the rest of the six-week trip Eric was 'Eric' to all of us. Little were most of us to know how close we were to become as a group in that relatively brief period.

My initial memory of our trip following the Calgary Stampede, was when our group disembarked at Canmore with our very meagre supplies, including our bicycles. We actually had only a few cooking and eating utensils distributed among us in addition to personal gear, including sleeping bags. We were to rely on finding adequate cover each night, having no tents.

Our initial objective was to cycle from Canmore to Banff, a relatively level route dominated by spectacular views of the mountains on both sides. En route, I was pedalling along on a down-hill gaining on a slow-moving car, when I tried unsuccessfully to brake, only to find that my chain had become disengaged. Accordingly, I lifted my left foot to clear the camping gear stored on a bracket above the rear tire, with the object of using my foot to brake my forward momentum.

Alas, although this embryonic plan should have worked, I caught my left leg on the camping gear and soon found myself in a ditch with a bloody left knee.

We biked on to Banff, where I had my knee stitched, the attending doctor advising against further biking for a day or so. Fortunately this coincided with a planned two-day stay in the Banff area.

Looking back and comparing the Banff recreational area in late 1995 to that in 1942, there is no resemblance. In our relatively brief sojourn at Banff, in 1942, we were 'based' out of a so-called park area which consisted of several fairly widely-dispersed A-frames covering table/bench eating areas with nearby 'rest-rooms' comprised of washing stands, toilets and urinals, all enclosed within bare, water-tight structures.

Although our stay at Banff was to be short, lasting only several days, we were exposed in that very brief period to several memorable events.

We were aware of the presence of bears - all black bears at that time - in the camp-grounds, and accordingly we took care to protect any edibles in containers elevated by straps or ropes in the supporting structures above the sheltered eating areas.

We were concerned enough about bears to sleep under the sheltered eating areas, underneath the tables. One morning a couple of us awoke to see a large black bear vigorously rubbing its back on one of the supporting timbers of the shelter, apparently oblivious of us.

On another occasion, while we were away from the camp-ground, we returned to find that some of our supplies contained in a pack-sack secured to the roof timbers of our shelter were strewn about on the ground. We followed a trail of ravaged supplies leading to a tree. Sure enough, there was a young black bear half-way up the tree. Logic suggested that we could easily dislodge the bear by climbing the tree and securing a rope beneath the bear - pulling hard on the rope would force it to descend one way or the other. I believe I readily volunteered to climb the tree with the rope - obviously not considering what might happen, should the bear decide to come down. Fortunately all went well - I secured the rope about two-thirds up the tree - the bear climbing higher, as

I ascended. Following my descent, we pulled on the rope with gusto and in spite of a couple of dogs barking below - the bear suddenly decided to come down, which it did with incredible alacrity, evading the dogs as it charged off into the bush. I don't know what I would have done had the bear decided to descend while I was still on the tree!

Another incident at Banff revealed my capacity to sleep relatively undisturbed at age 13. I had actually elected to sleep in the open in my sleeping bag as it was a beautiful clear night with no apparent threat of rain. However, it started to sprinkle before a couple of the older party members had turned in. They took a certain amount of pity on me and dragged me in my sleeping bag into the nearby men's washroom, depositing me immediately in front of the urinal area. I didn't stir until the following morning when I became aware of unusual activity in my immediate area. Fully awake, I arose pulling back my sleeping bag into a lower traffic area, while trying to adjust to my new surroundings.

From Banff we cycled to Lake Louise where we visited Moraine Lake and Lake O'Hara before cycling on to Castle Mountain and negotiating the road over the Rocky mountains during a 6-hour climb pushing our bikes. We rapidly descended into Radium Hot Springs, passing several cars en route. Following R & R at Radium Hot springs, we continued south along the Columbia River past Windermere and Canal Flats toward Kimberley. En route two of us cycling together suddenly encountered a huge bull moose astride the gravel road. We stopped to survey this magnificent animal, which quickly lumbered off across the road into the bush.

Although our plan was to cycle together, it was never rigorously enforced, and on more than one occasion we became separated from the rest of the party overnight.

The first of these occurred near Canal Flats when two of us were well ahead of the party, which encountered a prolonged rainstorm, so that they did not reach our objective for the night. We were also affected and sought shelter short of our target when we spotted an inviting home near the road with a sheltered veranda. The occupants readily agreed to our sleeping on the veranda and most generously provided us with a dinner that night and breakfast the following morning.

We travelled from Canal Flats to Nelson by train and then cycled / ferried to Penticton via Kaslo, Nakusp, Vernon and Kamloops taking in Kootenay, Slocan, Arrow and Okanagan Lakes partly by ferries over a two-week period. From Nelson we went by the S.S. Moyie, a sternwheeler to Kaslo.

Again, two of us got separated from the rest of the party at Kaslo, where we sought shelter among several evergreens in a park. We awoke next morning to the sound of singing voices. Unwittingly, we had selected a location where children of Japanese-Canadians, moved from the coastal area with their parents following Pearl Harbour had been shamefully interned. The children and their teachers were quite oblivious to our presence during preparation for their morning classes.

From Kaslo several of us rode our bikes into Kokanee Glacier Park and then climbed to the rim of a ridge encircling the glacier and a small lake within its cirque. We were able to dislodge several large boulders which we watched with awe as they bounded for several thousand feet down the snow slopes into the cirque, some actually landing in the lake.

From Kamloops we boarded a train for Boston Bar in the Fraser River canyon to overnight at the Spencer's Ranch on the west side of the river. John & Victor Spencer were both attending UCC at the time. Several of us got off at Lytton and cycled down to Boston Bar - a spectacular trip for us who had never been exposed to the area.

We took the ferry from Vancouver to Nanaimo and cycled from there to Victoria before returning to Vancouver where we spent a few days. En route back east we stopped at

Jasper where we spent several days cycling to nearby points of interest. Our final stop was in Edmonton at the invitation of an acquaintance of Eric Wiseman, then a doctor at the Edmonton General Hospital. One evening several of us accompanied Eric and his friend to the hospital where we were admitted to a poorly lit large room which turned out to be the pathological section. Eric's friend must have had a macabre sense of humour, for he quite suddenly pulled back a sheet on a table exposing a partly dissected corpse.

We had cycled some 1500 kilometres mostly on gravel roads through the southern part of B.C. and its spectacularly diverse environment. Life would never be quite the same and like my father before me, I was to eagerly return almost a decade later to this wonderful land with an opportunity to explore its remote regions.

My second year at UCC was not particularly eventful, although I fully participated to the best of my ability in the many extra-curricular activities available. By the time I left UCC, I had been a member of the chorus or held equivalent minor roles in the annual Gilbert & Sullivan musicals it presented, among them 'Pirates of Penzance', 'H.M.S. Pinafore', 'The Gondoliers' and the 'Mikado'. All roles, whether male or female, were undertaken by male students. One I recall quite vividly was my friend John Armour whose falsetto portrayal of *Katisha* (an elderly lady, in love with Nanki-Poo) in the *Mikado* was a highlight. My only 'solo' act was as the 'Drummer Boy' in the *Gondoliers*, which fortunately entailed no vocal lines.

I spent the summer vacation in 1943 on two farms north of Toronto. Both were dependent on obtaining temporary help from summer labourers - many, such as I, recruited from schools. This was related to the World War II Ontario Farm Service Force, an emergency program in Canada to provide up to 4000 summer workers from secondary schools to replace regular agricultural workers who had either enlisted or had been conscripted for the armed forces. Pay was $25 per month with free board and lodging.

My initial employment in 1943 was with a Scottish couple who ran a mixed farm near Langstaff, about 20 km north of Toronto, on behalf of the owner, Mr. Wheeler, a UCC associate. I was the only outside helper on this modest farm and I recall being accepted as part of the family. There were several memorable highlights. It was the first time I ever had an opportunity to milk a cow. Although I improved with time, I was never really adept. I was placed in charge of providing feed for the young pigs and quickly learned the basic chant "here pig - pig - pig" at meal time. I had a bad encounter when the sow somehow caught a chicken and I intervened, having to wring the poor chicken's neck as it was badly mauled.

Often, on those summer evenings, the family head would don a kilt and march slowly back and forth playing a variety of reels, marches and laments on the pipes. It was the only entertainment I had.

In the latter part of the summer I spent several weeks at another farm with two older helpers - both in their teens. By a remarkable coincidence I was to learn years later that Don Coburn, the younger of the two, at sixteen, dated Rene while at university. The older was a bespectacled young man who had tried to enlist but had failed his medical because of his weak eyes. We kept in touch for over a year. Eventually he was accepted in the army and prior to going overseas sold me his bike, a Raleigh I had prized, for $25.

By my third year in the Upper School, I was well entrenched and heavily engaged in sports - probably too heavily as my grades in maths began to suffer toward the end of the year.

Unlike most of the students at Upper Canada, I had not learned to skate before learning to walk! My ankles were extremely weak and I believe that they contacted the rink surface more than my blades. At a very early stage I was relegated to goalkeeping and my early hockey career did not advance beyond first form in the Upper School. However, I was to have a second chance, as it transpired, following my promotion to Eastern District Manager for Kennecott Copper's Canadian subsidiary in 1965, 24 years later, based in Toronto following a transfer from Vancouver. Shortly after my arrival I received a telephone call inviting me to come out for hockey with the Prospectors All Stars, a mining team which annually played a hockey game against the Evil Keevils of Teck Corporation at Maple Leaf Gardens during the Prospectors & Developers Association Convention in Toronto. My telephone contact was unwilling to accept my claim that he had made a terrible mistake until I checked in for my first practice one evening at the Richmond arena north of Toronto. I wasn't his only mistake - the other was Neil Hillhouse. That first evening was a total disaster as we tried holding each other up and wound up careening off the boards when we did manage a short flurry. However, I persisted and several years later my family were present at one of the annual PDAC games when I scored a fluke goal.

In the winter I concentrated on swimming and diving and was captain of my house swimming team in my third year and a member of the UCC swimming team in my last two years. In diving I was not overly proficient in a great variety of alternate dives and was most fortunate to win the Little Big Four diving title in my final year. This was an annual meet between UCC, Ridley College, Trinity College School and St. Andrew's, all private schools in Ontario at that time.

I also won my weight in boxing in my three final years with bouts against Jim Biddell, Bill Stewart and David Todd (see 'Boxing'). In preliminaries there was a bout against Bill Hewitt, Foster Hewitt's son.

Although I played end on the Senior Football team in my final year, I was not a regular as there were fleeter, taller and more experienced players available. In my fourth year, I regularly played right middle on a team which included several friends. Jim Biddell was a running half along with Allan Badley and Bill Hewitt was our quarterback. Between October 4 and November 8, 1944 we played twelve games, i.e. every 3 - 4 days, winning all but two, with a record of 168 points for and 83 against. In those days we regularly played both offense and defense, three of us having no spares. We also played our last two games, both of which we won, without Angus, Rennie and Watson, who were unfortunately suspended for smoking.

I became increasingly interested in field athletics in my last three years at Upper Canada. In those days we did not spend much time in training; it was not until I reached university that training was emphasized. However there were a number of events scheduled throughout the fall and early summer periods. These included intramural steeplechase and cross-country (harrier) runs, and intramural and interscholastic track and field meets, held mostly in May.

I did not win, place or show until my fourth and fifth years, when I obtained six firsts, three seconds and two thirds.

My first significant win was also my most memorable one while at Upper Canada. It occurred on Games Day in the senior mile run composed of six runners including Pete Bremner, an all-round athlete, Lt. Colonel of the school battalion and by far the most popular student in his final year, as evidenced by his being awarded the gold Herbert Mason medal.

At the end of the first half-mile I was running second behind Bill Leuty and ahead of Pete Bremner, the three of us about 20 feet apart. By the end of the third lap I had passed Bill and Pete was right behind me. He passed me at the start of the back stretch in the last lap, but did not maintain his pace. At the end of the back stretch I bit the bullet, trying to concentrate on maintaining my pace as I passed him. I won by about 10 yards.

Almost 50 years later in the 1993 B.C. Senior Games at Cranbrook, I entered the 1500 m, 5 k track and 10 k road races in my 65-69 year age group. I won a silver medal for second place in the 10 k road race. However, I recognized that my chance of placing in the 1500 k track race - at the lower end of my distance running capability - lay in overcoming an equal desire on the part of a John Harris at the upper end of his distance running capability who had won medals in the 100 m, 200 m and 400 m distances.

Harris was ahead of me for the first 3-1/2 laps of the 4-lap race. I had stayed within 2 metres of him for almost the entire event when I decided to try to break him. We ran neck and neck for the final 200 metres and he nipped me for third place at the end, our official times being 6:48.6 and 6:49.0 respectively. The first place finisher was Cliff Salmond, a former British Olympian at 5:57.3 and the fifth place Edmund Hamer at 9:40.8 minutes.

I was to win my second Magann Cup for the one mile in 1946 and my first MacDonald Cup for the Senior cross-country race in late 1945. It was at Games Day in 1946 when I also placed first in the pole vault that I was approached by Hec Phillips, one of Canada's outstanding sports figures and coach of the University of Toronto track and field team and the Harrier team, since 1933.

After identifying himself, he asked me about my future plans - indicating that I would be welcome at the University of Toronto (U of T), should I elect to continue studies involving an athletic commitment. Hec was true to his word. Although I had applied to both Queen's University and U of T, my preference was Queen's because of its pre-eminent position at that time in terms of engineering. Fortunately I had also applied to U of T, as my final grades were acceptable only to U of T.

I find that I am progressing significantly ahead of events as they occurred which directly affected my future. To regress, I intend to cover many incidents which occurred in my third to fifth year at Upper Canada and which I hope will be of interest to others more concerned with an earlier UCC perspective of events.

I have mentioned my early interest in swimming and diving on entering the Upper School. Our coach from the outset was Winston McCatty, a gentle but persuasive individual who rapidly gained the confidence of those he sought to instruct. I had little real aptitude for diving before entering the Upper School. Winston taught me the various dives which were in vogue in the early 40s. He spent a great deal of time with me and others, providing instructive confidence in both swimming and diving skills.

It was with some surprise and not a little dismay that our swimming team learned that Winston had elected to join the Royal Canadian Air Force in 1943. It was a great loss to the team, but nothing compared to the tragic news of his death in an accident during training with the RCAF a year later.

I have described the role of the UCC Cadet Battalion and its role in providing both leadership and guidance to those who might be called upon to serve all too soon with the armed forces. The UCC Cadet Battalion was under the overall guidance of Freddie Mallet, M.C. & Bar, during those years. I doubt that Mr. Mallet had ever developed much beyond the demands imposed upon him during World War I and that his role at UCC was about as much as he aspired to.

In my first year at the Upper School I joined the Band of the UCC Cadet Battalion where I gained only a minimum of proficiency as a bugler. As I advanced, I eventually rose through the 'ranks' to company sergeant major. During this period I earned a bronze medallion in shooting with the D.C.R. Association in 1945 and I also participated in a 10-day field camp at Camp Borden, attended by numerous other high school cadets. The training was rigorous, obviously intended to impress 16 - 17 year olds with what would be expected of them in the armed forces.

One of the memorable exercises at Borden was a demonstration of unarmed combat by two instructors - one unarmed and the other with a rifle with fixed bayonet. They were ringed by 200 - 300 students in the demonstration area. One of the 'ad-hoc' exercises included for realism required an unarmed expert to call out one of the students and urge him to try to hit him in the stomach region with a rifle and fixed bayonet. The instructor assured the student that he would not inflict any damage to the student in the exercise. The demonstration was concluded without incident, though I heard that in a later class, the same instructor was fatally wounded by one of the students during such a demonstration.

Our participation in cadet battalion exercises were never taken lightly and discipline was maintained at a level equivalent to that expected of armed services recruits. On the announcement of Wince McCatty's death, I was one of an honour guard which participated at his memorial service. The following anonymous tribute was provided.

>TO WINSTON McCATTY
>
>So soon we breathe farewell, O comrade true:
>A boy you were in manner light and gay,
>Your sunny smile infecting all who knew
>That open helping hand from day to day.
>Lover of living birds and fresh clean ways,
>Artist of health of body and of mind,
>Strength of the timid, friend of those whose days
>Were troubled, known to all as very kind.
>And there with birds you flew at duty's call,
>Bearing your part in building our new world ---
>Shattered your wings, your hopes and dreams, your all;
>Yet your clear spirit wings its paths unfurled.
>God rest you merry, gentle soul; and may
>The birds you loved still wing their happy flight
>Above eternity and keep you gay
>For ever in our hearts and in the Light.

Some measure of the importance ascribed to the UCC battalion in these years was the following excerpt from the Globe and Mail in late 1947 concerning an inspection of the battalion:

>"Twice postponed by inclement weather, the annual inspection of the Upper Canada College Cadet Battalion was held on the wind-swept campus yesterday, and the fury of the wind failed to prevent 300 cadets and a band of 60 from giving the usual cadet display of marching and manoeuvres.
>
>Inspection was made by Maj. Gen. Arthur Potts, D.O.C. for M.D. 2, who also took the salute. He was accompanied by L.M. McKenzie, school principal.
>
>Display included signals, mine laying and breaching, first-aid, Sten gun drill, respirator drill, weapon training, and Bren gun drill. Mass physical-training display and selections by the band rounded out the program.

I doubt that the developmental needs of 14 - 19 year olds, particular boarders, in the Upper School were ever seriously considered by the staff. There were certainly exceptions

- such as Jim Biggar, head housemaster at Seaton's, with whom I was most involved, and Eric Wiseman, junior housemaster. Few at the Upper School realized that Mr. Biggar, himself a UCC old-boy, had also been recognized as a popular individual by his peers, having been awarded the Herbert Mason medal as a student.

Mr. Biggar taught one of my geography classes, in addition to history. His desk was normally cluttered with many books and unfiled documents. I recall him describing the power of glaciers in one class, emphasizing his words using a yardstick to represent the face of the glacier. With his arms outstretched he slowly moved the yardstick across the face of the desk - sweeping all the stacked volumes onto the floor without hesitation as he described the catastrophic forces.

He also had a penchant for stories involving a humorous cat named Tobermory which he insisted on sharing with the class. Most of us enjoyed them - recognizing that they had no bearing whatsoever on history or geography.

Although he was never a father-figure to me, I always admired his forthrightness and honesty. Others probably thought me biased, because I could accept his consistency and attempts to provide both guidance and discipline to the extent he felt they were required.

As none of my masters were ever to meet my parents during my six years at Upper Canada, my parents welcomed letters they received, mostly from Jim Biggar as my senior housemaster, concerning my progress or lack thereof. In general my academic reports were satisfactory although rarely outstanding. As I advanced through the Upper School, I had particular problems with mathematics, starting in my third year, which reached chronic proportions in my final year, leading me to drop the position of editor of the College Times at Christmas in order to concentrate on obtaining a passing grade in my Senior Matric exams.

I never had an opportunity to read any of these letters until decades later following the death of my father in 1969. Also in my father's files were copies of his letters to various masters, mostly expressing appreciation for their comments or providing requested information concerning my vacation plans or sundry personal needs. Few people recognize the time, effort and consideration required by staff at an efficient boarding school to anticipate or arrange for the myriad of details affecting the personal needs of students. An example is the following excerpt from a letter written on May 27, 1941 by Alan Stephen, Headmaster of the Preparatory School at Upper Canada to my father:

    Dear Mr. Barr,

    I think Mr. Carson has written to let you know that David is spending part of his holidays with Mr. and Mrs. W.B. Wiegand, Ballwood Road, Old Greenwich, Connecticut. Thereafter he is going to Camp Onondaga.

    The Canadian authorities will not allow David to take more than $1.00 into the United States. This will provide him with a breakfast on the train. I know that Mr. Wiegand would gladly provide him with a little pocket money, but if there is any way by which you could send him five or ten dollars I thought you might like to do so. I would suggest that if this is possible, you send the pocket money direct to Mr. Wiegand.

    I know the Wiegand family very well and I am certain that David will have a very happy time.

    Yours sincerely,
    Alan G.A. Stephen

By cable dated June 19, 1941 my father requested remittance of ten dollars U.S. currency to Mr. Wiegand from his New York office thrift account. On July 1, 1941 Mr. Wiegand politely wrote Mr. Stephen, acknowledging receipt of the $10 as follows:

Dear Stephen:

Have received $10. by cable from D.A. Barr, Guayaquil, Ecuador; no doubt spending money for David. Since I do not know Barr's proper address, I am taking this means of acknowledging the receipt of this money, which will be doled out to David in regular weekly amounts and any residuum sent on to you for a similar purpose in camp.

Mr. Biggar's lengthiest letter to my parents was written on May 29, 1944, near the end of my third year in the Upper School when I was sixteen. He obviously had a very high opinion of me - which I was never really aware of, and his stated impressions of my scholastic abilities proved particularly prophetic.

Dear Mr. Barr,

Your son David, called Ditz by everyone here for no known reason at least none known to me, continues along as well as ever, that is very well indeed. His health is always good. He is always neatly dressed and carries himself very well. I do not [know] how recently you have had a picture of him but he has not changed in appearance since I have known him, he merely enlarges himself without changing. He is going to be a small man, but one of the most wiry. There is a splendid spring in his walk. In a mile run or a cross country he comes among the very first of his age group which includes scores of boys almost twice his size. It is the same in swimming and most noticeable in something like the long plunge. His splendid little physique is driven by more "heart" and "guts" than I have seen in anyone I have known. That sounds like exaggeration to please a parent but everyone here would agree.

In the class-room his work is always regularly, neatly and intelligently done, and with interest in it. But he is not an intellectual, though he seems to like reading. He stands up near the head of a medium calibre form. His studies will be perfectly successful I prophecy but not brilliant.

It is as a person he stands out most prominently. He is always merry, often mischievous, always friendly and enthusiastic. He has perfect natural manners. He is friends with everyone. I do not imagine he has ever been addressed an unfriendly word by anyone because he [is] so generally liked. He whistles all the time. Whenever a volunteer is needed for something he is the first volunteer. He is usually elected captain of any team he is on. He takes the duty seriously and though it seems a funny thing to say of a small young man like David one can see him inspire the other little boys.

His abundant energy makes it hard for him to waste a day and as soon as a holiday comes he looks for a job. I was glad to see that the employer he happens to have had last summer appreciated him and asked him to his house in town since.

For this summer he has obtained work at a lumber camp in Northern Ontario with and through another boy Eric Lubbock. The latter is English, the war-guest of an Old Boy of this school, J.D. Woods whom I know. Though the place is remote and the country rough, as the Woods are letting - have arranged for - Lubbock to go, I am sure that all will be well.

To the best of my recollections, Mr. Biggar never expressed a word of criticism to me - although this was by no means the case with several other masters. He caned me on only one occasion, which was an impersonal, but unsuccessful attempt at mass exorcism. It followed the annual Seatons House Pyjama Parade in my third year.

The Pyjama Parade was apparently a Seatons tradition of unknown initiation and involved all the boarders except the prefects. In the years in which I participated it averaged about 40 - 50 individuals under the direction of one or more seniors as to the date of the event. The outermost garments were pyjamas and the only accompanying items were numerous rolls of toilet paper.

In my second year the parade followed a typical route, starting at about 5:30 a.m. down Avenue Road to St. Clair, east to Yonge St., and around the Granite Club. We then boarded a St. Clair Ave. Street car with the consent of a good-natured operator recognizing that we were strictly local as we were obviously not intending to pay for our ride several blocks west. Before returning to the College we paid our annual early morning visit to Bishop Strachan School and its female boarders, hurling our rolls of toilet paper as best we could toward the windows above at our only audience. We were back at Seatons House within 3/4 hour.

Following our return everyone was called out by the prefects to the common room, where we were admonished by Mr. Biggar. We then paraded one by one up to him, flanked by two prefects to receive our punishment. It consisted quite democratically of one stroke with the cane on our pyjama-clad rear ends per year of recipient's Upper School form. Mr. Biggar was quite visibly fatigued by his early morning work-out and that concluded the 1943 Pyjama Parade.

Three years later, in my final year at Upper Canada, I was a steward as Head of Seaton's House. 'Butch' Mackenzie, my maths teacher, was principal, having replaced the highly respected Terry MacDermot who left the Upper School in 1942 for a wartime-related assignment with the government in Ottawa.

'Butch' Mackenzie not only had a fierce visage, even under normal circumstances, he also had little to no patience with wayward pupils, such as I, so it was with considerable trepidation that I became involved in an encounter with him over the Seatons Pyjama Parade in 1946. Immediately following the event I was approached by a senior in Seaton's who had been singled out as being personally responsible. He sought my help as he had been called in by Butch Mackenzie and expelled. I'm not sure why I didn't approach Jim Biggar - instead I went directly to Butch Mackenzie's residence which was on the school grounds and gained admittance. Butch said very little and only listened without comment while I recounted the tradition of the event and the accountability of all who attended as there was no enforcement involved. After leaving, I thought that it was probably a wasted effort; however, that evening the affected senior came in to thank me as his expulsion had been withdrawn. It was one of the most satisfying experiences I had at Upper Canada.

Dalton Mills, Ontario - 1944

One of my best friends at Upper Canada College was Eric Lubbock, a classmate, who like myself, had been evacuated from England to Canada in World War II. Through Eric and his uncle J.D. Woods, president of Harvey Woods & Company, and also an Old Boy of UCC, I was able to obtain summer employment at a lumber mill located at Dalton Mills, northwest of Chapleau, Ontario, during July and August 1944.

The experience was to be one of my more memorable summer vacations while at UCC which were to include the 5-week bicycle trip through the Rockies and British Columbia with seven or eight other students in 1942; partial summer labour on two farms in Ontario in 1942 - 3 and a part-time job cutting firewood for Pine Grove Inn at Dwight in the Lake of Bays area, Ontario in 1945.

The lumber mill was located in a remote area, about five kilometres west of Dalton, a whistle-stop on the Canadian Pacific Railway line, about 65 kilometres northwest of

Chapleau. From Dalton, a railway spur constructed by the company provided the only access to the mill, located on the edge of a small lake immediately north of Shikwamka Lake, both lying on the Michipicoten River drainage.

The existence of Dalton Mills was due to the resourcefulness of the Austin family which included four sons: Allan, Bill, Chuck and Jack who were half-brothers to Allan and Bill. Allan and Bill ran the lumber mill and Chuck and Jack founded Austin Airways in 1934 – a northern air service which was to prosper for over 50 years prior to being sold to Air Ontario, a subsidiary of Air Canada. Some years after 1944 I flew with Austin Airways on several exploration-related jobs and recall the air service as one of the most reliable I was to experience.

Allan had four sons: Jim, John, Alan (Mac) and Richard. In a recent letter Mac pointed out that Jim was enrolled at Trinity College School (TCS) in the fall of 1939 "and from that time on, for about seven or eight years, boys from the south came up to work at the mill (during the summer vacation). Most were from TCS, but word of the luxury in which we lived and the high wages we enjoyed got out, and as a consequence, boys came from other schools. Philip Ketchum, the Headmaster of TCS, was curious about Dalton Mills, and the summer you were there he came up to see for himself. He ended up being put to work shovelling coal out of a box car. Oddly enough, he never returned."

Mac also noted that it was difficult to get labour, however unskilled, for the mill during World War II. This accounted for the flow of students from UCC, UTS, St. Andrews, TCS and Lawrence Park Collegiate that worked at Dalton Mills during those summers of the war.

Apart from Eric Lubbock or "Yah" (pronounced 'yea'), as he was nicknamed, other students at Dalton Mills that summer included Dick "Cozy" Cole, also from UCC, John and Mac Austin, John W. "Bunt" Burrows, their next door neighbour in Chapleau and several others whose names have faded from my memory.

Mac Austin recalls that families lived in the two communities; about 400 men, women and children at Dalton Mills and about 200 at Dalton. Alas, I have no recollection of the community at Dalton Mills; however, I remember that I never strayed beyond the cookery. John has advised me that most of the permanent residents lived beyond the cookery, which may well have been *terra incognita* for other reasons. John believes that parents kept any young girls "well away from the imported, randy males in our part of town". Many of the employees were French-Canadians, several of them draft-dodgers, who were very conscious of the tentative nature of their employment. As mentioned, Mr. Austin had his home in Chapleau; however, it was not occupied during that summer, other than by week-end visitors as I was later to learn on a more personal basis.

He had very thoughtfully arranged to accommodate his youthful entourage in two uninhabited houses rather than a bunk house at the mill which could accommodate up to 100 men. From my meagre recollections, we spent most evenings in our abode following dinner, as there was little else that one could do as a diversion from the work load, Sunday excepted. Shortly after my arrival at Dalton Mills at a starting rate of 44 cents per hour, I found myself using a picaroon, a sharp pointed instrument with a lengthy handle, to snag cut railway ties on a system of motivated rolls and redirect them at a 90-degree angle of the belt system, to a loading docket. On one occasion I was unable to disengage the picaroon, attached by a cord to my wrist, and was dragged along the side of the belt until I could disengage myself. Fortunately, no injury resulted.

A few weeks later, "Harry", a long-term operator of a relatively small swing-saw used to cut timbers up to prescribed lengths, reached below the saw for a cut fragment and upon rising split his skull open from fore to aft on the still moving saw. He was attended by

Alan Richardson, the first aid man, a second-year medical student at U of T, who successfully restricted potentially fatal damage by taping his skull together. John tells me that Alan didn't finish his medical studies and acted as secretary for the lumber company for several years before teaching at Gravenhurst High School until he retired as Vice Principal. He remains a close friend of the Austin family, annually joining four Austin brothers and wives at the Shaw Festival in Niagara on the Lake. "Harry" was a Japanese Canadian, one of many from the west coast who were moved to the interior and much farther afield during the war. He survived the accident with no obvious problems. Coincidentally, John ran into him in 1945 in Brantford after enlisting in the Canadian infantry. Harry, who had also enlisted, recognized John and doffed his cap by way of recognition to show his scar.

The day after the accident Mr. Austin approached me and asked if I would be willing to take over the duties of the operator, to which I readily agreed. My hourly rate was immediately increased to 50 cents. I remember only a few incidents as a result. The "Cat", as we all called Mr. Austin, because of his watchful nature and rapid movements around the mill, was obviously keenly concerned about the welfare of his personnel. Mac Austin reminded me recently that "his presence at the mill was signalled by rubbing the right hand end of your imaginary wax moustache between your thumb and forefinger." Once I recall catching his eye while at the circular saw, only to have him let me know in no uncertain terms that I should pay strict attention to my work, for he had been watching my performance. I was somewhat subdued but most impressed. The job was incredibly monotonous. Not having a watch, I made a simple sun dial which I could refer to in order to estimate the time before the lunch break or the end of the shift. The monotony was relieved on one occasion when the head sawyer became ill and I was called on to take his place. This function involved operating a six-foot diameter saw, much in the same fashion as the smaller eighteen-inch diameter saw I was accustomed to. However, the timbers were fed by an operator on my left and the cut section removed by another individual on my right. The two men were large, husky French Canadians, both of whom chewed gobs of snuff incessantly, spitting the brown juice out at regular intervals. I knew I had been accepted when my partner on my right offered me some chewing tobacco from his can. Feeling I couldn't refuse his good-natured gesture, I accepted a small quantity and managed to survive. He smiled broadly when I passed, on his next and final offer.

Our work week was six days long. Breakfast was served in the mess hall at 6:00 a.m. and our work day officially started at 7:00 a.m. with a one-hour break for lunch at noon followed by an additional five hours terminating at 6:00 p.m. Dinner was served at 6:30 p.m. We were usually in bed by 10:00 p.m.

While at Dalton Mills, I was painfully initiated into contact with a member of the spruce beetle family, apparently a familiar and much feared resident of lumber mills in Eastern Canada. While loading freshly cut timbers into a boxcar, I felt a severe pain on my forearm and swatted off a black insect about two inches in length. It was my first and last successful attack by a mature spruce beetle, which I learned by the experience to avoid. Other workers later showed me how insensitive the insects were to their own body parts by holding them in gloved hands and pushing their lengthy feelers and stout legs within reach of their pincers to have them immediately amputated. I also learned like others to listen to their squeaky sounds while in flight over the mill site and to look skyward if the squeaks suddenly stopped. Much like a V-2 bomb in World War 2, this indicated that they were in sudden descent.

You might well ask: "what else did we do at Dalton Mills in our spare time?" Well, let me tell you that very little transpired because of the demands of the working week. Most of us were 16 - 17 years old. Some of our so-called leisure-time pursuits were innovative, although not necessarily successful as exemplified by the following incidents:

We found that one of our contemporaries had a proclivity toward sleep at an early hour. We had heard that if a sleeping person's hand was allowed to be immersed in warm water, that he/she could be expected to involuntarily urinate. We spent several convulsive hours trying to prove this with a particularly unresponsive individual, whom I suspect must have soon become aware of our malign purposes.

We spent many hours creating an alternative to the company jitney in the hope of providing our own transportation between Dalton and Dalton Mills, much to the delight of the older and wiser men who thought we were crazy. We had no problem in obtaining four railway wheels, axles, lumber and constructing a wooden box with a driving arm for our trolley. Our problem lay in the gear department. The only gears we could find appeared to be either too small or too large to provide for a smooth transition between starting and reaching an acceptable speed. Discouraged, we installed the only two that might work. The trolley's official launching coincided with its abandonment. The motion of the trolley using the smaller gear was barely visible and the operator's arms were almost shaken off. Three of us pulling as hard as we could with the larger gear engaged couldn't develop any sustainable speed. The trolley never left Dalton Mills and remained a constant reminder of our futile efforts in transportation design.

Because of our friendship with John and Mac Austin, Yah and I quite regularly had a real break from the demanding work week because of the over-night accommodation available at their parent's home in Chapleau, albeit parents in absentia.

I'm not sure we could have afforded the regular fare by the passenger train between Dalton and Chapleau, or whether scheduling was a problem. Perhaps it was a combination of both. In any event, there were many freight trains plying the main CPR line and they frequently slowed down or stopped at Dalton eastbound toward Chapleau. Following dinner on Saturday evening, we would either walk the five kilometres, or more frequently obtain a ride on the jitney bus from the mill to Dalton. We never had a problem catching a ride on one of the freight trains, our accommodation being either a box-car or loaded flat-car, many of them transporting war-time materials eastward.

By the time we arrived at Chapleau, it was generally beginning to get dark or had already reached that stage, and we always jumped off the freight car carrying us well out in the yards.

Returning to Dalton demanded our boarding a west-bound freight train by 4:30 p.m. on Sunday, in order to provide sufficient time for us to exit the train at Dalton and arrive at Dalton Mills in time for dinner, as we were generally half-starved following our modest breakfast and lunch at Chapleau.

On this particular occasion, John, Mac, Bunt and I arrived as usual at Chapleau, without incident. On the following day, we boarded a west-bound freight train late in the afternoon, as in the past. On approaching Dalton and preparing to jump off, we soon realized that this was not going to be possible. For the first time in half-a-dozen similar trips, the train did not slow down and we had to continue on. We considered Franz as the earliest appropriate location where we could expect to get off and re-board another eastbound freight train.

The train did not stop until we reached Franz, a divisional point, about 55 kilometres northwest of Dalton. We jumped off and waited for about two hours until an eastbound freight appeared and stopped. We had spent part of the time picking blueberries in a particularly well-endowed field at the margin of the railway yard. Full of confidence we boarded the train, only to be seen by the stationmaster who ordered us off. We had nothing more to do but rummage around for more blueberries until the next freight pulled in well after dark. It was about 11:30 p.m. No problem this time; the stationmaster was probably asleep.

We climbed into an empty boxcar, the lip of the car being almost four feet above ground level. Soon we were eastbound anticipating a slowdown at Dalton, where most freight trains did just that, in our experience. Unfortunately this did not appear to include trains passing by at midnight or later. We simply couldn't believe that we were once again bound for Chapleau!

We never seriously considered returning to Chapleau. We felt compelled to be back at the mill in time for our shift Monday morning. We knew that the eastbound freights leaving Dalton were on a gradual up-grade for several kilometres, their slowest speed being reached about 5 kilometres from the station. Because it was wartime, we had been exposed to aircraft bailout procedures through the media and films. We rapidly resolved to jump off at the top of the upgrade. For motivational purposes and to keep close together, we agreed to rap each other on the shoulder as the sign to jump. I would like to be able to say that it was an exhilarating experience. In truth, we were all terrified at the prospect of jumping into the darkness, not knowing where we were going to land. Fortunately, we had little time to reflect and we all jumped in quick succession.

On re-grouping we were incredulously relieved to find that none of us were injured, apart from some scrapes and bruises. We trudged off westward toward Dalton, about five kilometres away. Within about 100 metres, we encountered railway ties piled along the edge of the railway in small clusters for about 400 metres. Miraculously we had just missed jumping into them. On a later trip to Chapleau we noted the ties still at the same location and some distance to the east a lengthy railway trestle, which we had never considered as a potential hazard, although we had seen it many times in our earlier travels.

From Dalton we had to walk the additional five kilometres in to the mill. We learned at breakfast that the mill was shut down for the day because of a mechanical problem. We gratefully went back and slept.

Most of the students left Dalton Mills at the end of August to return to school. Four of us who had survived the epic train jump decided to spend some of our summer's earnings in a visit to New York, which we thoroughly enjoyed. Included were John and Mac Austin, Bunt Burrows and I. Even though our stay was relatively short, we managed to see most of the well-known attractions of the city at that time. We weren't flush enough to afford fancy accommodation and Mac remembered that we selected a flop house called Lindy's. John's memory was vivid enough to query, "Wasn't Bunt the one that punched a hole through the thin plaster wall between the two rooms we stayed in at that flea bag on Lexington Avenue?"

This and our epic return from Chapleau occurred sometime after Yah had received a telegram instructing him that arrangements had been made for his return to England, as his parents quite reasonably believed that the threat of bombing had significantly decreased following the Normandy landings by the Allies in June. Consequently he was not involved, as he had left to return to England by mid-August.

I remember our farewell to Yah quite vividly. Several of us went out on the jitney to see him off at Dalton. When the east-bound passenger train finally pulled in to Dalton, I was somewhat overcome by the realization that a good friend, whom I had expected to be associated with for at least another year, was leaving and that I would probably never see him again. Yah was more direct, and came over and gave me, not a handshake, but a hug, something not common at our age and time. And then he was gone. I didn't hear from Yah until I received a letter from him almost two years later. He had by then entered Oxford University. I later learned that he had successfully run as a member of parliament.

Although I made an attempt to locate him on one of our trips to England 30 years later, regrettably I had not anticipated the possibility of contacting UCC for an address, so we never did get together again.

Yah was a most appealing individual. He was good-natured, slight and wiry, but with a mischievous demeanour, always alert to a potentially provocative situation. His mastery of this was brought home to me most vividly on one occasion when we were about to enter a fourth year French class directed by Mr. Shearer, the most sadistic teacher that I and many others at UCC, particularly those in Wedds House (the other boarding residence in which he was a junior housemaster) ever had the misfortune to meet.

Mr. Shearer was known to derive great satisfaction in enforcing his so-called authority. With respect to those unfortunates in his residence, he administered caning with a bamboo wand, soaked in water and chalked prior to the caning, so as to leave a visible mark on the buttocks for aiming purposes in successive swats. On occasions he would require a victim to visit his room beforehand to see the cane soaking in a bucket.

Unable to administer such severe punishment in the classroom, which included both boarders and non-boarders, he had over the years developed a method of encouraging students to correctly parse French phrases, which I am sure he considered effective based on his experience. Yah was an extremely intelligent student; however, Mr. Shearer picked on him frequently. Yah was also one of the more courageous individuals that I have met and he was to demonstrate both this and his ingenuity in a most convincing manner at a French class in the spring of 1944.

Mr. Shearer's method of emphasizing parsing to students, which had stood for years, was to stand beside them and rhythmically pummel their upper arm most forcibly with his bare knuckles, as an accompaniment to the attempted parsing. I can vouch for his ability to benumb an arm.

Just before entering the class, Yah pulled me aside and said "Ditz, I'm going to cure that bastard from ever hitting me, or hopefully anyone else again!" We all wore blazers or jackets at the time. Yah pulled his jacket down on one side to expose a piece of cardboard impregnated with tacks at about one centimetre intervals. He had another equally well-endowed on his other arm, from elbow to just below his shoulder.

Whether Yah in some way encouraged Shearer to pick on him that day or not, I cannot recall. However, it happened. It would be difficult to adequately describe my reaction. I <u>knew</u> what was going to happen. I watched, fascinated, as Mr. Shearer said, "Ah, Mr. Lubbock - we seem to have a problem, will you try that again!" I just stared at Yah's smiling face directed at Mr. Shearer, and at the smirk on Mr. Shearer's face which changed dramatically to a look of astonishment as his fist struck Yah's arm. He emitted a slight gasp, his eyes dilated, and he mumbled some incomprehensible comment. To my knowledge he never struck Yah or anyone on the upper arm again.

Buzz Classey, another French teacher, also had a somewhat vicious reputation, but because he seemed to enjoy beating students. I think he was a rather short-tempered individual who believed that when appropriate, some form of physical inducement was required to provide effective education. Prior to our association, he had lost his caning privileges as a direct result of kicking a student down a flight of stairs. Unfortunately the student sustained serious injuries. During my tenure he used to whack us across the back of our heads with a copy of *Cours Moyen de Français*, our reference text. As his current copy deteriorated from this abuse, he would replace it with another. His mind was evidently a little warped as he would tack test results upside-down on the bottom of his door, making it most inconvenient for us to determine our marks. God forbid that you would dare risk untacking it and retacking it int a more convenient position!

The Old Boys volume contains many references to the teaching staff at Upper Canada, most of which seem highly critical. That was not my impression as I had all I could handle in maintaining a second-class grade. As Jim Biggar had observed, I would never be brilliant. However, even in the 1940s the academic record of graduating students based on Senior Matric examinations was better than that of most Toronto schools with the possible exception of UTS. My experience at University was to be different as I was to become challenged and deeply motivated by the courses leading to a degree in Engineering in which I finally achieved honours' standing in my graduating year.

My first direct association with girls at Upper Canada was in my second or third year of the Upper School when I became a dancing aspirant as a means of meeting those of the fairer sex in our limited environment. I was fortunate to meet several girls from Bishop Strachan School during this exercise which led to limited dates. Limited, because of the extraordinary time frames involved as a boarder when only part of a weekend was available without being 'out of bounds'.

In my fourth year in the Upper School my life changed when I had the good fortune to meet Nancy Lawson, a student at Havergal College, located north of U.C.C. above Eglinton Ave. Congenial and unsophisticated, pert and blonde, she was wonderful company although after several dates I realized that her background was probably almost as sheltered as my own, for we were both shy with those of the opposite sex. Nancy, however, had a brother whom I never met, although I was to know several of her cousins. Whenever we returned to her home after a date, her parents were always nearby and our relationship never progressed beyond that of very good and convenient friends. She wore my UCC pin for a while, but eventually returned it to me with a letter when I was working in the bush. I guess it was a "Dear John" letter, but I could fully appreciate her position. I simply wasn't around enough.

I did not maintain contact with her after my final year at University, when I spent the winter at Sudbury working underground as a miner at the Garson Mine. In the Spring of 1951 I was en route to British Columbia, when I once again met Nancy in Toronto for a final time. I don't think she had ever realized the nomadic nature of the vocation that I had chosen in mineral exploration.

My final year at Upper Canada was to demand several important decisions and to provide a few unexpected fulfilments. I have already mentioned my surprise at being appointed editor of the *College Times*. I had never aspired to this position and I believe it occurred only because of a falling-out between Gordon Adam, the previous editor and Michael Macklem, the very gifted academic who went on to become Head Boy and Governor-General Medallist among other accomplishments in this his final year. As should have been the case originally, he took over editorship at Christmas when I elected to concentrate on my final exams.

By far the most unexpected recognition was being the recipient of the J. Herbert Mason gold medal, Geoff Pringle receiving the silver. I still know little concerning the origin of the medal, except that it was instituted in 1888 and bears the inscription 'Awarded on the nomination of the masters & election of the boys of the Upper School'. I recall being completely overwhelmed by the announcement and wandering around in a daze as I walked through the school grounds afterwards.

I had many friends at Upper Canada, many of whom I spent time with during vacations and temporary jobs. One particularly close friend was Walter Bernard (Bert) Maxwell, a prefect at Seatons in my final year, a track and field enthusiast and my room-mate at my first and second year at the school of Practical Science (SPS), when ensconced at Ajax rather than at the University of Toronto, in 1946-47.

Bert's father was a mining man who was closely associated with the legendary Thayer Lindsay, head of the Canadian Ventures Group for many years. While I was a guest at the Maxwells' home over Christmas, 1945, Bert's father recognized that I was deliberating between architecture, civil engineering and possibly oil exploration, knowing little of the actual requirements for these fields. He quietly noted that there should be a great future in Canada for mining geologists. Because of his knowledge and enthusiasm for the field, I became interested and applied for a summer job in 1946 at the McIntyre Mine at Schumacher, Ontario. John Armour, another close friend from the graduating class, applied for a job in the electrical department at the nearby Dome Mine. Both of us were accepted and we arranged to share accommodation at near the mines.

We had little trouble locating a rooming house at Schumacher, following our arrival by train at nearby Timmins. Our room had floor dimensions of about 6 x 12 feet and was a little over 6 feet high with a small window, two single beds, a small vanity, single chair, two hooks for hangers on one wall and a bare light bulb suspended from a ceiling fixture.

We saw little of each other as we were generally on cross-shifts. My job had little to do with mining, as my principal occupation was as a mill-worker sitting beside the main conveyor belt extending over a large crusher, with a huge magnet suspended over the belt beside me. My function was to pick up pieces of wood, which came up the conveyor belt from underground and at intervals to remove steel fragments from the magnet. I wore no earmuffs nor ear plugs, as this era preceded the recognition that exposure to deafening noise could result in deafness! Interestingly enough though, there were several occasions when I found myself drifting off and dozing. My only relief on this job was when I spelled off the mill sweeper - a far more tiring job with no more sophisticated equipment than a sturdy broom and dust pan to work with.

I did not receive my Senior Matric results until mid-August and it was with considerable relief that I learned that I had literally squeaked by with one 1st, three 3rds and six Cs. I was ecstatic! I had applied to both Queen's University in Kingston and to the University of Toronto and was not at all surprised to be turned down by Queen's and thankful to be accepted by the University of Toronto.

Addendum: March 21, 1997

I wrote the above recollections in April 1995 – more than 50 years after the summer of '44. By one of those rather incredible coincidences that appear to occur all too often and support the observation that "it's a small world", I met Elizabeth Solsberg, (nee Austin) at the West Vancouver Recreation Centre this morning. Her T-shirt bore the unusual legend "DOWNTOWN MAGNETAWAN" which prompted me to ask if it was anywhere near Swastika. Although she knew of Swastika she spoke of Magnetawan's proximity to Chapleau. John and Mac Austin are two of her uncles. I have promised to send her a copy of these recollections with the hope that it will lead to a further contact with John and/or Mac Austin and an opportunity to correct any obvious errors and hopefully add recollections of their own.

Addendum: May 28, 1997

My above recollections, although faulty and hazy in many respects, have now had the benefit of being scrutinized, with corrections and most interesting input from both Mac and John Austin – Mac by letter and John by phone and letter. Although I had made an effort to locate both Mac and John when writing up the original draft in 1995, it was without success. I am now looking forward to an opportunity in the near future to renew our acquaintance, hopefully with all three: Mac, John and Bunt.

I have also learned that Mac and John's father passed away suddenly in 1945 and that the mill burned in 1949 and was not rebuilt. Richard Austin visited Dalton Mills in 1985 and said that, apart from the family cottage, which remains standing, there is no sign that man ever set foot there. "No tracks, no burner, no mill site, nothing."

Uncloistered Halls

My evolution from high school graduate to university freshman, though this term was never used then, was to be an unorthodox upheaval, for a variety of reasons.

Once again I found myself in a world almost totally devoid of females, but instead of being among the oldest boys in an age range of 13 to 18, I was suddenly transported into a world of men - real men. The average age of the 3300 engineering students in first and second years in the Faculty of Applied Science and Engineering, commonly called the School of Practical Science (SPS) at the University of Toronto (U of T), was about 24. There were five women. Over 80 percent of this horde were veterans from the armed services, only recently returned from the most personally threatening experience that most would face throughout their lives. For some the experience had been too traumatic and they would not survive into the second academic year.

Once again I was in an institutional environment with many similarities to that which I'd grown accustomed to at boarding school. It was not to be a leisurely immersion in high academia among cloistered walls. The return of veterans from World War II that was to give rise in subsequent years to the "baby boom" generation had a dramatic impact on university enrolments. The undergraduate population at U of T suddenly grew from about 10,000 to 17,000 almost overnight. Fortunately this had been anticipated by Dean C.R. Young, who solved the problem when he discovered the recently-vacated munitions plant at Ajax.

The 'Ajax facilities' were transformed to provide accommodation for the majority of students who wished to be on site rather than commuting 60 kilometres or more each day from the Toronto area. The 'annex' complex that developed included 'U'-shaped single-storied wartime buildings capable of accommodating 100 students, who shared rooms with an average floor space of 8 x 14 feet, equipped with two bunks, two desks, a wardrobe, one window and one door. They were cosy to say the least and occupied both sides of the arms of the building, accessed from a centrally located corridor. The main entrance at the bottom of the 'U' gave onto a large lounge, flanked by bathrooms containing toilets, wash basins and a large trough-like urinal.

I mention the latter as several ingenious individuals contrived to improve the appearance of the facilities at a dance held at our 'residence' with a courageous group of Bell Telephone ladies from Toronto. The urinal was thoroughly cleaned, its exits plugged and the trough filled with water to accommodate an assortment of goldfish for the occasion. A similar degree of acuity was demonstrated by others who installed temporary wiring in the ladies' lounge with disguised microphones in order to listen in on conversations concerning their dancing partners.

The complex had a large cafeteria and separate recreational and lounge buildings, store, medical and dental facility, etc. Lecture rooms were distributed some distance from the living quarters and consisted of separate buildings interconnected by covered walkways which extended almost to the living quarters. Transport between lecture rooms and the main part of the complex was provided by buses and large vans. The buses, equipped with seats, were used by faculty and occasionally by students whereas the vans, called Green Dragons, were designed to accommodate a mass of standing humanity.

The cafeteria was destined to be immortalized in verse, not so much for its essentiality as for a couple of its residents. One was 'that shaggy dog named Red' who ate everything going, and the other 'the maiden at steam-table three' who whetted engineers' appetites.

In February 1948 on a cold overcast day, I walked over to the Ajax water tower, over 130 feet in height, which was situated within the complex. With considerable trepidation I climbed the unprotected ladder leading to the top of the structure to take some photos of the bleak surroundings.

Our buildings were in the foreground, within the complex area. To the north the main highway to Toronto skirted the perimeter of the SPS subdivision and the village of Ajax with its block of about 100 almost identical houses. Lake Ontario would have been visible to the south, were it not for the low cloud cover; however, the string of lecture buildings with their interconnected covered walkways were well exposed. The general impression was reminiscent of photos taken of the concentration camps in Europe during World War II.

Considering the large number of individuals suddenly brought together and assigned to share accommodation for about eight months with an unknown roommate, the results were impressive. Our residence included two veterans from the disastrous Dieppe raid of August 19, 1942 when an Allied force composed of 5000 Canadians and 2000 British troops with small attachments of American Rangers and Fighting French landed on the French coast. The raid was launched with the object of gaining information for the subsequent invasion of Europe, though the invasion was concentrated over 100 km to the west at Normandy and took place almost two years later in June 1944. The two survivors of the raid in our residence had both been taken prisoner, but had never met each other before. They were thoughtfully assigned as roommates.

I was very fortunate to share a room with Bert Maxwell during our two years at Ajax. We rarely attended the same lectures, even though the curriculum for first-year engineering was essentially the same for all engineering students. Bert was studying chemical engineering. We would meet at dinnertime and spend at least two hours studying in our room prior to turning in at about 11:00 pm.

Our weekly workload was prodigious. In the second year the course content for the eleven degrees offered in engineering became specialized, with the exception of mining engineering and mining geology, which remained almost identical, specialization occurring in the last two years. In our last three years we wrote exams or were granted grades in an average of 21 subjects, including 9 labs and a thesis. Without evening or weekend study sessions, we averaged about 35 hours per week in lectures and labs.

A very high percentage of the students in both mining engineering and mining geology worked every summer, the earnings being sufficient in many cases to provide for room and board and tuition fees. In addition a degree required mandatory experience for a summer in a mining or geology related assignment. Most mining companies were cooperative in assisting geology students in this regard.

The motivation shown by the veterans in pursuing the opportunity for a university education was overwhelming. Their maturity was generally very evident. After what they had been through, they were impatient to get on with their lives, or what was left of them. One of the older individuals at one of our lectures hollered out to the professor "Jesus, we covered that in high school 20 years ago!"

S.W. Wright, a classmate in mining engineering who missed one of his years, persisted and graduated in 1951, the same year his son was attending first year engineering. Father was 61.

As I will soon reveal, my university experience was by no means all work and no play. However, our so-called social life was highly restricted and we rarely had the opportunity to get involved in university-related activities in Toronto in our first two years.

Much of my spare time was taken up with athletics, including boxing in my first year, and track & field and harrier (cross-country) events in all four years. Apart from participating in track & field events for SPS at the intramural level, I was on nine senior and intermediate intercollegiate championship teams.

In track & field our Canadian intercollegiate opponents were McGill, Queen's and Western universities in the senior category, and Ontario Agricultural College (OAC), Western and McMaster Universities in the intermediate category.

We had similar opponents in the Canadian intercollegiate harrier races, which were 4.5 - 7.0 miles (7.2 - 11.2 km) in length. The courses at McGill (Montreal), Queen's (Kingston) and OAC (Guelph) were the toughest.

The intermediate intercollegiate cross-country meet at Guelph in 1946 was my first on an intercollegiate 'harrier' team. The 7-mile course was also the longest I encountered during my four years at U of T. The race remains particularly memorable, not because I came first, which was not the case, but because the combined effort of our first four finishers earned us the championship over OAC, Western and McMaster. My competing on that day with the team was a matter of choice I made between two opportunities. In fact, I never seriously considered the alternative, although others certainly did.

Most of my classmates who graduated from Upper Canada College in 1946 and who went on to U of T were 'rushed' by various fraternities on campus. I was not particularly concerned whether or not I joined one and had really not seriously weighed the pros and cons. However, I was invited to become a pledge of Kappa Alpha, considered particularly prestigious, whereas I should probably have considered Zeta Psi, if I were really interested, as many of my friends, including John Armour, Jim Biddell, Rod Mclennan and others from UCC were to become members.

At my final meeting with KA representatives I was informed that my initiation had been planned for the following Saturday and in keeping with my athletic interests, would consist of carrying a suitcase from the Toronto Campus at Queen Park to Ajax, a distance of about 30 kilometres. The initiation didn't bother me, but the date, which coincided with our intercollegiate harrier meet at Guelph certainly did. Blindfolded at the time, I informed the group, perhaps too bluntly, that the date was unsuitable because of the meet. I was particularly chastised by one individual who hollered "Just who the hell do you think you are Barr? What's more important, one cross-country event or a lifetime association with your fraternity brothers?" I explained the team aspect of the event and the reliance the others placed on my being there to compete - but all to no avail with this one particular individual. Having made up my mind, I told them that there was no question of my not participating in the meet and accordingly, I saw no further benefit in our proposed association.

Later, a KA member whom I highly respected contacted me asking if I would reconsider continuing as a pledge with a more convenient initiation date and I thanked him but demurred. The acrimonious nature of the previous discussion had affected my outlook on fraternities - perhaps unreasonably.

The Globe and Mail carried the following summary of the race at Guelph on that Saturday:

> "In Guelph Varsity Seconds just edged OAC over a snow-covered 7-mile course. Varsity men finished 2, 5, 6, and 7 for a score of 20, while OAC placed 1, 3, 8, and 10 for a total of 22. Western and McMaster were well back in third and fourth

positions. Smylie and Shwitzer captured first and third spots for OAC, while Barr, McMullen, Hedwin, and McNeill of Varsity finished second, fifth, sixth and seventh respectively. Smylie's time was 42 minutes, 26.2 seconds."

We also competed in separate harrier events against Alfred University and Buffalo State Teacher's College in New York State and Wayne University near Detroit. Finally, as our Senior Intercollegiate Team had done so well in 1948, we were the first Canadian intercollegiate team to be invited to participate in the U.S. National Amateur Athletic Union Senior Cross-Country 10 K race, held at the Warren Valley Golf Club in Detroit, Michigan on November 27, 1948.

Our harrier team in my first and second year included Dave Preston, recognized at the time as the best distance runner in the Canadian intercollegiate circuit. He was closely followed by George Doull, who became the dominant Canadian intercollegiate runner in my third and fourth years. George was in the Faculty of Forestry - a short, compact individual who was to lose only two races in our third and fourth years in the mile and longer distances in either track or harrier events.

Our meet against Wayne University in Michigan was held on a moderately hilly 4-mile course in the outskirts of Detroit on a rainy day in the fall of 1949. It was to be my closest finish to George. At one point in the early part of the race I was running just behind him in about fourth place behind two runners from Wayne. Suddenly the grassy slope steepened and both George and I slid down some 20-30 feet on our backs and rear ends, emerging in first and second place! George completed the first mile in 4:48 and the course in 22:04 while my time was 22:25.

The U.S. National A.A.U. Senior Cross-Country 10K race, also held in the Detroit area in 1948, was to be a real eye-opener for George, who was accustomed to finishing in first place. Twelve universities, colleges and clubs with about eighty runners and eight unattached individuals competed. The course was laid out to provide the centrally located spectators with an almost continuous view of the event.

As in all harrier meets at that time, the lowest score for the first four or five finishers in each team determined the team winner. Appropriately, Michigan State College was the winner with 28 points, Syracuse University second (59 points), Shanahan Catholic Club third (73 points), New York Athletic Club fourth (96 points) and U of T fifth (123 points).

Our individual placings for U of T were George Doull, 13th in 32:23; Gordon Wilson, 22nd in 32:55; Dunc Green, 24th in 33:06; myself 31st in 34:21 and George McMullen 33rd in 34:41. The finish was remarkable as two runners, well ahead of the pack, sprinted the last 100 metres to finish in a dead heat. It required a delay of about 15 minutes to develop film and name Robert Black of Rhode Island State College 1st and Curtis Stone of Shanahan Catholic Club 2nd, both in a time of 30:00.2.

Hec Phillips, our coach died, at the age of 59 in the spring of 1950. Born in Scotland, Hec developed into an outstanding middle-distance runner and was asked to represent Britain in the 1912 Olympics. Instead he emigrated to Canada, serving overseas in World War I, during which he also won several army championships in England and France. He was a member of the Canadian team at the 1920 Olympic Games at Antwerp in 1920 and in Paris in 1924, running in the 400-metre and half-mile events. He held the Canadian 800-metre record for eight years until the mark was beaten by Phil Edwards in 1928.

Apart from coaching the U of T track, field and harrier teams since 1933, he also drilled the Toronto police track and field team and the West End "Y" runners. In 1948 Hec assembled a 10-man distance team composed of runners from both the West End YMCA and U of T to compete in the International Silver Company 30-mile Relay held at Burlington, Ontario. The team included Herb Tilson, Ken Nevin, Kit Kitagawa, Fred

Crompton, George Norman, Earl Phee, Dunc Green, George Doull and myself. We won the event, the trophy having been last won by West End YMCA in 1909.

At the time, Paul Poce was one of Canada's outstanding middle-distance runners. In 1950 I ran for U of T in the 2-mile open at the inaugural Eastern Canadian Indoor Track & Field Championships in Toronto with Paul representing West End "Y", both of us having been coached for the event by Hec. Neither Paul nor I had any illusions about the outcome of the race, for it featured Browning Ross, a U.S. Olympic team member from Villanova University. Ross won the event, setting a Canadian and track record of 9:40.6. Paul came second and I managed a very satisfying third. Ross went on to win the 1,000 yards open in 2:23.3, a track record, on the same evening.

That night the junior mile was won by 16-year-old Richie Ferguson of the North Toronto Track Club in a Canadian and track record time of 4:33. Richie was to distinguish himself four years later at the British Empire (Commonwealth) Games when he represented Canada in the famous Miracle Mile duel in Vancouver between Roger Bannister of Britain and John Landy of Australia. In a later chapter I mention Rene attending this event in my absence with a friend. It also featured the agony of Jim Peters of Britain trying to finish his marathon run, which was to be his last.

In the Intramural Indoor Track and Field competitions in 1950 I ran for SPS on two occasions in a one-mile event, setting a record of 4:29.2 one week and 4:29.0 two weeks later. Records are made to be broken and mine was no exception, lasting only a couple of years.

We travelled to various meets both in Canada and the U.S. generally by bus. In the fall of 1946 a group of us was en route to a harrier meet with Buffalo State Teachers College. On arriving at the U.S. border, a custom's official boarded the bus, asking each of us for our birthplace. Ted Gawinski was born in Poland and of course I was also a foreigner being born in Colombia. Both of us had to get off the bus where we were considerably delayed while our immigration status was checked. The delay was lengthy enough that we were provided with a police escort to the College in order to meet the scheduled starting time for the race.

From then on, during any of our events in the U.S. my birthplace changed quite dramatically to Toronto and Ted's to another Canadian location and we had no further problems!

While at university several incidents occurred which were not part of the curriculum. These ranged from simple pranks to some bizarre events, which in other environments might have resulted in civil action. They revealed the rather earthy or even infantile humour frequently ascribed to engineering students. One of these pranks in which I was directly involved as a last-minute assistant revealed the skillful planning, execution and timing of the older veteran members who participated.

"Engineering Drawing" was one of our lab courses during first year. The only implements involved in the course at the time were a drafting set, plastic triangles, French curve, black ink, set-square, thin aluminum perforated plate for eradicating minor errors and drafting paper. Our final major assignment was a 20-hour drawing, the excellence or lack thereof of the completed item weighing heavily on the final mark.

During one of our last labs, I noticed a group of my classmates clustered around a drafting table. Joining them I was stunned to see an upturned ink bottle on the side of the drawing with a huge blob of ink spilled across the drawing. We were soon joined by the drawing's owner, who almost collapsed when he saw the disaster. The perpetrator quickly reacted, picking up the prefabricated glossy blob, to be chastised by only a few of the onlookers.

The most widely publicized incident occurred while I was at Ajax in 1947. It involved the 'Skule Cannon', an unobtrusive implement consisting of a section of iron pipe about one foot in length, closed at one end and supported within a crudely crafted base. Gun powder rammed into the object could be ignited through an aperture resulting in an impressively devastating blast. The 'Skule Cannon' accompanied the SPS Toike-Oike band at appropriate functions.

Following a parade in the fall of 1947 in the university grounds in Toronto, the 'Skule Cannon', which only weighed about 10 - 15 pounds, was carelessly handed over to a nearby individual, assumed to be an engineer for safe-keeping. The recipient, a 'Meds' student, carted the item over to the Medical Faculty building.

Within a remarkably short period, considering the circumstances, the error was discovered, and a delegation of engineers entered the Meds building searching for its treasured cannon. Unable to locate it, the group left the building with a hostage in the form of Robert Hetherington, a future president of the Meds Student Administrative Council, who was transported to Ajax. He was held there for about a week while negotiations for his release and the Skule Cannon's return were in progress. I understand that he was adequately secured at night and attended engineering lectures with his bodyguards during the day. An abortive attempt to rescue him was made by three bus-loads of protectively clad Meds students. Finally, with due pomp and ceremony, the exchange was made on the university grounds the following weekend. The Engineers always thought that the Meds group gained on the exchange as a record of the Skule Cannon's capture was inscribed on the cannon. In 1949 a larger and certainly more traditional cannon mounted on wheels was obtained to replace its dishonoured predecessor. It was manacled via a chain to two attendants at future rallies.

Over some altercation between University College (UC) on the Toronto campus and the Engineers, UC students arriving for lectures one morning found that the arched entrance into the main building had been bricked-in overnight. Naturally and quite correctly the Engineers were accused of having committed the desecration and the Engineering Students Administrative Council was requested to pay $300.00 for the required repairs. Grudgingly the amount was raised and converted to very small denomination coins which were transported by wheel barrow to the university bursar's office where it was almost refused.

I later learned of another prank associated with the UC building. One of the lecturers had a Volkswagen, which he parked near the building. On arriving at his lecture room on the third floor one morning, he was greeted by a full classroom, including his vehicle balanced on his sturdy lab table. After the requisite amount of pleading the class carted the vehicle back down the spaciously accommodating hall and stairwell to its parking spot.

Pranks with vehicles, almost all Volkswagens according to my son David, are evidently appealing. While he was attending Engineering at the University of British Columbia (UBC), 'Engineers' cut a Volkswagen in half with a torch and re-welded the two cut halves together around a tree on the university campus.

Some years later, UBC Engineers managed to successfully suspend a Volkswagen with cables from the deck of the Lions Gate Bridge in Vancouver. Notwithstanding the justifiable criticism levelled at the event for safety reasons, the careful planning, execution and eventual retrieval of the vehicle were considered of sufficient merit from an engineering aspect that a paper covering the exercise was published in a journal of the BC Professional Engineer. I have never learned whether the write-up served as a useful reference to other undergraduate engineers.

My minor involvement in one of these pranks occurred in my final year when I was in residence at Devonshire Place in Queen's Park, Toronto. Apparently some minor dispute with nearby residents of St. Michael's College led to the event. I arrived at the rendezvous area near the front lawn below their residence to join a group equipped for retaliation, the details of which were completely unknown to me. Within a matter of minutes, I was provided with a mallet and some stakes and asked to place them at designated distances on large letters being outlined by lime on the grass. These letters "SPS" covered an area of about 10 x 25 feet. Another group quickly followed the stakers with gasoline, to be followed by others who lit the gasoline-soaked letters.

The fiery display was quite dramatic and very rapidly attracted spectators from the two upper stories, who shouted appropriately profane and threatening comments at us. Luckily we eluded capture or identification and fortunately the episode received little if any coverage in the media.

Prospecting in Northern Ontario

Near the end of my first year at University of Toronto I applied with other classmates for work with a mining company that would provide me with some geological experience. Because of the high number of individuals enrolled at mining engineering and mining geology courses at that time, I guess we were lucky to get any kind of summer job offering experience related to the mining industry.

I was accepted for employment by Wright Hargreaves Mines Ltd. in Kirkland Lake, Ontario and was hired on as a surface labourer on May 3. I managed to get accommodation at Mrs. Jacks' boarding house at an unbelievably reasonable rate for room and board. Her facility was well known and vacancies were rare because of the excellent meals.

My work during the first month included a considerable amount of time digging out frozen tailings ponds and skimming the bark off logs destined for cutting into mine timbers. Whenever I got a chance I would wander off into the bush searching for outcrops and mineralization of any kind. I worked with older men in addition to one student from Queen's University. Several weeks into the job I was met at the mine gate coming off shift by a bearded miner who had been working with me on surface and noted my eager but fruitless prospecting efforts. He motioned me to follow him and when we were clear of the property, handed me two samples from the nearby Lakeshore Mine carrying abundant visible gold. I was overwhelmed with his kindness and later gave one of the specimens to the Queen's student. In the late 1950s my sample disappeared from the mantle in our first home in Richmond.

In early June I worked for two weeks as a ventilation engineer's helper to G.R. Yourt, the work consisting mostly of drafting. It was followed by another two weeks on surface labour under a foreman named W.E. Duval. Part of my work included planting small birch trees around part of the Wright Hargreaves' property. Thoroughly discouraged by this time I asked Mr. Duval if I could take a day off to seek a job at Noranda which could provide me with some mining or exploration related experience. He agreed and I spent a day in Rouyn and Noranda visiting different mine offices, but to no avail. On my return I learned that Ned Nelson, the Chief Geologist at Wright Hargreaves had heard of my quest and arrangements were made for me to accompany two prospectors as their helper. I occupied this position from July 10 to August 27, returning to Toronto on August 29 to prepare for a supplemental exam in Calculus.

The 41 days with Larry Merrill and Sherman Tough were a great experience and a wonderful change from my earlier chores. We travelled mostly by canoe, prospecting in several townships north of Kirkland Lake with bare-bone camps set up at Beaverhouse, Splashwater and Burnett Lakes. The work included considerable back-packing, portaging canoes and rock-bashing. I couldn't have had more congenial instructors and companions. In addition, the fly season had essentially passed and the only real bites I suffered were from five hornets while on a traverse.

Student Geologist in North Quebec

Having had some experience with Wright Hargreaves in 1947, I was offered a job for the 1948 field season with the company as a student geologist. The work period lasted from May 1 - August 19 and included 111 days of camp preparation, line cutting, prospecting and geological mapping in the Waswanapi Lake area of northern Quebec.

Our second year exams finished in late April and I was invited to spend some time with Doug Bedford, a friend in second year mining engineering who was also going to be working with me during the summer. Doug lived in Grimsby, Ontario and we both enjoyed a welcome break prior to boarding a train for Amos, Quebec, our planned point of departure with other members of the crew.

Our party chief was Jack Harris, assistant geologist at Wright Hargreaves with considerable exploration and underground experience since graduation from McGill University in 1929. Joe Martin, an ex-underground mining captain was with the crew in its early stages as was Rusty Clark from Swastika who helped with camp construction and drafting. Another student geologist with the party was Jim Stewart from Queen's University at Kingston.

We spent about 10 days building a log cabin, a couple of tent frames and a dock. The cabin didn't really appear to be a necessity, however, it was deemed so and earned the name Martin's Folly. We used it principally for cooking and dining. We had no radio communication. However, we had two canoes plus an outboard motor and there was a fire ranger's tower about 5 kilometres away equipped with a radio and manned for most of the summer by two French Canadians. We were to use it on several occasions.

Doug, Jim and I all shook our heads in wonder and consternation when we helped unload the additional groceries purchased on the second flight by Jack Harris. There were umpteen dozen eggs in their cardboard containers which we stacked outside between two trees, much like cordwood. Whether this had been a mistake we never knew, however we were to eat them throughout the summer - the whites of the eggs totally green within a few months. Apart from that we had sufficient fresh meat for several weeks each month following supply trips, which was supplemented by canned meats, fish caught in the lake, reasonably abundant grouse and moose on several occasions purchased from a nearby Cree village. After Martin's departure we took turns cooking dinner, the designated cook coming in early from the bush. We usually had sandwiches in the bush at noon along with tea and fended for ourselves at breakfast with the exception of coffee. Martin insisted that there was only one way to make coffee, that being to add sufficient grounds to a very large pot at the beginning of the week for the first day, adding additional coffee during the following days until the pot was about half-filled with grounds by the end of the week. On one of my allotted turns to cook dinner, I had killed a couple of grouse which I cooked first, adding carrots, potatoes and onions to produce a mulligan. Dinner was almost ready when Rusty Clark came into the cabin just after I had spilled the entire contents on the floor. We swept the floor periodically, but being composed of axe-planed logs, it contained a healthy veneer of assorted origin. Rusty agreed it would be a shame to waste the meal so I gathered it all up, placing it back in the pot and swilled it out with boiling water. No one complained!

We had a number of unscheduled visitors who arrived either by plane or canoes, including Alf Townsen and Warren Mackenzie; Sherman Tough and Larry Merrill and Albert Zeemel, a well-known prospector and Bill Houston. Albert Zeemel brought in a couple of bottles of Seagram's VO and when adequately lubricated, he demonstrated his prowess with a canoe, raising himself into a steady headstand. I still have a photo of this performance.

We had a very large property to cover which required considerable line-cutting and mapping. I can only recall Jack Harris assisting in line-cutting on one occasion when we both were working together on one line and Doug and Jim were on another. Jack had forgotten his fly dope and it was about the worst day I can remember for black flies and mosquitoes. I loaned him my bottle as required. He finally had had enough of the flies and decided to head back to the camp and left me to carry on by myself. Sometime later when I went to add more fly dope, I realized that Jack had taken it with him. I didn't last very long without it. Every time I raised my axe before swinging to cut, my face was engulfed by flies. I finally left and joined Doug and Jim for the rest of the day.

In August we elected to set up a fly camp for Doug, Jim and myself to complete line-cutting and mapping in the most distant part of the property at a point about five kilometres from our base camp. While we were there, Jim cut his leg quite severely with his axe. As we had no adequate first aid supplies at the fly camp, I ran back to the base camp for the medical kit. On returning I applied a heavy coating of sulfa powder and a secure bandage. Doug and I supported Jim on the journey back to base camp the next morning. We then climbed into a canoe and went down the lake to the ranger's fire-tower. We both climbed the tower and explained the situation as best we could in broken French, as neither of the rangers nor their contacts on the radio spoke English. We asked them to have Gold Belt Airways send in an aircraft for Jim as soon as possible as the cut required stitching. Satisfied that we had done all we could, we returned to camp. That evening we heard the drone of a distant aircraft around dinner time and soon the Gold Belt Airways Norseman hove to, the pilot yelling "Get that injured guy down here right away - this is an emergency flight!" Jim was standing on the dock with Doug and me. The pilot was quite irate when he realized that his patient still had both his legs. Evidently the message as received was that one of them had been cut off!

We didn't get any news about Jim's condition until we arrived at Amos on our way back to Toronto. Apparently a doctor had examined the cut which was not infected and considered that it was too late to be stitched. Jim spent several days in Amos recuperating at the home of Scotty Stevenson, a local pilot. He was almost ready to come back to camp when he fell down a flight of stairs at a party one night, opening the cut up again. He eventually returned home to start his third year at Queens in September. Doug and I arrived back in Toronto just in time to attend a three-week third year survey camp at Gull Lake, just prior to commencing the fall session at U. of T.

We were required to attend the surveying course between our second and third years. Students in mining geology and mining engineering had the option of attending the three-week course, either immediately following the completion of the second year exams in late April and early May or prior to the start of third-year courses in late August and early September. Most of my geology classmates attended the spring survey camp with several mining engineering students, whereas I attended the fall session with four other classmates and most of the third-year mining engineering group.

Several good friends were among those at the fall survey camp, including Charlie Lockwood and Bob Ure in geology and Doug Bedford in mining with whom I had worked that summer at an exploration project in the Puskitamika Lake area of northern Quebec. Others were all in mining and included Jack Clark with whom I roomed on Church Street in Toronto during my third year; Bob Fahrig and Doug Williams with whom I drove to Calgary in search of a summer job in 1949 and Don 'Doc' Coates with whom I have maintained contact through the years on various work-related and other associations, including playing with the Prospectors All-Star Hockey team in Toronto in the 1960s. The 5TO Mining Engineering class included a most congenial group of individuals who were to remain in close contact with each other for at least 25 years. Unlike our group in geology with a similar size of graduating class, the mining group were to have several reunions, the 25th being organized by Don Coates, then living in Vancouver, who arranged for a meeting at nearby Whistler. It attracted 17 couples of the 32 who graduated in 1950 and the 29 still alive in 1975.

The university's survey camp located at Gull Lake northeast of Lake Simcoe in Southern Ontario had many appealing features. It lay adjacent to the lake in a lightly wooded region with partly cleared sections which offered sufficient space for simultaneous field surveys to be carried out by two to three-person parties. Several steep rock bluffs near the lake were used to simulate underground mining conditions where surveys might be required from transits to stations lying at high angles from the instrument.

The survey camp consisted of several wooden buildings, the largest providing sleeping accommodation for our entire group of thirty in triple-decker bunks. Fortunately there were two fairly large doorways at either end of the building, which permitted ventilation; otherwise it could have been stifling. An added threat at our fall camp was the presence of an extremely good-natured individual who will remain nameless. He had a common anal affliction which emitted noxious gases without any audible warning but with such nauseating strength that it pervaded the entire building, commonly clearing out all humanity within a matter of a few seconds through both doors.

Another building contained drafting tables and survey equipment. Other buildings included a staff unit for H. Macklem, our survey instructor and B.J. Haynes, his assistant; cook house and washroom facilities.

A lengthy dock with a diving tower at its end were welcome facilities for swimming. There were also several canoes available, although rest and relaxation time was quite limited.

The course was very thorough. Graduates having completed the course were eligible for an Ontario Surveyor's Certificate by completing a six-month apprenticeship with a qualified surveyor. Two of our classmates were to take advantage of this career opportunity.

We were to learn that isolated trees within the overall survey area were used on occasions to accommodate Mr. Haynes. He would climb one of these trees prior to the start of a test survey to spy on nearby survey crews in an attempt to ensure that results were not 'cooked'. This could be accomplished by a confrère from another survey party providing others with the correct latitude and longitude of the final destination!

One of our exercises was to lay out a railway curve within the cleared fields between designated starting and finishing points. Occasionally, one of the crews would err in its calculations and nearby crews would marvel at the faith of the unfortunate crew that its curve would eventually close, when it was clear that they were obviously heading straight for the bush.

All in all, the survey camp was an instructive experience and the close association between classmates prior to the start of another academic year, particularly in such a rural environment, was a welcome break and led to several lifetime friendships.

Jug Hustling In Alberta - Summer 1949

By the end of my third year at U of T, I was aware of the emerging importance of the energy sector in Alberta with the widely heralded results of the Leduc oil field discoveries in 1947. 1 resolved to try to gain some first-hand experience during the summer in support of my fourth-year thesis, which I was sure would propel me to dizzying heights as a sought-after geology graduate specializing in petroleum.

Being almost totally absorbed in preparing for my final exams in third year, in addition to other apparently important commitments, I had failed to assign any priority to job hunting for the summer months. My principal concern was how to reach Alberta where I was sure that the doors of opportunity would be more than open. My problem was my very limited resources.

From a few preliminary inquiries I learned that Bob Fahrig and Doug Williams, both in third year mining engineering, had made arrangements to travel to Vancouver for summer jobs by delivering a car with all transportation costs prepaid. Such was our 'fraternity', that they welcomed me as a non-paying passenger. Bob, whom I have mentioned in association with episodes both at Ajax and following graduation, was a former submarine commander and Doug a highly decorated World War II pilot.

Following an uneventful trip across Ontario, parts of the U.S. and Alberta, I left Bob and Doug south of Calgary, and arrived there by bus on a Friday morning with ten dollars to my name.

Seeking a field job, I initially contacted Westby Geophysical Corp., a Canadian subsidiary of Seismograph Service Corp., a major U.S. company, active in Alberta. Mr. Leedy, the manager of Westby was encouraging, indicating that there was a possible job opening with a seismic crew near Red Deer and that he would contact me on the weekend. Encouraged, but not committed, I decided to check some oil companies for possible field jobs with their exploration division. Due to my delicate financial situation I accepted a drafting job early that afternoon with Home Oil following a brief discussion with their Calgary geologist who could not promise any field work and who recognized that I might accept a job with Westby Geophysical.

On Saturday I received a phone call from Mr. Leedy at the boarding house I had located. The rent was to consume 50 percent of my available funds. He came right to the point. "Our crew at Red Deer needs a recording helper right away. If you are interested, you should be at our office by 8:30, Monday morning. I'll drive you to the airport and fly you to Red Deer where you'll be picked up and driven to the work site." Of course I was delighted and agreed.

I arrived at the Home Oil office with my suitcase at about 8:00 a.m., Monday May 9. Soon the company geologist appeared and I quickly apprised him of the weekend development. Although he appeared surprised, he insisted on my receiving pay for half-a-day's employment, which I accepted although I felt somewhat mortified.

I changed into work clothes during our flight to Red Deer in Westby's small aircraft, piloted by Mr. Leedy. We were met by George Schedler, the crew's party chief and by 10:00 a.m. I was on shift!

Seismic surveys as used in oil exploration rely on the detection by a seismograph of seismic waves propagated by a localized ground disturbance. The principle is similar to that used in the recording of earthquake phenomena. In oil exploration shock waves from the discharge of dynamite in shallow surface drill holes emanate downward, and in reflection-type surveys, such as we used, are reflected upwards from interfaces between different underlying strata at varying velocities to detectors placed on surface.

The physical properties of underlying strata are sufficiently different to cause detectable differences in the behaviour or acoustic properties, of strata or other bodies. Detectors, called 'geophones' or more commonly 'jugs' by seismic crews, are connected to a cable about 300 metres long at specific intervals, i.e. about 30 metres. The cables are generally laid out on a road both in front and behind a recording truck containing the seismograph. Holes for the dynamite charges are spaced about 600 metres apart and in our area were drilled to depths of about 20 - 30 metres through overburden to accommodate the overall length covered by the jugs, termed a 'spread'. Although my position on the crew was specified as a 'recording helper', I was recognized as one of two 'jug-hustlers'. We certainly hustled, for on an average day we each carried and connected 10 jugs on the spread, disconnecting them after the blast, covering about 10 spreads during a 10-hour shift.

Our party consisted of a dozen or more individuals including replacements during the summer and an L.M. & S. Contract drilling crew. A land man negotiated permission with land owners, mostly local farmers, for drilling holes adjacent to local roads. Drill hole sites were surveyed by "Smitty" Smith and his rodman. The company's Mayhew 1000' drilling rig was operated by Pop Reed and his helpers Gord "Slim" Philip, Gerry Rudiger, Ron Williamson and "Buttons" Petrunia. The L.M. & S. Contract Fahling 750' drill was operated by Elmer Golem, Slim Hoffman, Andy Cooper and Simian Hall. Included with the drills were two water trucks which were required to provide water for the drilling operation. The drilling operation generally preceded the shooting and recording sessions by a day or more. There were strict regulations concerning the shooting stage with respect to pre-loading of holes, for obvious safety reasons.

"Mitch" Mitchell, a rangy Texan, was the shooter and Bruce "Casanova" Weatherhead was his helper. A genial individual, Mitch however had a mischievous streak. As seismic surveys were a recent innovation in the 1940s, visitors were attracted to the operations, particularly on weekends. They would drive up, generally parking some distance from the shooting truck and commonly asking if their presence was acceptable. Mitch never volunteered any information to the contrary, his common retort being "hell no, you're not in my way". He was however acutely conscious of wind directions and velocities, for they governed the possibility of a vehicle receiving a "direct hit" from the mush blasted out of the drill hole when the dynamite charge was detonated. The August 1949 issue of the *Seiscourier*, an SSC company publication, notes that Mitchell had struck again. "The bag this time was one shiny 1941 Plymouth standing to windward which received a direct hit by assorted shale cuttings."

To my knowledge, no vehicles were reportedly damaged; however, I recall seeing one completely covered with mud. The woman driving the car emerged unscathed and pleaded with Mitch, "My husband will kill me if I bring the car back in this conditions. Can you help me?" Mitch could indeed. He got the water truck to drive up and wash off the debris.

Gene Sandburg was the observer or recorder of the recording truck and Howie Devon and his replacement George Samchuk and I were his helpers. Gene was from Minnesota and proud of his Swedish background. He once told me that the Sandburgs were a unique family, being descended from a class at a military institute in Sweden, where the entire graduating class was re-assigned the name Sandburg. I have never had occasion to confirm this rather unusual story, however, I was not going to debate it with Gene.

I mentioned working on weekends. Our time off was governed by the elements. Weekends off were not of much benefit because of the distance to most of the personnel's homes. Accordingly Westby had an arrangement acceptable to most employees, whereby the time required to complete an average of 220 hours per month of work governed the number of days taken off at the end of the month. Accordingly we generally worked a minimum of 10 hours per day. This was to provide me with a week off, one month, which I spent around Banff. On another occasion I spent some time with Elmer Golem and his wife Terry visiting the Leduc oil field.

During the summer we were based at several different locations including a lengthy period at the Sandy Cove Hotel at Pine Lake, about 30 kilometres southeast of Red Deer, followed by a period in Red Deer where I shared a room with Smitty. I recall our stay at Pine Lake as rather an austere existence, as we had little time off, missing many days of work due to unprecedented periods of rain; we had to be available on short notice when weather permitted.

I'm not sure whether the depressing weather motivated an unusually boisterous party one evening. However, as the night wore on, several of the more uninhibited members of the crew lost control - not that they ever had such an abundance of this quality that allowed for much loss. This resulted in considerable damage to furniture from being hurled out of a second-storey window. The consequence of this deplorable situation was that we were instructed to leave the premises forthwith.

The afore-mentioned issue of the *Seiscourier* is obviously not a source of reliable information. Schedler was to report our departure as follows: "Following a pleasant farewell party at the Sandy Cove Hotel, Pine Lake, the crew took up its new headquarters in Red Deer, the home of the World Wheat King, in the heart of Alberta's summer wonderland."

During my final two weeks in Red Deer prior to returning to Toronto, I met Laurie Willis, a waitress at a restaurant in Red Deer. Although we were not to spend much time together, we enjoyed each other's company. We attended several casual gatherings with the crew, one of which included a somewhat unusual event. Leaving a party with two of the crew in the front seat of a car in a freak snowstorm, Laurie and I were necking in the back seat when we suddenly found ourselves rolling over, out of control and coming to rest in a heap on the inside of the roof, still clinging together but not at all injured. Obviously unable to see, the driver had driven the car up an embankment beside the road and rolled it over. The car didn't weigh that much because we had no trouble rolling it back over on to its wheels and continuing on our way.

On September 14, I left by Greyhound bus from Red Deer to return to Toronto. On the bus I met a young lady from Winnipeg, which helped to while away the hours. The few hours when sleep seemed possible, it was good to have a shoulder to lean on. She invited me to break my trip in Winnipeg, but although tempted, I carried on, finally stopping in Detroit where I had arranged to spend a day with a friend in a dentistry course at U. of T. with whom Jack Clark and I had shared a couple of rooms at a boarding house on Church Street in our third year. I must have been great company, as I retired for a welcome bed at about 8:00 p.m., not to rise again until noon the following day.

I spent my final year in significantly improved student accommodation to that on offer in my first three years. I shared a two-room suite at 3 Devonshire Place in Queens Park with two third-year engineering students, having a room to myself.

My final marks with honour standing were more a result of compatible course content than anything else. I received 3 As, 17 Bs and one C for a third place standing in the class of 30 – but in a different world from Barney Peach, the only honours graduate in our class with 15 As, 3 Bs and 3 Cs. Barney went on to a career in petroleum geology with Shell Oil in Alberta and was one of the few graduates in our class to be offered a 'permanent' position with a company prior to graduation.

One Lucky Canuck

A review of *Torontoniensis* a 500-page yearbook dedicated to the members of the graduating class of the University for 1950 shows 1075 graduates in the various engineering disciplines. It is startling to see page after page of photos, only two of whom are women! No wonder one of them listed 'Ajax men' among her interests, associations or accomplishments.

A most unexpected entry in the 'Torontonesis' was the following item under 'S.P.S. Athletic Association':

Bronze "S"

> "Every year one member of the graduating class is voted to the exalted position of holder of the Special Bronze S for his outstanding achievements in athletics, both for the faculty and the University. This is perhaps the highest award any Skule-type athlete can receive and Dave "Ditz" Barr is fully worthy of the honour.

Throughout his four years' servitude under the Blue and White, Dave has amassed the fantastic sum of 85 points for athletics, in all forms of Track and Harrier both on the cinders and around the executive end of the job. He has been a star performer with the University Harrier squads in recent times, and led the SPS team to a very hotly contested second place in the indoor track meet at Hart House this season.

When his classmates went to the polls in February they made no mistakes about the location of the all-important X, with these pleasing results.

Horses & Packers

Although in 1942, as a 13-year old working for the summer on a farm near Thornhill, Ontario, I had obtained some experience around horses and had actually mastered the difference between 'gee' and 'ha', my principal responsibility was feeding the pigs. In 1952 I had my first experience with the use of packhorses for prospecting and geological parties largely dependent on these beasts of burden for most of their day-to-day transportation needs. The packers I was to meet and rely on prior to the transition to helicopter travel several years later were all true frontiersmen. Comfortable with the outdoors, self-reliant, they were at ease with themselves. They were to include Buster Groat, Cariboo John Bendicksen, Albert Alexander, Willard Freer and the legendary Skook Davidson. My time with them was far too brief.

In 1952, as a junior geologist working with Kennco, one of my responsibilities during the summer was to maintain monthly contact with a prospecting party working in the Sicintine Range, west of Bear Lake in north-central B.C. The party consisted of Axel Berglund and another prospector and was supported by Buster Groat as packer with about five horses. As the time Buster was in his early thirties, a handsome, genial, well-muscled six-footer. My liaison role was to visit them by float-plane at the nearest lake and check on their progress and ensure that they received their supplies for the following month when I visited them again. They had no radio communication. On my first visit I was horrified to see a large crescent-shaped freshly healing scar on Buster's forehead. He had been kicked by one of the horses and a wound that should have received hospital treatment and at the very least 20-30 stitches had been pulled together as well as possible with tape.

During the fall of 1952 I was to spend two to three weeks completing a bio-geochemical survey of the Dorothy-Elizabeth claims in the Omenica River area in north-central B.C. My companion was Albert Alexander, an Indian packer from Fort St. James. I was to learn the rudiments of the pack-horse business as I helped Albert cinch up saddles and loads. He taught me the packer's knot for tying off loads, which I have found useful ever since. I experienced my first numb-bum from riding on a pack-saddle softened by blankets and would walk as far as I rode in our journey from Germansen Landing into the property and back. If you have ever walked as a member of a pack-train you have probably noticed that horses have a habit which is particularly demeaning to the hiker, who feels subdued enough as it is, since he is walking while the packer is riding majestically on his appropriately saddled horse. On climbing up inclines, horses are occasionally given to measured ejections of excrement which are synchronized to their footsteps and completed with shattering resonance; the entire performance providing a severe test of the hiker's three principal sensory organs.

One of the more exciting events in that learning period was crossing the Omenica River on horse-back. The river is not particularly swift, but it was deep enough where we crossed that the horses had to swim part way.

In 1953 we used pack-horses throughout the summer during a four-month program in the Nina Lake area – again northwest of Germansen Landing in the Omenica River area. By this time the B.C. Dept. of Mines, predecessor to the B.C. Ministry of Energy, Mines and Petroleum Resources, was building the Omenica Road – a road to resources – northerly from Germansen Landing, and the Omenica River had been bridged. A rough

Jeep road provided access for six miles west on Germansen Landing and a pack-trail continued for 20 kilometres to our base camp.

I was the party-chief of a program designed to explore the lead-zinc-silver potential of a limestone belt traced for 50 kilometers northerly between the Omenica and Osilinka rivers. An average of 16 men were employed from late May to September in several capacities including geological assistant, prospectors, labourer, cook and a packer. Initial supplies and equipment were trucked from Fort St. James, about 200 kilometers north to Germansen Landing. These were shuttled by float-plane to Nina Lake and from there by a pack-train of nine horses to a base camp. During the season seven additional fly-camps were established at points along the limestone belt in the course of trail cutting and geological mapping.

The horses packed in our total requirements. The heaviest load I can recall seeing loaded and packed was a rock-drill weighing almost 200 pounds, which had to be balanced off with a load of equivalent weight. This was packed by our lead horse, a large gray mare, the undisputed leader of the group. I learned that packhorses walk in a pecking order recognized of course by the packer. Occasionally one of the horses would get disgruntled with its position and would start to quicken its pace, casting nervous sideways glances as it came abreast of the horse ahead. The threatened animal would invariably react by increasing its pace and attempting to bite or kick the intruder. Pack loads would get knocked about if the trail was narrow and in timber. The experience could be quite terrifying for an individual riding on a packsaddle with its makeshift rope stirrups if the horses started to canter. I never did learn how to spell 'not-heads,' or was it 'knot-heads' that packhorses were called? The first time I heard the term it was used by Paul Hammond, Kennco's western manager in the early 50s.

In May 1953 John Anderson, John Greenaway and I traveled by Jeep from Vancouver to Fort St. James to initiate our 1953 program. We arrived at the Fort on an unbelievably warm sunny day, grimy with dust and drove directly to a warehouse on the edge of Stuart Lake where we kept supplies over the winter. Stuart Lake was still ice-covered though the margins of the lake were ice-free for about 30 metres from shore. The three of us peeled off our clothing and dove into the lake. I actually reached and touched the edge of the ice before returning to shore.

We spent a couple of days in Fort St. James sorting out our gear – the horses we were to use already having preceded us to Germansen Landing. We stayed at the Fort St. James Hotel, which was nothing more than a large home owned by Dan and 'Ma' Fraser, which could accommodate 8 to10 guests. On our arrival Ma enthusiastically led us around to show us the modifications installed over the winter, including electrical outlets, a plumbing system providing running water and I believe a bathroom to replace the outside facility.

Paul Hammond arrived while we were there. The four of us were sitting at the dining room table for dinner when Ma Fraser brought in the meal. Unfortunately it included sliced SPAM, which we all loathed, as it had been, and remained one of our more frequent meat supplements. Gentility under such situations was not one of Paul's noted characteristics, and he bellowed, "Ma get rid of the God-damned horse-cock and bring us something edible!"

The pack-horses made weekly trips between Germansen Landing and base camp to obtain fresh supplies, transport mail and out-going samples for assay. Again, we had no reliable radio communication; however, Germansen Landing did, and it could be reached in a day's travel.

Our base camp area was in a natural clearing containing a small pond and surrounding meadows - ideal for grazing the horses. On one of my frequent trips between base camp and the fly-camps, I was tired after several hours of hiking with a heavy packsack, as I neared our camp I spotted a black bear walking on the trail toward me, about 50 metres away. I decided there was no way I was going to get off the trail and kept walking. Fortunately, he took off into the bush as soon as he spotted me. About 10 minutes later, near our camp, I had just emerged from the trail into our clearing when I was spotted by several of the horses about 50 metres away. En masse, they galloped across the clearing directly towards me, rearing to a sudden halt as I stood petrified. The leader looked down at me, snorted (probably in amusement), and the pack returned to their grazing without a further glance at me.

One does not expect to have to deal with a dead horse at an inconvenient location. One of the horses had eaten that-which-he-ought-not-to-have-eaten and succumbed in a draw containing a small creek. The site was about one kilometre from our camp and right on our main trail. The situation was potentially dangerous as the sudden depression where the poor beast lay was only visible within about 15 metres from the trail. We realized that the carcass would be detected and eaten by bears within a very short period. The most sensible solution seemed to simply move the half-tonne carcass downstream some distance. Not having anything stouter than pack rope, that was what we used. The task prompted laughs and curses as about twelve of us strained, fell, staggered and fell again as we dragged the carcass about 100 metres downstream. Three days later there was no sign of our departed companion.

I remain eternally indebted to Paul Hammond, as he had married (now the late) Sibyl Deveny, a friend of Rene's from Swastika, Ontario. We were to meet in January 1953 through a blind date arranged by Sibyl for Rene who had arrived in Vancouver in late 1952. Therein lies another story – Sibyl mistakenly judged me to be older than my friend John Anderson and selected me accordingly. John and I were to be each other's best men at our weddings in 1953 and 1956.

Paul had used horses extensively in the Omenica River area while working at the Dorothy-Elizabeth and Lorraine properties and in other parts of the area in 1949 to 1951. He later accepted a position as Exploration Manager for Conwest Exploration Co. Ltd. in Toronto. Early on in his new position he was confronted with the need to obtain a bulk sample from a property in northern Quebec which was within four kilometres of the nearest lake suitable for access by float-plane.

"No problem", thought Paul. "We will use a horse". A suitable candidate was selected for the task and taken to Amos, the nearest point with a suitable float-equipped aircraft. The pilot pointed out that there could be a certain problem in moving the horse into the aircraft and restraining it for the flight. "You'll have to sedate the horse once it's on the aircraft or we could be in deep trouble".

Their being no veterinarian at Amos at the time, Paul consulted a local doctor who agreed to administer the sedative once the horse was on the plane.

Well, you can imagine the interest that developed once the plan became public. In a festive mood, about 200 individuals, including the mayor, were at the airbase to witness the horse's departure. Paul enquired anxiously of the doctor, "How much of a dose are you going to give the horse"? The French-Canadian doctor thought for a while and replied, "Well, a grown man weighs about 150 pounds. The horse weighs about 1,200 pounds. We will give the horse eight times as much."

Following considerable effort, the terrified animal was loaded onto the aircraft. The doctor administered the injection and with satisfaction, saw the horse slump to the floor. He was later to emerge crestfallen and announce to Paul, "Da horse, she is dead"!

I have already told you of our experience in moving a dead horse... I never did find out how they manhandled the horse off the plane.

When I knew him in the early 1950s, Cariboo John Bendicksen was as hardy a soul as I have ever met. A good-natured Laplander, he had a stubborn trait, which may have been partly related to frustrating years of experience with his equine charges. John Anderson spent a summer with him whereas my experience with him lasted only for several enjoyable weeks. John insisted that if you wanted Cariboo John to accomplish a particular task, you would simply tell him you wanted the opposite, or vice-versa.

I met him at Fort St. James in the spring of 1954. We had arranged for him to take some horses from nearby Vanderhoof into northern B.C. for a summer exploration program. He spent the winter trapping with another couple based in their cabin near Uslika Lake. Uslika Lake was about 60 kilometres by trail northerly from Germansen Landing at that time. John took eight days to travel on snowshoes from Uslika Lake to Fort St. James, a distance of about 260 kilometres. He travelled light, carrying only a small tarp as a bedroll, a rifle and sufficient ammunition, salt, chewing tobacco and a few other essentials for cooking. He relied almost totally on living off the land for food. I asked him whether he had had a good winter.

"I got badly frost-bitten and almost lost one of my hands". He showed me the affected hand, which, although scarred, looked almost normal. "My fingers went black and I knew I had to cut out the dead stuff." And he did, cutting it back to the bone until it bled.

His friends at Uslika Lake brought him a nice blue sleeping bag to convert him to using a warmer, more comfortable bedroll. Although obviously delighted with the gesture, when we spent a week together on a trip I noticed that he never slept inside it, zippered up. He simply wound it around himself and slept as soundly as ever, with an ear always alert to the tinkle of bells on the hobbled horses. He rarely had to venture far to locate them in the morning.

In 1955 we had a large prospecting program in the Selwyn Mountains in the Yukon which also included a geological party examining previously discovered prospects in a detailed program of mapping and sampling. The program was supported partly by horses but principally by helicopter. Our packer was Willard Freer, a lean cowboy who never seemed to change appearance, because he was almost perpetually garbed in leather chaps and a weather-beaten, sweat-stained ten-gallon hat. He worked for many years for Skook Davidson who maintained a well-established coral of pack and saddle horses for exploration crews and hunters at his base in the Kechika River area.

The actual circumstances of Freer's death a year or two following our program are uncertain. He is believed to have been thrown off his horse and killed during an encounter with a grizzly bear.

At the end of the 1955 season, en route through Watson Lake, I was sitting at a table in the bar of the Watson Lake hotel with a group of our prospectors and two of Skook Davidson's young Indian packers. Skook walked in, surveyed the occupants in the bar, spotted one of the packers at our table and walked over. Talking in a pleasant enough voice, he picked up the packer's full glass of beer and saying, "Why, you mangy, no-good little son-of-a-bitch", and adding a lengthy list of other unmentionable names, he very slowly poured the contents over his head without ever taking his eyes off him. The poor packer sat, as if transfixed, staring mournfully up at Skook as the beer ran down his face. I never did learn the cause of Skook's irritation, but I have never forgotten the complete lack of reaction by the packer.

That memorable incident was my introduction to a remarkable and colourful individual, truly a legend in his own time. Years later I was fortunate enough to be referred to Moira Farrow's excellent book, *Nobody Here But Us Pioneers of the North*, with its 22-page Chapter I entitled, "'Skook', Guide, Packer, Horse Rancher". The chapter is particularly meaningful to me in describing many of Skook's exploits, as references include contributions from John and Nancy Anderson, Hugh Gabrielse and George Dalziel, all of them friends or former acquaintances.

Frank Cooke, a former close companion of Skook's who had shared many experiences with him and who, for many years, was a highly respected and well-known member of the Royal Northwest Mounted Police in northern B.C. told me several stories about Skook and himself while we were working in the Highland Valley area in the late 1950s. On one occasion Frank and Skook and two others, one of whom was only a casual acquaintance, spent a boisterous night at Manson Creek during which they had far too much booze and several fights. The next morning, all four were sitting around, trying to adjust to a new day when the casual acquaintance surveyed the other three and observed, "What a mess, each one of you has at least one black eye." Skook looked up and noting his relatively unblemished face, winked at Frank, who sauntered over and knocked the other off his chair. "There", said Skook delightedly, "now you have one too!"

Introduction to Babies, Helicopters, Perennial Snow and Ice British Columbia, 1954

As the title suggests, 1954 was a year of enormous changes in my life. By January 1, I had been married to the love of my life for almost three days and life would never be the same.

I met Margaret Irene McLellan through a blind date arranged by Sybil Hammond (nee Deveney) a friend of hers who had lived in Swastika, northern Ontario, Rene's birthplace. Sybil had married Paul Hammond, who in 1953 was exploration manager for Northwestern Explorations ("NWX") based in Vancouver. At the time I had worked for NWX since April 1951 and I was sharing a suite in Vancouver with John Anderson who was also employed with NWX.

Rene arrived in Vancouver in 1952 having just graduated from the University of Toronto in the School of Nursing. At the time her brother Bob was a Junior Intern at Vancouver General Hospital and Rene decided to accept a nursing assignment at the same hospital. In January 1953, Sybil met Rene and quite casually said, "Rene, there are two great young guys working for Paul, and I believe that Dave is nearer your age, being the older. Would you be interested in meeting him?" Rene agreed and the four of us went to a movie together on January 30. For me it was love at first sight! By the greatest of good fortunes, Sybil had erred, as John was about six months older than I. Within two months I had developed enough confidence to propose a lifetime union and had been accepted, neither of us being deterred by the date - April 1. Rene fibbed a little about the length of time we had known each other prior to our engagement in breaking the news to her parents, making it what she considered to be a more acceptable time span.

Our time together was all too brief, before I had to leave in early May for the summer to supervise an exploration program north of Germansen Landing in B.C., with no opportunity to return to Vancouver until the end of the season in mid-September.

We wrote numerous letters to each other that summer, most of Rene's originating from Swastika as she decided to spend the period while I was away with her family and friends in that area. How I looked forward to the welcome sound of those horsebells, the approach of our pack train with mail on its weekly supply trip from our base camp near Nina Lake to Germansen Landing. In spite of the workload the four-month program seemed endless. I was, however, fortunate in having excellent company, which included both John Anderson and John Greenaway. We had a number of interesting experiences including the wettest July I can remember when we got a daily soaking and spent each evening trying to dry out clothing at an open fire near our tents. John Greenaway taught both John Anderson and myself the rudiments of bridge, which, with cribbage, provided our principal source of entertainment. I was to lose the grand championship for cribbage to John Greenaway and as an agreed penalty, I suffered the indignity of a brutally cold dip in our nearby pond, which at the time was close to freezing.

Part of our entertainment was unintentionally provided by Joe Keenan, a lithe, more-than-60-year-old Irishman, who, like many other cooks, was a good storyteller. He had a store of fascinating remedies, some of them quite unusual. One of his favourite expressions was "all you need is coal oil for the hair, salt for the teeth (pronounced 'tathe') and raw onions for the piles". On one occasion he got an infection on his arm and after some searching near camp located some dandelions which he applied as a poultice. He was to awaken that night in agony with his arm still quite infected but burned in addition. His pillow was perpetually black from coal oil, for he practiced what he preached and true to his word he had a thick mass of curly hair, albeit of an odd colour, not quite grey, not black, but very oily.

From past experience with black bears around camp, he kept a bucket of kerosene beside the door of the kitchen tent, which contained a broom handle to which he had coiled some old rags around the end that sat soaking in the bucket. Sure enough, one day there was a considerable commotion as Joe chased a bear from the kitchen tent area with his flaming torch.

Most of us grew beards, John Greenaway's being a luxuriant thick growth of dark hair with a vivid brownish tinge. John Anderson's resembled Fu Man Chu and mine was a disaster, totally lacking any distinctive quality. I wisely shaved it off before returning to Vancouver but made the mistake of retaining a moustache to which I had become quite attached, or vice-versa! When I went out to the airport in Vancouver in September to meet Rene returning from Swastika, her eyes glazed over when she saw me, and it was quite obvious that she was thinking, "My God, what have I let myself in for?" I shaved it off within the hour.

Our reunion was all too short, as Rene had to return to Swastika to prepare for our wedding on December 29. She left Vancouver on December 1, so that our total time together before we were married was a little over six months.

Rene's bridesmaids were to include her sister Ann, university classmate Helen Langford (nee Lindley), close friends Shirley Wright and Joan Wyatt. My best man was John Anderson and the grooms were Gerry Noel, Ken Nevin and Duke Mullqueen. Gerry was working as a geologist with NWX at that time; Ken was a close friend from the University of Toronto who shared an interest in track and field and Duke was another close friend from my Upper Canada College days.

John Anderson and Gerry Noel very generously agreed to accompany me east for the wedding and we drove to Toronto in five days, arriving in Toronto on Christmas Eve. They carried on to Swastika after dropping me off at a used car lot where I purchased a blue 1949 Meteor for $1,000. The car with its stylish visor was to be our pride and joy for about six years. It was stolen for joy rides on no less than three occasions, the last time recovered by the police in one of those wide and deep drainage ditches in the Richmond area, south of Vancouver.

I approached Swastika with a fair degree of trepidation, for I had not met Rene's parents or any of her relations or close friends, other than her brother. I need not have worried, her family could not have made me feel more welcome, particularly when they went along when Rene insisted that none of the Christmas presents should be opened until my arrival!

In the two days prior to the wedding we had a rehearsal at the Swastika United Church, an underground tour of the Larder Lake Mine for the best man and groom and an informal get together for the wedding party one evening. I recall that this included a game where we sat around in a large circle and one of the party selected the first letter of a word, the object being that those following in succession were not to complete a word by adding further letters. Poor Duke provided an "f" after "wha" to everyone's consternation. Joan spoke up: "Duke, surely you are not thinking of the word waffle?" which alas was the case.

The wedding in the small, attractive United Church was a very warm affair, decorated pew ends containing sprigs of holly forwarded by my parents from Duncan on Vancouver Island, and other Christmas decorations abundantly displayed. The temperature outside the church was quite the opposite at minus 38º Fahrenheit and only the photographer ventured out with Rene and me for a numbing photo beside our car whose door had frozen closed during the ceremony.

The next day we drove off on our way south, Rene's father having kept the car inside the McLellan's busline service building overnight to ease our departure. We got about eight or ten miles down the road before the radiator boiled over and we somewhat reluctantly called Rene's father for assistance.

From Toronto we drove southwest via Detroit to St. Louis, where we had a casual encounter with another couple also on their honeymoon but in their sixties! On our way west we visited Carlsbad Caverns, the Grand Canyon, Petrified Forest, Kennecott Copper Corp.'s Chino copper mine in New Mexico, Yosemite Park and San Francisco; finally driving up the Pacific Coast highway, to arrive in Vancouver the day after the second heaviest snowfall the city has suffered in the last 50 years.

Our first 'home' was sub-let from our friends John and Gwen Greenaway and consisted of the second floor of a building at 2267 West Second Avenue which was owned by Joe and Mary Muscroft who occupied the ground level. We were to reside there until September, when we moved to an apartment on 71st Avenue in the south Granville area in preparation for the arrival of Susan on November 29.

My very early role as a father was eased considerably by the welcome presence of Rene's mother who had traveled out to Vancouver just prior to Susan's birth. With our limited furnishings, Rene and Nancy shared our bed while I was relegated to a bedroll on a safari cot in the living room. I would be awakened at night by the activity surrounding Susan's need for sustenance, to be provided a cup of tea by Nancy, without having to move from my rickety cot. This luxury didn't last long as Nancy was naturally reluctant to stay for the months of nighttime attendance Susan required. On returning to Swastika, the local paper reported "Mrs. W.A. McLellan returned from an expensive trip to visit her daughter Irene and son-in-law in Vancouver..." Although the trip was probably expensive the word was intended to be "extensive."

As part of NWX's 1954 program I proposed an exploration program designed to search for massive sulphide copper-zinc deposits in roof pendants within the Coast Crystalline

One Lucky Canuck

Complex in an area extending northwesterly from Toba Inlet to the head of Knight Inlet and northeasterly to the east margin of the Complex near Tatlayoko Lake. The area covered about 3,500 square miles of relatively inaccessible terrain, which included the Mount Waddington massif, with its large systems of glaciers and ice fields and many nearby unclimbed peaks. We felt that our target, a Britannia copper-zinc type deposit mined at Britannia Beach, about 85 miles to the southeast of Toba Inlet might be recognized with reconnaissance-type exploration and the aid of a helicopter, which was a new tool for prospecting at that time.

Mount Waddington with its 4000 metre (13,177-foot) summit is Canada's highest mountain outside the Yukon-Alaska border area and at the time of our program it remained unclimbed by a Canadian. Its northwest peak, only 20 metres lower than its formidable summit tower, was climbed in July, 1928 by the most persistent husband and wife team in Canadian history - Mr. and Mrs. Don Munday and Don's brother Bert. The first ascent of the summit tower was completed in 1936 by Fritz Wiessner and William House, two American climbers. In 1958, as Paddy Sherman was to write in his book, *Cloud Walkers*, British Columbia was celebrating its centenary, and a Canadian party finally climbed the main tower. Paddy Sherman himself was a member of a successful expedition to the northwest peak in 1962.

Our program benefited significantly from the advice provided by Charlie Ney, a long-time friend and associate, keenly interested in mountaineering, with whom I was to spend several days in a staking expedition in the Chutine Lake area in January, 1959, described elsewhere. Apart from Charlie, none of our party had any significant experience in safety requirements for parties operating in mountainous terrain containing considerable areas of perennial snow and ice. We purchased the bare essentials, which included flys and pup tents with built-in floors having dimensions of about 1.5 x 2 metres. They were designed for the few overnight stays originally envisaged and for emergency purposes, including adverse weather which might prevent pick-ups. Each two-man party was also equipped with ice-axes, crampons, about 30 metres of nylon rope, emergency food supplies, cooking utensils, propane stoves and fuel.

Our party consisted of myself, Charlie Ney, John Anderson and Kayce Campbell as more senior personnel, assisted by Bob Gifford, Bill Smitheringale, Sandy Dean and Joe Werner. Both John Anderson and Kayce Campbell had been with me in our 1953 program in the Omineca. The assistants were all attending geological courses at the University of British Columbia and I was to remain in contact with all but Joe Werner through the '90's. We had an extremely congenial English cook with us who was quite at ease with our spartan base-camp accommodation.

Helicopter-based exploration programs were in the pioneer stage in western Canada and we were fortunate in having direct access in Vancouver to Okanagan Helicopters, which was to establish a worldwide reputation in the rotary-wing aircraft field. Our helicopter was a Bell 47-D1, piloted by Bill Legge and maintained by his colourful assistant Dave McLeod - another relatively recent arrival from the British Isles. Although our exploration crew was completely green in experience in helicopter-supported programs, not to mention the specific requirements of such programs in rugged terrains, we worked well together, benefiting from our experience and capable helicopter personnel.

In the planning stage, I finally selected a site for a base-camp at Bear Bay on the west side of Bute Inlet, midway up this fiord, on the basis of its relatively central location and its proximity to a small logging camp run by Len Parker which contained an established dock. It also had a weekly supply service provided by a converted fish-packer named the "Patsco". We arrived at our base-camp location in early July and were ready to begin the exploration program within a couple of days following the arrival of the helicopter. Our program was to last about six weeks and we were to be blessed with incredibly clear weather for much of this period. We maintained the base-camp for the duration of the program, even though we extended our support programs to the head of Knight Inlet and to several other distant locations, mostly by the use of fly-camps. Our only visitors were Jim Scott, then manager of NWX and Bill Dean, Sandy's father and manager of Kennco Explorations (Canada) Limited, NWX's parent with offices in Toronto.

I will never forget my first helicopter flight with Bill Legge from our base-camp at Bear Bay. John and I planned a traverse[1] along a mountain ridge on the northeast side of Bute Inlet. On the approach to the landing site, Bill flew relatively close to the ground taking advantage of updraft. He deliberately overshot the landing site, suddenly exposing us to almost 1000 metres of abrupt dropoff on the other side of the ridge. We reacted as Bill expected from previous experience, rearing backward against our seat belts. Bill was an extremely cautious pilot and was to provide all of us with confidence in his ability and judgment. This degree of comfort was augmented by the helicopter being equipped with large rubber floats for safety reasons, rather than skids, because of the considerable amount of flying required in the immediate vicinity of the inlets.

The Bell 47-D1 could transport a 400-450 pound payload on full tanks under normal temperature conditions to elevations of about 2,000 metres. This was satisfactory for our program in the Mount Waddington area; however, a following program proposed by Charlie Ney with essentially the same personnel in the Dewar Creek area near Cranbrook, B.C. had to be aborted within several days because of the much higher timberline which averaged about 2500 metres. Unfortunately we could locate few landing sites below that altitude which restricted access within the area and required much reduced loads, including only one passenger per flight.

Compared with later models, the Bell 47-Dl was relatively delicately balanced in terms of loading of personnel and equipment fore and aft of the rotor-blade mast, its centre of gravity. With two passengers and the pilot, the helicopter's 12-volt battery sat on a bracket on the tail boom to compensate for the forward loading. When the passengers left the aircraft, the pilot had to move the battery on to a mounting bracket inside the bubble. Bill mentioned two cases that he was aware of when a fatigued pilot, dropping off loads to helipads on clifftops had taken off tail heavy, in one case with fatal results.

Our only incident causing damage to the helicopter occurred on landing in the Orford River area on a gravel bed, when the tail rotor was slightly dented by striking brush Bill could not see. This was to be a fairly common occurrence in later helicopter-supported operations.

We had many fascinating experiences in the Mount Waddington program and no real problems. The relatively swift transition provided by the helicopter from the densely forested slopes at water level at Bear Bay to the magnificent alpine areas inland was exhilarating. Frequently we traversed alone, although more commonly within earshot in pairs. Plodding along in the vastness of such inspiring mountain scenery, realizing that few others, if any, had passed this way before, produced a deep feeling of undeserved privilege. We were not required to travel directly across glaciers although I spent several days on large snowfields near Mount Waddington while camped near Mt. Gilbert with Bob Gifford.

One of my traverses took me across an extensive snowfield on a very sunny day. Although I was adequately lathered with snow-screen, after several hours without any shade traveling on unblemished snow I began to feel the effects of heat and intense

---

[1] Traverse: Reconnaisance-style survey

brightness. I approached the corniced edge of part of the snowfield and although I could distinguish the edge of the cornice, I badly misjudged the vertical distance below the cornice to the lower part of the slope still in snow. My slide took me over 50 metres instead of several metres expected, with no ill effects, but with a well-learned lesson.

Our principal break from an otherwise routine program was to provide helicopter assistance to a mountaineering expedition which included Paddy Sherman and which was attempting first ascents of Mt. Gilbert and Mt. Raleigh in the Southgate River area, southeast of the head of Bute Inlet. Both peaks had summits of about 3,000 metres and required ground access via Southgate River - a difficult two-week trek. The party had heard of our program and I arranged for the helicopter assistance in the spring. One of our fly camps was on the north side of Southgate glacier immediately to the south of Mt. Gilbert. Bob Gifford and I spent about 3-4 days in the area and on one of our traverses we spotted a greenish stain on the west buttress of the mountain. I had carried the company's 16 mm. camera with me on that particular day to film a typical traverse on the program. Although fairly weary, we decided to try to reach the apparent malachite-stained copper showing. We trudged off and finally gained the summit of the ridge, where sure enough the remnant of a 2 metre wide copper showing extended across the ridge. The elevation was about 2800 metres - only 300 metres below the summit looming ahead of us. We walked along the ridge for a short distance and came to an abrupt halt at a huge cleft separating us from the shear face. Later, when the mountaineering party arrived, I teasingly informed Paddy that we had been up Mt. Gilbert; his relief was obvious when he learned just how far! His party was to gain both objectives during their stay in the area.

While in the Mt. Gilbert area we ran out of fresh meat, having taken no more than a day's supply. Each of our fly camp parties was equipped with a .303 Lee Enfield rifle and an adequate supply of ammunition, principally for protection in the event of bear attacks. One evening I shot a mountain goat near camp. Although we enjoyed the goat meat at our base-camp, the experience totally curbed any further thoughts I may have had of developing into a hunter and my one lucky shot was my last.

On another occasion, Sandy Dean and I spent several days camped out on a gravel bar near the headwaters of Scar Creek. We were landed by Bill several miles downstream from its source area in what was considered a reasonably central location based on proposed traverses on both sides of the creek. On the first traverse, we planned crossing Scar creek near our camp, but felt that it was too deep and too swift so we walked upstream until we located a wider stretch. Even so, I had some doubts about the depth near the opposite bank a mere 20 feet away. I heaved my pack across the creek and tied my climbing rope to my waist giving the other end to Sandy so that I would be swept back to his side of the creek if I lost my footing. I didn't encounter any problem, although the water was ice cold, until I was within five feet of the cobble-strewn embankment. As I took my next step I dropped several feet and with a desperate lunge managed to grab some boulders on the opposite bank and pull myself out of the creek, completely sodden. We were to walk upstream for almost one mile before locating an acceptable crossing for Sandy.

Almost six months earlier, before our summer exploration plans had been finalized, I had obtained two seats near the finish line for the 1954 British Empire Games in Vancouver which was to feature the 'Miracle Mile' between Roger Bannister of Great Britain and John Landy of Australia. The race was the first time two runners were under four minutes in the same race. Overshadowing this event was the unexpected completion of the marathon twenty minutes later when favoured Jim Peters of Great Britain entered the stadium a full 20 minutes ahead of the next runner. Only nine of sixteen runners who had started the event on an unusually blistering-hot day were able to finish the race. It took Peters eleven agonizing minutes following his entry into the stadium to stagger, collapse six different times and finally crawl to what he thought was the finish-line, which was in fact still 55 metres away. It was Peter's last race and an ordeal which he was probably lucky to survive.

Of course I brooded as I realized that I would not be able to see the event. Rene agreed to attend and took Jim Scott's wife Jean with her. Like everyone else who watched the momentous duel between Bannister and Landy, she was also appalled at the sickening spectacle of Peter's struggle to complete the marathon watched by thousands none of whom went to his aid because of the disqualification which would follow if he was helped in any way.

Our departure from Bear Bay in mid-August was somewhat delayed by the projected arrival of a Stranrear aircraft which we had arranged to fly the group back to Vancouver in a single flight. With our camping gear all packed up, I located a fairly stout pole and a willowy branch and indulged in some pole vaulting. Several of the others had a go, including Bill, who had decided to wait with Dave for the Stranrear to arrive to ensure we were not stranded. Unfortunately his jump was a disaster and he sprained his ankle, which rapidly swelled. I fashioned a walking stick for him with which he limped around, until deciding that he had better take off for Vancouver in the helicopter before his foot became useless. He left his 'cane' which I delightedly delivered to the Okanagan Helicopters base in Vancouver when we eventually arrived back in the overdue Stranrear, with a label marked "Limpy Legg". The name stuck for a while.

I spoke briefly with Bill in a telephone conversation in the early 1980's while he was still active in a managerial capacity and learned only recently that he died of a heart attack in 1986.

In my career I have had the opportunity to report on examinations or visits to well over 400 metallic and non-metallic mineral properties and/or mines, principally in Canada, but also in the USA, Ireland, Zambia, Mexico, Guatemala and the Republic of South Africa. Some were of short duration, others such as Galore Creek (6 years) were extremely lengthy. Because all mineral occurrences are unique, each offers new experiences and challenging opportunities in assessing their potential at various stages of their evaluation, the ultimate goal of an explorationist being a viable producing mine.

In the late summer of 1954 I received another opportunity for travel into mountainous glacial terrain in the examination of a disseminated copper[2] prospect called the Lucky Lady group which was situated about 3 kilometres north of the Granduc Mine site on the north side of an arm of the South Unuk River glacier. This particular prospect combined a variety of unusual experiences in the short time span of about 4-5 days, which made a marked impression on me as a young geologist seeking all the experience I could possibly acquire. The Lucky Lady claims were owned by Wendell Dawson, a former Lieutenant Commander in the U.S. Naval Engineering Corps. in the early 1950's and a part-time prospector in Alaska and the U.S. One of his associates was Howard Fowler, also interested in the development of mineral properties and a fairly well known pilot in the northwestern part of British Columbia and Alaska. I had not met Wendell before and both he and Paul Pieper, his young assistant, were on the property at the time of my examination.

I was accompanied by Bill Smitheringale and we were flown from Wrangell, Alaska into the mouth of the Chikamin River in a Cessna 180 on floats. There we were met by

---

[2] disseminated: Scattered as minute particles.

prearrangement by Don Ross, a pilot known to Wendell from Ketchikan, Alaska with a Super-Cub named the 'Copper Queen'. He flew us with our pack boards, sleeping bags, etc., one at a time from a wide gravel bar on the Chikamin River to the junction with the Leduc River and up the Leduc River to the International Boundary. I recall looking down from an elevation of about 600 metres above the river to the postage stamp-sized landing strip which had been hacked by hand out of the large alders and willows which covered the banks of the river. The airstrip was actually 250 metres long and about 15 metres wide but believe me, it didn't look that large on approach! After removing some supplies and my pack from the aircraft I waited for the pilot to return with Bill and filmed the tight landing with the company's 16 mm camera. After the pilot left, Bill and I hiked up to the mouth of the Leduc Glacier - a distance of about 3 kilometres. En route we had to ford a major tributary, and several small ones, making use of hip waders and stout poles, which had been left at the crossing by predecessors.

On reaching the Leduc Glacier we found that it afforded safe and secure footing. It was generally free from any wide crevasses for the 2-3 kilometres we walked to gain access to the Granduc Mine camp where advanced exploration of the 2B ore body was still in progress in preparation for the major mining development that was to follow. The operation was being managed by Jack Crowhurst and arrangements had been made for us to be accommodated there overnight.

The Granduc camp consisted of about 20 widely-scattered tent frames perched on a 30 degree talus slope above the north side of the Leduc Glacier. The camp was located in the immediate area of the large copper deposit, accessed by adits spaced about 30-50 metres apart in elevation. Transportation into the area was provided either by fixed-wing aircraft on skis or large balloon tires, depending on the surface conditions of the glacier, or by tractor trains plying the glaciers from north of Stewart at the head of the Portland Canal some 30-40 kilometres to the east.

Jack Crowhurst gave us a tour of the operation before we left the next day for the Lucky Lady prospect. To reach the property and its camp-site we travelled north on part of the Leduc Glacier to its merging with the source of the South Unuk River Glacier system. We then climbed up to the camp site lying within a grassy meadow at the top of a mountain ridge, about 600 metres above the glaciers.

There we met Wendell, a tall, rangy and deeply tanned individual with strong features and a most engaging disposition. He was 58 years old at the time. His partner Peter Pieper was at least 25 years younger, also incredibly fit and equally good-natured. They both obviously enjoyed their Spartan existence for they were completely at ease with only the barest essentials.

We were blessed by clear skies during our stay and were able to complete a fairly thorough examination of the prospect, which had few workings, consisting mostly of well-exposed outcrops on the very steep, south-facing slope below the campsite extending at an average slope of about 55 degrees to the glacier below. The well-developed fracture system across the trend of the slope with its glacial sculpturing had produced a very hackly surface, which afforded safe passage over most of the mineral-bearing exposures. We collected a number of representative samples for assay, but unfortunately the results were to be insufficiently encouraging to recommend further interest.

One evening we sat and listened to Wendell describe his early work in the area, commencing in 1931 when he and his partner, Bill Fromholz discovered the copper mineralization in the nearby area which later became the focus of the large mine planned by Granduc Mines Ltd. I was very impressed by the sincerity of Wendell's account of the major lawsuit, which had been in progress for less than one year, to determine whether

Wendell was entitled to a prospector's traditional 10% vendor interest for having brought the location of the prospect to the attention of Karl Springer, president of Helicopter Exploration Co. Ltd. through his friend Dr. Des Kidd, also of Vancouver.

Wendell described the history of the discovery and the subsequent lawsuit in simple terms. Over a year or more later I was to learn from Wendell that the lawsuit had eventually been settled in his favour by the Supreme Court of Canada in October, 1955 and that the terms of settlement had been referred back to the Supreme Court of British Columbia to determine the extent of damages. Unbeknown to Wendell's wife, Grace, as I was to learn almost 40 years later, Wendell had immediately forwarded 5000 shares of Granduc Mines Ltd. stock to Amy Formholz, the widow of his former partner. A letter to Wendell from Amy dated January 21, 1956 from Juneau, Alaska reflected the extent of her gratitude with the comment, "It paid off the mortgage on my home. I did a lot of repair work, and sent my granddaughter out to school."

By coincidence I had become interested in the ruling of the Supreme Court of Canada in the case in 1992, having obtained a copy from my friend Lisa Holmgren, a former summer employee of DuPont of Canada Exploration Limited who subsequently earned a legal degree and was a solicitor with the legal firm Lang Michener Lawrence & Shaw in Vancouver.

The case between Wendell Dawson (Plaintiff) and Helicopter Exploration Co. Ltd. (Defendant) is of particular interest to explorationists, assuming they are discoverers of mineralization which could be of economic interest, particularly in terms of negotiating a deal with another party. Contingent on the nature of the discovery, an accepted term of agreement for over a century has been that the discoverer is entitled to receive, as a minimum, a 10% non-assessable interest. The discoverer need not have title to the deposit in question, merely the knowledge of its location and agreement with an interested party willing to acquire legal title to the lands containing the discovery.

While preparing the first draft of these recollections in May 1993, I wrote Grace Dawson, enclosing a copy and requesting any comments or information for clarification purposes – particularly as to the final settlement. She provided considerable information by return mail including a front page of the October 18, 1956 issue of the Vancouver Province with the headline, "Engineer wins $333,450 award". The amount was based on the value of 10 per cent of 750,000 shares which Helicopter Exploration Co. Ltd. received from Granduc Mines Ltd. on the sale of the Leduc River property, as of the date of trial, less advances made to Wendell.

In reading the March 14, 1994 issue of the Northern Miner, I was dismayed to read an article crediting the discovery of the Granduc copper deposit to prospectors Einar Kvale and Tom McQuillan, reportedly in 1946. I wrote the following response, which appeared in the April 4, 1994 issue and forwarded a copy to Mrs. Dawson.

Claims staked years earlier

The Odds 'n' Sods article "An outstanding exploration team" (*T.N.M.*, March 14/94) refers to prospecting carried out in northwestern British Columbia by Einar Kvale and Tom McQuillan, reportedly in the summer of 1946. The article suggests that their exploration led to the discovery and staking of the copper deposits which became the Granduc mine.

Both men were not only exceptional prospectors, but outstanding individuals. Out of respect for the memory of Wendell Dawson and Bill Fromholz, I wish to clarify the events in which all four were ultimately involved in the Granduc discovery.

One Lucky Canuck

The copper-bearing outcrops on the north side of Leduc Glacier are recorded as having been discovered and staked by Dawson and his partner in the summer of 1931 and recorded on Oct. 19 of that year. The claims subsequently lapsed during the Depression.

On Dec. 28, 1950, Dawson wrote to his friend Des Kidd (who was actively exploring in northwestern British Columbia), recalling his discovery and seeking his interest. Kidd forwarded Dawson's letter to Karl Springer, an associate. Springer then wrote to Dawson expressing interest and proposing to "carry you for a 10% non-assessable interest in the claims." Dawson replied on Jan. 22, 1951, that the proposal appeared fair and suggested a meeting.

Dawson, a U.S. commanding officer in the Pacific during the Second World War, was recalled to active duty in February, 1951, and was unavailable for a proposed site examination.

Springer's company, Helicopter Exploration, retained Kvale and McQuillan as prospectors and in July, 1951, they were flown into the Leduc area by helicopter. On Aug. 2, 1951, on a return visit to the area, they staked a number of claims on behalf of the company.

Dawson did not learn about the staking until June, 1952. His claimed entitlement to a 10% interest led to a dispute and an action against the company filed Nov. 23, 1953, in the Supreme Court of British Columbia. Dawson lost the case and subsequent appeal but was successful in a Supreme Court of Canada decision in 1955.

I had the good fortune of meeting Dawson while examining some of his claims north of Granduc in 1954. I was later to learn that, prior to his successful settlement, he had forwarded 5,000 shares of Granduc stock to the widow of his former partner.

D.A. Barr, West Vancouver, B.C.

While examining the Lucky Lady claims, Paul Pieper and I arrived at the head of the glacier below the showings one evening en route back to camp. Paul was familiar with the route back and as I gathered my gear he asked:

"Can you glissade?"

"Somewhat, " I replied.

"Well follow me and jump when I jump."

I thought the instructions somewhat strange, but fortunately they were sufficiently clear and, as he slid off, I followed as instructed.

The slope of the broad head of the glacier steepened as we travelled downward at a reasonable speed, then I noticed Paul shooting high above the glacier some 30 metres ahead of me. All too soon the reason was obvious -- a sizable crevasse which I cleared without difficulty because of the warning. There were several others before we reached the turnoff point to camp.

Following completion of our examination, Wendell and Paul decided that they would come out with Bill and I and we left for a prearranged shuttle from the airstrip at the head of Leduc River back to the mouth of Chikamin River. En route we met two individuals on the Unuk Glacier, one quite old and obviously completely worn-out. During a brief exchange we learned that they were in the area to restake some mineral claims that had been sold and later contested for improper staking. The Gold Commissioner had generously allowed the original stakers to return and complete the staking according to the Regulations, which I thought was most considerate under the circumstances. Apparently in the heat of the Granduc claims taking spree, the posts with duly attested tags attached had been dropped through a hole in the clouds from a Beaver aircraft 600 metres above the level of the glacier. The pair had not located the posts, which was not surprising.

Staking claims in the rugged northwestern part of British Columbia by dropping claim posts from aircraft is probably rare for obvious reasons besides safety, not the least of which being that the claims posts may never be located on the ground and if they did not shatter on hitting the ground they would be unlikely to be within several hundred metres or more of the desired location. Several years after our trip to the Lucky Lady group I visited Charlie Ney at Kinskuch Lake, about 10 kilometres northeast of Alice Arm where he was supervising a diamond drilling program for Kennco. The lake lies within a bleak, rocky area above timberline. During my visit Charlie mentioned that a couple of days earlier he had heard an aircraft flying high overhead on the other side of the lake from his camp. On investigating he saw an object drop from the aircraft like a bomb to land in a small snowfield. On investigating he located a bundle of claim posts tagged with a description of the area claimed, which as it turned out was well within the boundaries of the property he was drilling.

On arriving at the airstrip Bill, myself and Wendell were successfully flown in succession back to the Chikamin River. The final trip with Paul Pieper was the final flight of the "Copper Queen". We learned that it hit a sudden down-draught on take-off at Leduc and crashed into the alders and willows at the end of the airstrip. Apparently the aircraft fell apart gracefully, strewing itself in such a sufficiently accommodating fashion that it imparted no severe injuries to Paul and the pilot. Both were cut and bruised but managed to walk back to the Granduc camp-site where they received first aid and a later flight back to civilization.

Win Some - Lose Some

Flat River, Pelly Plateau and Selwyn Mountains Prospects

Yukon & Northwest Territories 1955 – 6

What John Sullivan, Kennco Explorations (Canada) Limited's president, referred to in the early 1960s as the "Flat River Disaster", had its roots in the earliest known geological reconnaissance completed in the Nahanni River area of the Northwest Territories by Dr. E.F. Roots of the Geological Survey of Canada, in 1953.

The South Nahanni River region, which now lies within Nahanni National Park Reserve, was recently identified as a World Heritage Site. It includes spectacular Virginia Falls with a drop of 316 feet through its rock-strewn gorge among limestone cliffs soaring as high as 2,000 feet above the river. Long a fabled region because of the early prominence and publicity surrounding Deadmen's Valley, it has retained its wild character since first explored in the early 1800s.

In 1905 Frank and Willie McLeod, two brothers from Fort Liard, and a third man made a second expedition into the region in search of gold. They failed to return. A search was eventually made and the bones of the brothers were located in the valley between the Lower and Second canyon of the Nahanni, since dubbed Deadmen's Valley. The newspapers of the time, ascribing great significance to the fact that both skulls were detached from the skeletons, undoubtedly improved their circulation by referring to the location as Headless Valley and suggesting the possibility of foul play. The general public would of course not know of the dismemberment of most carcasses following discovery by carnivorous predators such as grizzlies.

Kennco's interest in the mineral potential of the region undoubtedly stemmed from exploration carried out in the Yukon under the direction of Gerry Noel in 1953. A prospecting party consisting of Axel Berglund and Walter Cannon, working under a prospecting agreement with Kennco's Western Canada subsidiary (Northwestern Explorations Limited) was active in the area in 1954 from June 1 to September 15. The party was supported by six packhorses under the direction of Buster Groat, acquired from Skook Davidson's establishment at Terminus Mountain in the Finlay River region far to the southeast. An area of about 600 square miles on both sides of Flat River, a tributary of the Nahanni paralleling the Yukon boundary, within a few miles of the border in the Northwest Territories, was covered by basic prospecting with little to no geologic guidance. Gerry Noel described the area as follows in his 1954 report:

The area prospected includes parts of both the Logan and Selwyn mountains. These ranges consist of long northwest-aligned chains of rugged mountains with summits reaching 10,000 feet, although in general the topography ranges from 3500 feet in the main valleys to 7000 feet on top. Numerous mountain glaciers occupy the higher cirques. Flat River valley, a long northwest directed trench about one mile wide, separates the Logan and Selwyn mountains. Except very locally, good timber is scarce in the region, and is confined to spruce and balsam. Timberline is about 4000 feet.

Snow did not leave the upper levels (6000 feet) of the region until mid-June in 1954 and began to accumulate for the winter in mid-September.

The results of the 1954 prospecting program outlined a promising copper-zinc-lead mineral belt along the southeast margin of the undetermined extent of the Nahanni batholith[3]. Although none of the showings were staked, Noel recommended further work on all three showings.

I will have mentioned Gerry Noel on more than one occasion; an excellent geological scout, and good friend, of quiet disposition, he was respected and trusted by one and all in the exploration community. I recall Buster Groat saying, "In all the time I've spent with him, I never ever heard him say anything bad about anybody." After leaving Kennco, he joined Utah Copper, the predecessor of BHP-Utah and was instrumental in negotiating an agreement in 1965 with Gordon Milbourne, the owner of the claims on the northern part of Vancouver Island, which later became the Island Copper Mine, for a time one of B.C.'s principal copper-gold producers.

My first involvement in exploration in the region occurred in early 1955, when I was asked to plan for an exploration program in support of Gerry's recommendations. Gerry had also noted in his concluding statement of his 1954 report that "the presence of mineralization west of Hyland River suggests that further reconnaissance work should be directed to include a wider northwest-trending belt southwest of this area." Accordingly, I began to plan for a combined prospecting program of areas in a large region in S.E. Yukon and an adjoining portion of Northwest Territories, plus a geological/sampling assessment of the three known showings located by prospectors in 1954.

In preparation I contacted Fred Roots and spent a fascinating evening with him and his wife in Ottawa during which he discussed his impressions of the geology of the Nahanni area. He also described what must have been an indelibly etched experience he had with the International Geophysical Year Program in 1951-52. In that period, Canada established a very small group of geoscientists in the subarctic, who were resident there for a 5 - 6 month period with no guarantee of physical support or supply other than what had been originally provided. Fred was one of the participants.

Sometime during their long entrenchment, one of the group complained of failing vision in one of his eyes. Fortunately, the party had radio communication with a doctor at a whaling station on an island off the coast of Argentina. The only medical assistant on the program, a third-year McGill student, well-equipped with texts, but lacking expertise, spoke to the doctor, who eventually stated that every effort should be made to preserve sight in the unaffected eye, which meant surgical removal of the damaged eye.

The IGY team had no surgical instruments on hand nor anaesthetics. Fred and the McGill student decided that it would be in the best interests of the patient to work rapidly and privately in order to comply with the doctor's recommendations while reducing anxiety problems for the patient. Accordingly, they fashioned surgical instruments and read up on surgical procedures. They were so confident that they decided to film the operation for posterity. Fortunately, the entire event was successful and the patient retained the sight of his unaffected eye.

As part of our preparation for the 1955 program, I checked on the availability of adequate maps for the area. I found that in one of the three larger National Topographic Survey (NTS) grids covering our proposed prospecting area, each covering two degrees of longitude by one degree of latitude, or about 5000 square miles, the largest-scale topographic map available was only on a scale of one inch to 8 miles. This was obviously not adequate for ground control purposes or aerial reconnaissance work. The map area in question (NTS 95-E) when completed, was to be designated "Flat River" produced at a scale of 1:250,000 or about one inch to 4 miles.

---

[3] Batholith: A body of igneous rock at least 40 square miles in area.

I scurried around and was delighted to learn that air photo coverage of the Flat River sheet had only recently been completed by the federal airborne surveys and that photocopies were available from Ottawa, although no map had yet been prepared.

I also learned that the University of British Columbia was offering a short course on photogrammetry which included radial-line plot techniques. I enlisted and subsequently produced a complete drainage-pattern map on a scale of one inch to 2 miles for NTS 95-E, which was more than adequate for both helicopter reconnaissance and prospecting purposes.[i]

In planning the 1955 program, I only had three seasons of background experience to draw on under the kind of conditions contemplated for a combined prospecting and property evaluation program in the Cordilleran region. This included a helicopter-supported reconnaissance program in the Coast Mountains in 1954; a horse-supported property examination and reconnaissance program in the Omineca Mountains of Central B.C. in 1953 and various property examinations in 1952. Based on this limited experience, I was convinced that helicopters would be the work horses of the future rather than their four-legged predecessors. Accordingly, I had noted that:

"Previous experience by our prospecting parties during 1949-53, on programs in northern B.C. and Yukon which were supported by pack-horses, indicated coverage of areas varying from 500 to 1 100 square miles in a four-month season. The average amounted to about 700 square miles at an average cost of $8000 per party. Statistics were not available for the number of camps occupied by these parties."

I was planning for coverage of about 7000 square miles in 1955.

After considerable research, both with acquaintances in the fixed-wing and rotary-wing sectors, and study of the proposed exploration area, I reasoned that four to five two-man prospecting parties plus a 2-3 person property examination crew could be serviced from centrally-located base camps consisting of the helicopter pilot and engineer, a cook and myself with radiotelephone communication to Watson Lake, our principal supply point.

In preparing the Flat River map (NTS 95-E) from aerial photographs, I also completed a photogeologic interpretation for the Nahanni sheet (NTS 105-I) and the Flat River sheet.[ii]

During the winter period I also obtained work commitments for the summer program from various prospectors, students and a graduate geologist to direct the property examinations. In addition a contract for a Bell 47 D-1 helicopter was arranged with Okanagan Helicopters Limited of Vancouver.

By May 31 the following individuals had been assembled at Watson Lake for the 1955 program: Axel Berglund, Wilf Christian, Pat Cook, Buster Groat, Jake Hunderi, Ed Kasmer, Len Maley, Jack Marshall (prospectors); Ed Freeman (geologist), John Greenaway, Dave Fletcher, Ken Darke (assistants); Oscar Schmidt (cook); Al Smillie, pilot and Jack Rich, engineer (Okanagan Helicopters).

For most of us, this was our first visit to Watson Lake, situated on the Alaska Highway in Yukon, near the border with British Columbia on the north side of Upper Liard River.

Watson Lake is named after Frank Watson, an Englishman who ventured into the area, married a native girl from nearby Lower Post and lead the life of a prospector and trapper, settling into a home on the shores of the lake now north of the town in 1898. The community known as Watson Lake came into existence in 1942 as one of the accommodation and supply points on the Alaska Highway. This wartime project, designed to link Alaska to the lower 48 States by a connector highway constructed between Fairbanks, Alaska, and Dawson Creek, B.C., totalling 1,522 miles, was roughed out in the remarkable time of 8 months and 12 days.

In 1955 Watson Lake was a community of no more than 200 residents, dominated by the Watson Lake Hotel, run by Rita and Jim Lund. It included a mining recorder's office, post office, general store, garage owned and operated by the remarkably-bearded Johnny Friend and of course the air-base at the lake for Watson Lake Air Services owned and operated by George Dalziel, which was to be our essential supply link between our base camp and civilization in the 1955 and 1956 programs. Watson Lake immediately impressed all of us as a real frontier community.

The Sign Post Forest is Watson Lake's unique attraction. It was initiated in 1942 by Case K. Lindley of Danville, Illinois, a homesick U.S. Army corpsman with Company D, 341st Engineers. While working on the Alaska Highway, he erected a sign indicating the mileage and direction to his hometown. Others followed suit and by 1990 when the population of Watson Lake had increased to 700, the 10,000th sign had been placed.

Before losing the opportunity I must compare this supply point on the Alaska Highway with another at Rancheria, also on the B.C.-Yukon border, about 80 miles to the west. I had the great misfortune to obtain accommodation at the local motel in the early 1980s while examining a nearby property. Arriving late one evening, I checked in and quite exhausted and grubby from the long day in the bush, I decided to have a cleansing and relaxing shower. Sure enough it worked, but all too soon I realized that I was being covered with an oily solution smelling strongly of diesel fuel. I slept soundly enough, but at breakfast I had the worst cup of coffee imaginable - again overpowered by diesel.

A short walk after breakfast revealed that the motel was constructed on a large levelled-off dump, with oily seeps at the base. It was one of the huge dumps at supply points on the Alaska Highway where diesel fuel had been stored, and 30 years later the seepage was still affecting the water supply.

While at Watson Lake I asked Oscar Schmidt, our base-camp cook, to review a tentative supply list for our first base camp. I had prepared it with Rene back in Vancouver. Prior to leaving Watson Lake we had to plan for provisions at our base camp at Hyland Lake for 16 people, recognizing that we could be delayed in establishing our proposed fly camps. Oscar made a few changes, but one vital item must have been misinterpreted, not only by him, but by the local store which prepared our supplies.

Soon after arrival at Hyland Lake, we realized that no toilet paper had been included in the supply list. Fortunately, this gross omission was quickly rectified and an adequate supply was provided as a priority item on a subsequent supply trip. A few days later, in checking all his supplies, Oscar asked me "What the hell did we order 48 cans of tomato paste for?" On reviewing the original supply list the answer was obvious; Rene was used to referring to toilet paper as T.P. and that's what I had written down!

I have mentioned that in 1955, Watson Lake was a pioneer settlement. In our relatively brief stay lasting five days we were fully exposed to its Wild West environment. I would have to say that the most influential person in maintaining law and order at Watson Lake at the time was Rita Lund. Because of its prominence as an accommodation hostel plus bar, the Watson Lake Hotel was not only *the* place, it was the *only place* for socializing. Rita, as proprietor, was a no-nonsense type of woman. As long as you toed the line that was fine, but she was more than capable of enforcing order. A fair and gentle person at heart, her physical attributes were also more than impressive, they were awesome! Weighing nearly 300 lbs, she was accustomed to handling beer kegs, delivered by truck to the hotel, which she would roll off the truck and cart to their destination.

One evening her husband Jim, weighing only about 250 lbs, got into an altercation with another individual and a fist fight broke out on the verandah of the hotel. The verandah was accessed by a single set of stairs rising about six feet from the ground and was enclosed by a railing about three feet above the floor. Noting that her husband seemed to be in trouble, Rita stepped in, picked up his adversary and threw him over the railing. He did not return. I was impressed that that this was not considered an unusual event. Patroness, keg handler, bouncer, Rita handled the hotel as required.

On May 30 Al Smillie and John Greenaway made a flight with George Dalziel to Flat Lake, where I had planned our first base camp. Observations indicated that the area was still covered with a heavy blanket of snow and that break-up at Flat Lake was not expected for at least several weeks.

As Hyland Lake, one of our proposed base camps situated about 90 miles north of Watson Lake, was in the break-up stage with its marginal areas already ice-free, I decided to establish the initial base camp there next to a gas cache which had been placed at a potential site location during the winter. On June 2 a portion of the crew was flown to Hyland Lake to clear a landing strip for helicopter approaches and a landing platform was erected for the helicopter.

My progress report for the period ending June 10, 1955 included the following observations:

"Although a reconnaissance flight was attempted on the evening of June 5, mist prevented any reconnaissance from being accomplished. A flight on the morning of June 6 was made along the eastern side of Hyland River to attempt to locate landing sites on a mountain range extending 20 miles to the north. Snow during the evening of June 5 and early morning of June 6 had blanketed the area to camp level at 2800 feet, and the rugged nature of the mountains did not offer abundant landing areas. It was later found that the range was largely composed of granitic rocks, considered not favourable for prospecting purposes.

"A preliminary aerial examination of the area lying within a 25-mile radius of Hyland Lake was completed by June 9. Although the area is at this date still snow-covered above the 4000 foot elevation, and covered by patches of snow in the more sheltered areas to valley level, a period of four days of sunny weather has helped to remove much snow. Suitable landing and camping sites have been located for prospectors' camps throughout the area, but these are generally not of value at present as snow prevents any travel.

"The first prospecting team consisting of Berglund and Groat was moved into a low-lying area located 15 miles north-easterly from Hyland Lake on June 8. A belt of limestone located by aerial reconnaissance lies to the northeast of the camp location and several miles northerly of a granitic stock."

A significant portion of my time throughout the 1955 season was taken up with providing preliminary geologic information based on my aerial reconnaissance from the helicopter covering favourable prospecting terrains and adequate fly camp sites. Much of this reconnaissance flying was completed between 4 - 7 a.m. at the request of Al Smillie, our pilot, to maximize lift potential of the aircraft under cooler air conditions and enhance the selection of prospective landing site locations. This routine was to persist throughout the summer season."[iii]

By June 17, three prospecting parties had been established at initial fly camps by helicopter, two of which had been moved to second camps.

In this early exploration stage I frequently completed reconnaissance surveys of potentially favourable prospecting areas accompanied by one of the prospectors. On one of these, Jake Hunderi was my companion. We got dropped off at the same location with an agreed pick-up by helicopter high above a series of cliffs. Jake agreed to a circuitous route around the cliffs, whereas I felt that a direct ascent would provide an uninterrupted stratigraphic[4] sequence. Although I was right, I ran into problems at the extreme top of the series. By extreme I mean within an arm's length of the top of a cliff, above which we had agreed to a pick-up. In this precarious position, I realized that it would be extremely hazardous to try to lower myself. I waited, yelling frequently, and eventually Jake's anxious face appeared above me. Without further ado, he reached down and with locked arms, hauled me to safety.

By June 24 all four prospecting parties were active and the geological crews were employed on combined prospecting and geological traverses as they waited to access the first of the three properties to be evaluated. [iv]

On June 24, Al Smillie flew me to the most northerly of the three showings discovered in 1954 by Axel Berglund and Walter Cannon. It was to be named the Axel Showing and lies about 15 miles southeast of Flat Lake on the southwest side of Flat River Valley. The Fitzsob Showing lies about 12 miles further to the southeast and the Buster Showing is an additional 25 miles to the southeast, both lying west of Flat River.

As none of the showings had been staked by the prospectors in 1954 and even though there was no obvious competitive activity in the area, we were anxious to protect the Axel showing as we were not planning to map and sample it for at least a month.

My staking spree was without doubt the easiest I have ever had. Northwest Territories regulations at the time permitted the posting of a notice stating that application had been made to prevent further staking within an area of four square miles for one month around the mineralized area. I staked the Axel No. 1 claim by notice on June 24 and returned to assist in staking an additional three claims on July 22.

In late June we moved our base camp 60 miles east across Mackenzie Mountains from the Yukon to the most easterly of the Skinboat Lakes in Northwest Territories. The intriguingly named Skinboat Lakes, whose derivation I have not attempted to research, lie at the head of one of the many tributaries of Nahanni River in the southern part of our proposed prospecting area.

The site enabled us to move our prospecting parties through unmapped and unprospected terrain to the south of the area prospected by Axel Berglund and Walter Cannon in 1954. We were also able to provide adequate support to our geological crew who were moved on June 29 to the most southerly of the three showings discovered in 1954, to be named the Buster claims.

The crew consisting of John Greenaway, Dave Fletcher and Ed Freeman was to be joined in late July by Jim Rutherford, another recently graduated geologist.

Apart from mapping and sampling the deposits, the geological crew staked seventeen claims covering the mineralized occurrences and extensions of the deposits which were located up to two miles from the discovery area.

Although I had visited the site at the initiation of the program, I did not return until about a week later accompanied by John Sullivan and Jim Scott, Kennco Western's manager. I recall that we did not have use of the helicopter at the time and had to walk about six miles northwest from Lucky Lake to access the showings. We walked through scrub willow most of the way, without the benefit of any good game trails, returning the following day after examining the showings.[v]

.Prospecting by our four crews while we were based at Skinboat Lake was not encouraging, and no significant discoveries of potential interest had resulted. I decided

---

[4] Stratigraphic: Pertaining to the composition, sequence and correlation of stratified (layered) rocks.

that Seaplane Lake, lying near the big bend on Flat River about 30 miles to the north, would be more serviceable for our crews. The entire base camp and seven drums of gas were moved to Seaplane Lake on July 13 in seven flights by Beaver aircraft. We were to be assisted in the move by an unexpected visitor.

Just before our move our rather mundane base-camp existence was stimulated by a totally unexpected event. I had read Raymond Patterson's book *"Dangerous River"*, which recalled his early adventures in the Nahanni River area. A few days before our proposed move to Seaplane Lake, Vic Maguire, who had made a belated service trip, took me aside and said, "Dave, there is an individual named Raymond Patterson in Watson Lake who has planned to link up with Albert Faille[vi] on the South Nahanni, but so far has been unsuccessful. Would you consider having him stay with you for about a week at Seaplane Lake, as he would dearly like to get back into the area?" I had no hesitation in agreeing.

Vic arrived early on the morning of July 13 in his Beaver with his distinguished passenger, just before a welcome breakfast. Afterwards, Raymond immediately volunteered to assist us in our camp move, prophetically destined for the largest island on Seaplane Lake, which was also the site of one of Albert Faille's cabins in the Nahanni area.

Seaplane Lake is one of those rare beauty spots enhanced by vivid blue and green hues from algal deposits. I was to take many photos of this particular lake under different lighting conditions with sunsets ranging from golden through orange to deep red. No wonder Albert Faille chose this location for a cabin, apart from its proximity to abundant game.

Raymond eagerly accompanied the first load hoping to find Albert at the lake and agreed to stay and help unload the aircraft's initial loads. I am describing this one of many base camp moves we made in this and other seasons in reconnaissance work in the north; Raymond devoted eight pages to this single day in a chapter called "A Cook's Tour" in *Far Pastures*, a novel published in 1963. He recorded his disappointment in not locating Albert as he approached his cabin and noted the following written message left pinned on the door: ". . . as might a man two hundred years ago - exactly as it suited him:

>Albert Faille. Yun. 3. 1955. Gone up Flat.
>
>Bike here Yulay. 1. 1955. Stay till 10th.

But something had gone wrong with that plan; he had probably failed to run on to any game up the Flat, because he had come and gone again before the end of June:

>Here on Yun. 25. 1955. On the way to
>
>Ervine Crike no grub cant stay.
>
>Albert Faille.
>
>PS stay at ervine til 10 Yulay 1955.

If only he had been able to stay here till this day of grace, 13 Yulay, it would suddenly have rained grub from the skies."

Raymond was with us for about one week and in that period we provided him with some helicopter time to visit several nearby places of interest. He was obviously deeply concerned about Albert's welfare, even knowing that he had existed on his own for many months without outside help.

During his stay, Raymond spent considerable time in the cook tent chatting with Oscar Schmidt and as it turned out falling under Oscar's spell for story-telling. As a general rule, Raymond retired for several hours each afternoon to a small pup tent which he had brought with him. When not out in the wilds, he would appear refreshed for dinner and more of the cook tent banter dominated by Oscar.

Almost ten years later I received a Christmas present from my parents ~ it was a copy of *Far Pastures* in which Raymond vividly describes his return to the Nahanni and includes two final chapters, the first being "A Cook's Tour" which describes his stay with us and an Asarco camp to the south in 1955, and the second, one of Oscar's stories. I phoned Raymond after receiving his book to compliment him on his description of our operation based on such a limited stay. "Well", he said, "you may have wondered what I was doing during those afternoon naps in my tent - as you can guess, I was trying to accurately record the day's events".

While based at Seaplane Lake, I managed to make a helicopter landing at two of the numerous hot springs in the Nahanni river area. At a nearby hot spring on McLeod River I recorded its being shared by a calf moose, perhaps attracted as much by the warmth of the water as the lush edible plants at the bottom of the shallow pools. A brief landing at Rabbitkettle hot springs with its spectacular well-preserved stacked tufa-formations[5] was to be remembered years later when I visited similar hot springs in Yellowstone Park with Rene.

In early August we moved to Flat Lake, 75 miles northwest from Seaplane Lake at the headwaters of a northwest flowing tributary of the South Nahanni River and the Flat River. We made the move, conscious of the need to provide sufficient time to adequately support the geological crew at both the Axel and Fitzsob properties prior to mid-September, with the possibility of snow at higher elevations and the impending loss of students returning to university.

One of our priority activities after establishing the base camp was to provide accommodation in the cirque basin on the Axel claims for the geological crew. I had arranged for a shipment of sufficient lumber for construction of two tent frames on the bare rocky ground in the cirque.

When it came time to fly the first load of lumber into the property we decided to take in shiplap. I had arranged with Al, our pilot, to fly with him on the first load and remain on site to help offload the lumber during the day. Al knew by then that my weight with minor gear was about 140 pounds. He looked somewhat dubiously at the load of shiplap and asked Jack Rich, our engineer, what the weight of each of the eight-foot boards would average. Jack hefted a couple and said, "Oh, about five pounds". Al said, "Okay, load on about 25 on each side". We took off from the camp across the lake, and immediately started to lose elevation, as we were tail-heavy and descending toward the water. Al yelled, "Get into the bubble!" I didn't need any elaboration and immediately unhitched my seatbelt and complied. Later, our compatriots on shore were to say that the skids dripped water as we very slowly turned and marginally gained altitude heading towards the Axel showing. As the ground level fortunately dropped gently to the south down Flat River Valley, Al was able to maintain his elevation. Also by good fortune, the elevation of Flat Lake was essentially the same as that at the cirque on the Axel property, 12 miles from Flat Lake but about 1200 feet above the floor of the valley.

We approached the cirque with barely 100 feet of elevation to spare and Al picked a convenient snow patch near the proposed campsite and made a direct-approach landing, the snow cushioning the impact. On returning to our base camp at Flat Lake, Al rigged up a simple scale, balancing off several boards of shiplap against some of our bulkier supply items with established weights. As expected, each board averaged almost ten

---

[5] tufa formations: Sedimentary rock composed of calcium carbonate, or of silica, deposited from solution in the water of a spring or lake, or from percolating ground water.

pounds in weight and the 60 percent difference was far beyond the safe carrying capacity of the machine. Even with this experience, it was to take several years before accurate weights were to be established for <u>most</u> helicopter loads at remote locations.

The Axel claims were examined between August 11 and 27 by the same geological crew moved from the Buster claims. I was on site initially and periodically, also accompanying John Sullivan, Jim Scott and Gerry Noel for a visit in July and Jim Scott again in late August.[vii]

During the 1955 program, the surface showings were mapped by plane table at a scale of one inch to 50 feet and extensively sampled. In addition, a belt 7 miles long by 2 miles wide centered on the mineralized area, was prospected and geologically mapped on a scale of one inch to 200 feet.

I also arranged for completion of a detailed mineralogical study following the program and the results were available prior to a drilling program completed the following season. Alas, no other ore-bearing minerals of economic interest apart from copper were recognized! The drilling program was completed on my recommendation to test for "the possibility that the exposed showings could represent fringes of a large replacement occurring to the southwest."

Following the completion of work on the Axel showing, the geological crew were moved to the Fitzsob showing. The derivation of the name of the showing eludes me. It seems completely incompatible with the Axel and Buster references. Surely Walter Cannon, Axel Berglund's partner, did not have such a nickname, ancestor or affliction![viii]

The party was initially provided with supplies sufficient to last for a 4 - 5 day period. Shortly after arriving at the new camp site, a yearling caribou was spotted near camp. John ruefully recalled that "Jim dispatched the caribou with his Husquvarna rifle, which he highly prized, but was not particularly adept with." It was butchered and two quarters were sent back to base camp by helicopter. One quarter was hung on a tree near camp and the other eaten following a brief period. "That fat yearling was one of the tenderest meals I've ever had," John said, while describing his recollections.

A couple of days after their arrival, a snow storm prompted the group to move the camp further down the mountain slope to a spot with a nearby landing area. It was about 1000 feet below the point where the other quarter of the caribou hung near the remnants of the carcass.

When the crew returned to the camp one evening from the work area, Dave spotted a grizzly eating part of the carcass. He mentioned it to John who commented, "If Rutherford sees that, he'll want to shoot it," which neither relished because of his questionable marksmanship. They kept quiet, but about thirty minutes later Jim let out a "whoop" and they knew that the jig was up. Jim decided on the spot that he had to kill the grizzly and so he and John, who had the camp 303 Lee-Enfield rifle, both cautiously approached to within 150 yards of the bear. Without warning, Jim started firing. The grizzly immediately charged down slope in their direction and John joined in the firing. The grizzly was hit by several shots before falling about 40 - 50 yards from the pair. John recalled it as a very nerve-racking experience. Considering John's World War II experience, in which he was severely wounded in Holland, it must have been just more than memorable! Jim skinned the bear out.

While at the camp, our helicopter became unserviceable and I called B.C.-Yukon Air Services on the radio and arranged for their Beaver to come to our assistance. As virtually all of the prospectors' fly camps were only directly accessible by helicopter, I arranged for air drops of supplies from the Beaver. Never having been involved in this procedure, I benefited from advice provided by George Dalziel, the famous "Flying

Trapper" of the 1930s and then owner of B.C. Yukon Air Service. He supplied us with gunnysacks and arranged for us to tie supplies tightly in one sack and place them in a second sack tied loosely. We picked marshy spots for the drop near the camp, buzzing them several times to alert them to the drop site. A note was attached with each drop indicating the predicament and an anticipated one to two-day delay in further service trips by our helicopter. About 75% or more of the items dropped were sufficiently undamaged to be of use.

No sampling had been completed in 1954 when the showing was discovered. The 1955 program included geological mapping and sampling and the results were not encouraging with very low copper and silver values returned in assays. The property was not staked and on September 8 the party was moved out to Flat Lake en route back to Vancouver.

The results of the general prospecting program were also most disappointing; no showings that were discovered warranted staking. However, I recommended a similar program for 1956 in the Pelly Plateau and Selwyn Mountains areas covering a similar-sized area of about 7000 square miles immediately to the northwest of the region prospected in 1955.[ix]

Preparation for the program during the winter months was similar to that for the 1955 program, with the exception that it was not necessary to produce any topographic maps. However stereo-coverage of the proposed working area was obtained and proved of value to both geologists and some of the prospectors during the season. Geologic information which could serve as a guide to the disposition of prospecting parties was limited to several government reports on the fringes of the region.

In the early spring months fuel caches were established at the proposed base camps which included Finlayson Lake, Summit Lake and Sheldon Lake. These lie between 130 to 200 miles northerly from both Watson Lake and Whitehorse.

The Canol Road, completed in 1944 to service pumping stations along a pipeline route designed to carry oil from Norman Wells on the Mackenzie River to Whitehorse crossed the northwestern part of the project area but was only passable in the summer months and then only from Johnson's Crossing on the Alaska Highway to Ross River Post on Pelly River. Accordingly we planned on being serviced by fixed-wing aircraft chartered either from B.C. & Yukon Air Services in Watson Lake or in Whitehorse. Aircraft we eventually used included Junkers, Beaver and a Cessna 180 with payloads of 1400, 1200 and 580 pounds respectively and costs ranging from $90 to $60 per hour.

The field program was supervised by Jim Rutherford who was assisted for certain periods by Dick Campbell and myself. Dick was preparing for a doctorate in geology in California at the time and was later to join the Geological Survey of Canada and have a distinguished career based mostly on his contributions on the geology of the northern part of the Cordillera.

The prospecting parties selected included Wilf Christian, Pat Cook, Ed Kasmer, Jake Hunderi from the 1955 crew and Jake Copeland, John Hewitt (replaced by Mrs. Copeland in August), Al Mackillop, Don Bragg, Jack Marshall and Len Millar. Additional personnel on the company payroll included a cook and a part-time camp-attendant.

We again contracted for a full-time helicopter with Okanagan Helicopters Limited. The program was to benefit significantly by having a Bell 47 G-2 with an effective payload of not less than 500 pounds at sea level compared with about 400 pounds for the Bell 47 D-1, which we had used in 1955. In addition, it had greater power available at elevations over 5000 feet above sea level. Finally, John Porter, our pilot, had outstanding experience and qualifications with a keen interest in cooperating in every way possible to make it an efficient operation. His engineer on our program was Ray Williamson.

John held an instructor's certificate with Okanagan Helicopters Limited. I recall discussing his experience in training prospective helicopter pilots with him. He mentioned that one of his more demanding requirements as an instructor was to decide on a convenient time and place during advanced training to test the pilot's ability to react to an emergency. John would reach over and turn off the key providing power to the engine. He would then say, "O.K., this is an emergency and you have to make an auto-rotational landing. Just pick a spot and put it down." His capability as an instructor was evidenced by the fact that he had survived many of these tests without having to turn the ignition key back on again.

In late 1955 Kennco had become interested in a lead-zinc-silver prospect on Haskin's Mountain near Cassiar and about 100 miles southwest of Watson Lake. It was owned by Ray McKamey of Atlin, and Glen Hope and J.W. Thompson of Lower Post. Glen Hope had been a long-time resident of a cabin which he had built adjacent to a creek a few miles east of Cassiar. He had lived there for many years, eaking out an existence from recovery of placer gold[6]. Most importantly he and his wife had raised a family of two girls, Glenda and Diana and two sons in this remote location, with schooling administered according to provincial standards by Mrs. Hope. Glenda and Diana were both artistically inclined and talented and went on within a few years to the Banff School of Fine Arts.

While in Watson Lake in 1956 one of my tasks was to have an option agreement signed providing for Kennco to complete an exploration program on the Haskin's Mountain property that summer. John Anderson was to supervise the program and had arrived with me at Watson Lake. Following our arrival in Watson Lake I wrote a letter to Jim Scott on May 30, portions of which I have excerpted:

"The Haskin's Mountain agreement was signed last night by Thompson, McKamey and Hope. McKamey arrived from Whitehorse yesterday morning and as he was in some hurry to return, we drove down the Cassiar road to Glen Hope's home in the evening. As expected, Glen Hope was on Haskin's Mountain staking, and had his two sons with him. We reached their tent at 10:30 p.m., finding that snow in the upper areas is melting quickly. The agreement was read and signed by candle-light, with no major objections raised. You will note that McKamey has initiated changes covering his middle initial, which I thought should be done to check with his signature. He also wished to initial each page of one agreement, and I would suggest that this particular agreement be returned to him.

Ralph Hope, who is nine years old, made the observation that our company "sure is a fly-by-night outfit". It's the second time that we have interrupted the Hope's sleep in four days."

The significance of Haskin's Mountain will become evident in the latter part of this saga as will my reasons for introducing this seemingly irrelevant episode in our 1956 program.

In preparing for the 1956 program we also had to consider the recommendation I had made concerning a diamond drilling program on the Axel deposit. Because of the distance between our proposed base camps and the Axel deposit, I had opted for a diamond drilling project supported by both helicopter and packhorses. This was to be provided by Skook Davidson at Terminus Mountain through Willard Freer, his packer designated for the project, who was to lose his life some years later while packing alone. He was assumed to have died after being bucked off his saddle horse, possibly during an encounter with a grizzly.

---

[6] Placer gold: Native gold occurring in more or less coarse grains or flakes and recoverable by washing the sand, gravel, etc. in which it is found.

Our project team had arrived at Watson Lake by May 30, within one day of the proposed initiation of the 1955 program! It was, and remains a deja-vu experience. Furthermore, our crew was again delayed in being placed out to their initial prospecting sites by weather conditions.

I was particularly pleased with the support of the diamond drilling crew at the Axel showing by a combination of the helicopter and Willard Freer and his horses. The drilling program, which included seven EX-holes totalling 2193 feet, was completed to test the lateral extensions of known mineralization, particularly to the southwest, as I had previously recommended.

Recognizing that we would be carrying out drilling programs at both Haskin's Mountain and at the Axel showing, both supported from Watson Lake, I arranged for a small core shed to be constructed at Watson Lake to provide for partial storage of diamond drill core from both the Haskin's Mountain and Axel showing drilling programs. Although Dick Campbell and I logged the Axel showing drill core on site, it was transferred for storage to Watson Lake during the season. At the same time drill core being obtained from the Haskin's Mountain program was being transferred for logging by John Anderson at Watson Lake.

Work previously completed at Haskin's Mountain by Aaro Aho, later co-developer of the Faro lead-zinc-silver deposits in the Yukon, had recognized minor contents of scheelite, a tungsten oxide, with the lead-zinc mineralization. Accordingly Kennco had arranged for an ultraviolet lamp to be available during core logging of the Haskin's Mountain drill core. Although the Haskin's Mountain core was lamped for possible tungsten content, not a single box of rather similar Axel showing core was ever lamped at the time either accidentally or otherwise. I reluctantly accepted Murphy's Law years later to account for this abysmal oversight.

To cut a long story short, our prospecting teams in 1956 found no showings worthy of staking except for two named the Oke and MacKillop showings, later determined to be of no economic interest, after more detailed examination.

John Porter prepared a report for Okanagan Helicopters on the results of the season's operation.

Included were the following statistics:

"Days lost due to weather (5), mechanical problems (15.5), not required (5.5), no gas (3).

Flying time: May - ferry to Watson Lake from Vancouver (17:10 hours), June (95:50 hours), July (77:25 hours), August (70:45 hours), September (49:50 hours); ferry to Vancouver (19:00 hours), for a total of 330:00 hours.

Freight carried: 57,935 pounds

Number of passengers: 307

Number of geological reconnaissance landings: 205."

John also provided the following additional information on general performance of the operation which explained the relatively high number of days lost through 'mechanical problems':

"The machine... was found very good for the work involved. Outside of fire damage there were two mechanical failures: a scissor arm bolt failed on a landing and a crack developed in one of the tail rotor drive shaft support brackets - the latter necessitating a welder being flown in from Whitehorse. Some trouble was encountered in starting the machine with any degree of consistency, but this was eliminated with practice.

On July 25 an unfortunate ground fire damaged the helicopter, necessitating replacement of the main rotor blades and all rubber components of the engine compartment. The machine became serviceable again on August 9, our commitments being looked after for a week by a P.W.A. helicopter flown in from Whitehorse. "

I had arrived at the base camp by fixed-wing on one of my visits only minutes after the 'unfortunate ground fire' occurred to find it was caused by the engineer washing down the bubble of the helicopter with gasoline while smoking. This had probably happened many times before; however on this occasion the helicopter unfortunately caught fire. I felt extremely sympathetic to the engineer at this misfortune and happily, from what I can recall, it did not adversely affect his relationship with Okanagan Helicopters, who I am sure were all too aware of the incredible learning curve that was developing in their organization.

No further work was carried out by Kennco in the general area of the Selwyn Mountains and Pelly Plateau in 1957 or 1958.

As project manager for the 1955 and 1956 programs, and as a responsible professional engineer, I had written reports on the prospecting programs and the results of the work done on the Buster, Fitzsob and Axel showings. The latter were filed in the Northwest Territories for assessment work purposes.

As a result of the disappointing results of the diamond drilling program on the Axel showing in 1956, no further work was recommended. Abandonment of the Axel claims was approved in December, 1956 and of the Buster claims in March, 1957. However, sufficient work had been applied in 1955 to maintain the key Axel claims (1-4) in good standing until expiry dates falling in the summer of 1958.

In late 1958, I learned that a confidential announcement had been made by the Mackenzie Syndicate, an exploration group including Dome Mines, Leitch Gold, Area Mines, Highland-Bell and Ventures (later Falconbridge Group) indicating that its 1958 prospecting program had resulted in the discovery of several occurrences of skarn-bearing scheelite in the Flat River area. Among these was the Axel showing, which was staked shortly after the expiry of Kennco's claims. The lack of recognition by Kennco of the tungsten content of the deposit is what Sullivan later referred to as the Flat River Disaster.

On December 11, 1958 I wrote a memorandum concerning "Tungsten in Flat River Area, N.W.T." describing various showings in NWT/Yukon and other areas which could have a tungsten potential either as a tungsten property or as a valuable by-product, also noting the current poor market for tungsten and potential access problems to the Flat River area. Accordingly a proposal was made on December 18, 1958 to the company "to immediately stake the more important previous finds and to evaluate the tungsten potential of the known lime-silicate (skarn) occurrences in 1959".

The proposal was not approved.

Almost a year later, in November 1959, it was formally announced that the Mackenzie Syndicate had made a significant discovery of tungsten in the Mackenzie Mountains region of Northwest Territories.

Being back in Vancouver at the time after my first season on the exploration program in northwestern British Columbia, I received several phone calls and personal enquiries from both corporate associates and others, who were aware of my past involvement in the exploration programs in the Flat River area. I answered all of these forthrightly, perhaps accepting more responsibility than was required for the subsequent recognition of the tungsten potential. I consistently stressed our corporate exploration objectives, which did not include an interest in so-called minor metals, including tungsten.

It's difficult to appreciate that in the mid 1950s, in contrast with current exploration practices, most companies, including Kennco, did not spend the extra amounts required for specific element detections. These costs were significantly reduced when multi-element analyses were implemented.

Fortunately for myself and perhaps others, particularly more experienced peers who had examined the Axel showing, I never named individuals who might have been expected to recognize the tungsten potential of the Axel showing and share in my misfortune. This was deliberate and after almost 40 years, remains a personal satisfaction.

I found that I was unable to share my bitter disappointment with others, including my family, in what was for me a particularly tormenting event. In addition I had to contend with a corporate claim that a directive had been issued earlier to exploration personnel requiring provision for the detection of minor elements, such as tungsten, in properties under examination. None of Kennco's personnel were aware of such a memorandum and a subsequent search of corporate files failed to reveal any such document.

During this very disheartening period, I received a telephone call from Lyle Dunn, then a geologist with the Mackenzie Syndicate, asking if he could meet with me concerning the tungsten discovery. I agreed and he arrived with Len White, another acquaintance, who had directed the 1958 program in the Mackenzie mountains for the Mackenzie Syndicate.

My most important recollection of our meeting was their extremely sympathetic reaction to my misfortune. They had not arranged the meeting to obtain any information from me on the deposit, but only to make me aware of the details of the discovery and to indicate how unexpected it had been.

Recalling their comments in 1994, while preparing the initial draft of this episode over 35 years later, and seeking to describe the circumstances of the discovery as accurately as possible, I made enquiries and was fortunate enough to finally contact Charlie Aird, a good friend, who had been a member of the Mackenzie Syndicate in 1958 as a student.

According to Charlie the main exploration focus of the Mackenzie Syndicate at the time was prospecting for beryl. The prospecting party consisted of four prospectors, among them Hugo Brodell, Tony Aarea and Jack Campbell, and their assistants which included Cam Ogilvie in addition to Charlie, both of whom were attending the University of British Columbia.

The party had just arrived at Flat Lake when they were requested to move into the Churchill Copper area, some 200 miles to the north. Brodell was left at Flat Lake with his dogs while the rest of the party departed. Lyle asked Hugo to walk to the Axel Showing while he was at Flat Lake and obtain some samples. Later when Lyle had an opportunity to examine the samples, he noted that they appeared similar to ore at the Emerald tungsten mine in B.C. where he had previously worked.

It was common practice for the prospecting party to grind up samples they obtained using a mortar and pestle to provide material suitable for panning. For readers not familiar with the principle of panning, by adding water to a pan containing finely ground up material composed of minerals of differing specific gravities, gold for instance being 19.6 times heavier than water, swirling of the minerals with water provides a crude separation of those from low to high specific gravity in the pan.

Tony Aarea ground the material up and noted a white tail among the sulfide minerals which he could not identify. It was later examined with an ultraviolet lamp, and as Lyle had surmised, it proved to be scheelite, i.e. calcium tungstate. Although the Axel claims were scheduled to lapse in mid-summer of 1958, the Mackenzie Syndicate apparently did not stake the property until 1959. Accordingly, the members of the 1958 prospecting team were sworn to secrecy concerning the discovery of tungsten at the property until it could be staked.

Canada Tungsten Mining Corporation Limited was incorporated in 1959 to develop the tungsten showings on the property. Funds for production were provided in 1961-2 by Dome Mines (14.52%) and Falconbridge (14.52%), two of the original members of the Mackenzie Syndicate and Northwest Amax Ltd., a subsidiary of American Metal Climax Inc. (70.96%). By good fortune the federal government, already constructing one of the early development roads in this part of Yukon, agreed to pay two-thirds of the cost of a branch road to the property.

Production commenced in November 1962 at a rate of 300 tons per day using open-pit methods but was suspended in July 1963 because of low tungsten prices. Operations resumed in September 1964 only to be curtailed on December 26, 1966 when the concentrator and crusher caught fire. A new 350-ton per day plant was constructed and in operation in late 1967.

By 1980 the mine was capable of producing 8 million pounds of tungsten per year, making it the largest producer of tungsten in the western world. Although it was forced to curtail production because of low metal prices on several occasions, it eventually closed indefinitely in May 1986 with reserves of 1.4 million tons grading 1.2% tungsten trioxide.

What I have attempted to describe in this section is an account of the conditions under which exploration was carried out in the mid-1950s in a relatively remote and unexplored region in western Canada. Hopefully it provides an insight into the impact which rotary-wing aircraft and their pilots were to have in exploration in such remote areas during a transition period where reliance had previously been placed on the more limited capacity of horse and fixed-wing supported programs.

In addition this epic provides one more example of the serendipitous nature of many significant mineral discoveries. In the 1960s I recall attending a meeting at the Engineers Club in Vancouver where a presentation was made by a well-known Canadian 'mine finder,' in which he described his discoveries. Following his presentation I commented upon the fortuitous aspect of some discoveries, which was certainly not brought out during his talk. Quite obviously he did not believe in good luck, or being 'at the right place at the right time.' Conversely, as in our exploration programs in 1955 and 1956, one can be at the right place, but at the wrong time, be it caused by unrecognized mineral values or a host of other influences which can deny a successful discovery.

After 40 years of exploration in the region we explored in 1955 and 1956, it is prophetic that the Axel showing, the first significant deposit recognized in the 14,000 square mile region explored by Kennco is the only one to have reached production by 1994. But then I have already mentioned Murphy's Law.

Exploration in Northwestern British Columbia – 1959

By 1959, Kennco had again become interested in the copper-bearing potential of northwestern British Columbia, but this time in a regional rather than a local context like the abortive examination of the Bronson Creek copper deposit in 1949. Then a senior geologist with Kennco, I planned a program designed to provide reconnaissance geochemical coverage of 10,000 square miles, within a 20-30 mile wide belt on the eastern margin of the Coast Plutonic Complex as a two-year program. The survey area, as proposed, extended from 60 miles north of the British Columbia-Yukon Territory border north of the town of Atlin, 425 miles south to the vicinity of Nass River.

The 1959 field program began in early May when our party flew from different parts of British Columbia - mostly Vancouver - to Whitehorse and then drove by truck to Atlin, a small, historic, attractive community on the east side of Atlin Lake about 100 miles southeast of Whitehorse.

Atlin lies in a setting of outstanding natural beauty, nestled as it is within the broad valley containing Atlin Lake and dominated by the Coast Mountains to the west. The community lies within the dry belt on the lee side of the Coast Mountains with an average precipitation of about 11 inches.

On February 10, 1898, Fritz Miller and Kenneth McLaren reached Atlin Lake from Skagway, Alaska with six companions. They staked claims to cover a gold discovery on Pine Creek near Atlin. Two years after the discovery of gold in the Klondike, about 350 miles to the northwest of Atlin, hordes of venturers were still travelling to Dawson. When news of the Pine Creek discovery was publicized, miners bound for the Klondike turned to Atlin instead and before the end of 1989, over 3,000 persons had visited the booming camp. By 1898, the population of the immediate district reached about 5,000 but dwindled thereafter and in 1959, there were about 200 people resident at Atlin, mostly dependent on placer mining and prospecting for their livelihood. By 1994, the population of Atlin had increased to about 430 reflecting the increasing numbers of individuals attracted by small, remote northern communities with their relatively modest living conditions in serene pioneer-like environments.

Our party in 1959 included Dick Woodcock and myself acting jointly as party chiefs. Like myself, Dick was a senior geologist with Kennco and since then he has enjoyed an outstanding career as a geologist working principally in western Canada and, in recent years, as a consultant specializing in molybdenum and other deposits.

In order to provide for sample collecting, I allowed for five two-man teams on our proposed program. It is safe to use that male terminology as there were few women involved in exploration in 1959 and none in our party. I was to become an advocate of heterosexual field parties after joining DuPont in 1974, being partly influenced by Marshall Smith, one of our geologists, but mostly by the outstanding work of Louise Eccles, our first permanent female geologist. By 1980, our field parties commonly consisted of young men and women, about equally divided but highly skewed toward women in terms of university enrolment by students in geological sciences. However, in the 1950's our silt sampling crews were mostly male university students and our party included Ross Blusson, Don Bragg (a prospector), Harley Goddard, Bela Gorgenyi, Ray Gullison, Denis Hale, Hugo Laanela, George Lechner, Jim McAusland, and Bob Patterson. In addition, we had a two-man prospecting party composed of (now the late) Wilf Christian and (now the late) Gordon Davies, Jr. Of the above individuals, not all of whom were enrolled in mining or geological courses, Ross, Don, Harley, Hugo, and Jim were still active in Canadian exploration or mining-related ventures in 1991.

Because of the need to provide for follow-up of any anomalous sample sites while we were in the general area, I had arranged for a mobile analytical laboratory at our base camp. This was most efficiently run by our chemist, Rudy Jalbert and his assistant, Baird Palmer. Our cook was Bob Russell.

Although we used trucks and boats for sampling part of the project area, our principal transportation was provided by a Bell 47 G-2 helicopter on contract from Okanagan Helicopters and piloted by Mike McDonagh, who was killed several years later in a helicopter accident. John Gill acted as engineer. Our base camp was periodically moved and frequently supplied by float-equipped, fixed-wing aircraft, principally by Herman Peterson of Peterson's Air Service in Atlin and also by "Wild Bill" Harrison, then based at Telegraph Creek.

Claim Staking - Chutine Lake, January 19, 1959

I had previously flown with Herman Peterson on several occasions and was comfortable in the company of this expert pilot, who incidentally still lived in Atlin in 1992. In January, 1959, on the release of information by the Geological Survey of Canada on the results of their 1958 program in north-western British Columbia, we moved quickly to stake a reported occurrence of molybdenite in the Chutine Lake area, about 30 miles west of Telegraph Creek in the Coast Mountains. Our party consisted of Charlie Ney, John Anderson and myself.

John and I were to complete similar short staking trips in February and March of 1959 to stake claims covering similar reported occurrences. One was the Barrington River molybdenite report and the other a tungsten occurrence in the Dragon Lake area of Yukon Territory. Our equipment and provisions were similar in all three cases as was our dependence on Herman Peterson's air service in Atlin for our two trips in B.C.

It was our first venture into a relatively hostile area; hostile only because of our inexperience in glacial environments under winter conditions. We flew to Whitehorse from Vancouver with our equipment and drove to Atlin where we met Herman who was well prepared for our hastily arranged trip.

I recorded our trip into the Chutine Lake area in diary entries which are reproduced verbatim in the following summary.

January 19, 1959 Left Vancouver 3:30 p.m. flight for Whitehorse - DC6B stopping at Prince George, Fort St. John and Whitehorse. During flight moonlight gave good visibility of land. The flight follows the Alaska Highway almost the entire way. The temperature gradually decreased as we flew north reaching -36°F at the air field in Whitehorse and being estimated at -40°F in the townsite. We were met by Hertz U-Drive (agency of Bill Drury) and stayed the night at the Taku Hotel. It is common practice to plug the battery into a heater to permit easy starting on cold nights, and to ensure no delays; we left the vehicle - a 1959 Pontiac at Bill Drury's garage.

January 20, 1959 Bill Drury dropped the car off at the hotel at 7:00 a.m., and having obtained all our baggage and equipment from CPA the previous night, we were able to leave for Atlin at about 7:20. The drive took about 3 hours to cover 120 miles, winding for a long way along the east shore of Atlin Lake. At Atlin we met Herman

Peterson who owns Peterson's Flying Service, and his wife Doris. We obtained all our groceries at one store and picked up some odds and ends at a general store. Atlin is a picturesque community, underpopulated for its size as a result of time's passage from the boom period when placer gold fields were in operation. The houses are almost all constructed from wood frame and among newer buildings are the government agents' offices and a school. Apparently only three white children were attending school last year.

Herman had kept a heater attached to his plane since the following day. As we had postponed our trip by one day, he had not been sure of our arrival time. The only part of the message he received which we telegraphed advising of our delayed arrival time was the word "delayed". We left Atlin at around 12:30 after buying sufficient supplies to last us about 10 days - we planned being at Chutine Lake for 4 days initially. The temperature at Atlin was -20°F at take-off, but as we gained altitude it increased to 0°. Our course was via Nahlin Lake and the Sheslay River across a divide into the headwaters of Chutine Lake. The distance in air miles from Atlin is 130 miles and we were in the air for about 1½ hours. The area becomes more and more rugged as the Coast Mountains are approached. East of Sheslay river the mountain ranges are broad and bevelled, consisting of Tertiary volcanics in flat lying flows. Near Chutine Lake, relief increases and jagged crest lines appear, divided by numerous glaciers. We recognized the approach to Chutine Lake by a towering peak 9660 feet in height which forms the east wall of Chutine Lake. This probably represents one of the greatest amounts of steep relief on the Coast as the summit is attained within a horizontal distance of 2½ miles, rising from 850 feet - the elevation of Chutine Lake. From our elevation Chutine Lake looked like a questionable landing field, although it is four miles long,, as we lost altitude its size became apparent. Herman was worried about possible slush ice - a condition caused by the weight of overlying snow on blue ice causing cracks which permit underlying water to seep through. He made one pass at the lake where he permitted his skis to touch the snow, and then finding conditions satisfactory we landed, stopping near the north end of the lake. We packed our supplies about 100 yards onto the shore - a flat terrace covered with willow and very sparse balsam. The temperature was -20°F and about 8 feet of dry powder snow covered the ground. We pitched our Logan tent after stamping out a site with snow shoes. We placed two silk flies on the base of the tent and then packed in our sleeping bags and immediate essentials including a Coleman stove, Coleman lantern, cooking equipment and some food supplies. The Logan tent measures only about 8' x 8' on the floor with a central pole reaching a height of about 7', so there is not too much room for supplies. One of the continuous problems is trying to keep excess snow out of the tent, which is mostly brought in on clothing, particularly foot gear. With regard to the latter, we all used leather (moose skin) moccasins with a felt insole and two pairs of socks. Our feet were continually warm as long as not wet and kept moving. We went to sleep about 9:30 p.m., and I enjoyed my best night of sleep - awakening about 4 or 5 times during the night, but able to return to a good sleep.

January 21, 1959 We rose at 6:00 a.m. and were not ready to leave camp until about 9:00. All our water was obtained from melting snow, and cooking took time. Washing dishes added almost ½ hour. The weather was partly overcast but moderately bright. We took about two hours to reach the toe of the glacier which was to provide access to our staking area. We made one false start - where we were stopped by unstable-looking ice over the Chutine River near open water. We retraced our steps and crossed the river at a safer spot. On the way we cut and broke off about 50 small willow branches with which to mark our route, in the event of later drifting. We roped up at the base of the glacier and slowly pushed our way around the ice-fall area with its crevasses, reaching an elevation of 500 feet above the lake level in about one hour.

Of the three of us, Charlie was by far the most experienced in mountain travel, although we had all had several seasons of exploration work in glacial environments by that time. Charlie's experience led us to pack the willow branches to mark our path up the glacier so that we could return by the same route. He also had arranged to bring sufficient candles with suitable can containers should we need them as lights in the restricted daylight hours. In spite of his experience, he was terrified of going to sleep because of the cold temperatures. Both John and I had difficulty in reassuring him that he would wake up shivering if it got too cold rather than drift off into a final sleep.

We alternately broke trail from then on, stopping for lunch at 1:00 p.m. Lunch consisted of frozen salmon which leaves much to be desired. After lunch we continued on through an increasing snow fall which caused white-out conditions and necessitated occasionally throwing bits of our willow ahead of us to indicate sudden drop-offs. We reached the start of a small mill-hole area at about 3:00 and then decided that we had reached our end for the day, as we had to allow time to return to camp. This was accomplished in about 2 hours covering 4½ miles, and the last part was done in darkness with the aid of one of Charlie's 'bugs'. This is a can containing a hole through which a candle is inserted to permit a protected light.

That night we prepared all the forms that we would require for staking the Nat claims. These were inserted with accompanying tags into small aluminum cans, all to be left at a cairn, marking the position of a witness post for the inaccessible claims. We had calculated that we would be able to reach two witness posts approximately 3000' apart on the west side of the glacier, and witness posts to the west over a high mountain. We hit the pit at midnight.

Preparing for bed was an exercise in itself, requiring careful placing of various garments below the sleeping bag in an attempt to provide some insulation from the hard packed snow below. This was not wholly successful and toward the end of our stay, we slept less because of the cold. I was surprised to find following the third night that perspiration from sleeping - or warmth from my body had melted the snow below three layers of canvas soaking all the canvas, two layers of an oil skin parka and the bottom of the sleeping bag. An insulating medium between body and ground is essential for a warm dry sleep and short of carting in mattresses, boughs from evergreen trees are probably most suitable. We cut some of these and used them on the fourth night, but by this time there was too much moisture around to prevent a repetition of the soaking.

We had no water supply in spite of having our tent pitched on the northern end of the ice-covered lake. We melted sufficient snow for our needs and I do not recall washing during our three-day stay. Because of the humidity created by steaming water, ice crystals formed around the lower part of the tent reaching lengths of one inch or more. We jokingly noted that our stay would be restricted by the increasing length of these crystals. The bell of the tent was hot, there being a 100°F range from the floor to the highest point of the tent. One evening, I hung my damp socks over a string attached to the sides of the tent near the top of the tent pole. We were anticipating our evening cup of cocoa, which was brewing in an open pot immediately below. Somehow the socks got dislodged and fell into our brew. Charlie calmly reached over and fished them out, ringing them as dry as possible into the pot. The message was abundantly clear: we would be drinking the cocoa. It still tasted pretty good! Bill Matthews, a University of British Columbia professor, was to recall this story in a eulogy at a memorial for Charlie at his untimely death at 55.

January 22, 1959 To our horror we overslept by one hour - getting up at 7:00 instead of 6:00. We rushed through breakfast - which consisted of mush, bacon and bread dipped in bacon fat plus coffee. We left our dishes unwashed to take advantage of daylight conditions and left by 9:00. By 11:00 we had reached the lunch spot of the day before, as a result of more rapid travel on our broken trail. By 12:00 we were extending the trail across the glacier, and it was around this time that the sky began to clear and visibility increased. A small plane flew overhead shortly after making a circuit of the glacier and then flying off over Chutine Lake. This plane returned about 2 hours later and made several passes over Chutine Lake - leaving again. We later learned that it was piloted by Bud Harbotle who was bringing in Gordon Dickson to stake the same area that we were after. Apparently they did not see us on their first trip and returned to Telegraph Creek to consider the chances of staking on the glacier. On their return trip they noticed our tent and tracks and decided that since someone had already arrived, they would not be able to get the ground, and so they left. Another independent staker attached to a syndicate flew up all the way from Vancouver to Whitehorse, rented a truck and went to Atlin and asked Herman Peterson to fly him in to Chutine Lake. Peterson dissuaded him as he would also have been too late.

At around 1:30 we reached a point near the west side of the glacier after having plodded up and down numerous snow-covered holes which as a result of white-out were barely visible. We called this "the point of indecision" because we debated whether to make for one of our sites and stake from only one witness post - altering all our tags accordingly, or attempt to cross a highly-crevassed area to the second of our two selected points. We decided to spend not more than half an hour trying to cross the crevassed area, and finally had to return back as conditions became progressively worse.

Our other point was reached without too much incident - except that white-out prevented one from seeing the edge of a prominent moraine on which we walked without prodding ahead with an ice-axe or pitching snow balls. We ate our lunch - sardines and raisins as we changed the tags, and finally left the site at 3:00. Our trip back took 3 hours and we were all tuckered out - looking like three of the seven dwarfs with our 'bug'. We finally bedded down at about 11:00 full of anticipation for the arrival of Peterson to take us back out on the next day.

January 23, 1959 The area was all misted over on the following day and it started to snow at mid-day. We spent the morning tramping out a landing spot for the plane which had two long poles which we cut for the skis to rest on. We also placed about a dozen small trees in a line at intervals of 200 feet as a landing guide for Herman. We then explored part of the lake-shore, knocking off several hunks of rock for specimens. John and Charlie collected the first of a load of boughs while I made a late lunch and started completing staking forms. During the afternoon we tramped around some more to avoid using up too much of our fuel for the stove in keeping warm. The inside of the tent was always coated with ice when not in use, and we judged the temperature by the drop in the 'frost line' down the side of the tent. The creased portions of the tent usually retained ice as a result of pockets with poor circulation of the warm air.

We had frequent trouble with our stove during lighting. It had a tendency to leak and the generator would become coated in flames. On one occasion the base of the tent caught fire - but it was so wet or frozen that it was not a serious threat. We finally changed the plunger on the Coleman light, greasing the new leather with butter to create friction on the wall of the tube, and it worked like a charm from then on.

Frequently we heard the rumble of snow sliding down the mountain-sides. After hearing the sound, we could see the source and watch tons of snow come spewing down the near-vertical face at the bottom of the mountain. Several bright green splashes of frozen run-off made a beautiful curtain of icicles at one point where they draped across an over-hang. Charlie was most impressed with this mountain because of its outstanding relief.

By night fall, snow was falling heavier and on the following day we estimated that a total of 8 inches had fallen during our four-day visit, most of it dropping in the final 24 hours.

January 24, 1959   Our initial view of the weather led us to speculations of how long we were to be imprisoned at Chutine Lake. It was completely misted over. Following a leisurely breakfast, I poked my head out the tent flap and saw several patches of barely perceptible clouds which broke the overcast. Within an unbelievable half-hour the whole of the valley was clear. The visibility gradually increased and became crystal-clear by around 1:00 p.m. It was one of those almost rare days when even the most distant peaks are so sharply etched that they appear drawn into a common plane. While Charlie went out after a water sample John and I moved our landing site away from a section of ice which had become slushy.

Around 1:30 with Charlie returning I started to make lunch, and had beans all thawed out in a pan when they both announced that Herman had just pulled into view. I took the frying pan outside the tent and we gobbled our lunch out of the pan with three forks and then commenced to tear down our camp and pack our remaining gear. We had already packed up some equipment during late morning in anticipation of Herman's arrival. We finally left the lake at around 2:30 and made a sweep over the glacier on our way back to Atlin. We found that Herman had been in the air almost all morning after Telegraph Creek weather, so that he could pick us up, as conditions at this time of year are so uncertain weather-wise. He had flown above ground fog flying to Chutine Lake and ducked into a hole near his destination. By the time we reached Atlin almost all the fog had been dispelled. We landed at around 4:30 and were invited to sandwiches and coffee at the Petersons.

Herman and his wife have lived in the Atlin District for about 15 years. Both of them hail from the east and we spent quite a while talking about Quebec with which John was familiar, as he had spent a winter in Montreal.

After settling a couple of bills, we left Atlin and arrived in Whitehorse at about 9:30. One of the startling facts about road travel in Yukon and Northern B.C. during winter and to a lesser extent summer, is the speed with which distances are covered. The road from Atlin to the junction with the Alaska Highway, 61 miles long, is of second-grade calibre with several winding sections. It is maintained partly by the B.C. Government and partly by the Territorial Government, as the 60th parallel passes midway along the route. We stopped to look at this feature, a swath cut through the bush about 30 feet in width. The Alaska Highway in winter and summer is well-maintained, and travel at 50 to 60 miles per hour is normal on the hard, crushed gravel surface. In winter travelling conditions are better - there is no dust and less chance of blow-outs as a result of the colder temperatures. During our trip a low of -50°F was recorded at Whitehorse.

January 25, 1959   We spent the day around Whitehorse. John visited several of his friends, many of whom he saw at the curling rink where the Yukon bonspiel was in progress. We both knew Al MacKillop, one of the members on the Whitehorse team, and were very happy that his team won the bonspiel following an exciting final game in which the final rock determined the result of the game.

We drove around Whitehorse, down by the old wharves where several of the paddle-wheelers which used to ply between Whitehorse and Dawson on the Yukon River, are gradually decaying on dry-dock. A large stock of asbestos in neatly piled gunny sacks was waiting for shipment via the Whitehorse and Yukon Railway to Skagway. This originates from Cassiar where eight varieties of fibre are currently being produced. The fibre is mined during the summer months, mostly May to November, and excess is stored in a covered shed capable of storing about 100,000 tons. The fibre in the broken rock is crushed, screened, dried, sorted and packed at Cassiar. It is then trucked in large trucks about 100 miles north on the Cassiar Road to the Alaska Highway and then about 280 miles northwest to Whitehorse. The trucks contain large tanks below the units which are filled with diesel oil on the return run. The trucks are equipped with a recording tachometer to prevent the drivers from exceeding a safe speed limit. Even with this deterrent, they drive very fast and make the run in around 11 hours.

We saw a recently 'erected' nugget of native copper, weighing 2700 pounds which was hauled out of the White River area by bulldozer to Whitehorse. The nugget, a flat slab about 10 feet in height and 4 feet in width is standing on its edge outside the Whitehorse Museum.

We drove around to a recently constructed dam on the Yukon River. The dam of rock fill construction crosses the Yukon river at a point slightly above the Whitehorse Rapids, approximately 2 miles outside of the city. We used a road which gives truck access to the dam to cross over the dam itself and stopped to gaze at the water thundering through green ice-coated caves at the bottom of the spillway. The gate control structure was coated with ice and reminded me of one of those pictures you see of destroyers or other vessels coated with ice in the winter. We drove around to the power house and I found a friendly type on duty who showed us around. The power house is one of the neatest I have ever seen - all of the equipment being the latest available, mostly from Sweden. Construction was started about two years ago and the cost to date is $3½ million dollars, with another million dollars to be spent on the erection of fish ladders.

Inside the power house the lower portion of the walls are painted blue to simulate the height of the water outside. The gross head averages 60 feet and the two units currently in operation each have a rated capacity of 6900 kw, but are now generating 1500 kw. One of the interesting side-lights was an electrically operated clock which was running at 60.2 cycles in order to make up an 11-second deficit per 24 hours existing by comparison with an adjacent spring-driven clock which has a tolerance of ±½ second per week. Since all electric clocks in Whitehorse are geared to the power house clock, the standard must be closely controlled.

The master control panel contained a unit which could be called a trouble panel. It was subdivided into various sections on which were marked possible power failure causes. These become lit by a small bulb until the particular trouble is corrected.

In the afternoon Charlie and I climbed up the local ski hill and watched the skiers. A Dutchman told us that the best skiing temperature is about zero degrees as below this temperature the snow becomes sticky and slows up skiing.

In February 1959, John Anderson and I made a similar staking trip into the Barrington River area immediately west of Little Tahltan Lake with access again provided by Herman Peterson. The trip was shorter and the only excitement occurred when our stove suddenly flared up and I pitched it out while John held the tent flap open. We soberly reflected on the consequences of a burned tent and sleeping bags.

May 1959

We returned during our 1959 summer program under more congenial circumstances to learn that both staked properties contained only minor molybdenum mineralization of no further economic interest.

Our stay in Atlin in May was brief and provided us with an opportunity to consolidate our equipment and basic supplies before establishing our first base camp at Palmer Lake, a small lake on the east side of Atlin Lake, about 10 miles to the south of Atlin and accessible by road.

While in Atlin, we had occasion to set up a laboratory tent on a flat area near the edge of the lake immediately in front of the Atlin Hotel. The tent had been specifically designed and made in Vancouver as a mobile laboratory for the analyses that would be required on samples collected by the fly-camp crews. While we were inspecting the finished product with some satisfaction, we were visited by the disgruntled owner of cabins next to the hotel who mistakenly assumed we were to become on-site residents. Without any specific introduction or enquiry, he complained that he had perfectly adequate cabins for rent and that it was "tough enough to make a living as it was". This was a very isolated incident. Our presence in the area was welcomed by the residents we met; they recognized that we would be supplied from Atlin for several months.

Shortly after our arrival, Herman Peterson removed the ski-wheel landing gear from his Beaver aircraft at the Atlin airstrip to replace it by floats for his summer operation. Several of us watched with considerable fascination and not a little apprehension as he prepared to fly the aircraft down the gravel airstrip and take off from a wheel-equipped dolly. The operation went very smoothly and we were all much impressed - particularly with the velocity of the apparently flimsy dolly as it careened down the airstrip after the aircraft became airborne.

Following our move to Palmer Lake, we completed sampling of the area accessible by road northerly to the Yukon border and beyond, augmented by helicopter flights. Within a relatively short period, we had our six fly-camp parties deployed at low elevations; the higher regions were still largely snow-covered.

We relied principally on firewood for fuelling our tent's airtight heaters and on propane for our cook stove. Obtaining firewood was a continuous chore and one morning I was stunned to see that the back of one of our pick-up trucks was filled with blocks of neatly bucked up cottonwood. I enquired as to its source and Bob Patterson delightedly answered, "We found it on the roadside between here and Atlin" I recalled having seen some individuals cutting trees in the general area and had to say very regretfully, "Well, unfortunately you'll have to take it back, as it was obviously piled there for later pick-up by the woodcutters." Gloomily, Bob drove off to return our easily acquired bounty. A day or so later, on driving into Atlin, I was surprised to see a neat pile of freshly cut cottonwood stacked by the road within an area containing only spruce and about two miles from the nearest area of wood-cutting activity. I laughed, anticipating the astonished reaction of the woodcutters to the changed location of the wood pile when they returned to pick it up.

While at Palmer Lake, we met an American who had wintered two or three horses near Atlin. He had brought them across country, partly via the Telegraph Trail, all the way from Death Valley in California and was heading up to Mt. McKinley in Alaska. His object was to be the first to ride horses from the lowest (282 feet below sea-level) to the highest (20,320 feet above sea-level) points in North America. He couldn't have been too aware of the access problems to Mt. McKinley on foot, let alone by horses.

Routine sampling was completed by the five collecting teams, each composed of two samplers based at fly-camps at distances of up to 40 miles from base camp. Their sampling patterns extended for up to 4 miles from the fly-camp and they were moved to a new location at four- to five-day intervals by helicopter. In addition, the two-man prospecting party composed of Wilf Christian and Gord Davies was used to explore areas of high geochemical background values obtained by the samplers where no obvious target areas for metallic mineralization appeared evident.

The fly-camp crews were moved in a southerly to south-easterly direction from Palmer Lake to new camps about 8-9 miles from their previous locations. During the move, they were provided with additional supplies, mail and a new set of air photos with film overlays on which they would plot new sample sites and sample numbers. Outgoing items included their mail, samples and reference material. Samples were processed in the lab and the results were plotted on reference maps. The results were interpreted by Dick and myself and anomalous areas were quickly defined as we wanted to determine the potential importance of the anomalies while they were still reasonably accessible from base-camp.

We generally worked alone in following up anomalous findings. We would get set out by helicopter at the nearest point to an anomalous sample site, armed with a portable cold-extractable chemical kit, prospector's pick, hand-lens and emergency gear in our pack-sacks. After verifying the site, which had been previously flagged and numbered by the samplers, we traversed upstream taking samples and checking creek boulders for signs of mineralization. We also completed very preliminary geologic maps of any mineralized areas detected and took preliminary samples for assay.

The follow-up process was very effective. During the 1959 program, a total of 112 anomalies were defined, of which 33 contained patterns or values of sufficient interest to warrant immediate follow-up. Time permitted follow-up of 28 of the anomalies and relatable mineralization was discovered at 25 of the localities. Additional analyses in the 1959-60 winter provided further data which assisted in defining less distinct anomalies, which were followed up on a priority basis in the 1960 program.

In addition to the follow-up activity, Dick and I generally worked together in supplementing our database by sampling relatively inaccessible sites which did not warrant the establishment of a fly-camp. For this purpose, we would fly out with Mike and be dropped off independently to complete the sampling as rapidly as possible, either while the helicopter waited nearby or to be dropped off and picked up on a hop-scotch basis. Between us we sampled about 600 sites that summer.

Follow-up work was always exciting as there was the challenge of locating a source area and the hope that significant mineralization might result. Many of the sites we visited had probably not been occupied by others, as much of the mineralization discovered occurred in relatively remote areas difficult to access on foot. Helicopter-supported exploration was still in its infancy.

As would be expected, the time we spent on follow-up work was quite variable and was influenced by the size of the anomaly and the number of sample sites requiring further investigation upstream to their source areas. These source areas could be miles away or quite close to the anomalous sample site. Rugged topography was always an inhibiting factor as detours might be required around inaccessible cut-banks or falls on drainages. Two examples will illustrate this time factor.

In June, I was dropped off by helicopter one morning on a northwest flowing tributary to Sutlahine River, about 80 miles southeast of Atlin when we were based at Trapper Lake. I was to investigate two of three anomalies obtained on the creek, the third requiring more detailed sampling over a broad area on both sides of the main creek which was completed in July by Bela and Hugo. After completing follow-up on one of the anomalous sites, I began on the second which required following a steep creek gully for about 3,000 feet with an elevation gain of about 1,200 feet. I kept getting stronger reactions with my Holman copper kit as I worked my way up-stream and also noted rusty massive pyrite fragments in the creek. Late in the afternoon, I had to retrace my route for a pick-up as there was no other landing site available nearby. After an aerial reconnaissance, I flagged a landing spot above timberline and beyond the head of the anomalous gully. I returned the next morning and worked my way back downstream, obtaining negative reactions with the Holman copper kit until I suddenly found myself at the obvious source - a massive pyritic lens[7] exposed over an area of about 100 feet square. After traversing more than one mile and descending some 1,500 feet in elevation, I was within 100 feet of the point where I had turned around the previous day! A sample which I chipped for 120 feet on the showing assayed 2.5 ounces silver per ton and 0.34% copper. More work was recommended in the area in 1960. The property, to be named Thorn, was re-discovered by Julian Mining Company Ltd. in 1963 and since then, considerable exploration, including several drilling programs, have been carried out by various companies, exploration activities continuing through 1989. I had the unexpected pleasure of completing a qualifying report on the Thorn property for Shannon Energy Ltd. in May, 1989 through arrangements made by Dr. Gerald Carlson who had acquired an interest in the current claims.

One of the fastest and most direct follow-ups we ever made was also one of the most distant from the anomalous site to the source area. In early July, Dick and I flew to the mouth of a tributary to Sittkanay River near the Alaska-B.C. border, about 20 miles east of Juneau. Sampling by Hugo and Bela had indicated a strongly anomalous molybdenum content and the pair had noted molybdenum in float fragments at the toe of a glacier about one mile up-stream. Dick and I examined the toe of the glacier and noted that the molybdenite occurred in fractures and veinlets in rusty-weathering fragments of quartz monzonite porphyry strewn over a medial moraine at the toe of the glacier. We climbed into the helicopter and followed the medial moraine up the glacier for three and one-half miles to a juncture of two arms of the glacier.

There the rusty weathering medial moraine became a lateral moraine to an ice field at its head immediately east of Mt. Ogden (7,441 feet) lying on the border. We climbed out of the helicopter about one mile above the split in the medial moraine and walked up to the margin of the ice field where the quartz monzonite porphyry could clearly be seen apexing into overlying sedimentary rocks. In my report on the 1959 program, I noted *"there is a strong possibility that a partly scoured molybdenite deposit occurs in the floor of the glacial basin"*. Although we returned again to complete preliminary mapping in August, our examination on July 6 took only about two hours. The molybdenite occurrence was later independently recognized by the Geological Survey of Canada in a mapping program of the Tulsequah and Juneau map-sheet headed by Dr. Jack Souther with fieldwork completed in 1958-1960. Dick and I both felt that the possibility of discovering a viable deposit was unlikely because of the location and physiographic problems for a bulk-mineable deposit. The prospect became known as the Mt. Ogden Molybdenum property and was explored principally by Julian Mining Company Ltd. and Anaconda American Brass in a joint venture in the 1960's.

We moved to Trapper Lake in June where we established our base camp at the mouth of a river draining into the lake from Tunjony Lake, about three miles to the west. The site had been a base for earlier explorationists including Jack Souther's Canadian Geological Survey party.

Camp moves were always a hectic, physically draining activity because of the need to complete the move in such a short time. After scouting out a potential site, we arranged for a preliminary fuel cache to be established with only a minor amount of prior physical

---

[7] Pyritic lens: A body of pyrite (iron sulfide), thick in the middle and thin at the edges, ie lens-shaped.

preparation. Essential requirements for the site were good access for float planes, preferably dry ground conditions, a suitable landing site for the helicopter and readily accessible drinking water and firewood. We always prayed for good weather during the move which was essentially completed in five to six flights on a single day, the distance between base camp locations being about 70 to 75 miles. Not infrequently we completed fly-camp moves with the helicopter on the same day. These would be located at maximum distances to the north of the new base camp and would gradually be moved southerly past the base camp location. Where possible, we would arrange for the fly-camp personnel to enjoy some R & R at base camp but such occasions were rare and one party at least spent the entire summer being moved some 20 times from one site to another without a base camp break.

Although we all saw game frequently, we never had any serious incidents as a result of encounters with wildlife. Wilf and Gordon, our prospectors, were to count 12 different grizzlies during the summer. One night, a moose got tangled up with the ropes of their Logan tent while they were sound asleep. They awakened as if into a nightmare knowing only that they were being battered around the tent as it moved along the ground. The moose finally became disentangled but not until it had dragged them for almost one hundred feet. Thereafter, their tent ropes and an outer roped off area were generously festooned with small bells.

One of the favourite recollections of my companions at Trapper Lake was my encounter with a grizzly bear at our Johnson bar where I was caught with my pants down. Our latrine was quite conventional, consisting of a stout branch securely attached to two trees between which a pit of suitable dimensions had been constructed to satisfy the requirements of six man-months duration. This essential structure was situated about 200 feet from the cook shack on the edge of a cottonwood stand. The location could not have been more poorly selected as unbeknown to us wandering grizzlies are more sensitive to odours from a garbage pit than those from a latrine. One morning while cogitating restfully on the Johnson bar, I became aware of brush rustling behind me. Concerned, as I had asked Baird Palmer to cut some firewood, I turned to warn him off and was absolutely appalled to see not Baird, which would only have been embarrassing, but a grizzly standing on the other side of the pit about six feet away sniffing the air and looking at me.

My reaction was swifter than I ever dreamed myself capable of, and I had pulled up my pants and ran the 200 feet to the cook tent in what seemed like 20 ten-foot bounds, almost certainly approaching a world record for that distance under those circumstances! My companions were not overly sympathetic as I believe they doubted my story. However, that evening Mike was subjected to the same experience. The next day, Baird, who was a sharp shooter, dispatched the grizzly at the garbage pit.

We were to see abundant game in the Trapper Lake area, which lived up to its name. One day, on returning to camp in the helicopter, we watched a cow moose trying to encourage her twin calves across the mouth of the swift flowing creek emerging from lake, which was well beyond their depth. She was eventually successful and we admired the calm and gentle way in which she gave them confidence.

In the nearby Tatsamenie Lake area, Jack Souther recalled witnessing a pathetic encounter between a grizzly and a young moose while its mother stood nearby. With one measured swat, the grizzly knocked the poor calf's head from its body.

Trapper Lake was the scene of a tragic accident in 1963 when three or four explorationists working for Julian Mining Company and the pilot of a Beaver aircraft owned by Trans Provincial Air Services were killed shortly after take off from the lake.

According to witnesses, there was a strong wind at the time and the aircraft suddenly went into a spin and crashed into the lake. There were no survivors.

From Trapper Lake, we moved 55 miles southeast to Little Tahltan Lake, alias Pete or Valentine Lake, situated about 20 miles west of Telegraph Creek and only 12 miles northwest of Stikine River. Our campsite was particularly scenic, lying in alpine meadows just above the lake, which provided excellent fishing. The lake is one of the most distant spawning locations for salmon making the long and arduous ascent of Stikine River for 100 miles from tidewater near Wrangell, Alaska to the mouth of Tahltan River above Telegraph Creek and a further 30 miles up Tahltan River to Little Tahltan Lake. From Tahltan Lake a short brook provides a link to Little Tahltan Lake. The summer of 1959 had been quite dry and the water in the brook was too shallow, running partly underground, to permit the salmon to reach the lake. The Department of Fisheries had two young men carrying the fish across the inaccessible portion for several weeks during the spawning period.

On one of my early flights in this part of the Stikine River area with Herman Peterson, we were flying south from the Sheslay River area on a very hot day. As we approached Stikine River near Telegraph Creek, Herman checked his altimeter and gained altitude. A short time later we began to lose altitude and crossed a pass near the river, a comfortable 1,000 feet above it after dropping 2,000 feet due to the down drafts created near the river. I was later to learn that Herman had lost a Cessna 180 aircraft with its pilot and two passengers, one a nurse tending an injured individual, because of a down-draft in the area. A similar Cessna 180 with several passengers aboard spotted smoke from the wreckage and descended to investigate, only to meet the same situation but with miraculous results as the pilot was able to crash land the aircraft into trees without injury to himself or his passengers.

While in the Stikine River area, we arranged to complete stream sediment sampling of portions of the river system accessible by riverboats. Ross Blusson and Harley Goddard spent five weeks, principally on the Stikine and part of the Iskut River, its major tributary, completing this task with the assistance of two different boat operators, both of whom were familiar with the rivers. One of these, Mr. Wigglesworth, was drowned several years later on Stikine River near Boundary, a former customs post on the Alaska-B.C. border.

Almost 30 years after our 1959 program, I wrote a letter dated October 28, 1988 to the Honourable William Vander Zalm, Premier of British Columbia, urging the recognition of the multiple resource-use potential of the region and the significance of this major travelway which was being threatened by a proposal to convert its status to that of a recreation corridor where access for other purposes could be denied. The following excerpt from the introductory portion of my letter discussed The River, Historical Markers, and Mineral Resources:

The River

Worldwide, waterways are the principal travel corridors which have influenced cultural and economic development. In Canada probably in excess of 95% of the nation's population reside along waterways and their influence in early exploration and settlement is legend. The Stikine River drains an area of 50,000 square kilometres (5.3% of the Province) and is one of British Columbia's principal waterways. Its potential importance as a waterway in modern times was recognized over a century ago when it and the Porcupine and Yukon rivers in Yukon Territory were declared international waterways in 1872. By terms of the treaty these rivers were to be kept open to navigation and, until recent times, a snag-clearing crew funded by the U.S. and Canadian governments was active in this pursuit on the Stikine River.

Stikine River rises on Stikine Plateau, flows a distance of 500 kilometres to the B.C.-Alaska border and an additional 35 kilometres through the Alaska panhandle to the delta near Wrangell, Alaska. The Iskut River is the largest tributary of the Stikine and discharges into it several kilometres above the B.C.-Alaska border.

Stikine River is navigable by shallow draft vessels from tidewater in Alaska to Telegraph Creek in British Columbia, 260 kilometres upstream A Russian sailing ship is recorded to have reached the boundary area in 1863 and weekend excursions by paddle-wheelers from Wrangell in the early 1900's accessed the Great Glacier whose front at that time lay on the west side of Stikine River some 15 kilometres north of the International Boundary.

In 1962, while in the area, I travelled up Stikine River from Wrangell in the 'Judith Ann', a shallow draft stern-wheeler that maintained a weekly service from Wrangell to Telegraph Creek. Its travel time up river was about three days and it returned to Wrangell in ten hours. The river is considered dangerous for travel by inexperienced persons as its channels are constantly changing, the current runs with great velocity, the water is cold with a high possibility of hypothermia occurring in the event of a fall into the river. There is constant danger from underwater snags not readily visible or evident because of the silty nature of the water. In 1959 two geologists working under my direction spent six weeks with experienced guides collecting stream sediment samples from the mouths of tributaries to the Stikine from chartered boats. Several years later one of the guides was drowned near the boundary. In 1964, I met the survivors of a small float-equipped aircraft who hit a submerged sandbar on Iskut River and spent six days on a protruding float until rescued by a concerned supply pilot. The same year, as a passenger on a small float-plane, we struck an underwater snag during take-off from Stikine River near Anuk River, opening a two-metre long gash in one float. The pilot was able to land the aircraft on the unaffected float at Wrangell before the damaged float became completely water-filled.

Historical Markers

Prehistoric cairns along both sides of Stikine River, from the International Boundary to Chutine River, 40 kilometres south of Telegraph Creek, provide evidence of early human activity in the area. These are most abundant near Scud River and are generally situated on prominent locations overlooking the river at elevations of about 1,000 metres. Although their purpose is not known, they may represent a border zone between the Tlingit tribe of the coast and the Tahltans of the interior. They are difficult to access from the river and would seldom be seen by boat travellers. They would not be threatened by road development and would be more accessible to those interested.

Mineral Resources

In 1861, placer gold was discovered a short distance below Telegraph Creek. One of the main routes to the Klondike area following the discovery of gold in 1896 and at Atlin Lake in 1898 was via the Stikine River to Telegraph Creek and overland to Teslin or Atlin Lake.

The most intense exploration in recent times was initiated in the mid-1950's during the search for porphyry-copper and molybdenum deposits in British Columbia and since 1975 in the search for precious metals.

The principal results of this exploration were the discovery of the previously described **Galore Creek** copper deposits and the **Liard Copper** deposit near Schaft Creek which contains 330 million tons averaging 0.40% copper, 0.036% molybdenum. Both deposits are currently undeveloped but are potential producers of major importance.

In the Iskut River area, exploration for copper near Bronson Creek between 1906 and 1930 and intermittently thereafter, failed to yield production. With the increase in gold prices in the last decade, exploration in the same area has resulted in significant discoveries with **Skyline Exploration** commencing production in 1 988 at a rate of 400 tons per day with reserves of 680, 000 tons averaging 0.67 ounces gold per ton. Nearby **Cominco Ltd. Delaware/Resources Corp.**, with drill indicated reserves of 1.2 million tons averaging 0. 75 ounces gold per ton, promises to be one of British Columbia's significant gold producers. These developments presage expectations that this area will develop into another gold camp, the first in 30 years, with only five other gold camps having accounted for over 75% of the lode gold produced in British Columbia to 1980.

The record of mineral exploration in this highly mineralized but remote area provides an example of the persistence required for significant discoveries, the importance of affordable infrastructure in attaining production and the vital need to preserve transportation corridors.

Although the emerging Iskut gold camp is not directly affected by the LSRRC[8], any future decision to provide access to the Iskut River area via the Alaska panhandle, could be compromised. The same could apply to any development utilizing the Stikine Corridor or for the Galore Creek copper deposits or other mineral deposits which may require that access in the event of a future production decision.

Other parts of my letter discussed hydroelectric power development potential, fisheries and timber resources and aesthetics which are not relevant to the 1959 program.

Ross and Harley were to make several memorable traverses while accessing stream sediment sample sites on the Stikine and its tributaries from points on the river which could be reached by the boat. The late Forrest A. Kerr who, with others, completed geological surveys of the area in the 1920's to be published in the geological memoir "The Lower Stikine and Western Iskut River Area", noted that:

"The river flats above high-water level are a veritable jungle, in which large cottonwood trees and a few evergreens tower above a tangled mass of willows, cranberry bushes, devil's club and other shrubs. "

Devil's club, *oplopanax horridum*, may be unfamiliar to many, but not to any who have traversed the lower slopes of the Coast Mountains. A good description is contained in *Wild Flowers of British Columbia* by Lewis J. Clark, who writes:

"Under three different genera this ferocious yet handsome shrub has retained its specific "horridum", suggesting a unanimity of opinion among many authorities. The current generic name comes from the Greek hopton, weapon (the root also of the Hoplite, the heavily armed soldiers of early Greece), and Panax the name of a large-leaved member of the Araliaceae, related to the ivy. The springy stems are 4-10 feet long, often bent to the ground, the ends then rising again to an erect position. Many a hiker or forester pushing his way through the thick growth of stream-beds and wet ground, has been wounded by tripping on these procumbent stems so that the upright portions lacerated his face or hands. Every part of the stems and leaves bristles with long yellow spines. The punctures and scratches are poisoned, and soon become swollen and very painful - a fact we can ruefully confirm. "Devil's Club" is appropriate."

From personal experience, I can vouch that the most tender areas affected were the knees and the back of the knees, as one had to keep walking while impaled by broken off thorns which were difficult, if not impossible, to remove until several days later when the wounds festered and the thorns popped out with ease. During traverses, we rarely wore

---

[8] Lower Stikine Recreational River Corridor

gloves and many times one would regretfully grab a thorny stem while attempting to regain balance on wet, slippery ground. Some masochistic botanists actually consider the plant 'beautiful' and recommend its inclusion in the corner of gardens - no doubt behind barbed wire.

During their traverses in the Stikine River area, Harley and Ross were to have many encounters with Devil's Club. When the two were near the end of their sampling program in the Iskut River area, Ross was to soberly comment that, "Traversing conditions had noticeably improved as the Devil's Club is so tall that you can walk under the canopy of leaves and see where you're going".

Ross and Harley not only covered the Stikine watershed, as described in the stream sediment program, in May of 1959 they completed a similar program of the Portland Canal area while working from the *Haida Queen*, a fishing boat rented for a six-week program.

In August, we once again moved our base camp, this time to Mess Lake, about 35 miles south of Telegraph Creek and about 45 miles southeast of our previous camp-site at Little Tahltan Lake. This was to be our final base camp in 1959 as most of our crew had to return to university in early September. The final month of our program was to provide us with some spectacular anomalies in the Jack Wilson and Galore Creek areas, which led directly to our acquisition of a large ground position in Galore Creek in 1960 and the subsequent discovery of a major copper-gold-silver discovery to be known as Stikine Copper or Galore Creek.

By the end of the 1959 season, a total of 172 fly camps had been established from which 3,800 steam sediment sites had been sampled covering about half of the originally proposed survey area. The sites included those sampled by Ross and Harley from river boats on the Stikine River system. Also included were the 600 more inaccessible sites sampled by Dick and myself using the helicopter. We also collected an additional 500 samples during detailed follow-up of anomalous sites, for analysis at the base camp laboratory.

I left Mess Lake before the end of the season for an event of major importance - the birth of David which was scheduled in late August. On arriving in Prince Rupert on my way south, I phoned Rene that evening to let her know that I would be arriving the following day - August 31. She indicated that everything was fine and that she would meet me at the airport. She started in labour the moment she put down the phone. By 10:00 the next morning, she checked with our doctor who told her to get to the hospital right away. I was met at the airport by Ivy Barnes, one of our neighbours in Richmond, who brought me up to date. About an hour later, I was contacted by the doctor with the announcement of David's birth.

Exploration in Northwestern British Columbia and Northwest Territories - 1960

Nahanni Butte, Northwest Territories
Galore Creek, Stikine River Area, B.C.
Completion of Helicopter-supported, Geochemical-based Reconnaissance Lower Stikine to Nass River

A previous chapter entitled 'Exploration in Northwestern British Columbia - 1959' describes various activities which directly resulted in the recognition that Galore Creek, a tributary to Scud River in northwestern British Columbia, contained by far the most spectacular copper contents of any stream sediments from watersheds sampled during our 1959 program.

In 1959, we were aware of copper showings on claims held by Hudson Bay Exploration and Development Company Limited on the West Fork of Galore Creek and those owned by American Metals Company Limited on the East Fork. These discoveries had been made in the mid-1950's during aerial reconnaissance from helicopters. In both cases vivid areas of green malachite-stained outcrops were visible from distances of several hundred metres. Initially we believed that the values obtained on the West Fork probably reflected the known copper showings. However, more detailed analyses from our sampling and their distribution relative to the known mineralization on the West Fork of Galore Creek, where values were significantly higher than those on the East Fork, suggested that the extent of the known copper mineralization was far too restricted to explain the outstanding geochemical results and that a much more extensive source area remained to be discovered.

I proposed that the geochemical exploration program in northwestern British Columbia which was initiated in 1959 should be continued in 1960, as originally planned, with similar coverage being extended southerly through the Lower Stikine and Iskut Rivers to the Nass River area.

In the late fall of 1959 we checked the status of claims held in the West Fork of Galore Creek by Hudson Bay at the Mining Recorder's Office in Vancouver. We learned that most of the mineral claims were scheduled to expire by the early summer of 1960, providing that no cash-in-lieu of work payments were made in order to extend the expiry dates. Assuming that the claims lapsed, the holdings in the West Fork of Galore Creek would be reduced from over 100 claims to four claims covering the known showings in the vicinity of the assumed principal area of interest plus an additional sixteen claims covering two other copper showings which I subsequently named the Junction and North Junction deposits.

The four claims were subsequently assigned in 1960 by Hudson Bay to Murray Morrison and William Bucholz, two of the company's prospectors credited with discovery of the showings on West Fork in 1955.

As part of our 1960 program I also proposed a staking program in the West Fork area to be followed by preliminary evaluations of the known copper deposits in the Galore Creek area.

The preparatory work for the 1960 program absorbed considerable time during the 1959-60 winter period. Little could I know that work in the Galore Creek area would become my principal assignment with Kennco for the next six consecutive years. Nor could I envisage the personal demands that would be required during that period, not only in time on site but more importantly in time away from my family each year from May to September, or later.

Quite unexpectedly, a new development affected our planning schedule. Because of my experience with helicopter-supported exploration programs, I was asked to assist Bob Hutchison, then district manager of Kennco's Eastern division, in planning and managing a six-week program in late May and June, based out of Nahanni Butte, Northwest Territories. The project was planned to investigate the potential for bedded-copper mineralization in sedimentary formations in the southeastern part of Mackenzie Mountains in the Northwest Territories. This had been reported by Blake Brady, a geologist working for Union Oil Ltd. to his friend Bob Hutchison. In his reconnaissance work, which included measuring and describing stratigraphic formations in the general area, which could be of potential interest for oil and gas accumulations, Brady noted that the malachite-stained outcrops containing casts and fragments of fossils had obviously been receptive to copper-bearing solutions, coating and partly replacing the fossiliferous remains. Brady had not determined the grade of copper mineralization within these formations, but as an astute observer and recorder had noted the association.

Nahanni Butte, gateway to the Nahanni, was still a very small settlement in 1960. We had contacted Dick Turner, a pioneer trapper in Canada's north, river man on the Liard and pilot who had settled at Nahanni Butte in the 1950's. He indicated that we would be welcome to have a base camp on his property which faced the Nahanni River immediately below the impressive bluffs at Nahanni Butte which tower above the river valleys.

We arrived in early June somewhat surprised to find that Nahanni Butte was a popular centre for helicopter-based oil exploration parties. However, there was plenty of room in the grassy fields near the river and we soon had a tent-camp established, complete with unreliable radio communication. Our camp supported about twelve personnel, including Charlie Weir, pilot and Keith Williams, engineer - both helicopter operators with Okanagan Helicopters. Apart from Bob Hutchison, our exploration group included Dominic Cecire (cook), Jim McAusland, Harley Goddard, Gord Davis, Bela Gorgenyi, Gerry Rayner, Ed Lawrence, Ed Banas, Ted Carpenter and myself, all with Kennco's Western Division, who were to be involved in the ensuing program in the Stikine River area. Personnel and residents, apart from the Turners, also included an RCMP resident officer, a couple of nurses and a small missionary group. At the peak of our four-week operation there were four or five helicopter-supported parties based at Nahanni Butte.

From Nahanni Butte we completed reconnaissance surveys ranging up to 50 miles to the north, within the southern part of Mackenzie Mountains, covering portions of the Canyon Ranges, Granger River and Ram River with its spectacular deeply incised canyons. Prominent lakes within the area are Little Doctor Lake, Cli Lake and Trench Lake.

We rapidly recognized that the copper-bearing stratigraphic horizon reported by Brady could be readily identified from the helicopter and projected from ridge to ridge over many miles. It really was quite astonishing during this familiarization period to take off from a recognized copper-bearing site and fly on a compass-bearing with the aid of excellent air photos and the helicopter's altimeter to a predicted projection of the equivalent horizon several miles away, land nearby, locate and sample it. Unfortunately, our assays were extremely low and the results led to an earlier termination of the program than expected.

One of my earliest traverses was in the Trench Lake area. I was dropped off by helicopter on a bright sunny day in a broad valley which supported a heavy grassy growth, with widely spaced and scattered clusters of alpine fir. Following my landing in this peaceful scenic setting, I extracted a couple of air-photos of the surrounding area from my backpack to orient myself with my pre-selected traverse route. As I surveyed my surroundings, I noted a huge bull-like animal in the middle of the valley several hundred yards off. My first reaction was to wonder how such an animal came to be grazing in such a remote place. Within a split second I started to run at full tilt across the meadow to the nearest timber some 200 yards distant and gained access, my heart thudding, to the second or third branch of a sizeable tree, 10 to 20 feet above the ground. I had realized that the animal was a magnificent and sizeable grizzly bear. I am somewhat embarrassed but considerably relieved to have to relate that the grizzly did not charge after me, in fact it is doubtful that he was ever aware of my presence. He eventually wandered off, as did I, each of us, thankfully, to complete our respective missions.

I recall a couple of recreational interludes during our time in camp. One evening, Charlie Weir took off in the helicopter with a full roll of toilet paper. After gaining an acceptable altitude, he opened his side door, threw the roll out, retaining a portion of it in the helicopter. He then proceeded to cut the long ribbon which had developed, with incisive thrusts of the helicopter's rotor blades. Charlie, at six feet with finely chiselled features, curly blonde hair and an athletic physique was a ladies' man and it was fairly obvious that the show was for the benefit of the nurses rather than the large male audience!

On another day we gathered at a radio to listen to a sputtering broadcast of Ingemar Johansson's unsuccessful defence of his heavyweight boxing title against Floyd Patterson, the previous titleist. Such was our limited recreation.

In late June, I chartered a Cessna 180 aircraft to fly me into the Galore Creek area, some 400 miles to the southwest. The object was to determine snow conditions for our proposed program in that general area and to see whether there were any signs of competitor activity. This latter possibility had been a growing concern to me during the previous winter. Consequently, I was apprehensive as we approached Galore Creek in the aircraft.

This was my first trip into the area and I was immediately awed by its dramatic setting. Several rugged snow-covered peaks rising to elevations of about 7000 feet dominate the south rim of the wide basin of the West Fork of Galore Creek. The West Fork emerges beneath a glacier, which with its source ice fields occupies much of the southern portion of the basin. It flows northerly, dropping from an elevation of about 2500 feet, to its confluence with the East Fork some four miles to the north at an elevation of about 2000 feet. From its juncture with East Fork, Galore Creek flows due north for about six miles, dropping to an elevation of about 600 feet on reaching Scud River.

Much of the drainage in the Galore Creek watershed area of 60 square miles is derived from meltwater from several ice fields and glaciers occupying the head of the drainage basin. The dominant East and West Forks drain an area of about 30 square miles of which 30% or more is covered by perennial ice and névé[9].

As was evident from the aircraft, much of the area contains relatively good exposures of bedrock because of its higher relief and recent glaciation. Talus slopes and moraine deposits obscure much of the lower sections of the basin of the West Fork except in certain parts where stream gullies are deeply incised into bedrock. Forest cover below timberline elevation of about 4000 feet is mainly balsam with some spruce and hemlock

---

[9] névé: Granular snow, also called 'firn.'

and poplar at lower elevations. As we were to learn from many traverses within the area, dense growths of alder, willow, huckleberry and devil's club cover slide areas and valley slopes.

During our reconnaissance flight we made several low passes within the basin, and much to my relief there was no sign of competitor activity. It was apparent that the depth of snow in much of the basin would have inhibited the start of surface exploration prior to late June or early July, which satisfied our modified schedule.

Our 1960 program was coordinated to provide support to personnel operating periodically in July through August from two small fly camps established on the west side of West Fork, and on nearby Jack Wilson Creek. The balance of time was spent in completing the geochemical sampling program which terminated in early October at Alice Arm. It was a very demanding program which involved helicopter-support by no less than three pilots for the single machine because of a variety of scheduling problems and the eventual need to replace Charlie Weir, our anticipated pilot for the duration of the program.

Apart from myself, those directly involved in our Galore Creek and Jack Wilson Creek programs in 1960 were the same individuals previously mentioned in our Mackenzie Mountains program. All but Gerry Rayner and Gord Davis had been with me on the 1959 program which had led to the discovery of the highly anomalous geochemical copper contents of the Galore Creek drainage and similar values in portions of the nearby Jack Wilson Creek. I believe that all of the crew were either attending courses at the University of British Columbia or had completed courses before joining Kennco.

Gerry Rayner and Gord Davis had both worked for Kennco in the Highland Valley area in British Columbia in 1958, while I was supervising an exploration program at the Krain property in the same area. At the time Gerry was involved in a large regional exploration program supervised by Joe Brummer, while Gordon Davis was working with a geochemical exploration party collecting silt samples in a program headed by John Fortescue. Gerry was to be active in the Galore Creek project through 1964 while Gord left after the 1961 season, during which he completed a B.A.Sc. thesis on "A Magnetite Breccia, Galore Creek, Stikine River Area, B.C.". Ross Blusson, cheerful and dependable under all conditions was a notable companion on the Galore Creek program from 1960-1965, with the exception of 1961. Similarly, Bela Gorgenyi was on site from 1960-1964, again with the exception of 1961.

Following graduation from U.B.C., Jim McAusland was to return to Galore Creek in 1966-67 to supervise the completion of a 30-mile road into the area via Scud River and Galore Creek Canyon and the driving of adits on two of the mineral deposits (North Junction and Central Zone).

One could not have asked for more enthusiastic, committed and competent associates during the early stages of exploration on this project with its remote location and physical demands. As I write in 1994, recalling those days, all still reside in the Vancouver area and I have maintained contact with most of them through our common interest in the mining industry.

In early July, 1960 our crew and most of our camping equipment were flown from Nahanni Butte to Telegraph Creek, situated near the head of navigation on Stikine River, about 60 miles upstream from its juncture with Scud River.

Telegraph Creek has an interesting origin tied to the mid-nineteenth century challenge of establishing direct telegraphic communication between Europe and America. In 1856 a syndicate headed by Cyrus W. Field succeeded in laying a submarine cable across the Cabot Strait to Newfoundland and establishing a telegraphic link between New York and

St. John's at a cost of over one million dollars. After five abortive attempts in 1857 and 1858 steam and powered ships successfully laid and connected cables to provide a short-lived period of communication lasting twenty days. In 1866, after several other abortive attempts, a single cable, 2300 nautical miles in length, weighing 5000 tons was laid by the Great Eastern Cable ship between Ireland and Heart's Content in Newfoundland.

In the mid-1860's a rival company, the Collins Overland Telegraph Company, also in the U.S., considered that the 2000 miles Atlantic Ocean crossing was impractical, based on the record and conceived of a much longer crossing, but with less ocean, via the Bering Strait and across Russia to Europe. By 1866 when it learned of the successful Atlantic crossing, the company had cut out a trail north from Hazelton for about 200 miles to Telegraph Creek which became known as the Telegraph Trail. Although the venture was abandoned, the village established on the Stikine River survived.

By coincidence, as my father's career had consisted of 33 years in the telegraph field with All America Cables, a subsidiary of ITT, one of his prized possessions which I can still view on my den wall beneath a commemorative plaque, is a 6-inch long section of the remnant of the 1858 transatlantic cable sold in 1866 to Tiffany's in New York by Cyrus Field!

Telegraph Creek and more particularly nearby Glenora, 15 miles downstream, was to enjoy several periods of resurgent activity as a result of gold rushes and the major discovery at the Klondike in 1898, when prospectors sought the Stikine as an alternate route to the goldfields. In support, the government extended the Telegraph Trail north to Atlin and later to Whitehorse and Dawson.

By 1960, only vestiges of the older villages at Glenora and Six Mile remained, whereas Telegraph Creek contained a Hudson's Bay store, Catholic and Anglican churches, nurses' residence, primary school, R.C.M.P. residence, the Diamond 'C' Cafe and scattered residences, all clustered on the steep north slope of Stikine River. In 1960, Telegraph Creek supported a Tahltan community which by 1994 had grown to about 200 residents.

The very brief sojourn of our party at Telegraph Creek, coincided with a tragic accident in which one young native resident was killed and two others were severely injured when their vehicle went out of control, crashing on the steep slope facing the Stikine River. Both Jim McAusland and Gord Davis were among those providing aid to the injured lads who had to be flown to Whitehorse for treatment.

Our preliminary supplies, camping equipment and personnel were transported from Telegraph Creek to the mouth of Scud River on a small scow pushed by the *Judith Ann*, a diesel-powered river boat, sixty-four feet long, owned by Ritchie Transportation Co. of Wrangell, Alaska, on one of its weekly trips. Within several days our very modest camp, complete with its geochemical laboratory tent, supervised by Rudy Jalbert and his assistant Ed Banas, had been established in support of our continuing reconnaissance program on the broad dry floodplain at the mouth of Scud River, a large braided tributary of the Stikine, lying about 45 miles downriver from Telegraph Creek.

Our first priority however, was to complete the staking in the West Fork area of Galore Creek and also to evaluate the importance of the significant but more localized anomalies, obtained in nearby Jack Wilson Creek. This activity had to be coordinated with the continuing regional geochemical reconnaissance program.

Initially we established a fly camp near the original showings on the west side of the West Fork of Galore Creek at the old Hudson Bay camp site with the assistance of Charlie Weir and his Bell 47 G-2 helicopter. Armed with air photos of the area and copies of claim maps of the original Hudson Bay staking, I elected to simply re-stake along the old claim lines which ran east-west, rather than adopt a new staking plan. At the time we were using two-post staking, i.e. claim posts with descriptive information on aluminum claim tags attached to the posts were placed no more than an estimated 1500 feet at either end of a blazed location line. Each claim was described as lying 1500 feet to the left or right of the location line from the No. 1 post to the No- 2 post.

Four of us completed the staking of 110 claims in a couple of days, each of us working alone on designated location lines. One of my lines ran across the general area of the old Hudson Bay camp and the four claims that were later transferred to Murray Morrison and William Bucholz and which were still in good standing. Approaching a small, but well-incised narrow creek valley on the south side of Dendritic Creek I gazed across the valley to the east looking for the extension of the location line. I was surprised to see an apparent outcrop area just off the assumed trend of the location-line, rimmed by trees. Continuing blazing of the claim line on the other side of the creek, I traversed to the north and soon emerged on an outcrop about 20 x 30 feet in area composed of breccia[10] with well mineralized chalcopyrite[11] rimming the fragments. I was extremely excited by this find as it was the first evidence of copper sulfides I had seen in the area. The outcrop lay about 1000-1500 feet west of the malachite-stained exposures near the Hudson Bay Camp which had attracted the prospectors. Several years later it would be included in the reserves of the Main Zone, yet to be recognized and delineated.

My re-staking in the old Hudson Bay camp area also revealed an interesting situation. As I plotted the location of claim posts as nearly as possible on an overlay of the air photos, it became evident that the distances between claim posts on all four transferred claims appeared to exceed 1500 feet and in addition the distance between the two parallel east-west location lines was well in excess of 3000 feet, the maximums allowable. Although it was highly unlikely that a competitor might recognize these differences, I staked a fractional claim covering only the area between the two location lines as a safeguard.

Later in 1960, on learning of the transfer of the four claims by Hudson Bay I recommended immediate negotiations to provide for an option of the claims by Kennco. Unfortunately Kennco were too late and the claims were dealt by the prospectors to Consolidated Mining and Smelting Co. (Cominco). In 1961 when we were continuing our exploration of Galore Creek and were accommodated in tent frames at a new location on a prominent morainal bench on the east side of West Fork, Grant Gibson, a geologist with Cominco arrived in the area with an assistant to carry out preliminary investigations of the optioned claims. The two were based at the more modest Hudson Bay camp and we arranged for them to make use of our shower tent facilities. More importantly I realized that Grant was required to perform assessment work on the four claims, so I informed him of the gaps not covered by our fractional claim, which enabled him to complete work on only two of the four claims to cover the assessment required on the adjoining claims by staking additional fractional claims.

By 1962, following negotiations between Cominco with its four claims, Hudson Bay Mining & Smelting Company Ltd., who retained sixteen claims covering the North Junction deposit and Kennco who, by now, held a total of 268 claims following additional staking completed in 1961, the holdings were incorporated into Stikine Copper Limited with initial ownership Kennco (Stikine) Mining Limited, as operator (76%); Hudson Bay (19%) and Cominco (5%).

---

[10] breccia: Rock composed of angular fragments cemented together.
[11] chalcopyrite: Sulfide of copper and iron.

Lacking a fulltime helicopter, which was otherwise occupied on the regional geochemical program, we spent considerable time bushwacking to and from work areas, commonly taking up to 1½ hours to access a work site, compared to 5-10 minutes of helicopter time in later years.[x]

One of our more daunting tasks was sampling the Butte deposit, so named because of its precipitous shape as a well-mineralized copper-bearing remnant protruding above talus slopes on the steep west margin of Galore Creek basin. Our initial forays around the deposit were cautious as we carefully grasped for footing and hand-holds. After having mapped and sampled this and similar areas and having gained considerable confidence in exposure we were sympathetic to the few visitors who felt compelled to complete a 'hands on' examination of the deposit in later years.

Our preliminary evaluation of the Jack Wilson Creek area required only about one week from a fly camp established near the head of the creek in a timbered area. The setting was far different from that at Galore Creek and there was much less evidence of visible copper mineralization.

While in the area Gerry and I 'traversed' one day for about one-half mile through dense alder and Devil's Club, mostly squatting and partly crawling to get through the tangled undergrowth. It was so dark that Gerry did not notice that his glasses had been knocked off his face until we came into a relatively open area on a river flat. We retraced our steps as best we could but were unsuccessful in recovering them. Because of the irregularity of our mails and the problem of checking his prescription, Gerry decided to wait until he got back to Vancouver before getting another pair. During the six-week period, his eyes adjusted so adequately to conditions that he decided not to bother getting his eyes re-tested and as far as I know, he has not required glasses since then!

Before leaving the area we decided to stake some claims covering the apparent area of interest, to provide for some trenching in the following year, which was completed but without sufficient encouragement for further work. In 1991 Gerry returned to the area when many old prospects and districts were being re-visited and re-evaluated for their gold potential following encouraging developments in the Iskut River area of the lower Stikine and other parts of B.C. At one of our brief meetings he mentioned that he had something of interest to show me. The object was a badly mauled fragment of one of our aluminum claim tags bearing the identification "JW" claim and my signature as staker. He had found it near the mouth of the Jack Wilson Creek some miles from its original location, and assumed, probably quite correctly, that it had been torn off its claim post by a bear.

During July and August we were to range far and wide in our regional geochemical sampling program while based out of our camp at the mouth of Scud River. My 35 mm slides include coverage of items considered of interest at Yehiniko Lake, Hankin Peak, Dokdaon Creek, More Creek, and Snippaker Creek, lying up to 60 miles to the east and southeast and Barrington River and Chutine Lake about 45 miles to the northwest.

Our camp at Scud River was in an enviable setting. The broad flood-plain of Scud River was composed of sand and gravel, generally bare, but containing various low-lying shrubs and forested mostly by cottonwoods on its margins. Within a few years it would become the terminal point on the Stikine for access by road to Galore Creek, but that had not been envisaged in 1960. Its flat overall surface and lack of trees made it ideal for the construction of an airstrip, which became a reality in 1964 in support of the Galore Creek project.

From our small base camp, composed of several tents, one had magnificent views to the west of a mountainous skyline, rising above several glaciers and ice fields to peaks at elevations of 10,000 feet or more near the boundary with Alaska, only 20 miles away.

Ed Banas, our geochemist, was an ardent member of the Vancouver Outdoor Club at UBC at the time and had pointedly told me of his familiarity with glaciers and ice fields. I can imagine the strain he must have endured gazing up at the distant ice fields, so near, yet so far. One day he asked me if he could have the helicopter drop him off on part of an ice field near the head of Patmore Creek in the morning and be picked up after the end of the work day. Impressed with his background, I agreed and eventually a convenient day arrived with clear blue skies.

Charlie Weir later described the drop-off on the glacier. "No problem, I flew him over the edge of this ice field containing a few crevasses, dropped him off and arranged to pick him up after completion of crew-change, probably at around 5:00-6:00 pm." Charlie returned on schedule, put down at the drop-off point, turned the motor off and looked around, as there was no sign of Ed. Sometime later, as Ed had not reappeared, Charlie cast his eyes around the general terrain and noted a dark object sitting about 100 yards away on the surface of the snow. Knowing Charlie, he probably wandered casually over to the dark object, which was a backpack sitting within a few feet of a large crevasse. Charlie looked into the crevasse and saw Ed, standing shivering about 10-15 feet below in the middle of a 3-foot wide snow platform forming the floor of the crevasse.

Charlie, always nonchalant, greeted Ed and told him to remain cool, or words to that effect, as he would return to the helicopter, lower a line and extricate him. Charlie accurately placed his skids on the edge of the crevasse and with the looped line attached to the skids, hauled Ed out of cold storage.

We were all most sympathetic. Being experienced with crevasses, Ed had approached his ice field as an old friend and armed only with a prospector's pick, had dropped 10 feet into the crevasse, envisaging no problem in the challenge of scrambling out with the aid of the rock pick. Alas, after hours of trying this was not possible and he suffered the chilling indignity of about 10 hours ensconced in the crevasse, but fortunately with no serious after effects.

It's probably a testament to the risks we had anticipated at that time that I accepted the responsibility of permitting Ed to be dropped off under those circumstances. As I have mentioned elsewhere, notably in the Safety Manual, 'Mineral Exploration in Western Canada', which I arranged to have printed in 1982 by the B.C. & Yukon Chamber of Mines, I am personally aware of four fatalities related to falls in crevasses on glaciers by explorationists with inadequate rescue equipment available, all of which were of course not anticipated.

In August we were moved by Herman Peterson in a Beaver aircraft from Scud River to a new base camp on a small island in Bob Quinn Lake, about 60 miles to the southeast near the Stewart-Cassiar highway. Gerry was the last to leave with Herman, the rest of us having left on earlier flights to set up camp. By then the weather had deteriorated to the point where Herman seriously considered the alternative of flying north to Telegraph Creek to await better weather, or flying down the Stikine and up the Iskut to Bob Quinn Lake, a more circuitous route than he had been taking during the day. Herman elected to head down the Stikine, but as the weather conditions were obviously becoming marginal to the south he turned east up Porcupine River into Sphaler Creek. As Gerry recalled:

"We circled into the Porcupine a couple of times and then we ducked into Sphaler Creek. Once you got into Sphaler heading up the valley you felt you were committed. We could

look ahead and you could just about see your way through there, but it was very shaky. We got to the moraine at Round Lake and the ceiling was right down to the deck.
I had no idea what Herman was going to do. I didn't think that there was any room to turn in there. Just as we came up on the moraine, Herman pulled back on the stick and we went up into the clag[12]. We went along in the clag for what seemed an eternity, but it was only about 1½ miles (according to Herman), straight line, straight forward and then down and we came out on the other side."

Both Ross Blusson and I were with Gerry when he described his recollections of that flight. We remembered similar situations with other pilots and I have documented a similar incident which Herman described in the Atlin Lake area during a winter flight.

Our base camp at Bob Quinn Lake was in an idyllic setting and in sharp contrast with that at Scud River. Instead of sand and gravel we were among stands of fir at the edge of the lake. Occasionally we could hear the distant rumble of trucks working on the construction of the Cassier-Stewart highway, a rarity indeed at our basecamp sites in 1959-60 in northwestern B.C.

Shortly before our move to Bob Quinn Lake, I finally bit the bullet and requested a pilot change from Okanagan Helicopters. In the more than ten years of intensive helicopter use that I experienced, supported by several dozen pilots, it was the one and only time I felt compelled to request a replacement. It was some time before I realized that Charlie Weir suffered from migraine headaches, which made him unavailable for flights on several occasions. However, a combination of factors contributed to my decision, which included a number of dropped loads which had not been properly secured and work days lost through his unavailability coinciding with camp moves. In addition several of the crew with considerable experience with other pilots simply felt that he was not particularly interested in making the effort to pick drop-off or pick-up points which could have significantly reduced their traverse time to target areas.

Our temporary replacement was Don Poole, a highly experienced, congenial individual who immediately contributed to improved morale among our crew. One evening three of us, including Don, were reminiscing near the lake beside a fire, when the conversation turned to life's most embarrassing moments. Only Don's was memorable. As a sixteen-year-old, he had been captivated by a girl in his class and had finally generated enough confidence to ask her for a date. When he called for her at her home he was about to ring the doorbell when he felt a growing urge to break wind. Faced with this dilemma he decided "better now than later", and moved along the verandah to distance himself from the door. His task completed he started back toward the door, when a voice called out "We're here, Don! " Around the corner of the verandah mother and daughter were sitting on a chesterfield. Although the girl could not have been nicer, he could not get over his acute embarrassment, and never dated her again.

Our work progressed well, out of Bob Quinn Lake, and we were soon ready for our next move, which was to Meziadin Lake in September. A few days before leaving we had our only 'night-out' incident of the year when a scheduled pick-up failed to take place. Ed Lawrence had been dropped off near Tom McKay Lake, near an old mining property 25 miles to the south to complete a day of sampling. He failed to turn up at the rendezvous location in the evening and Don Poole reluctantly returned at dusk empty handed. Ed had misjudged the time and distances involved but managed to spend the night in one of the old buildings at the property and was picked up next morning.

---

[12] clag: Low cloud-cover

Our complement dropped following our move to Meziadin Lake when several of the crew left to return to university. However, we now had Jim Murphy as our third pilot, replacing Don who had been provided only until Jim became available. In addition we acquired Bob Lackie, engineer, to replace Keith Williams who was required elsewhere.

Our final move was to Alice Arm where we planned only a brief stay of about one week in October to complete our proposed coverage to the Nass River based at the Alice Arm Hotel.

We had about two days of sampling left to complete the program when the weather suddenly deteriorated to the point where flights to the remaining samples sites were not possible. We were all quite tired of the long program and anxious to get back to Vancouver. Alice Arm, with only a few residents, was quite depressing. We spent a few days reading, with brief walks in the rain to check for signs of a possible break in the weather. We also located a couple of packs of playing cards and spent hours clustered around a table playing hearts. However, this diversion, in our bushed condition, was not enough and I recall seeing Bob Lackie on one occasion roll his eyes upward, heave a deep sigh and leave the table. It wasn't the cards, it was simply tedium at the end of a long season.

My first of more than half-a-dozen trips to Alice Arm was in 1952 to examine a property near the toe of the Kitsault Glacier at the head of Kitsault River about 20 miles north of Alice Arm. The property was owned by "Preacher" Smith, one of the few remaining residents of Alice Arm, who had built a small cabin on his property and was on site at the time of my visit. I was fortunate, or so I thought at the time, in securing a ride up the Kitsault River in a truck returning to the Torbrit mine site. The narrow road constructed along the west side of the river hugged the canyon wall some 200 vertical feet above the river. Sitting on the exposed side, I began to wonder whether I really was "fortunate" in hitching a ride.

There had been a heavy rainfall the night before and the road was muddy in sections and pooled with water. Conversation between the truck driver and myself had been desultory, he was obviously tired and I was apprehensive. Suddenly the trucker put on his brakes in the middle of a large pool of water covering the roadway. Without comment, he backed up slowly, turning the wheel and came to a stop. He then drove directly toward the edge of the roadway and I found myself suspended over the river with a magnificent view as he rocked the truck back and forth. Without another word, we were on our way again. Knowing how the weight of the vehicle was distributed he had created a run-off ditch for the water, apparently seeing no need to reassure his passenger! As we drove on, our conversation became far more animated.

At Torbrit, I was provided with a tour of the silver mining and milling operation and also took on some nourishment before hiking off on the trail up the west bank of the river toward the glacier, about 3 miles away.

Preacher Smith was in his cabin when I arrived late in the day. We spent a pleasant evening together during which I learned much about the early history of the district. One particularly memorable event that he mentioned, which I have never forgotten, was his watching an eagle dive into the salt chuck at Alice Arm after what the eagle must have believed to be a near-surface salmon. As its talons contracted and it sought to fly upward with its prey, it floundered and disappeared beneath the surface. Preacher was convinced that the eagle had unwittingly grappled with a seal, and due to its inability to release its talons in upward flight with a load, had been drowned.

Preacher urged me to visit a nearby acquaintance who had been driving an adit with handsteel for many years, without a companion, in quest of what he was convinced was a bonanza. At the time he was over 70 years old, but his physical condition was remarkable. He asked me, as many less educated, but far more experienced prospectors are wont to ask geologists, whether I thought he should continue. Where with younger "clients" or others I have rarely hesitated to state my humble, but best impressions, in this case I couldn't. I could see his search was what was keeping him fit and alive. In the search for minerals, miracles do happen. There was no way that I was going to discourage his persistence in pursuing his long-time goal.

This seems like an appropriate time to relate another episode in the Kennco era in which I had little or no involvement, other than as an active witness to the vicissitudes of mineral development at this particular time.

Molybdenite mineralization in the Alice Arm area was first recognized in 1911. The first molybdenite production from the area was a World War I supported venture on the Tidewater property in 1916. Intensive exploration for molybdenite deposits was not to occur in British Columbia until the late 1950's following an abrupt increase in the demand for molybdenum in the stainless steel industry.

Kennco acquired its first interests in the Alice Arm district in the late 1950's on the Tidewater and Roundy Creek properties. In 1950 Kennco optioned claims on Lime Creek from Gunn Fiva, an Alice Arm resident, and carried out considerable exploration including diamond drilling on the property between 1959 and 1963.

My good friend Dick Woodcock was project geologist on the property during the exploration stage. I recall him describing an incident which occurred in the early 1960's while he was on site. Walter Ramsey, a helicopter pilot based at Terrace, highly-skilled, experienced and respected, with his Korean War background, was outward-bound from Lime Creek to Terrace with two surveyors on board. Having gained an altitude of about 5000 feet above the steeply inclined walls of Alice Arm, the northeasterly extension of Observatory Inlet, he was unable to convert to forward flight from altitude-gaining mode. Checking and re-checking the controls while trying to maintain a cool composure for the benefit of his passengers, he watched with increasing concern the continuing gain in altitude. Finally exasperated, he exclaimed to his hushed passengers. "Man, oh man, are we in trouble now!". Then, in an exquisite demonstration of expertise, he shut off the power and from his lofty vantage point selected what was probably the only potential auto-rotational site in the area. This was the least-inclined of the widely-separated talus fans at the base of the deeply-incised creeks draining into Alice Arm.

Many people have recalled his comment, and I hope as many know that he successfully landed the aircraft with only minor damage to the rotor blades and that passengers and pilot were eventually rescued without serious injury.

Kennco's subsequent exploration at the Alice Arm molybdenite property was to lead to a production decision in 1964 by British Columbia Molybdenum Limited, a wholly owned subsidiary. However, that is another story.

In mid-October the weather finally broke, and we rapidly completed our program in the Alice Arm area before returning to Vancouver.

Galore Creek, B.C. 1961-65

Mineral discoveries have occurred over the past 5000 years under a variety of conditions during our society's evolution. In the dawn of human history our early predecessors encountered many surface showings of well-exposed minerals, of great value today, which would however not be of any use for many millennia.

The broad stages in the development of human culture (Stone, Bronze and Iron Ages) as defined for archeological purposes by the materials used principally for weapons and utensils, etc. reflect the importance of minerals in our evolution. However, it is only within the latter part of the last century that we have witnessed the proliferation of exploration methods that have emerged to supplement basic prospecting and which are based on the physical or chemical characteristics of minerals, or the geological environment of their settings.

By the early 1960s most of the exploration methods still in vogue for exploring for porphyry-copper type deposits such as those occurring in the Galore Creek area, had generally been adopted in western Canada.[xi]

Exploration of the Galore Creek deposits followed this general pattern, albeit highly influenced by such factors as the remoteness and difficult access of the area, the maximum practical exploration season of about 5-6 months and the need to control exploration expenditures.

It is not generally recognized by the western Canadian exploration sector that the 1965 program at Galore Creek was the largest ever mounted in western Canada, with fourteen diamond drills employed on the project. It is highly unlikely that any future project in the Canadian cordillera will require more drills. The reason? All of the drilling completed in the 1965 project predated the use of wireline drilling, a modification of standard diamond drilling practice developed in the 1960s which was in the formative stage in the early 1960s.

Rather than describe this memorable period of the property's exploration history year by year, as in 1959 and 1960, I have elected to focus on the personal contributions of those involved on the project at the time while describing the major factors which had to be considered in planning and achieving objectives. In retrospect, I realize that the many individuals involved were among the hardest-working explorationists that I have encountered. Throughout the period contractors who contributed their services on the project did so not in isolation, but as part of a co-operative team effort. Many of them spent 2 - 3 years at the property, even though it was not general policy for some companies, e.g. Okanagan Helicopters, to have their personnel spend more than one season at a project without a break.

Transportation

As the Canada geese fly on their migratory routes over Vancouver and up the Pacific Coast to their summer breeding grounds in the Arctic, it's about 640 miles from Vancouver to the headwaters of Galore Creek, where our base camp lay.

Fortuitously, with the introduction of jet service in the early 1960s by certain airlines between Seattle and several communities on the Alaska panhandle, it was possible to leave Vancouver on a morning flight and arrive in Galore Creek in the late afternoon of the same day. All that was required was confirmed reservations with both commercial and charter airlines; advanced notice of estimated arrival time at the mouth of Anuk River in a float plane from Wrangell, Alaska to provide for helicopter shuttle into the Galore Creek camp, a mere 10 miles east from our staging area at Anuk River, and finally, favourable weather conditions in the Stikine River area. This was the most reliable passenger service, although we used alternate routes on occasion, including chartered fixed-wing service, mostly Trans Provincial Airlines from Terrace, North Coast Airlines from Prince Rupert and of course Ellis Airlines at Wrangell. The latter were also supply routes where fast service was required for freight or express items.

By contrast, the all-important pre-planned shipment of most of our major camp requirements, including building materials, drilling and fuel supplies, mobile equipment, etc., was provided for annually, via ocean barges loaded at Vancouver and Prince Rupert and transshipped to river barges owned by Ritchie Transportation Co. at Wrangell for the trip up the Stikine River to staging areas at the mouth of Anuk River and later at the mouth of Scud River, some 90 miles north of Wrangell. These spring shipments grew from about 30 tons in 1961 to 600 tons in 1965.

Shipment via waterways provided for cheaper transportation of the bulk of our annual requirements than air transport. All our shipments on the Stikine through 1964 were handled by the "Judith Ann" and the "Totem" river boats and their 20-ton barges, owned by Ritchie Transportation.

In 1965, the last year of my direct involvement as Project Manager for the Galore Creek operation and by far our largest operation at a cost of about $2 million of the $5 million spent between 1961 - 5 on the project, North Arm Transportation of Vancouver, who had provided shipping to Wrangell from Vancouver in 1964, made a significant commitment to our operation.

North Arm was a family business operated by three Stradiotti brothers: Napoleon, the oldest; Ricardo, who handled finances and Alda, the youngest, responsible for daily operations. Originally involved in the fishing business, the company hired Doug Dewar as General Manager to grow the business.

My contacts were mostly with Doug, a tall, rangy and highly motivated individual who generated an air of enthusiasm and confidence in meeting potential challenges. Doug was on the 1964 operation on three separate occasions. This involved the transshipment of equipment and supplies at Wrangell to the Judith Ann and Totem river boats and their barges. He also spent time on the Stikine portion of the operation as an observer. Based on his observations he was later to ruefully comment, "The Stikine goes up and down as fast as a toilet seat at a mixed party."

Early in 1965 I met with Doug to discuss our proposed program for the year. I then learned that he had proposed to the Stradiotti Bros. that a tug and barge, partly under construction, should be committed to the 1965 campaign proposed on the Stikine. Doug rationalized that the equipment could also be used on future coastal operations in support of the logging industry.

Early in April I received a phone call from Napoleon. Rene and I were invited to attend the launching of the tug 'North Arm Prospector' on April 24. The tug and barge 'North Arm Explorer' were specially designed for shallow draft river operations. The tug only drew two feet of water and with two 250 hp engines was, according to Doug, "One of the most powerful in Canada." In 1994, it is still owned and operated by North Arm and I note in the B.C. Telephone directory that it can be contacted through the Marine Service Call

Radio Operator through 111334 on Vancouver 25 Channel. I must give them a call one of these days!

The North Arm Explorer was designed to draw six inches with no load and could carry 90,000 gallons of fuel with a four-foot draft. Its dimensions are 28.5' wide, 105' long with a 7' moulded depth. Both vessels are still operating (1994) although the North Arm Explorer was sold to another company several years ago.

On our arrival at the Fraser River for the launching, we were delighted to learn that Rene had been selected to complete the traditional send-off. Following a very informal ceremony, we were invited to lunch with the Stradiotti Bros. and Doug at the Fraser Arms Hotel at the foot of Granville Street. Rene was presented with a silver-covered box engraved for the occasion. I remember the congenial and very informal atmosphere. At one time several of the young Stradiotti Bros.' offspring were being more rambunctious than was considered acceptable and Napoleon waved them off with the comment "Go on, why don't you go outside and play on the white line." They abruptly left, but I very much doubt that they followed his instructions!

Few people realize that the Stikine, which drains an area of 19,500 square miles, is British Columbia's fourth largest waterway behind the Fraser, Columbia and Skeena rivers. The Stikine is an international waterway, navigable during the summer months by river boats from tidewater near Wrangell, 89 miles upstream to Telegraph Creek, where the river emerges from the Grand Canyon of the Stikine, only recently navigated by kayaks, albeit with umpteen portages.

The river flows in the summer months at a rate ranging from 5 to 8 miles per hour in its lower portion, influencing the time river boats such as the Judith Ann take to make the 3-day upstream run from Wrangell to Telegraph Creek versus the more potentially hazardous 10-hour trip downstream. Potential hazards, as for many braided waterways, include the shallow depth of relatively inactive channels and the number of snags in the river. For many years a snagging crew worked on the river to clear stumps and snags.

My direct involvement with this hazard included several experiences in 1964 affecting both the Judith Ann and the Totem river boats and their highly skilled captains, Al Ritchie Jr. and Ed Callbreath, the two operators most experienced with navigation on the Stikine at the time.

I arrived at our base camp for the season's operation via the Judith Ann from Wrangell on May 12. Shortly after arrival I was informed that both the Judith Ann and the Totem had been damaged and that no further barge loads from the ocean-going barge at Wrangell could be expected for about three days. I decided to check out the details to see if there was any way that we could help. I took off in one of the Hiller 12-E helicopters around 8:00 p.m. that evening and spotted the Totem near Great Glacier, about 30 miles south of camp. On landing, Al Ritchie said that the Totem had struck an underwater snag which bent a fluke on the propeller and also bent the propeller shaft. Apparently the Judith Ann had simultaneously struck bottom and had twisted the shell of the drive shaft. Al's father had already delivered the shell for repair in Wrangell. We flew the damaged propeller into Wrangell in the helicopter and left it for repair at a machine shop. The parts were repaired and delivered to both vessels the following day. On May 17, the Totem was en route back to Wrangell with both barges, having delivered two more loads to the Anuk River when it struck another underwater snag and sank in about six feet of water. Preliminary reports indicated that it might be a complete write-off. However, it was pumped out and repaired and was back in operation within a week.

Later that season, I was the only passenger on an Ellis Airlines Cessna 180 on floats when we struck an underwater snag during take-off for Wrangell from Anuk River, opening a six-foot long gash in one float. The skilful pilot was able to land the aircraft on the undamaged float at the air base in Wrangell and run the aircraft up the landing ramp used by wheel-float equipped aircraft.

Prehistoric cairns occur near timberline at elevations of about 3500 feet on both sides of Stikine River and are particularly abundant on the west side of Pereleshin Mountain just south of Scud River. They are generally situated at prominent locations overlooking the Stikine River. Although their purpose is not known, they may represent a border zone between the Tlingit tribe of the Coast and the Tahltans of the Interior.

The earliest references to modem-day activity on the river include exploration by a Russian sailing ship which is recorded to have reached the boundary area near the juncture of the Stikine with the Iskut River, its principal tributary, in 1863. In the early 1900s paddle-wheelers from Wrangell accessed the Great Glacier on weekend excursions. At that time the front of the glacier lay on the west side of Stikine River, some 10 miles north of the international boundary although it has now receded some 3 - 4 miles. It must have been a magnificent sight.

In 1861 placer gold was discovered a short distance below Telegraph Creek at Glenora which led to prospecting activity and moderate recoveries of gold on various bars. The Stikine also provided access to the Cassiar gold rushes of 1873 - 5 and most significantly to the Atlin and Klondike gold rushes in the late 1890s. One of the main routes to these later discoveries was up the Stikine to Telegraph Creek, and then overland to Teslin or Atlin Lake. Traffic on the river was particularly heavy in 1899 and 1900, to the extent that it was considered necessary to maintain a watchman with a signal at Little Canyon, below Telegraph Creek, where the river is too narrow for boats to pass.

I have previously described the activities by the Collins Overland Telegraph Company in the U.S. in the mid-1860s, which led to the establishment of a telegraph line from the U. S. through British Columbia, destined for a cable crossing of the Bering Strait. Its crossing at Stikine River led to the development of the small community at Telegraph Creek.

While working on the Galore Creek project I became interested in the background of the Stikine as an international waterway and the provision for a jointly-managed snagging operation with the object of keeping the river navigable. In Vancouver I visited the Public Library and obtained a photocopy of Article XXVI of the Treaty of Washington dated 1871 which states in its entirety:

"The navigation of the river St. Lawrence ascending and descending, from the forty-fifth parallel of north latitude, where it ceases to form the boundary between the two countries, from, to, and into the sea, shall forever remain free and open for the purposes of commerce to the citizens of the United States, subject to any laws and regulations of Great Britain, or of the Dominion of Canada, not inconsistent with such privilege office navigation.

"The navigation of the rivers Yukon, Porcupine, and Stikine, ascending and descending from, to and into the sea, shall forever remain free and open for the purposes of commerce to the subjects of her Britannic Majesty and to the citizens of the United States, subject to any laws and regulations of either country within its own territory, not inconsistent with such privilege of free navigation."

In 1988 and 1989 1 was to write two letters to the Honourable William Vander Zalm, Premier of British Columbia, objecting to the proposed designation of the Lower Stikine River Valley as a recreation corridor with restrictions to access which could adversely impact on any future mining or hydro-electric development in the area. I noted that Article XXVI of the Washington Treaty appeared to be in direct conflict with the objectives of the December 1988 draft of the Lower Stikine Recreation Corridor Management Plan, with its recreational bias. The Management Plan was approved in modified form in 1989, pointing out the importance of mining, but still contained many controversial and unclear statements concerning rights of access.

My good friend Bob Stevenson, became the Chamber's representative on the Lower Stikine Management Advisory Committee, the only industry representative in a committee composed principally of government agency and environmental association representatives. In March 1994 1 succeeded Bob as the Chamber's representative.

At various times in my long separations from Rene and our family, I managed to write letters which were more like descriptive reports of some of my more interesting projects. Of course these are in no way to be confused with the torchy love letters that passed between us in our earlier years where events, no matter how potentially earth-shattering, naturally played second fiddle to our far more absorbing emotional needs. I have excerpted portions of several of these although they tend to convey more of the excitement of the times, only some of the frustrations, but little of the yearning for family, which time has never dimmed.

As an early geoscientist, I can recall on more than one occasion when a visiting superior or 'outsider', ebullient with his visit to one of our remote operations under the best of conditions, i.e. clear skies, helicopter transport, good food, fantastic scenery, good progress, etc., while sitting on top of a particularly advantageous viewpoint, suddenly being overcome with the grandness of it all having observed "Dave, you have to be one of the luckiest guys around. All this and you're getting paid to be here!" Of course I always mumbled some acceptable comment. I never ever pointed out that he could be included as one of the more privileged having it both ways, a status I was not to achieve until my appointment as Eastern District Manager for Kennecott in 1965.

I have excerpted the following portions of letters written to Rene in early May.

"May 9

We spent the morning of May 8 getting together the rest of our supplies and had these down at the dock for loading by noon. We actually left Wrangell at 3:30 p.m. in order to hit the mouth of Stikine River at high tide. The following excerpt from the *U.S. Coast Pilot -Alaska, Part I, Dixon Entrance to Yakutat Bay* describes navigational controls as follows:

Stikine River empties by two mouths. One, the North Channel, following the mainland westward, enters the head of Frederick Sound, the other follows the mainland southward and forms the only navigable entrance to the river The North Channel can be navigated only by small craft at high water The southern entrance has a least depth of about 2 feet at mean low water The mean range of tide is about 11.5 feet and the diurnal range, i.e. the range from mean lower low water to mean higher high water, about 14 feet. The channel is from 0.2 to 0.5 mile wide, and changes with every freshet. The river freezes in winter, and with the spring freshets the current runs with great velocity. Small river vessels from Wrangell navigate the river in summer and early fall months for a distance of 162 miles to the mouth of Telegraph Creek.

There was a light wind blowing across Wrangell Straits when we crossed to the mouth of Stikine River. This billowed out the large polyethylene sheets which we used to cover the barge loads, and from the pilot house the power of the wind appeared formidable. We followed the "Totem" which weaved a seemingly aimless course, but which in effect follows the irregular unmarked channel, discerned by Ed Calbreath and Al Ritchie only from countless previous trips.

As we turned up Stikine River it wasn't long before snow started appearing on the banks of the river. We saw several bald eagles along the banks, and these are very common throughout the entire coastal area. After a while we started noticing seals popping up here and there. They are currently following the King Salmon which are starting to run. I photographed some of these that had been caught by the Commercial fishermen and were being landed at the cold storage plant at Wrangell. They averaged out at between 20 and 30 pounds.

It's 11:00 a.m., May 9, while I am writing this and we have just come to our first stop due to low water on Fowlers Riffle, approximately 20 miles north of Boundary. In these shallow sections one of the bargemen takes soundings with a pole at the head of the barge. The water is very low at this time of year and averages 4 - 5 feet in the shallows. The "Totem" displaces about 2½ - 3 feet of water, so that there isn't much leeway.

During the latter part of yesterday afternoon, the two barges were tied up together, parallel to each other, and have run that way up until now. The skipper has decided to leave the large barge being driven by the "Totem" here at Fowlers Riffle, and continue up with the "Judith Ann". We will return for the other barge tomorrow.

The large steel barge has surface dimensions of 90 x 20 feet, and the smaller wooden barge 80 x 16 feet. Both the "Judith Ann" and the "Totem" are each about 60 feet long. The "Totem" is powered by two 225 hp engines while the "Judith Ann" has only about half that power. The "Totem" was used up to several years ago on Taku River to haul concentrates from the Tulsequah Chief and Polaris Taku mines. The "Totem" pushed a 110 ton barge load about 35 miles to tidewater on that operation.

Recently Ritchie has added an engine to his smaller barge and this additional 80 hp is turned on in fast water or shallow sections. We are currently using full power trying to negotiate Fowlers Riffle. Progress is very slow being measurable in inches per second in the worst sections.

A little while back we passed by Great Glacier, which is no misnomer. This glacier which extends right down to Stikine Valley from the west flows 15 to 18 miles and then enters a narrow canyon about a mile wide to re-emerge as a symmetric fan 4 miles wide. From the air it is a beautiful and spectacular sight as the front is highly crevassed and contains impounded bodies of water between terminal moraines.

May 10

Yesterday we proceeded to within a ½ mile of our landing at Anuk River. The water at the bend before camp sounded out at a minimum of 3 feet during a trial run with the "Judith Ann's" skiff. The skipper decided to wait until morning for better lighting conditions, as it was almost dark before sounding was completed.

We took about 7½ hours to run up river from Fowlers Riffle. I decided to check Anuk River landing to make sure that our helicopters had not arrived. There was no sign of any helicopter having recently landed, as this would have been evident in the snow. As it later turned out, one of the helicopters had actually touched down on the other side of Anuk River on May 8 after making excellent and uninterrupted progress from Vancouver. Not seeing us around, they continued on to Telegraph Creek.

On our way up river yesterday afternoon, we lost about one hour trying to turn a sharp curve in the river in which the shallow fast water kept forcing the boat and barge over to one side of the river. The skipper tried just about everything including using the skiff powered by an 18-hp motor to try and turn the prow of the barge. We finally got into the centre of the current by a combination of using the skiff and four of us pushing on aluminum poles against the river bed.

We saw several moose at various points on the bank of the river, and one grizzly bear feeding on a recently killed moose. Although he moved off a little way into the timber, he still remained well in view while we passed by. On arrival at Anuk river landing we also saw the remains of a moose - which had been dragged from the river about 100 feet up onto the bank.

We arrived at Anuk River landing this morning at around 6:30. There is an average of between 4 and 5 feet of heavy snow on the ground, the only bare spots being where the river has dropped exposing a 10-foot shore line in places. By 9:00 a.m., we had had breakfast in two shifts starting at 8:00 and moved 35 tons of equipment from the barge to shore. Most of this was done by hand, although several of the heavier pieces, e.g. 1300 pound drills were winched by a tractor.

The tractor also proved very useful in clearing out two places for tent frames. It is a small John Deere. Four of our men returned with the "Judith Ann" to help shift another load from the "Totem" barge onto the smaller barge, as it is the only one that can cope with the shallow water. Six of us remained behind to start setting up two tent frames. One of these was completed by nightfall and five of us are currently comfortably bedded down for the night. The sixth man has returned to the "Judith Ann" with the two helicopters for accommodation purposes. They should intercept the "Judith Ann" about five miles downstream on its way back here.

The first helicopter arrived with Jim Keen, pilot, and Bruce Elwyn, engineer, around 2:00 p.m. When I found that they did not have their cargo sling or parts box with them, and that it had not been delivered to the barge at Wrangell, there was no choice but to head back to Wrangell and hope that it had arrived. The round trip with lunch thrown in, supplies obtained and a phone call to Vancouver took less than 4 hours.

It rained all day today, in contrast to the fine weather we have had for the last five days, and the only trip we attempted over the Anuk River pass to Galore Creek was unsuccessful. Perhaps tomorrow.

I wrote similar travel logs to Rene in 1964 during our spring migration into Galore Creek. There are so many similarities in my 1964 narrative to our 1962 trip that it gives me confidence that the experiences I have previously described were not as unusual as might be suspected. The major variations were only in scale as we were mounting a much larger program with the objective of completing about 45,000 feet of diamond drilling using eleven drills, at an overall cost of $1,094,000 compared with 15,471 feet achieved with four drills at a cost of $357,000 in 1962. Much of the program focused on drill testing of geophysical anomalies obtained from the 1993 program. Accommodations at the property in the form of prefabricated buildings were expanded to provide for the 85 personnel on site which had grown from 42 in 1962 and 70 in 1963.

Camp preparation in 1964 commenced April 28, and the main body of Kennco and Midwest personnel started to arrive at the property on May 11. Barges operated by North Arm Transportation Co. Ltd., and loaded with approximately 550 tons of fuel, drilling equipment and miscellaneous supplies, arrived at Wrangell from Vancouver on May 10. During the latter part of May, supplies were transported on barges attached to the Judith Ann and Totem river boats to Anuk River from Wrangell, in spite of unusually shallow water caused by late runoff. Included in the new equipment was a Land Rover, which I had ordered to use on roads to be constructed to various drill sites by a 24-ton D-7E bulldozer, also shipped on the barges in palletized loads which could be handled by a Sikorsky 58 helicopter which could transport loads of up to 4000 pounds. The use of the Sikorsky was prompted by discussions I had during the winter with Okanagan Helicopters Ltd. It was agreed that the aircraft should handle shuttling of our initial barge loads in the spring, supplemented by three Hiller 12 E helicopters.

On reaching Wrangell on May 10, we were brought up to date on the imminent arrival of the North Arm barge which had been spotted 20 miles south of Ketchikan by Al Ritchie

Sr., General Manager of the Alaska Ferry System, at 2:00 p.m. on May 9. Our sources were Ned Zenger and Fred Belmas, respectively the U.S. and Canadian customs officers at Wrangell. For many years the Canadian government had maintained a customs post at Boundary on the international border near Iskut River. The lack of traffic prompted the much more practical location. I had befriended Ned and Fred on my initial trip into Wrangell in 1960 and they both made us most welcome on our annual visits.

Another friend and booster of our efforts to develop a mining property which might benefit Wrangell through concentrate shipments was Frank Murkowski, then a bank manager in Wrangell. In later years he was elected Senator for Alaska and throughout his career has been a strong promoter of Canadian mining efforts in the Panhandle region. One of our memorable meals at Galore Creek consisted of three large sack loads of Alaska King Crab delivered by helicopter to the site by Frank on one of his rare visits to the Stikine.

I spent the evening of May 10 in Wrangell in the company of John Barakso, Kennco's geochemist, and Paul Smith, an Amfab Building Supply Co. employee. Both were on their way into Galore Creek, John to supervise the construction of an assay laboratory to reduce sample shipments to Vancouver and provide for rapid analytical data on drilling results and Paul to supervise erection of the prefabricated buildings by Kennco personnel.

The following morning I took Paul Smith with me in a fixed-wing aircraft to the North Arm barges which were then anchored off the mouth of Stikine River and left him on board the tug to ensure that the building supplies were marked for shipment in the correct order on the smaller service barges.

One of my travel logs in 1964 contains the following description of an incident during off-loading of one of the barges at the Anuk River which involved Don Jackes, the remarkably talented Chief Pilot of Okanagan Helicopters at the time and pilot of the S-58 helicopter on our operation. One of Don's achievements was co-piloting a large helicopter to the Far East via the Atlantic Ocean which required crossing through three separate war zones in the Middle East. He was also an efficient ham operator, his helicopter being equipped with radio-telephone equipment which enabled him to maintain contact with his family during his many assignments in remote locations. Tragically, Don and Merv Hess, another Okanagan pilot who had flown briefly with us at Galore Creek, were both killed in the fall of 1975 during the installation of the second gondola towers on Grouse Mountain, in Vancouver. "Rocky" Pearson, Chief Engineer for Okanagan for many years, investigated the crash for which no cause was ever determined because of the destructive force of the crash; four full fuel barrels in the aircraft were found flattened by the impact.

We spent May 14th and 15th on camp construction work. This consisted primarily of snow shoveling, John Anderson and I have been working on our radiotelephone installation. We were on this again today, although much of my time was also devoted to checking the location for loads to be dropped by the three helicopters operating currently on the transfer of loads from the Anuk River landing to our camp area. It snowed fairly steadily on both the 14th and 15th so that we only had about four or five Hiller trips up to camp. However today was moderately clear, with no wind and ideal for flying, and the pilots established a record. Both of the Hillers flew 19 round trips and our large helicopter, the Sikorsky (S-58) flew 26 round trips. Between the three of them they moved in a total of 45 tons.

S-58 managed to knock several heavy components of a bulldozer off the barge into the Stikine. These are lying in about six feet of icy water which is also quite swift. We have tried grappling for them but have been unable to hook them so far, although the locations have been determined.

May 17

This morning I flew back down to the Anuk landing to see how things were going. One of my worries concerned the missing parts of the bulldozer. A barge hauled by the Totem had just finished being unloaded when I arrived. All the equipment had been moved over to the side of the bank with a crane and I felt that if we could hook a line around the sunken parts, we would have no problem in pulling them out with the crane.

I had been advised that the pilot of the S-58, Don Jackes, had volunteered to try and hook a cable to the parts himself. This was not intended as an act of penance - it so happens that he is a scuba diver, and if you can imagine any greater coincidence, he had thrown his wet suit into the helicopter before leaving Vancouver, although he had no air supply. After checking in the locations of the pieces I asked Don if he was prepared to make a try at fishing for them and we arranged to try it after lunch.

Don was flown across the river from our Anuk camp in his wet suit and he looked every inch the pro. We jabbed a pole down a long side one of the cat parts, secured a rope around Don for safety purposes and after a great gulp of air he let himself hand-over-hand down the pole. Needless to say the water is very cold, and swift. The suit is insulated so Don didn't feel any discomfort. After several similar dives he found a satisfactory anchoring point and we handed him the end of a ½ inch steel cable which he threaded through part of the component. In no time at all we had both Don and the pallet on the barge. The process was repeated once more satisfactorily. The only objects that we actually lost were some parts books and possibly some tools.

I came back up to base camp around 2:00 p.m. and went back to the construction business. We got the radio working around 6:00 p.m. Everything takes so much longer to accomplish with all this snow. We had to dig a 10-foot hole to secure our generator and an 8-foot deep hole to accommodate one Land Rover which was flown in this morning by the S-58. That sure was a sight to see - the vehicle 'riding' serenely through the air at an elevation of about 1000 feet above the ground. Don took about 5 seconds to center the vehicle above the hole and then dropped it down. The buildings that we are erecting all have to have snow removed from foundation sites prior to construction. The largest snow excavation measures 60 x 22 feet and varies from 6 to 7 feet in depth. The parts for our bulldozer were all laid down by the S-58 on a mat of logs which our boys cut and hauled over to an open area near camp. A total of 20 logs about 20 to 25 feet in length were required to contain the major components, and many unimportant items were allowed to fall in the snow.

While hauling the palletized components of the bulldozer off the barge and into the Galore Creek base camp, the crane operator moved two pallets aside, preparing them for a single load for the S-58. Unknown to him the two pallets contained two portions of the blade of the bulldozer which had been cut into two halves, to meet the anticipated upper limit of 4000 pounds per load for the S-58. Don was unaware of the overall weight which, later, I found out was only about 3000 pounds and he flew the load into the camp without incident, of course, much to the chagrin of the two Finning Tractor mechanics on site to re-assemble the D7E, as they had to re-weld the two halves of the blade together; they had apparently been involved in cutting them apart!

One evening, shortly after Don's epic salvage job in the cold waters of the Stikine, I presented him with a 'medal' in recognition of his heroic services, well beyond those required of a helicopter pilot. The 'medal' was one of those circular metallic plaques, somewhat modified, which hung by a chain from an expensive bottle of Scotch that I had somehow acquired. The brief ceremony took place in the mess hall before his many admirers!

By now the dependence of our Galore Creek operation on a combination of water- and air-borne transportation will be evident. The medium used impacts on costs as has been demonstrated world-wide for many years with water generally being the cheapest, followed by land and then air, contingent of course on availability and local conditions.

In 1960 during our claim-staking and preliminary evaluation of the Galore Creek deposits, we were essentially totally dependent on helicopter transportation on-site, requiring much travel on foot within the general area to access and evaluate mineral occurrences.

Even in 1961 with our first modest drilling program totalling 1239 feet in six BX holes completed by T. Connors Diamond Drilling Co., we were without helicopter support on many occasions requiring drillers not only to spend 3 - 4 hours accessing drill sites, but actually carting fully loaded drill core boxes back to camp.

It was not until 1964 that we had the equipment on site to construct air strips for fixed-wing access. The first strip was built on a glacial moraine on the west side of West Fork Glacier. It was 1700 feet in length and in the final stages of completion when John Anderson and I made an inaugural flight to the property in Trans Provincial's new Otter aircraft piloted by Doug Chappell. The landing and subsequent test take-offs and landings by Doug were all highly successful, although it turned out the airstrip had marginal use in subsequent supply operations.

Late in 1964, a 5000-foot airstrip was constructed by Stikine Copper at the mouth of Scud River, principally to provide access for a proposed road 20 miles up the south side of Scud River to the mouth of Galore Creek and thence about 10 miles up Galore Creek to the base camp.

Another temporary airstrip was constructed on Scud River near the mouth of Galore Creek and finally an additional airstrip was constructed on the west side of Galore Creek in a heavily timbered area near the confluence of the West and East Forks. This strip was eventually lengthened to about 4000 feet and years later, in 1990, it was the principal access point into Galore Creek by fixed-wing aircraft operating from Bob Quinn Lake on the Stewart-Cassiar highway, 40 miles to the east.

In spite of the relatively rugged terrain and confined approaches in the West Fork area, no aircraft accidents were recorded from fixed-wing operations on these airstrips. However in late 1964, gale force winds of 50 - 60 mph, which can be expected several times a year, were responsible for damage to aircraft, tents and Parcall buildings.

At the time the Otter was sitting on the West Fork airstrip secured with ropes to large boulders as a precaution against the rising wind. During the night it was blown about 200 feet off the airstrip on to coarse moraine, sustaining significant damage. The hull was not flown out until January 1965 during a visit to the property by Ross Blusson in a Bell 204 helicopter, which also retrieved Pat Clay, caretaker and his dog Cheeko.

Restraining blocks had also been attached to the main rotor blades of the two Hiller 12E helicopters on site at the base camp. During the night one of the Hillers suffered extensive damage when the rope attached to the restraining block snapped and the main rotor blade auto-rotated at a sufficient speed to lift the aircraft off the ground before it tipped over and crashed. The following morning I took a picture of Gary Carstensen, the engineer, 'sobbing' as his head rested on his arm against the bubble of the machine.

In the 1961 program, with $134,000 in expenditures, we upgraded our camp facilities and Gerry Rayner, Gord Davis and I initiated plane-table mapping over the entire basin -

a program which was to take three years. In addition 1239 feet of diamond drilling were completed in six holes by T. Connors Diamond Drilling Co. of Vancouver through arrangements made with Dunc Chisholm. We also completed trenching and sampling, more claim staking, induced polarization[13], airborne and ground magnetic surveys and test AFMAG surveys on the West Fork Glacier. The geophysical surveys were contracted out to Hunting Survey Corp. of Vancouver and McPhar Geophysics Limited of Don Mills, Ontario. Fortunately we had the benefit of the highly experienced services of Hal Fleming, Kennco's Chief Geophysicist, for part of the time. An average of about twenty company and contract personnel were involved.

Most importantly we also had helicopter support provided on contract by Okanagan Helicopters Limited for most of the four-month season. The pilot of the Hiller 12E was Bill McLeod, one of Okanagan's original helicopter pilots whose experience at the Kemano project I have previously described. His engineer was Art Johnston.

Sufficient information on the 1962-5 programs has been provided to show how in each year since 1960 the scale of succeeding programs was significantly increased on the basis of the encouraging and successful results obtained. In each case it was critical to maintain adequate helicopter support. The flexibility provided in 1964 and 1965 with the partial-to-total availability of an S-58 paid dividends. An overall reduction resulted in the cost per ton mile flown and is also reflected in the lowering of the helicopter support cost per foot drilled from $4.00 per foot in past programs to $2.55 per foot in 1965.

The much larger carrying capacity of the S-58 at 4000 pounds versus that of the Hiller 12E at about 800 pounds impacted very favourably on drill crew rotations, where instead of two drillers and their core boxes being moved in a single trip from drill site to camp, the Sikorsky accommodated 12 drillers and their core boxes in a single trip. It was always quite a sight for myself - let alone visitors - to see the drillers pouring off the S-58 at camp during crew changes.

As is evident, our helicopter-supported programs at Galore Creek relied heavily on the pilots flying the machines and their engineers who were responsible for maintenance. The project would not have been possible at that time, or even now, 30 - 35 years later, without their capable support. Nor would it have been as enjoyable without their enthusiasm, good humour and experience. All were employees of Okanagan Helicopters Limited of Vancouver. Included were the following:

Year	Helicopter(s)	Pilots	Mechanics
1960	(1) Bell 6-2	Don Poole, Jim Murphy, Charlie West	Keith Williams
1961	(1) Hiller 12E	Bill McLeod	Art Johnston
1962	(1-2) Hiller 12E*	Peter Moore	Gary Carstensen
1963	(2-3) Hiller 12E*	Hughie Hughes, Greg Walker	J. Pawikowski
			Gary Carstensen
1964	(2) Hiller 12E	Harvey Evans, Hughie Hughes, Greg Walker	
	(1) Sikorsky S-58*	Don Jackes	Gary Carstensen
1965	(1) Hiller 12E	Harvey Evans	
	(1) Sikorsky S-58	Glen Soutar, Denis Light	

---

[13] induced polarization: A geophysical survey used to detect disseminated sulfide mineralization, as opposed to massive sulfide.

*Denotes partial use of one helicopter to supplement heavy transportation requirements from Stikine River in the early part of the season

They enjoyed their work and were used to assignments in remote places involving difficult living conditions and the somewhat demanding expectations of project completion objectives encumbered by seasonal restraints.

By mid-1965 exploration had given sufficient encouragement to consider construction of an access road into Galore Creek from Stikine River. Frontier Construction and Development Limited with its expertise in mine development work was invited to tender a bid. I was on site when Herb Donaldson, the president of the company visited the property to consider access possibilities. He spent the first day being flown by helicopter over the general area which included the Anuk, Jack Wilson, Scud, Porcupine, Sphaler and Galore Creeks, all situated on the east side of the Stikine and covering an area of about 500 square miles.

The first evening of his visit I asked him for his preliminary impressions and he shook his head saying, "There's no obvious route without one hell of an expense." At the end of the second day he was more cheerful and optimistic and on day three he announced, "There's only one obvious route in all things considered and that's up the Scud with a temporary road subject to local washouts which could be maintained on an annual basis, and then up Galore Creek." Frontier was subsequently granted a contract at a maximum cost of $470,000 for 1965. By September the road up the Scud had been roughed out and was passable to the mouth of Galore Creek where the approach for the road up Galore Creek was also completed. This necessitated blasting of the precipitous outcrops on the east side of the mouth of the creek to heights of up to 200 feet above the roadbed in order to maintain an acceptable grade. In September, a supplemental budget of $1,290,000 was approved for completion of the road, including bridges; the driving of an exploration adit on the Central Zone and drill testing to bedrock along the adit route.

During our 1959 - 60 exploration program in the lower Stikine, we located several prospective areas warranting further exploration, in addition to Galore Creek and Jack Wilson Creek. One of these was in the Sphaler Creek area, a tributary of Porcupine River, southeast of our base camp at Galore Creek.

It was not until 1965 that a small party headed by Gerry Rayner staked claims on Sphaler Creek covering some copper showings and completed a preliminary prospecting, geological mapping and sampling program. Gerry established a small fly camp and maintained radio contact with our base camp at Galore Creek and also had supplies delivered by a pilot, who must remain nameless, for pickup by helicopter at the mouth of Anuk River. Two interesting situations developed as a result of this program.

Gerry recalls one of these quite vividly:

"This would be in 1965, when I was doing reconnaissance work in the Stikine. I had my gas cache at the mouth of the Anuk and I was working out of there. George Cargill, geologist with Hudson Bay, was also working out of there with their S-55 helicopter and also had a sizeable gas cache. Quite often the pilot would come into the Anuk in his aircraft servicing Kennco's Galore Creek camp.

"For some reason or other, I either assumed or was informed by the pilot that he had some sort of arrangement with George for the use of gas from the cache. So whenever he came in and I was there, he would throw a drum of gas down from the cache and I pumped it into the aircraft for him and away he would go.

"It wasn't until many years later I got to comparing notes with George and I found that the pilot had no rights whatsoever to any gas from the Hudson Bay cache. In fact, when he was there and I wasn't, George informed me that it was a drum of _my_ gas that was being pumped into his aircraft!"

Should you think that helping yourself to fuel in isolated places was an unusual affair, it wasn't. Many similar, but less blatant incidents occurred. In several cases the fuel was replaced, but it was by no means the rule.

Shortly after staking the Sphaler Creek claims, Gerry was surprised to see a helicopter drop off a couple of men in the main valley of Sphaler Creek, not too far from the high grade copper showings in a fairly well-exposed breccia which had attracted his attention. Later, he was fascinated to find that claims had been staked on behalf of a mining company covering the showings. As Gerry had staked his claims using claim posts and location lines following the ridge crests on both sides of the valley, claiming ground for 1500 feet on either side, he reasoned that the stakers had not seen his claim posts and believed the ground to be open.

Aware of the flurry which would be caused on the company's shares if they announced assays from the showing on newly staked claims in the very competitive Galore Creek area, he foresaw an opportunity for a rather quick capital gain. He very generously decided to share his information with us, but in a surreptitious way, because of the difficulty of informing us by radio. He indicated that if he sent us a "go" message, we should immediately buy shares in the company.

Sure enough, a short time later, we were alerted and without wasting any time, Ross, Gerry and I bought some stock. Within a week the stock had risen sharply and we bailed out. Later we learned that the company had never recorded its claims, had not made any announcement and that the rise in the stock was caused by intersection of high grade values at its own mine!

Diamond Drilling

By the end of 1965 a total of 154,234 feet of diamond drilling had been completed at the Galore Creek project with direct drilling costs, i.e. paid to contractors, accounting for about 50% of total costs of $4,950,000 on the project since 1960.

All of the diamond drilling, except for 1239 feet of BX core drilling done in 1961 in six holes by T. Connors Diamond Drilling Company of Vancouver, was completed by Midwest Diamond Drilling Co. Ltd. of Winnipeg. At the time Midwest had no previous diamond drilling experience in western Canada. Apart from their head office, they had an office in Toronto, which included Rod ("Bud") McDonald as Contract Sales Manager. Don Coates, one of my classmates from the University of Toronto was General Manager based in Winnipeg. Dave Marshall, another classmate joined Midwest in 1964 and ultimately became the Manager of the Drilling Division.

By late 1961 rumors about the potential of the Galore Creek deposits had begun to leak out and Bud had several discussions with John Sullivan, Kennco's president in Toronto, to confirm the rumors. He informed his head office that, "Galore Creek has got to be one of the biggest copper deposits in western Canada - we should try and get the 1962 contract."

I was impressed by the enthusiasm of Midwest. Their expertise, innovative ability and growth-oriented interests were factors that attracted us to accept their 1962 bid and all subsequent contracts through 1965.

Prior to our move to Toronto in late 1965 as Eastern Division Manager for Kennco, I had few opportunities to meet Bud. A former lineman for the Winnipeg Blue Bombers, he was of imposing girth, robust visage and in addition extremely good-natured with a great sense of humour. On one of his rare visits to Galore Creek, I happened to be in

Vancouver and learned of his visit only on my return. Shortly after arriving at the property he had marched up to the office and introduced himself as Jack Gower, then Western District Manager for Kennco, to one of our summer students, the only occupant at the time. By contrast, Jack was older, grey-haired, and slight of build. Not having met Gower, the student complied to the best of his ability with Bud's request to be brought up to date on progress at the project, aided by maps and other references.

My most vivid personal contact with Bud occurred several years later when we were both involved in weekly hockey practices with the Prospectors All Stars in Richmond, in preparation for our annual game in Maple Leaf Gardens with Teck Corp.'s Keevils and their ringers at the Prospectors and Developers meeting in Toronto. I was a 135-pound winger and I had received a pass and was advancing for a shot on goal with only Rod in front of me, admittedly blocking any view of that part of the rink! I thought I had passed him when I was suddenly hurled to the ice: Rod had tackled me!

Some time later I met Rod and his wife one weekend when Rene and I were walking through a mall in the western extremities of Toronto. I hadn't seen him for a while and after an exchange of some common pleasantries, he gave a deep sigh and commented "You never know what's ahead. With seven kids, all active in sports, I found myself sitting on the John this morning and reflecting that for the next 9 - 10 years I would have to be getting up at 6 - 7 each morning, to take them to and from some practice." I provided what I considered was an acceptable level of sympathy before we parted - but Rod's right - these are the sort of obligations that most couples obviously never consider in family planning!

Midwest encountered all the challenges that they could handle at the Galore Creek project. Apart from the many logistical problems, the highly fractured nature of bedrock in the upper 600 feet adversely affected drill core recovery and progress. Up to 100 closely-spaced sub-horizontal fractures, which we called 'fracture cleavage', could occur within one foot of depth. At depths below 600 feet the fractures were sealed with anhydrite, which had been removed by meteoric waters[14] near surface.

Recognition of the extremely adverse conditions encountered in near-surface drilling commencing in 1962 led Midwest to grudgingly commit to heavy shipments of drilling mud, which was not considered essential in the initial bid. Experimentation provided significantly improved conditions.

During the start of the 1962 program I had collared one hole in the southern part of the Main Zone where fracture cleavage conditions were severe. I knew that there was virtually no overburden at the location from observations in 1961 even though the site was covered with about six feet of snow. After setting up, the drill night shift collared the hole and commenced drilling. The runner reported back to me the next morning enthusiastically that he had no problems drilling but had not reached bedrock after drilling an unbelievable 85'. I was stunned, realizing that he must have ground all core as none was retrieved. We moved him off the hole and re-drilled it later after the snow had gone and when we had a good supply of mud on hand. With the mud the drillers had excellent recovery but the rock could be crushed easily with your fingers. We stored the core boxes from this section in core racks for several days before logging the hole. When we retrieved them for examination each tray held nothing but mounds of silt as the core had disintegrated on drying out.

Don Coates was to comment about the drilling conditions: "In normal drilling, once you get into bedrock, the deeper the hole the more expensive it gets. At Galore Creek, it costs

---

[14] meteoric waters: Water derived from rain, water courses or other bodies of water.

less to drill at 1,500 feet than at 500 feet because of the problems in coring near surface." This was later to be reflected in Midwest's drilling costs charged at different increments of depth.

Standard diamond drills were used on the project, almost exclusively, by Midwest. These were BBS-14 and BBS-20 diesel drills, which were operated seven days per week on two 12-hour shifts. In addition a rotary overburden drill (Mayhew) under sub-contract, weighing about seven tons, was used in the latter part of the 1961-65 period to facilitate drilling in areas of deep overburden. Its progress through overburden composed mainly of glacial till, averaged 73 feet per 12-hour machine day compared with the average of 43.3 feet achieved in the best year (1965) in core drilling. In 1963 Midwest also experimented with a Boyles wireline drill, still in the development stage, which unfortunately had many problems. It was the Longyear wireline drill which was eventually developed that provided the major breakthrough in improved diamond drilling practice.

Prior to the development of wireline drilling, standard drilling practice required the successive addition of 10-foot long drill rods preceding drill rods as the drill hole deepened. During drilling, the drill bit, impregnated with diamonds on its leading edge and on a reaming rim to maintain the gauge of the hole, cut out a washer-like hole in cross-section as drilling advanced, leaving behind a continuous core of undisturbed rock within the central part of the bit. The core eventually entered a core barrel attached to the leading drill rod behind the bit from which the drill core was retrieved. This required pulling all of the drill rods from the hole to access the barrel. In fragmented ground, common in the surface portions of the Galore Creek deposits, blockages could occur before the core barrel was entirely filled, necessitating a frustrating increase in the average number of pulls required with reduced footage per shift.

Following retrieval of the drill core from the core barrel, the string of rods was lowered back into the hole, requiring re-connecting of all the rods.

The wireline method revolutionized diamond drilling performance. The governing modification was the reduction of rod-pulling to retrieve the core barrel. When the core barrel is full, or if a blockage occurs, an inner tube containing the core is detached from the core barrel assembly. The tube and contained core are then pulled to the surface by a wire dropped down within the string of drill rods. After the core is removed, the inner tube is dropped down into the outer core barrel and drilling resumes.

Among several thousand photo slides taken of the Galore Creek operation, I have one showing core recovered at a depth of 85 to 120 feet in one of the drill holes completed by Midwest Diamond Drilling Company in 1962 - their first year on the property. There are 23 blocks in the core box designed to contain 25 feet of fully-recovered core. It would have taken several days to drill and recover that 25 foot interval in the hole.

In 1965 up to 14 drills were used in completing 57,636 feet of drilling in 81 holes. A percussion drill, used to penetrate deep overburden, accounted for 7,865 feet of total drilling with an average of 43.3 feet achieved per 24-hour machine day. By contrast, in 1991 progress with wireline drills averaged 230 feet per 24-hour machine day based on 44,900 feet of drilling completed in 48 holes with an average of three drills.

Although most of Midwest's diamond drillers had not had any previous experience in rugged mountainous terrain, let alone working in some cases on glaciers, all but a few individuals rapidly adapted to conditions far different from those in the Canadian Shield.

One in-bound diamond driller actually got as far as the Anuk River from Winnipeg. However, he refused to leave the float plane and was returned to Wrangell without having set foot on the project. Another reached the base camp by helicopter but that 10-mile shuttle terminating in the dramatic setting of West Fork basin was his Waterloo and he was shipped back to the flatlands.

The 1962 drilling program included drilling of a vertical hole by Midwest about 1500 feet above the toe of West Fork Glacier. The hole was drilled as a preliminary test of an airborne magnetic anomaly in the central part of the glacier. Our previous drilling and sampling programs had noted a strong, but inconsistent relationship between copper mineralization and magnetite, so we were excited about the possibilities of discovering an additional deposit in the area. In addition, our stream sediment surveys indicated just as strong copper contents emerging beneath the glacier as the values obtained at the mouth of the West Fork several miles downstream, which we had initially believed reflected the marginal areas of the Central Zone.

The first hole passed through 251 feet of ice, which was cased in order to provide recovery of an additional 48 feet of sludge from poorly recovered bedrock. This section assayed 3.23% copper with an associated content of 0.02 ounces gold per ton.

Midwest actually designed and created a variety of ice bits faced with tungsten carbide in an effort to improve drilling through the glacier and in 1963 gained improved core recovery. The West Fork Glacier deposit was never completely delineated; however subsequent drilling indicated by 5 - 6 holes involving 3031' of drilling in ice defined a zone up to 400 feet in width with a core area about 100' wide grading 1 - 2% copper.

Snow Conditions

I have mentioned the snow conditions encountered each year when mobilizing crews and equipment into Galore Creek to start the annual programs. Years later in comparing average annual precipitation encountered at existing open-pit mines, I obtained data from Climax Molybdenum's operation in Colorado, which has operated for many decades at an elevation of about 10,000 feet above sea level. I was heartened to learn that the conditions were very similar to those at Galore Creek.

At Galore Creek estimates indicate that precipitation ranges between 50 - 80 inches annually with 60 - 70% falling as snow. Cumulative snowfall data for winter months were obtained during the period January 9 to December 31, 1965, during which total snowfall amounted to 267.6 inches and total rainfall to 22.93 inches. The data for the winter snowfall conditions were accumulated by a reclusive prospector, Pat Clay, whom I had met many years earlier working in the Yukon.

Pat spent the summer as caretaker and serving float planes and helicopters at Anuk Landing. In the winter he stayed at the base camp supported by rare visits by helicopter. Cheeko, his long-time companion, half-wolf and half-husky, shared his duties, as the two were inseparable.

Some indication of the actual conditions existing at Galore Creek during the annual mobilization stage can be garnered from the following excerpts from my letters to Rene dated May 21 and June 9, 1962 - this of course was Midwest's first experience in the area.

"May 21

So much has happened since I last wrote on our progress up here that it is hard to know where to start. At present there are only two men left at the Anuk River landing to assist in unloading another barge~load of supplies which is due to arrive from Wrangell tomorrow afternoon. There are thirty-six other men in the Galore Creek area at two camps. Our base camp has greatly expanded and now includes thirteen separate units including three very elegant Parcall buildings. All of these units are either on frames, or in the case of the Parcall buildings, rest on logs on the ground. The amount of snow which has been removed to date in diggings would easily exceed 600 tons.

Our largest single 'excavation' involved digging out a space for the Parcall units. an area 24 x 54 feet was dug out to an average depth of about 9 feet. The Parcall units are prefabricated box-like units 8 x 2 x 1 feet. They hinge out to form a floor space 16 x 2 feet, and 6 inches deep. The inside of the plywood containers (the floor sections) contain enough formed and pre-cut aluminum tubing, with a thick insulated nylon cover to span an aluminum arch four feet wide. As sections of flooring are added together, the aluminum frame is extended to form a quonset-like building, with doors at either end and four windows, one on each side of the doors. Actually one door is closed in, but the space is available for linking units end to end.

The units were purchased by Midwest at an approximate cost of $2000 each. We have a total of six of these units on the property, three at this base camp, and three others at a drillers' fly-camp, which is still in the construction stage. I am fortunate enough to have one of these for our Kennco office, and as I have a bed in one corner, I can assure you that I have never had a better bush accommodation. The interior is all grey, with grey floors and grey nylon. the exterior of the unit is composed of a vivid orange nylon fabric - being toned this eye-catching hue in order to permit ready visibility of the units from the air

Midwest does not believe in stinting on help. Our base camp personnel include a foreman, a clerk, cook, bull-cook and cookie. The latter makes the difference between acceptable accommodation and completely unaccustomed comfort. He visits every unit daily to fill up pails and basins with water, fills the stove up with oil, and sweeps out the units. Twice a week he mops the floors.

Two days ago I went off with the fly-camp foreman to select a site for the drillers' fly camp which would be centralized for their initial drilling period. This includes about twelve holes to be drilled on the southwest side of the basin about 1½ - 2 miles west of base camp. After flying around for a while we were let off at the only available landing spot in the area. We walked downslope and then climbed out of a creek onto a broad ridge near timber line. We located a likely looking bench on the slope which looked as if it might be suitable. We then ploughed around in the snow and cut out a landing area for the helicopter and built a heliport out of the fallen trees.

Yesterday a total of thirteen men were flown into the spot to start digging out space for the Parcall units which were flown in during the day from Anuk river, I had suggested that one of the units should be put on skids right on top of the snow to provide temporary accommodation, in the event of deep snow. So while the men were digging, others set up one unit, which accommodated seven men last night, the balance having returned to base camp. You can imagine our surprise when the hole was completed today for the first Parcall unit, and varied from 17 to 19 feet in depth. Needless to say they are going to set the remaining unit up on the snow for a while also.

To date, there have been approximately 165 flights over the Anuk River pass, and most of the essential equipment from the first barge-loads is already distributed in the various camp and drill-site areas. We have spotted enough drill holes to hold Midwest for a while, and today I went down with Frank Polkosnik, the Midwest foreman, and saw the first drill in operation".

"June 9, 1962

Right now we are in the process of moving drills to new drill sites. I made a final check on one this morning and surveyed two during the last two days. Our survey work is all of a preliminary nature at this stage, as it is impossible to locate any of the survey hubs from our last year's grid survey. The best we have been able to do is to detect cuttings near the base of trees which define the position of a picket line and then chain to the nearest known point. We had one exception where our chaining came to within six feet of a picket which was just visible at the base of a large tree which had shielded the surrounding ground from deep snow accumulation.

Both yesterday and today we have had light snow flurries, and we are still travelling by snow shoes to all points outside the immediate vicinity of the camp area.

This morning, while checking on a drill site location near our fly camp, I found myself looking unbelievingly at the fly camp setting. The snow level is still above the level of the tops of the Parcall units, being at least 9 feet in depth. Unless we get a prolonged rainy period or warm spell, that snow will still be around there in July. "

Avalanche Hazards

From an early stage I had been aware of an old avalanche path which disintegrated at the base of a northwest-trending spur several hundred feet north of the location originally selected for our base camp on the northwest extension of a prominent glacial moraine on the northeast side of West Fork Glacier. Based on tree growth, it appeared to be at least 20 years old.

We were also aware, however, of the dangers of traversing outcrop below the upper portion of West Fork basin because of the avalanches that frequently occurred in that area in the late spring.

In the 1961 - 64 period, when our exploration activity was rapidly accelerating, we had occasion to site a couple of proposed drill holes in gullies in the North Junction area - about 1½ miles northwest of our base camp. Because of the precipitous nature of the terrain in that area, the only practical drill-site locations were in subsidiary drainages leading southeasterly to Dendritic Creek - a subsidiary of West Fork.

I had surveyed in a drill site location in early 1964 just south of the North Junction deposit in one of these gullies. Several days later on a relatively warm day in June, the Midwest drillers occupied the site and started to excavate their set-up, which required removal of about six feet of snow above the dormant creek. That evening I learned that one of the Midwest drillers at the site had been buried by wet snow from a slide which started on the steep snow-covered bank of the creek several hundred feet above. Three of the four drillers had decided to take a coffee break, leaving the fourth driller alone shovelling out the site. The snow slide, with a head originating some 100 feet above the creek, maintained sufficient momentum on reaching the valley floor to continue downslope and of course funnelled into the hole being created for the drill site. Fortunately the other drillers heard a sharp cry as their fellow was bowled over and buried. They ran up the gully and immediately extricated him beneath over a foot of compacted snow. The driller suffered no after-effects; however we immediately deferred drilling at similar sites pending melting of snow cover.

My earliest experience with an avalanche occurred in 1954 while working in the Mt. Waddington area in B.C., a project which I have previously described. One bright sunny day I was traversing alone across a broad snow-covered basin rimmed by several lofty peaks. It was near noon and I decided to stop at a *nunatak*[15] rising about 50 feet in the middle of the basin where I could sit on some rock for lunch. It was an unbelievably fortuitous decision. While sitting down admiring the spectacular scenery below the basin from my vantage point, the stillness, which is so characteristic of these high alpine areas

---

[15] Nunatak: Mountain top projecting above an ice sheet

devoid of trees or any signs of life was suddenly broken by a whooshing sound. I looked around for the source and was stunned to see that the floor of the valley below the *nunatak* was in motion. A large avalanche had apparently occurred at the head of the basin and was passing by at a rate of about 10 miles per hour. It was some time before I gathered sufficient courage to continue across the basin.

Although we were aware of the potential dangers from avalanches at the head of West Fork basin, we avoided these areas, particularly in May and June, but we had never made any detailed studies of potential hazards from avalanches or snow slides which could affect winter operations at any of the satellitic deposits at higher elevations around the rim of the basin. The large Central Zone, which was the principal focus of all our drilling lies in the floor of the basin, mostly covered by overburden and timber, and bears no evidence of avalanche paths.

Little did I know when I visited the west terminal of the Granduc Mine operation in 1954 that eleven years later it was to be the site of the worst avalanche disaster affecting miners in western Canada since 1915 when 57 persons were killed in a rock and snow slide at Britannia, North of Vancouver.

On February 18, 1965 at about 10 a.m., an avalanche passed over a portion of the Portal Camp of Granduc Mines causing the loss of 26 lives and total destruction of part of the camp. By 1965 the camp had been relocated about one mile west of its original site near the mine workings, which I had visited in 1954. A few years prior to 1965 it was located in the middle of the Leduc Glacier which would have been in the central part of the avalanche path, leading to almost total loss of life among the camp's occupants.

At its new location the camp lay on glacial moraine at the juncture of the north and south forks of the Leduc Glacier, about 300' vertically above the north edge of the south fork and 2200' vertically below the origin of the avalanche. Just south of the camp on the edge of the glacier work had commenced on the camp adit which had been driven easterly for about 800' towards the Tide Lake Portal 11.5 miles distant and the eventual site of the Granduc mill and camp.

On the day of the avalanche there were 154 men in the Portal Camp area, with 94 men working on day shift. Of the day shift, 21 men were working underground and at least five men were in a safe portion of the camp, which was bypassed by the avalanche. Of the 68 men directly exposed to the avalanche, 42 were saved, of whom 20 received injuries ranging from bruises to fractured arms and legs.

The Portal Camp contained four 40-man bunk houses, a large recreation hall, warehouse, a first aid building, a temporary hospital, and a nearly new mine dry <??>, as well as a small helicopter hangar with a workshop. There were ten other small buildings in the area. A large power house, compressor building, workshop and the former mine dry were also located near the portal of the adit. The only buildings not destroyed were the bunk houses, the mine office, warehouse, first aid building and the temporary hospital.

By good fortune the avalanche spilled over the portal of the adit but did not block it, so that the 21 men working in the adit were immediately available for rescue work. In addition, Dr. H.B. Veasey, the Stewart area medical practitioner was in the first-aid building with the first-aid man at the time of the avalanche and being unharmed, both were able to assist in the rescue, receive the injured and give medical aid.

The mine shift boss who had recently been on surface, knew the positions of all 15 men working in the portal area and immediately directed his parties to probe these areas. Six men were found alive.

Although the avalanche immediately knocked out the camp power, within minutes auxiliary power was connected to the radio transmitter and a distress signal was sent to Stewart. The call was heard by the Alaska State Police and was also relayed to the RCMP. Immediate assistance came from many quarters. Lieutenant-Colonel Mathews of the Queen's Own Rifles of Canada was appointed search and rescue co-ordinator and soon numerous groups were on the scene including an army group from Chilliwack, an RCAF helicopter from Comox, the Snow Mountaineer Rescue Squad from Vancouver plus the Alaska State ferry "Taku".

Rescue dogs were also brought in but were not of much assistance as everything in the disaster area carried human scent. Bulldozers were used, working along with a probing crew.

Most of the survivors were quickly located; however, there was one miraculous exception. Eino Myllyla of Winnipeg, a 32-year old carpenter of Finnish descent was carried about 50 feet from a bunk house by the avalanche and buried under 3 to 4 feet of compressed snow "lying face down with his arms wide apart and rolls of plastic pinning his knees" as described by *The Vancouver Sun*. The location was a flat spot in the camp area which was selected as a landing site for helicopters. Myllyla remained conscious to semi-conscious for 79 hours before his discovery, which occurred when it was realized that the area had not been probed. He was partly exposed by the blade of a bulldozer. He had severe frostbite on hands and feet, and although amputation was predicted, it was not required. His survival was partly attributed to his low metabolism, extremely calm disposition and the snow conditions, which enabled him to breathe and have partial movement of his arms.

Granduc Mines engaged M.M. Atwater of Tahoe City, a recognized snowslide and avalanche consultant to investigate the cause of the avalanche. He concluded that it was a "Climax Avalanche", a rare condition caused by high winds transporting snow into a slide area to form a hard slab snowpack on an unstable snow base. Later large quantities of new snow and soft slab snow were deposited on the unstable base and the pressure caused a movement in the top section of the snowfield which culminated with the release of other portions of the snowfield over a one mile front. He estimated that about 50,000 tons of snow moving at a speed of over 100 miles per hour struck the camp. Dr. Atwater emphasized that the build-up of conditions can be recognized and that there are methods of breaking the sequence and destroying the conditions necessary to the formation of this type of avalanche.

At an inquest held on March 4, 1965 the jury attached no blame to any person or persons for an act of God. It recommended that either the camp site be abandoned during the winter months or precautions taken to prevent the build-up of snow on the slopes immediately above the camp.

Although much coverage was provided of the avalanche in the media, the most authoritative account is contained in the 1965 Mines and Petroleum Resources Report of the B.C. Ministry of Energy Mines and Petroleum Resources, which I have used as my source for most of the above description. The report further notes that "snow slides have taken their toll over the years, but in one's and two's, except in 1935 at the Taseko Motherlode mine, north of the Bridge River district, the connected bunk and cook house was demolished by a snow slide which killed the entire crew of seven men."

As a result of the avalanche, I made enquiries concerning short courses in avalanche control. John Anderson and I attended a course in Colorado which opened with a film showing a broad snow-covered valley disappearing into the distance beneath high mountain slopes. Attention was drawn to a white puff which suddenly appeared on the side of the mountain slope several miles away. As the film continued the puff broadened filling the valley as it grew and rapidly approached the camera site. The coverage ended abruptly in a solid white mass obliterating the scenery. Apparently the photographer and assistant had deliberately visited the site to film an avalanche from the road crossing the valley, but which had never been struck by an avalanche. The filming provided a tragic exception; both individuals lost their lives.

I also contacted Dr. Atwater describing our program at Galore Creek and asking for his recommendations. Unavailable for several months, he recommended that a competent pair of skiers try to initiate a snow slide in the basin of the West Fork when we arrived on site in May or June, 1965. We didn't have any competent skiers, so John and I spent several hours being flown by helicopter to points around the basin from which we had some marvellous runs but were fortunately not involved in any slides. Our most dramatic run was from the top of the Saddle deposit area on the east side of the basin at an elevation of about 6000' down to the base of the mountain slope at an elevation of about 2500'. We started in powder snow, ran through a long section of corn snow and finished in slush.

Dr. Atwater visited the property in early February 1966 and concluded that open-pit mining on the Central Zone would be difficult and expensive in winter because of heavy snowfall, drifting and stormy weather. Needless to say we had been aware of this since day one! However he made no specific recommendations concerning avalanche control.

Geological Mapping, Sampling and Core Logging

Although geochemical and geophysical surveys aided us in defining target areas warranting trenching or diamond drilling, an overall appreciation of ore controls at Galore Creek, as at many other mineral deposits in the exploration stage, was largely the result of painstaking geological mapping at the property at a scale of 1" = 200', sampling of exposed mineral occurrences and trenched areas exposing bedrock and even more detailed studies of drill core and sampled split drill core sections from over 150,000' of diamond drilling in 231 drill holes completed between 1961 - 5.

Geological mapping was completed by plane table surveys covering about 3.5 square miles, much of it in rugged terrain under diverse weather conditions. This form of survey, not commonly used today, employed an instrument person who worked over a mapping surface attached to a horizontally inclined board about 3' x 2' set on a tripod over a selected station. The instrument used is an alidade, consisting of a telescope pivoted to swing through a vertical arc atop a vertical stand attached to a steel rule, one edge of which is parallel with the line of sight of the telescope. The alidade is moved over the map surface and aligned between a point representing the station and a number of reference points up to several 100 feet distant selected by the geologist for plotting purposes. The geologist holds a graduated measuring rod vertically above his reference point and both the horizontal and vertical components of the distance are calculated by the instrument man who records the position of the reference point and its elevation for contouring purposes.

In 1960, our first year of mapping, Gord Davis acted as instrument man and Gerry Rayner and I described the geology of the reference points as rodmen. Gord was to complete a detailed study in his spare time of an area of magnetite breccia in the West Fork Glacier area which became his B.A.Sc. thesis project at the University of B.C. in 1962. In 1960 we were frequently without a helicopter and each day the three of us would trudge off from our fly camp, clambering through the bush and climbing up 2000 feet or more vertically to reach our working area. It took us over three hours to hike back and forth compared with a 5 - 10 minute helicopter trip when the machine was available.

While working in this area on the west margin of the basin, we had to kill a black bear which kept returning to our camp despite all our efforts to drive him away permanently. One of the students had skinned him out and the carcass remained in the camp area. I asked Jim Murphy, our helicopter pilot, to haul the carcass out of the camp and dump it a safe distance away so that it would not attract more bears.

The three of us were at work mapping on a bright sunny day, when we heard the helicopter approaching our area long before we were due for pick-up. We looked up as it neared us, flying about 300 - 400 feet above, and could see what appeared to be a body hanging underneath the machine. Of course it was this horrible carcass hanging by two legs with the others dangling underneath. Overhead, Jim veered away toward the steep snow-covered western slope of the basin and about half-a-mile away he suddenly released his cargo - probably chuckling all the way back to camp!

In 1962 our plane-table crew included Cyrus W. Field, Gerry Rayner and myself on an alternating basis, with Ross Blusson as instrument man. The three of us also shared core-logging duties, with Ross and others doing core splitting. One of the areas to be mapped, which I vividly remember, was the top of the South Butte, a precipitous *nunatak* at the head of West Fork basin, accessible to us only by helicopter and certainly not climbable with our limited mountaineering skills. For emergency purposes we left a tent and food ration at the site in the event of our helicopter becoming unserviceable. One of the dreariest photos I ever took was of Ross and Cyrus beside the plane table, all bundled up in wet gear in a heavy fog which eventually restricted our mapping. In 1994, I submitted this photo to the CIM Special Volume 46 on 'porphyry coppers', hoping that it would qualify as part of a proposed collage.

Our plane table mapping had eased off in 1963 and 1964 when Ross and John Anderson also participated in the mapping as the project neared completion.

Surface sampling, mostly within the mapped areas, included 11,739 lineal feet of continuous chipped samples collected over portions of about a dozen deposits. A considerable amount of trenching with pick and shovel, augmented by portable drills and powder was mostly completed by John Nuppunen and Frank Matic. John Nuppunen, a genial jack-of-all-trades, an excellent carpenter, mechanic and even bulldozer operator, was one of the most essential field men in Kennco's organization. He was much sought after on our projects and with his resident Finnish contacts, he recruited many of them in camp construction at Galore Creek and several other camps, notably the Chappelle property which later became the Baker Mine, with many of its original 14' x 16' buildings still in use.

John had established his own mastery of the English tongue and much to the consternation and delight of his peers, he produced many challenging variations by commonly dropping the first or second letters of nouns and occasionally by adding other letters in a frustratingly inconsistent and undecipherable code. Thus the much feared and maligned bear family would be referred to as "lack, rown and rizzly bears". To those city-slickers unaccustomed to winter travel in the far north, a "no-shoe rail" might be a challenging object.

He was a frequent visitor to our core-shack where up to 8 - 10 of us would be engaged in logging core according to its lithology, alteration, mineralization and fracture patterns; splitting core to produce a split retained for reference and a split retained for assay at our adjacent assay lab designed and equipped under the direction of John Barakso.

One morning while we were all engrossed in our particular chores, John Anderson let out a righteously wounded bellow, which immediately attracted our attention. Apparently a day earlier he had left a banana obtained from the mess-hall on one of the core-shack's cross studs above his work bench for a future snack. Let me make it perfectly clear that as eating and sleeping were considered essential pastimes at Galore Creak, and given high priority even for limited periods, you could basically eat as much as you could possibly consume without being subjected to criticism by anyone. However, over-sleeping was simply not allowed!

When John, possibly frustrated in trying to determine the ancestry of a particular lithologic form, grasped his banana, i.e. the one on the stud - it collapsed in his hand, obviously devoid of its essential entrails. To put it another way - the edible portion of the banana was missing. All of us were accused of being the guilty party. Gathering around John in a consoling group, we noted that he had previously taken pains to designate his claim to the banana by inscribing his initials 'JMA' on one of the shredded peels with a black felt marking pen. However, as we collectively inspected the banana peels, John denied any knowledge concerning the letters 'PS' following his initials and quite fortunately, none of us had those initials.

Sometime later in the day, John Nuppunen made one of his frequent visits to the core shack area. Suddenly John Anderson roared knowingly "Nuppy, you bastard, you've eaten my banana!" The accused's face reddened and with cheeks puffed out with the effort, he yelled back "pull sit!"

Immediately we all collapsed in hysterics, while John Nuppunen looked around in bewilderment. (For any readers unable to decipher this development at such short notice, we all recognized that in Nuppunese "pull sit" clearly meant "bull shit").

John Nuppunen had attracted his friend Frank Matic to accompany him in the trenching programs at Galore Creek. Bela Gorgenyi, a student at the time who had spent the 1959 season working in the upper part of the Stikine area with his partner Hugo Lanella, had been in far closer contact with Frank than I was able to achieve. Years later when Rene and I met Bela at a shopping centre in Richmond, B.C., and he and I started to reminisce, he recalled some of his principal incidents. He remembered Frank, a good-natured individual who claimed to all newcomers that his brother Otto Matic had developed the first repeating rifle!

Frank had an extremely hairy chest and as he wore the same T-shirt day-in and day-out, eventually the hairs protruded through the material. He strenuously denied that he had to go to the barber at the end of the season to get his chest hairs cut so that he could remove his T-shirt!

Bela also remembered an incident involving Frank and his trenching party which had been supplemented by several relatively inexperienced apprentices at Galore Creek. Due to extremely heavy rainfall, the party was forced to wade waist-deep across one of the tributaries to West Fork to get back to camp. One of the less experienced helpers refused to cross the creek. Frank then pulled out a stick of powder from his pack, stuck a short fuse into it, lit it and said "get going or I'll blow you across!" According to Bela he filmed this event and claims that the concerned individual "only had his feet wet by the time he scooted across."

When one is recalling events that occurred 20 - 30 years earlier, it's obviously important to obtain the lingering impressions of individuals directly involved at the time. From personal experience, depending on the life-threatening aspects of the situation, I have found that generally those types of events are so well recalled, that no embellishments are necessary. However, I have problems with second-hand recollections.

In 1959 while engaged in the first year of the 1959-60 Stikine Project, an incident occurred which could have been life-threatening but with which I was not directly involved. Remarkably after 30 years, when I contacted individuals engaged in the project, two of them immediately isolated one incident that I barely remember. In July 1994, I talked to Dick Woodcock, my associate in the 1959 program who recollected a helicopter crash at the mouth of Ball Creek, tributary to Stikine River and some 4 miles distant.

Both Dick and I shared the stream sediment sampling and subsequent follow-up of significant, but isolated geochemical values in areas not warranting establishment of one of our 2-person sampling crews. These were often difficult to access and a disproportionate amount of time was required to obtain samples; and if they proved significant, to follow them up.

Dick had indicated to Mike McDonough, our helicopter pilot, that there was an unmarked power line above the mouth of Ball Creek that Mike was not aware of. They subsequently landed and obtained a silt sample at the location, but during take-off Mike forgot about the power line and the helicopter struck it at a height of about 50 feet. Miraculously he managed to make a hard landing on a small sandbar near the mouth of the creek. Dick recalls that Mike said "Don't move." as Dick tried to get out of the helicopter while it was still in motion. Even after Dick exited he recalls that Mike stayed for some time and then without a word, felt under his seat cushion and retrieved pad and pencil to document the damage to the helicopter. Then they had to walk about 4 miles into Telegraph Creek along a trail which Dick recalls exhausted Mike as "he hated to walk". Bela recalls that in this emergency, when the helicopter was unserviceable for several days, I had dropped a message from a fixed-wing aircraft at his camp, plus similar messages to other fly camps, indicating that there would be a delay in their pick-up for transfer to their next camp. At the time, Bela recalls that they were low on food but were fortunate to be located near a small lake where they had been able to obtain sufficient grouse and fish to tide them over.

Many helicopter pilots have had fatal or serious accidents following airborne collisions with unmarked or even marked power lines. As I have described in a separate section relating incidents involving these valiant individuals, Mike was ironically killed in a similar accident in southern B.C. several years after our 1959 program. Don Roadhouse, a student who had worked with Kennco on exploration projects in the 1960s, later became a helicopter pilot and was killed while lifting a load from a truck and being guided aloft by an individual on the ground. The helicopter struck a power line which neither appeared to have noticed.

Visitors

In spite of its remote location, laudatory reports concerning the developments at Galore Creek and its potential soon spread along the grapevine and as early as 1962 we began to be besieged by 'visitors'. The term as quoted needs some elaboration. Visitors at promising exploration sites can be divided into at least three categories, all determined to learn as much as possible about the project in the short time available. They include a variety of senior management and affiliated corporate personnel requiring a first-hand examination of the project. Closely related are contractors bidding on development work. The next legitimate category includes representatives of government agencies, some of whom, such as Dr. W.G. Jeffery in 1965, then a staff geologist with the Mineralogical Branch of the B.C. Dept. of Mines and Petroleum Resources at Galore Creek, provide independent studies for the government on the possible importance of an evolving mineral district. The final category are the competitors, anxious to determine whether to stake adjacent mineral claims in the area, who are considering farm-in proposals or seeking to learn about the geology and exploration techniques being used which might be applied in similar projects.

John Sullivan, president of Kennco Explorations, could hardly be called a 'visitor'. He was the first senior company executive to visit the property in 1961. 1 have verification of this in a photo of him at the Butte deposit. John was impressed with the potential of Galore Creek from the outset and was a staunch supporter of the increasingly high appropriation requests which we submitted in subsequent years.

In this same period we had a visit from Lowell Moon, Exploration Manager for Kennecott Copper Corporation (Kennco Exploration's parent), based in New York, who visited with John. Where John was extremely spry and accustomed to fairly strenuous property examinations, Lowell was in his mid-50s at the time and walked with a pronounced limp from an early injury.

At the time we had no roads on the property and only one sturdy footbridge near the mouth of West Fork, spanning the creek. About two miles up-stream near West Fork Glacier, we had limbed two trees, each about 25 feet in length and placed them side-by-side spanning the creek which at this point was chest-deep at its deepest point, icy cold and about 20 feet across.

Approaching the crossing, Lowell lagged behind and I realized that he was very apprehensive about crossing, having watched the others reach the opposite side.

"It's okay, Lowell, I'll give you a hand," I said, wading ahead into the creek on the upstream side of the 'bridge', where I could press one arm against the pole and provide sufficient support with the other to steady him. We crossed without incident. However, on returning to the crossing later in the day, Lowell suddenly went ahead and into the water in the upstream side, as I had done. Half-way across, he suddenly twisted and losing his footing, reached up grabbing the pole with both arms as his legs were swept underneath. I ran into the water on the downstream side, having great difficulty keeping my feet on the rough stream bed, but managed to reach him and grab his shoulders over the two poles. 1 somehow managed to continue to the other side keeping him afloat.

It was a bad scene. Lowell felt very embarrassed and I was equally distraught, realizing that I should have anticipated a problem with the crossing that we were accustomed to using, but that others would consider totally inadequate. Lowell thanked me for my efforts and apologized for his decision to try to cross the creek on his own. Needless to say, an improved crossing was immediately prepared following our visitor's departure.

In 1963 we had 32 visitors at the property, several visiting on more than one occasion. Included in the first two categories described above were several other Kennecott executives, namely, Frank R. Milliken, President; Charles H. Burgess, Vice-President Exploration; Ed Rache, Vice-President Mining; Lowell Moon and John Sullivan; Hon. W. K. Kiernan, Minister of Mines, British Columbia; E.R. Hughes, Senior Inspector of Mines; Hartley Sargent, Chief, Mineralogical Branch, B.C. Department of Mines.

Of course, having Frank Milliken at the property was like a visitation from God and we were apprehensive about the logistics which required such senior personnel to be exposed to the risks inherent in travel by helicopter and small fixed-wing aircraft. So were Kennecott, for it had been a long standing corporate policy that senior management executives should not travel together even on commercial airlines. This policy stemmed from a tragic accident in 1946 in the St. Lawrence River area when Kennecott's vice-president, president elect and another senior officer were the victims of an insurance-generated plot by Albert Guay, in which all 26 occupants in a commercial aircraft were killed in a crash caused by a bomb planted in the suitcase of his wife, who was also aboard the aircraft.

Rache was not particularly enthusiastic with the setting of the Main Zone, Galore Creek's principal deposit lying in the floor of the valley of the West Fork. As we were sitting in a group near one of the many showings, Rache pointed at the mountain spur which descends from the Saddle showing and forms a backdrop for the camp site. "You should concentrate your exploration along that ridge", he said, "if there's an ore body there, it would sure give a helluva good stripping ratio!" He was comparing the location to that of the Main Zone, our principal deposit, where the waste-to-ore ratio would be considerably higher, i.e. the amount of waste rock which would have to be removed to access ore-bearing deposits at depth. The ratio commonly ranges up to 5-6 tons of waste per ton of ore at many mines. Some near-surface deposits in elevated terrains have had 1:1 ratios or less.

While at the camp, we arranged a game of horse-shoes to entertain our distinguished visitors. Ross, who was selected on one occasion as a partner for Frank Milliken recalled, "He always used to stick his tongue out every time he took a shot." For some reason we all laughed at this recollection. Perhaps it was due to the recognition that he was mortal after all! I was to remember Ross' fascination with this not-so-rare habit of many athletes when concentrating; the most memorable to my recollection being Pete Sampras, the great tennis player in the early '90s.

In 1964 our visitors totalled 42 and included M. O'Shaughnessy, Manager, New Mines Division for Kennecott Copper; Larry Ogryzlo, Director Explorations & Development for Hudson Bay Mining & Smelting Co. Ltd.; Neely Moore, General Superintendent of Exploration for Cominco; Dunc Heddle and Peter Sevensma, Senior Geologists with Cominco, the latter all representing Kennecott's partners in Stikine Copper Limited.

Included in the visitors list were a number of Japanese geologists from various Japanese mining companies who were so supportive of porphyry copper developments in British Columbia in the mid-1950s and 1960s and who initiated financing for Bethlehem Copper, B.C.'s first porphyry copper mine, when Canadian financial groups were reluctant to accept the risk. The Japanese visitors rarely indicated how many representatives would be arriving at the property and because of their relatively slight stature we hypothesized that up to six might squeeze into a Hiller 12E helicopter.

While I was off site, John Anderson attended to a Japanese geologist whom he found one evening in the office tracing one of our maps without permission on rice-paper. He confiscated it somewhat apologetically; we had to be careful about the details on developing reserves and other confidential data. I recall that Ken Northcote, formerly a Kennco Geologist who later joined the British Columbia Department of Mines in Victoria, had a similar experience in that period. A Japanese geologist was in Ken's office discussing moderately classified information about B.C. porphyry copper deposits. Ken informed him that he was unable to release copies of certain technical data. He had to leave the office momentarily and on his return he found the geologist astride his desk photographing copies of the data under review!

Rest and Relaxation

Those involved in exploration at Galore Creek in the 1960s, whether employees of Kennco, summer students or contractors, were not concerned about amenities such as rest and recreation. Essentially everyone involved recognized that their mere presence was evidence of their own objectives: experience and/or summer wages, or superior wages with overtime depending on status as so-called permanent personnel, temporary or contract. For the record, permanent and temporary Kennco personnel received no overtime pay even though they might be expected to work 7 days per week and up to 10 - 15 hours per day. This was an industry norm at the time.

In the mid-1960s, as in earlier years, exploration personnel were accustomed to relatively little rest and relaxation. The most that was generally expected by individuals willing to work 12-hour days or more, was an acceptable bed and above all an acceptable cuisine.

From an early stage I concentrated on these requirements, as did Midwest Diamond Drilling, the principal contractor. In retrospect our recreational facilities were poor to abysmal, consisting of table tennis, horseshoes, cards (mostly poker) and movies, as long as the generator functioned. However, most, if not all, had no trouble with an acceptable sleep and as all the diamond drillers worked a 12-hour shift, this was the principal perk in addition to the excellent meals offered by the catering staff. By good fortune we had a free-lance journalist visit the project in 1965. He produced a diversified array of drafts of articles for specific journals covering common themes, none of which, as I can recall, were ever published. However, he did record the fare offered to the residents, which quite evidently appealed to him, even as a visitor.

Having said that, we were fortunate in having this individual on site as an outside observer with a mission sanctioned by management. It is evident from the copy that he was inclined to exaggerate the conditions under which we performed our daily functions; however, he recorded much factual data that we were too busy to document because of our own priorities.

With regard to our meals, he noted quite consistently the high calibre of bush meals served:

'I've worked in mining camps all over North and South America, " says Martin Menard, a foreman with Midwest Diamond Drilling, the company doing the drilling for Stikine, "but this camp has the best food I've ever tasted." Here's a typical day's menu.

Breakfast: choice of dry cereals, fruit juices, ham or bacon and eggs, sausages, wheat cakes and syrup, hot rolls and breads, jam and marmalades, tea, coffee, milk. All breakfast orders are individually cooked and six eggs and half a pound of ham doesn't raise any eyebrows.

Lunch: A smorgasbord of cold cut meats, salads, raw vegetables, cottage cheese, macaroni and cheese, potato salad and pickles. This is followed by a main course such as porterhouse steak (trimmed of fat and bone) parsley potatoes, curried rice, apple sauce, hot raisin bread and biscuits, gravy followed by strawberry shortcake and canned fruit and cream. Some beverages or ice cold water from the 475 ft. thick hanging glacier on Saddlehorn Mountain behind the camp.

Supper: Smorgasbord, main dish of pineapple ham or mountain salmon, with two or three vegetables, side dishes and always choice of two desserts.

Louis Stratmoen, the camp cook, bakes fruit pies, cakes, biscuits, all kinds of bread (including raisin, served hot), several times a week,. these are stacked under the smorgasbord counter so the men can help themselves. The drillers make their own lunch from a mountain of cold cuts, cheeses, pickles, vegetables, fresh fruit, set out at 6 am and 6 pm every day for the two shifts.

As John Nuppunen, Stikine Copper's bearded, barrel-chested carpenter says: "The biggest danger here is getting a big pot," though the writer saw "Nuppy " as he's known in the camp, pack away a plate of smorgasbord, five beefsteaks, Spanish rice, followed by a heaped dish of strawberry shortcake and washed down with two mugs of black coffee."

Because of the heavy work and 12-hour shifts put in by the drillers, their appetites were prodigious. I recall being in line behind a large driller at breakfast following his night shift. When he reached the cook to place his order he simply said "eight, seven and six" which meant eight rashers of bacon, seven fried eggs and six hot cakes. I gasped in astonishment, but soon learned that such orders were not considered uncommon.

Mineral Reserves

One of the most important roles for a geologist working for a mining company is to provide realistic estimates of potential mineral reserves at various stages of a project's exploration. Many factors are involved in the determination of a project's ultimate viability, but none are generally so important as its mineral reserves, for they are the basis for all the subsequent calculations which must be made to determine economic feasibility - represented by a 'Feasibility Report'.

In the case of Galore Creek, there is no doubt whatsoever that it would have been in production decades ago were it not for its relatively inhospitable location dictating high capital and operating cost estimates which have increased over the years with increasing metal prices.

At Galore Creek, these estimates were recorded annually at the end of an exploration season in progress reports. For the record, in the 1961 - 65 period they were as follows:

In 1961 we concentrated on surface mapping and sampling of the various mineral deposits, as exposed. The report indicated the following estimates, supported only by a minor amount of diamond drilling:[xii]

Assuming an equivalent spatial distribution of deposits in drift, talus, or ice-covered sections of the Complex, a potential copper-bearing area of 5,000,000 square feet might be inferred to exist."

The H.B. deposit as enumerated became recognized as part of the Central Zone, in which over 80% of the subsequent diamond drilling was confined.

Subsequent estimates follow:

1963 Report
"The potential at Galore Creek, of at least 100,000,000 tons of 1.5% copper to a depth of 1000 feet, is one of the highest known for a copper property in Western Canada. The potential reserve of the Galore Creek area might well be 150 - 200 million tons of this material."

1964 Report[xiii].
In 1963, the potential reserve of the Galore Creek area was estimated at between 150 - 200 million tons of 1.5% copper. The reserves of 72 million tons of plus 1.5% copper indicated and inferred by drilling to relatively shallow depths, i.e. a ratio of 7:1 for length to depth of deposit in the Central Zone, would indicate that such potential reserves are realistic objectives."

1965 Report
The results of the major diamond drilling program in 1965 provided few changes to the overall potential reserve estimates, which again were principally based on additional drilling in the Central Zone. The overall estimate indicated a total of 59.5 million tons grading 1.41% Cu to a depth of 700' in the Central Zone with an additional 12.0 million tons grading 1.79% Cu to a depth of 500' in satellite deposits (North Junction, Junction and Southwest).

1976 Estimate
A paper on the Galore Creek deposits was published in CIM Special Volume No. 15 'Porphyry Deposits of the Canadian Cordillera', in 1976. The paper described the geology of the deposits and maintained that by 1973 a total of 807 metres of tunnelling had been completed in two adits and 78,516 metres of diamond drilling had been completed in 346 holes.

The Central Zone was estimated to contain 125,000,000 tonnes of ore grading 1.06 percent copper 0.40 ppm gold and 7.7 ppm silver.

1994 Estimate

A draft of a paper on the Galore Creek deposits scheduled for publication by 1995 in CIM Volume No. 46 provides a further update on statistics and tonnage estimates. By 1991 a total of 99,474 metres of surface drilling had been completed in 462 holes and 163 metres of underground drilling in 7 holes. No further underground work had been done.

The Central Zone was estimated to contain 233,900,000 tonnes grading 0.67% copper, 0.35 ppm gold and 7 ppm silver at a 0.27% copper equivalent cut-off and the Southwest Zone was estimated to contain 42,400,000 tonnes grading 0.55% copper, 1.03 ppm gold and 7 ppm silver, also at a 0.27% copper equivalent cut-off.

It will be noted that the much higher tonnages and lower grades stated are attributed partly to a much lower cutoff at 0.27% copper than the 0.6 - 1.0% copper cutoffs used in earlier estimates.

However, increase in the estimated size of the Southwest Zone is significant compared to that envisaged in the mid-1960s with only a few drill holes for guidance. The much later focus on the deposit was based on its much higher gold credit than that in most of the other zones and of course the increase in the price of gold from about U.S.$35 per ounce in the mid-1960s to over U.S.$400 per ounce in the early 1990s.

In late 1993 I learned that Cominco had sold its 5 percent interest in Stikine Copper to Kennco (Stikine) Mining Limited, but I did not obtain any details of the transaction. In mid-1994 in a conversation with Bill Wolfe, Exploration Manager for Cominco, he mentioned that Kennco did not like the corporate structure as it existed, whereby a deadlock could be created by Hudson Bay with its eligibility to earn a 45 percent interest combining with Cominco and its 5 percent interest.

Apparently Kennco wished to disband the existing interest structure and convert to a joint venture. Kennco's insistence upon obtaining a satisfactory modification to the agreement prompted Cominco to reconsider its position, particularly from the view of potential liability, as it was dealing with two companies whose parents were not Canadian. Cominco reasoned that although unlikely, it could be exposed to 100 percent liability with only 5 percent interest. Accordingly, Cominco suggested a buy-out of its interest, which was later agreed to at a price of $2.6 million (Canadian).

When I learned the amount of the settlement, I mentioned to Bill that the amount placed a 1993 value of $52 million on Galore Creek and suggested that the exploration expenditures over the 33-year period could well be in that bracket in 1993 dollars.

In my position as the B.C. & Yukon Chamber of Mines' representative on the Lower Stikine Management Advisory Committee, mentioned earlier, I phoned my friend Mark Fields, geologist with Kennecott Canada in mid-1994 asking if I could obtain the actual annual expenditures at Galore Creek by Stikine Copper through 1993. My object was to emphasize the potential significance of the property in terms that could be appreciated by the other committee members in supporting my purpose of promoting future access rights for mining operations along the Stikine Corridor.

I mentioned my estimate of about $50 million in 1994 dollars since inception of exploration activities in 1960. Mark laughed and said "You'll be surprised at just how close that is". He sent me a copy of an estimate he had prepared in 1991 using the Consumer Price Index, which adjusted actual costs to $50.3 million in 1991 dollars!

Galore Creek Revisited

I had written many letters to Rene from Galore Creek or its environs in the 1960-65 period and I had always regretted, following my transfer in 1965 to Toronto with Kennco, that she would probably never have the opportunity to visit that remote site in its remarkable setting, which had absorbed so much of the energy of many dedicated individuals who had nevertheless been well paid in return by their memorable experiences.

Following our return from Toronto in 1971, I happened to meet the head of Harrison Airways at a Christmas party in Vancouver. He was aware of my past involvement with Galore Creek and quite casually said "Dave, if you and your wife would like a trip into the Galore Creek area, we have a weekly scheduled flight from Vancouver which currently stops at Burns Lake, Hyland Post, Schaft Creek (on request) and Eddontenajon. You would be most welcome to be our guests." I thanked him for his generous offer and when an opportunity arose early in the following fall, I made arrangements for a visit to the property which at the time was being further explored by Hudson Bay Mining with Ray Freberg, a former classmate from U of Toronto, acting as project manager. We arranged for a helicopter to meet us at Schaft Creek, weather permitting, to fly us over the relatively short and direct route to the property.

We left Harrison Airways' base at the Vancouver airport one morning in early fall, flying with about five other passengers in a DC-3. It was a moderately clear day and we had a good opportunity to view the landscape as we flew at about 12,000 feet altitude northwesterly, initially over the Coast Range toward Chilko Lake. Leaving the mountains in the western portion of the Chilcotin, we caught glimpses of occasional ranches and dirt and gravel roads near the remote settlements of Kleena Kleene and Anaheim Lake en route to Burns Lake.

Apart from the rare settlements in the Tatla Lake area and just south of Burns Lake, a passenger aboard an aircraft could easily miss seeing any sign of human habitation throughout the 600 km. of this route This was in sharp contrast to the experience of traveling easterly across British Columbia where there are continual signs of roads, settlements and the familiar patchwork of clear-cut logging areas and their access roads.

At Burns Lake we waited for about twenty minutes for the flight attendant to obtain sandwiches for our approaching lunch. The ongoing flight provided views of logging operations far below us, lying to the west of Babine Lake, but within about 70 miles these terminated and the landscape below appeared to consist of virtually undisturbed bush and mountains. Near Bear Lake at the head of the Skeena River system, the right-of-way for the B.C. Railway extending northwest from Fort St. James, 170 miles to the southeast, and destined initially for Dease Lake, over 200 miles to the northwest, was a welcome sign of human activity. This 'Road to Resources', the ill-fated gamble of the W.A.C. Bennett regime, was to be a costly disappointment; rail eventually laid northwest of Bear Lake having to be retrieved because of lack of any foreseeable traffic.

To the northwest the skies darkened perceptibly during the afternoon and clouds increased with rain squalls just before we landed on the grass strip at the isolated settlement of Hyland Post to pick up some freight. Wet weather had affected the landing area and one of the wheels of the aircraft suddenly dropped into a muddy section. The pilot immediately applied power and managed to regain more solid ground.

From Hyland Post, which lies near the head of Stikine river on the east margin of Spatsizi Plateau, we flew westerly for about 120 miles toward the rough landing strip at Schaft Creek, constructed in support of the major exploration program at a nearby copper-molybdenum property being supervised by Harold Linder on behalf of Hecla Operating

Company. The aircraft contained considerable freight required at the site, and we had planned to use the company's helicopter to fly about 30 miles southwesterly to Galore Creek. By the time we reached Schaft Creek, visibility was marginal for the approach, and heavy rain pelted against the aircraft windows from the darkened skies. When we landed, Harold came aboard the aircraft and told us that based on weather reports, it was unlikely that we would be able to get through to Galore Creek the following day, as the route required considerable flying through the mountains. We rapidly elected to continue on to Eddontenajon with the hope of chartering a helicopter which would provide a far safer, but longer, route to follow down the Stikine River to the junction with Scud River and finally to its confluence with Galore Creek and the nearby campsite, a distance of about 130 miles.

I was disappointed at this turn of events, as I had looked forward to showing Rene the area around Schaft Creek where Dave Miller, Gord Davis and I had spent a week at a fly-camp while carrying out some preliminary geological, geophysical and geochemical surveys in June 1959. This was about two years after discovery of the copper-molybdenum deposits at Schaft Creek in 1957 by Silver Standard. Our access into the property from nearby Mess Lake, some 7 miles from the property where we had landed in a Beaver aircraft with our supplies, was quite memorable.

An earlier reconnaissance flight had revealed a small meadow near our selected campsite close to a mountain ridge on the property. The meadow appeared quite moist and I felt that it was feasible to airdrop many of our camp supplies from a small aircraft rather than have to make numerous backpack trips from the lake. "Wild Bill" Harrison was obviously the pilot for such an undertaking and he appeared as requested on schedule and after a brief reconnoiter of the drop site, agreed to our bombing mission. This was accomplished with a minimum of problems, the only concern, mostly to myself as the individual dropping the items, being when a wash-tub filled with rather light items twanged off one of the aircraft's guy wires. However Bill seemed unperturbed.

Our DC-3 flight continued on the short hop to the small airstrip at Eddontenajon without incident and we arrived there in the early evening. We were met and transported to the nearby motel where we were fortunate to obtain a pleasant room without a reservation. The pilot and engineer were not so lucky and had to fly on to Dease Lake for a room.

Years later, Dave McAuslan and I arrived at Eddontenajon unannounced after fruitlessly trying to obtain overnight accommodation nearby and we spent the night in the motel parking lot, Dave in the cab and myself in the bed of the pick-up - fortunately with sleeping bags, and no rain.

Following our arrival I was able to arrange for a helicopter charter from Dease Lake for the next morning - weather permitting. As we were scheduled to return to Vancouver via Harrison Airways two days later, we were most concerned after checking the weather the next morning to find the area completely overcast with low clouds, and no noticeable breaks. However, at about 7 a.m. we heard the drone of a helicopter, which subsequently landed near the parking lot.

We were soon buckled into our seats and airborne. The pilot skillfully made his way down Stikine River, flying only several hundred feet above the water because of the cloud cover and veering around the occasional sections where clouds extended almost to water level. We caught glimpses of Telegraph Creek and the Grand Canyon of the Stikine as we flew by toward the mouth of the Scud River, but it certainly wasn't a scenic flight by any stretch of the imagination. Turning up Scud River, we had no real problems until we reached Galore Creek and headed up the steep valley, wondering whether our visibility would be closed off before we could reach the broad basin of West Fork, some 1500 feet higher than the mouth of Galore Creek. Miraculously we got through and reached the campsite at about 8 a.m. much to the surprise of the residents as the ceiling was so low that no aircraft had been expected. As the rotor blades slowed following landing, Ray Freberg emerged and we climbed out to greet him. Realizing that our time was limited to about a two-hour visit, Ray suggested that we proceed to the dining hall for breakfast while he arranged for a trip to the underground workings, which I had never seen as they were completed in 1966 - the year following our departure for Toronto.

As we left the helicopter, Rene looked uncertainly at her handbag and asked "Should I take this with me?" "No", I said, "It'll be perfectly safe here." I could just imagine the stir she would have created walking around camp and possibly underground toting her handbag! After all, there was nothing to buy at Galore Creek except the very limited stock in the commissary. We walked up to the dining hall and I showed Rene the cafeteria-style routine for breakfast and a spot where she could sit and left her briefly to consult Ray. On my return, Rene giggled and told me she had looked up to see a tall individual walk through the doors toward the serving area. He suddenly spotted her and with a look of utter astonishment, quickly removed his cap from his head and breasted it! I can imagine his bewilderment for there were no women working in the camp and he may not have heard the helicopter or, if he had, would not have necessarily attached any connection. To the best of my knowledge Rene was the first woman to visit Galore Creek.

After a quick hike through the camp area, we drove over with Ray to one of the two adits, which had not been accessed for some time. Both Ray and I were concerned about the air situation as there was no assurance that the ventilation was adequate near the face of the adit, about 800 feet in from the portal. I went ahead striking matches, and sure enough about 300 feet or so from the portal the matches began to flicker out shortly after striking, so I immediately turned around and we returned to the vehicle.

We had sufficient time to fly over to the toe of West Fork Glacier where the ceiling was only several hundred feet, but it gave Rene an opportunity to walk up to the edge of the glacier and of course I was curious to see how far it had retreated in the seven years since my last visit.

With the totally restricted visibility, it was not possible to see the rim of the basin or even much of the surrounding mountain slopes. I wondered at the impression that Rene would take with her after all she had heard abut the site, and the numerous photo slides that she had seen of the impressive setting under more ideal conditions. It would be far different from mine, having worked in the area for such a lengthy period and experienced Galore Creek's many moods, some of which were so turbulent and many so awe-inspiring. I suppose the situation was comparable to the unlucky visitor to Vancouver during one of its really dull, rainy periods, when the clouds hang low over the mountains and the proud residents offer profuse apologies, all to no avail.

Our return flight followed the same route, however the ceiling had lifted slightly and we had improved views of the Stikine River section and Telegraph Creek.

The Berg and Lost Claims

In exploration, as in other risk ventures, the losers far outweigh the winners in terms of participating in successful discoveries - though fortunately not in the enjoyment of the challenge and all the ancillary benefits. Many are called, but few are chosen. As an inexperienced but eager mining geology graduate in 1950, I looked forward to success - not failure, or disappointment - in my chosen field. In the demanding four-year university program emphasizing engineering disciplines rather than the arts, which required passing at least 88 final exams, laboratory courses and survey schools, I considered that to be one of 27 graduates out of 54 original enrolees was an achievement in itself. Mind you, in the following winter, with no employment opportunities as a geologist and working as a miner at the Garson Mine in Sudbury, I felt lucky to have a job.

One of several potentially important courses excluded in our university curriculum, which could have been particularly beneficial to explorationists would have been one covering the legal issues that we might encounter in our careers. By 1992 I had appeared as a witness for two companies with which I had been employed in four separate legal disputes and I had also provided expert witness evidence in two additional hearings. The growing demand placed on geologists by legislation enacted to support rapidly changing socio-political-economic developments particularly as they impacted on tenure consideration was simply not envisaged in the late 1940's. Nor was the alarming growth of society's tendency to rely on litigation in settling even minor disputes. Most of the early federal and provincial mineral acts in force in Canada were relatively simple documents. Environmental issues, employee health, welfare and safety requirements affecting explorationists were embryonic compared to those of the 1990's. Early explorationists, like others involved in wilderness careers and pursuits, had to be self-reliant and tended to be trusting. In the late 1940's, as a prospector's helper in northern Ontario and later as a geologist in the 1950's in British Columbia and Yukon Territory, I recall several occasions when either alone or with a companion, I encountered an unlocked cabin in the bush. One of these had a simple sign which read "Use what you need, but please replace the firewood". Others apparently felt that even this request was unnecessary.

Compare this attitude with one incident in the early to mid 1970's and its consequences. On being promoted to Vice President of Kennco Explorations (Canada) Limited and Kennco Exploration (Western) Limited in March 1971, with a transfer from Toronto to Vancouver, one of my early and more time-consuming involvements was a claims dispute concerning the relative location of Kennco's claims covering the Berg copper-molybdenum deposit and the nearby uncertainly located Lost 1-4 claims owned by Sierra Empire Mines Limited. Little did I realize how emotionally engrossed I was to become in this absorbing dispute, which had originated in 1967 and possibly earlier, and its ultimate consequences and resolution in 1977. At one time I even felt that the topic could produce an interesting book. In retrospect, I feel it warrants a lengthy chapter.

I am indebted to Bill Guiler, office manager and Russ Babcock, Director of Exploration for Kennco Explorations (Canada) Limited for their willingness to provide me in August 1989 with photocopies of the principal legal records of the subsequent trial and appeal in the lawsuit, between Kennco and Sierra Empire Mines Ltd., William Henry Patmore and Godfrey Frederick Sanft. The documents include the Charge to the Jury at the conclusion of the trial and the Reasons for Judgement in the Court of Appeal from which I have liberally quoted. It is most important for the reader to recognize that the quoted testimony contained in Justice Dryer's Charge to the Jury is not necessarily a direct quotation unless so specified. As Justice Dryer partially described it in his Charge to the Jury:

"At the moment I am going to go on and deal with the evidence.

Now, some of this you may find rather boring, but some of this evidence you heard over two months ago, and we have spent so much time on this case now that I do not think it would be inappropriate for us to spend a little more time and go through the evidence.

You will remember what I said about this: I do not put everything down, and what I think is important you may not think is important, and in any event your recollection is more important than mine. However, I do have notes, and you do not, and this may be helpful to you. At least, I found it helpful when I read through it, so maybe you will too."

The Berg property lies between elevations of 4,900 to 7,500 feet on the east fork of Bergeland Creek in the Tahtsa Range in west-central British Columbia, about 55 miles southwest of the town of Houston. The area lies in the transition zone between the Coast Mountains and the Nechako Plateau and is characterized by relatively rugged topography with permanent snowfields and remnants of glaciers at higher elevations.

The property area was not accessible by road until about 1965. In previous years access, apart from that provided by helicopter, was on foot from Tahtsa Lake six miles to the south or from Nanika Lake the same distance to the west.

The earliest known record of prospecting activity in the property area is of that by Cominco Ltd. in the late 1920's. Prospecting of a large rusty-weathering zone in the vicinity of the present Berg copper-molybdenum deposit led to the staking of peripheral lead-zinc-silver veins in 1929. The claims covering the showings were allowed to lapse and were restaked in 1948 by George Seel as agent for Fred Sanft on behalf of the Lead Empire Syndicate, a predecessor to Sierra Empire, initially as the Lost 1-4 mineral claims. William Patmore accompanied and assisted Seel during the staking and was actively involved in the subsequent exploration program.

Northwestern Exploration Ltd., a predecessor company to Kennco Exploration (Western) Limited, initiated a geochemically-oriented exploration program directed principally to the discovery of porphyry-copper and porphyry-molybdenum deposits in Western Canada in 1957. The program was organized by Joe Brummer, a Senior Geologist with the company. This was the first regional stream-sediment geochemical survey carried out in the Canadian Cordillera and the area surveyed was the Guichon Creek batholith in southern British Columbia.[xiv] The results showed that the batholith was copper-rich and established the effectiveness of the method as a reconnaissance tool for assessing the mineral potential of large areas.

While this program was in progress, I was supervising an exploration program at the Krain property in the northern part of the centrally located Highland Valley area of the Guichon Creek batholith. This area was to become the major site of copper production in western Canada commencing in the 1960's with the Bethlehem Copper, Lornex and Highmont mines and later Valley Copper. I recall how difficult it was to locate John Fortescue and his sampling crew (which included R.E. Gordon Davis a co-discoverer of the Faro orebody in Yukon). Because of the competitive advantage of the survey method possessed by Kennco in this period, it was considered essential to maintain a very low profile. Access to the party's base camp was of necessity by four-wheel drive vehicle as camp sites were selected off travelled bush-roads with entry ways concealed by freshly cut branches. The samplers maintained a torrid but stealthy pace in their traverses, pack-sacks laden with up to 70 seeping Kraft paper bags, which contained their silty treasures.

In 1959 Kennco committed a major part of its exploration budget to systematically sampling large areas of the Canadian Cordillera by similar stream-sediment surveys and to a lesser extent water surveys directed specifically to molybdenum detection. Up to 60 personnel were involved annually in separate surveys and by 1970 when the level of these programs was reduced, about 41,000 square miles had been sampled in British Columbia and Yukon Territory. The programs contributed both directly and indirectly to the discovery of six potentially significant mineral deposits, containing lead, zinc, silver and gold in addition to those predominantly mineralised by copper or molybdenum. Three of the deposits (Chappelle, Lawyers, Sam Goosly) subsequently reached production as the Baker, Cheni Gold and Equity Silver mines respectively.

The Berg deposit was discovered in 1963 by Kennco by following up a single-station stream-sediment sample site sampled in 1961 which contained anomalous copper and molybdenum contents. The source, which coincided with the well-exposed gossan[16] previously discovered by Cominco, lay several miles upstream from the site on Bergeland Creek. Although the source of the copper anomaly was recognized and staked in 1961, initial exploration work at the property was not encouraging. Later work revealed that barren leached areas exposed at surface masked underlying mineralization of potential economic significance.

By 1971 Kennco had spent over $1 million in exploring the Berg deposit, the property consisting at that time of 98 full-sized and fractional mineral claims wholly-owned by Kennco and covering about six square miles.[xv] All but three of Kennco's claims were in good standing until the years 1980-2010 through the filing of assessment work and obtaining excess work credits which at that time were unlimited, but subsequently reduced to a maximum of ten years per claim.

As porphyry copper-molybdenum exploration intensified in Western Canada in the 1960's the frequent occurrence of lead-zinc concentrations in the peripheral portions of mineralized porphyry systems was more commonly recognized. I recall visiting Kennecott Copper's awesome Bingham Canyon copper-molybdenum operation near Salt Lake City, Utah during the period. The major open-pit at that time was at a depth of about 5,000 feet from the original surface, through mining at a rate of 300,000 tons of ore and waste rock per day with ore grading about 1% copper and 0.05% molybdenum. Occasionally exposed on the walls of the pit were portions of old underground workings from lead-zinc mines in the periphery of the copper molybdenum deposit.

In 1971 with overall responsibility for Kennco's exploration program and increased exposure to the parent's exploration objectives, I became increasingly convinced that joint venturing of Kennco's mineral property assets would provide the only guarantee for some future return on its substantial investment in Canadian exploration since 1946. After so many years of concerted effort by such a large number of individuals and with the technical successes which our team had enjoyed, this was a depressing decision. Reluctantly, I drafted a farm-out proposal and submitted it to Lowell B. Moon, then Kennecott Copper Corporation's director of exploration. I was not surprised when the program was approved but without any specific guidelines as to the type of deals to be negotiated. Twelve properties were proposed for farm-out and by 1972 seven, including the Berg, had been farmed-out. In January 1974, I was also to negotiate an option of the

---

[16] Gossan: Rusty-cloured decomposed rock or vein material resulting from oxidized iron pyrites.

Chappelle property with my successor at Kennco, Bob Stevenson, but under different circumstances, as in the same month I left Kennco to become Vice President, Exploration, for DuPont of Canada Exploration Ltd.

By 1976 following additional drilling Canex Placer which had optioned the Berg property from Kennco, estimated geological reserves at about 440 million tons grading 0.4% copper and 0.05% molybdenite using a 0.25% copper cut-off.

Principal terms of the agreement negotiated in 1972 with Canex Placer, later to become Placer Dome, provided that Placer could earn a 51% interest in the property by spending $2 million with further financing on a pro rata basis or alternately reduction to a 30% carried interest by a non-participating party. The agreement recognized the uncertainty of the outcome of the pending lawsuit by providing several conditions which were contingent on the outcome of the trial and the portion of the Berg deposit that might fall on Sierra Empire's Lost claims.

I have mentioned earlier the ever-changing regulations which explorationists must consider in carrying out their work under differing jurisdictions. In Canada, mineral rights may be acquired to leases occupying hundreds of square miles, but most commonly the basic claim unit prior to metric measures was about 40 acres (1,320 x 1,320 feet) or 51.65 acres (1,500 x 1,500 feet). Claims may be acquired by staking on the ground or by applying for leases or claims based on areas defined by latitude and longitude.

In 1948, when the original Lost 1-4 Claims were staked for the Lead Empire Syndicate, the British Columbia Mineral Act required that the area claimed for lode claims, as opposed to placer claims, be defined by two posts, rock cairns or mounds of legal size with the required identification at each post. The claimed distance between posts was not to exceed 1,500 feet with a suitably marked location line between them. Part of the identification on the No. 1 post designated the bearing to the No. 2 post and stated whether the area claimed, being 1,500 feet in distance, lay to the left or the right of the location line.

The Mineral Act provided specific requirements on a legal form Affidavit which was required to be filed for a full-sized claim, which inter alia included the following:

The name of the Mining Division.

The name of the staker and his address and, if acting as an agent, the name of the party and his Free Miner's Certificate No.

The staker's Free Miner's Certificate No. and its date and place of issue.

The date of locating the mineral claim, its name and a description of its approximate position, including the names of any mineral claim or claims it may adjoin.

Confirmation that a No. 1 and No. 2 post has been placed on the legal dimensions of the claim with the legal notices (descriptions) on each post.

The actual wording written on the No. 1 and No. 2 post. In the case of the Lost No. 1 Claim staked by George Seel on September 8, 1948, the following words appeared on the form:

"I have written on the No. 1 cairn the following words: Initial Post No. 1 of lost No. 1 M.C. No. 2 Post lies 1,500' in a northerly direction. 1,500' lie to the right of location line. Located the 8th day of Sept., 1948 by Geo. V. Seel, agent for G. Fred Sanft.

"I have written on the No. 2 Post the following words: No. 2 Post of lost No. 1 M.C. located the 8th day of Sept., 1948 by G.V. Seel, agent for G. Fred Sanft."

Certification that the location line between the No. 1 and No. 2 posts has been marked as required.

Certification that to the best of the staker's knowledge the ground within the claim is unoccupied by any other person as a mineral claim, nor by any building, dwelling house, orchard, land under cultivation, naval or military reservation.

Certification that the staker does not hold more than seven other claims staked by himself, or on his behalf within the past 12 months within a distance of 10 miles of the claim.

Certification that the staker has affixed or caused to be affixed to the legal posts of the claim, metal tags marked or impressed with the number stated.

A sketch plan on the reverse side of the affidavit showing as near as may be the position of the adjoining mineral claims.

The Mineral Act provides for staking claims over such areas as glaciers, lakes or other inaccessible terrain which may be considered to contain mineralization of potential interest to the staker by the use of 'witness posts'. I was involved in the past in staking at least three or four areas by 'witness posts', several of these being in mountainous terrain in the winter above timber line where cairns could not be constructed because of the depth of snow and others being to protect claims staked over parts of a lake.

The requirement in staking such ground is essentially identical to the procedure used in ordinary staking with the exception that the staker places legal posts for the area claimed at a safe location and designates the distance and bearing to each of the posts of the witnessed claims which could be a mile or more from the witness post.

An amendment to the Mineral Act in 1962 specifically provided that one post or one cairn with properly affixed tags and inscriptions could be used to identify common posts for more than one claim, i.e., a group of four claims covering a square area 3,000 x 3,000 feet could have a single centrally-located post rather than four separate posts.

Prior to the amendment the Mineral Act was not specific and it was common practice during staking to take the less demanding interpretation. As will be described later, whether one post or cairn could be used as the legal post for more than one claim was to become one of the issues at the trial and in the later appeal.

Although exploration activity was carried out each summer at the Berg property from 1962 to 1968 and again following 1969, the origin of the dispute between Kennco and Sierra Empire can be traced back to at least 1965.

By 1966, Kennco had spent almost $500,000 on exploration at the Berg property and had sufficient information to recognize its potential significance. A preliminary evaluation by Sam Smyth in 1967 indicated a potential annual operating profit for a 3 million ton per year operation ranging from $5 - $13 million at a copper price of $0.40 - $0.45 per pound.

Evidence presented at the trial in 1973 indicated that Kennco geologists George Stewart and the late Charles Ney had unsuccessfully attempted to locate the claim posts for the Lost 1-4 mineral claims in 1965. By 1966, there was an obvious difference of opinion between Sierra Empire and Kennco as to the locations of the claims. The Sierra Empire position was that the centrally located No. 1 posts for the claims lay on the southeast trending ridge on the northeast side of Bergeland Creek which forms the divide between Bergeland Creek and the northerly flowing Kidprice Creek about 2 miles to the east. A copy of the survey map completed by John Tustin, surveyor, on behalf of the Lead Empire Syndicate in 1951 which showed the position of the claim posts, the orientation of the location line, and trenches and pits on the claims failed to support Sierra Empire's claim as to the location of the Lost 1-4 claims when it was used by Stewart and Ney in attempting to locate the claims posts in 1965. The map had been shown to Ney, probably in August 1965, according to his recollection and testimony as shown in the following excerpt from the court records:

"... I was first on the Lost Claims around September, 1963, and I looked over the area very generally. I first saw Exhibit 154, the Tustin map, in the Burrard Building. I had a discussion with Sanft about it in Kennco's office in the Burrard Building. The late Jack Gower was in and out of this but no one else was there. I believe this conversation took place in the fall of 1965. We were under the impression that this map, Exhibit 154, represented correctly the position of the Lost claim posts. Sanft said that the posts were shown correctly on this map. He also told us that the posts were mounds and that they were near the ridge between Kid Price Creek and Bergland. We, Stewart and I, took a copy of it into the field and tried to reconstruct it on the ground. Our object was to show that the mounds we knew were on the ridge coincided with the No. 1 post shown on the map. We tried to reconcile the mounds with what were shown on the map but we couldn't do so. There was clearly something wrong because the series of pits on the map, instead of being lower that the posts, were actually higher. Later we found a No. 1 post had been discovered and it fits in well with the map as to elevation and distances and direction. The No. 1 post is about 700 feet east of the ridge where the mounds are. You go down a fairly gentle slope which becomes considerably steeper beyond the position of the No. 1 post. I have a little doubt as to the time I discussed Exhibit 154 with Sanft but I am sure it was between '65 and '66. After examining the two mounds we came away concluding they were the claim posts. In the fall of '65, I spent two hours up on the ridge. Stewart was there longer. I didn't go back until 1966. That could be in July. I didn't do any looking. I, myself, never looked again until 1966."

Charlie Ney had been one of the stakers of additional Berg claims in 1963, the initial claims having been staked in 1961. As such he was an essential witness for Kennco as the counter claim against Kennco in the lawsuit challenged the validity of certain Berg claims which he had staked using witness posts. Charlie, like most of us, assumed that more than one claim could be staked from a common post, as was to be emphasized in the 1962 amendment to the Mineral Act. In giving evidence at the trial, I stated:

'I have staked in B.C. and the Yukon and one cairn was frequently used for more than one post in the period prior to 1962. The same apply to one cairn containing a number of tags and identification numbers."

At one location in British Columbia where I noted the use of witness posts, stakers had erected about 20-30 cairns which were scattered over an area of several hundred square feet. The resulting setting in the barren alpine area resembled some primitive ritual site. I recall musing about the nightmare that the surveyor would face if a significant discovery resulted.

At the trial Charlie was also asked why his Application for Claims, i.e. the recording affidavits he filed did not concede the existence of the Lost 1-4 claims, as required under the Act. His response was:

"You can't show on a map something of which you do not know the location."

Because of the uncertainty of the correct position of the Lost 1-4 claims, Kennco in 1966 commissioned Underhill and Underhill, B.C. Land Surveyors to complete a photo control survey of the area containing the Lost claims and to include all the adjoining Berg claims. Testimony provided by Jack Parnell, a partner in Underhill and Underhill, at the trial indicated that he arrived at the property with his crew on July 7, 1966. The next day they located the initial post of the Lost 1 - 4 mineral claims.

"On July 8th we had a helicopter all day. We looked for posts and established a chain of title and looked for the number 1 post. We had searched the records of the Mining Recorder's office first, and we had copies of the important records with us."

The records referred to were photocopies of the affidavit for Full-Sized Claim prepared by George V. Seel, who as agent for Fred Sanft, staked the Lost 1 - 4 claims on September 8, 1948 and recorded them in Burns Lake on September 14, 1948.

"We found the initial post of Lost 1 to 4 mineral claims. I found a cairn with a can and four tags. Four notes in the can. Later I went up there with the sheriff and brought these down, and these are now Exhibit 125. 100 feet east of the cairn it dropped steeply to a glacier. West of the cairn it dropped at about 45 degrees to the valley. I just checked the cairn for notes and tags, copied down what was on them, and left them there. I put in a survey post.

We saw two cairns further west. Then we went south to try to find the number 2 post, but we could not find it."

Leggett then referred to two other cairns.

"I opened them up, found nothing in the cairns or in the ground below them. There were two or three more tins around the edge. I had with me a copy of the plan which was given to us by Kennco."

The plan provided by Kennco was the copy of the survey map prepared by William Tustin in 1951.

"It helped me to find the number 1 post. There were holes 4 or 5 feet in diameter blasted into the ground. We realized they were the holes shown on the plan. Some had pits between them. We found the pits before we found the posts."

Because of the problem encountered in locality of the No. 2 post, Mr. Parnell then attempted to use the services of Mr. Sanft in locating them.

"I had a discussion with Mr. Sanft in 1966 by phone, in August, and I met him twice on the site at the Berg camp . . . I went in on August 16th and he arrived on the 30th of August. I didn't tell Sanft where the number 1 post was, because I wanted to test his memory."

Data provided by Fred Sanft to George Stewart and Jack Parnell while on site at the time failed to lead to the discovery of the No. 2 posts of the Lost claims.

"I told Sanft that we didn't find the posts, and that I would search again, and ask as to where the Lost number 1, 2, 3 and 4 posts would be, and he pointed to the two cairns on the Bergland Creek side, that is to say, where I marked the "Z" on Exhibit 127. Then I told him I'd already found the number 1 post, and it was 800 feet east of where the two cairns were. He said, 'If you've found that, okay. I'll give you an affidavit on the number 2 post.

"I do a survey on the basis of all kinds of information. I decided to position the number 2 post on the basis of that information. That is to say, I had what was written on the notes and the check on the back of the Seel affidavits -- the description of the Seel affidavits. What I drew up was on the basis of that, and my discussions with the Gold Commissioner. I wanted Sanft to stay over, and tried to get the number 2 post, but he couldn't stay."

The survey map completed by Mr. Parnell became known as the "Parnell Plan". Because of his inability to locate the No. 2 posts of the Lost 1 - 4 mineral claims, and relying partly on the information in the affidavits which indicated 'northerly' and 'southerly' trends for the location lines, and partly on the nature of terrain to the north and south of the No. 1 posts, he assumed location lines trending north 7 degrees east and south 7 degrees west respectively from the No. 1 posts in his survey plan. Importantly for Kennco, although the orientation of the claims were not identical to that of the "Tustin Plan", the position of the centrally located No. 1 posts were similar, being about 700-800 feet east of the two cairns on the ridge crest above Bergeland Creek - where Sanft and subsequently several other witnesses for Sierra Empire had claimed they lay.

Just as Kennco had elected to have an independent survey completed of the Lost 1-4 claims, so did Sierra Empire. S.R. Leggett, a British Columbia Land Surveyor, was retained by Sanft in 1967 to locate the No. 1 and No. 2 posts of the Lost 1 - 4 claims. In describing the survey work at the subsequent trial in 1973, Leggett commented:

"I drove in with Sanft and Patmore to Houston, met Dr. Black (Phil Black, Kennco Senior Geologist) there and went in by helicopter to the site. The helicopter could not put down on top of the mountain because of the wind so he set down in the camp. Only Dr. Patmore and I were inclined to climb up -- I think this was Labour Day weekend. Patmore and I climbed to the top to search for the cairn. We found several rocks that could have been the cairn. I found another old can, or what was left of it, and another can beside it –

"They were rusty old cans. I thought one was a milk can but I cannot remember now why. Then we went on and found Parnell's cairn. I don't know the direction, it was downhill. It had an aluminum post and a pick on it.

"Patmore and I were not together. I think Patmore reached the Parnell cairn first. I took the can out and checked it, the tag numbers. I was there again in 1969 and at that time I wrote down what I found in the can in the cairn. The can was maybe a foot down in the cairn; we did not have to tear it apart; it was like a brand new cairn with a survey post in it. I would not be able to judge as to whether there was any aging of the rock."

Phil Black was supervising a drilling program in 1967. He had also directed the original regional stream sediment survey program in this part of B.C. in 1960 when the anomalous site on Bergeland Creek was sampled.

Leggett then searched unsuccessfully for the No. 2 posts traversing initially east and west based on verbal information provided by Patmore as to his recollection of the correct line orientation, and later north and south.

"I didn't report what we had done to Sanft except it was discussed that we had found another cairn. As far as he was concerned the Parnell cairn was the only cairn but we said we had found another cairn. I didn't tell Mr. Black what I felt was the No. 1 cairn. I don't claim to be an expert on the Mineral Act."

The survey done by Parnell in July-August, 1967 included a survey of Berg 15 claim for the purpose of proceeding to lease. As part of the leasing procedure, the survey was submitted and accepted.

In the fall of 1970, Leggett returned to the property and completed a survey of the Lost 1 - 4 claims ("Leggett Plan") which showed considerable overlapping by the Berg claims. Kennco complained of unauthorized work that had been done by Sierra Empire in 1969 on its Berg 18 claim, just south of its mineral lease covering most of Berg 15 which had been granted to it by the Crown after the 1967 Parnell survey. Sierra Empire stated that it was having new surveys prepared, but Kennco later learned that on October 25, 1971 Sierra Empire filed notice of its intention to apply for a Certificate of Improvements covering the Lost 1 - 4 claims based on the 1970 Leggett survey. According to the provision of the B.C. Mineral Act, had this Certificate of Improvements been issued and leasehold title granted, then the title of the claims could only be challenged thereafter on the grounds of fraud.

As a result of the apparently irreconcilable nature of the conflict which had developed, and the possibility of losing title to an important portion of the Berg deposit, Kennco commenced adverse proceedings under section 85 of the Mineral Act, R.S.B.C. 1960, Ch. 244, as combined by S.B.C. 1961, Ch. 39, without which its claims would, under terms of the section, be deemed to have been waived. Under subsection (4), a jury consisting of eight persons was ordered, this being a provision of the Mineral Act at the time in an action as an alternative to trial only by a judge.

A Writ of Summons was filed with the Vancouver Registry on January 12, 1972 by Kennco's solicitors naming Sierra Empire Mines Ltd., William Henry Patmore and Godfrey Frederick Sanft as Defendants in an action brought by Kennco Explorations, (Western) Limited as Plaintiff.

A Statement of Claim was filed on behalf of Kennco with the Vancouver Registry on February 25, 1972. The 14-page document's principal issues are succinctly described in the Reasons For Judgement of the Honourable Mr. Justice Bull dated March 18, 1975 following the Appeal of the earlier judgement.

"The appellant's claim was, firstly, that the four Lost claims were void and invalid because of non-compliance with provisions of the Mineral Act and were neither properly located, recorded or maintained, and for a declaration that the Leggett survey plan upon which the respondent was seeking the issue of a Certificate of improvements leading up to leasehold title was invalid. Secondly, it claimed a declaration that the fifteen Berg claims (each of which purportedly in whole or in part covered the Lost claims) were valid mineral claims, or Lost alternatively, should the Lost claims be held valid (in whole or in part), the fifteen Berg claims were valid insofar as they were not within the boundaries of the Lost claims as those boundaries were determined by the Court. Lastly, the appellant claimed damages against the respondent and the defendants Sanft and Patmore for fraud and conspiracy with respect to the alleged alteration of the original locations of the Lost claims causing encroachment on several of the Berg claims".

A Statement of Defence and Counterclaim was prepared on his own behalf by Dr. Patmore and filed with the Vancouver Registry on March 1, 1972. A Statement of Defense was also filed on June 14, 1972 on behalf of Mr. Sanft.

I was strongly opposed to making charges of conspiracy and fraud against the defendants. However, the solicitors considered that it was the only practical method of ensuring that the principals involved could be examined prior to trial under provisions of discovery.

The trial commenced in the old Vancouver court-house on Georgia Street in mid-April 1973 and ended on June 15, 1973 after more than nine weeks and 24 actual court days. Within the nine-week period, court appearances were recessed on two separate occasions while awaiting an appearance by Mr. Sanft who was ill or hospitalized. The claims against him were subsequently dropped by Kennco.

I recall having considerable sympathy for the seven men and one woman on the jury. The dispute they were being asked to settle involved a moderate amount of technical data and conflicting testimony, presented quite convincingly by most witnesses. The length of the trial was exacerbated partly by each of the defendant's having their own counsel, all of whom had opportunities for cross-examination and partly by the number of witnesses called - eight by the defendants and seven by Kennco.

Although there was almost a 15-month period between Kennco's decision to proceed with a lawsuit and the commencement of the trial, there was a colossal amount of preparatory work required. Kennco was indeed fortunate to have assigned as its counsel Larry M. Candido of Lawson, Lundell, Lawson & McIntosh of Vancouver, a former prosecuting attorney for the City of Vancouver with considerable court experience. He was assisted principally by David L. Rice, a younger, less experienced lawyer.

In preparation for the trial, I combined some business with pleasure. In July 1972, Rene's parents visited us in Vancouver and we elected to take a drive through central British Columbia using both the company four-wheel drive Blazer and our car. Part of our route took us west from Prince George to Smithers. My son, Rob and I then drove south from Houston in the Blazer along gravel roads to near Tahtsa Reach and then via a very rough muddy road to the Berg property which, at that time, actually required partial travel along an active creek bed whose waters were fortunately at a relatively low level. I had never been to the property before although I had visited several projects in the general area in previous years. At the time there was a diamond drilling program in progress by Canex Placer.

Rob was 15 years old and he had developed a keen interest in hiking. His enthusiasm was infectious and led me to join him in many interesting hikes and climbs in the next ten years. Rob kept a record of his more than 100 hikes from August 1971 to December 1984. The hike in the Berg area on July 30 and 31 was No. 10 for Rob and he recorded it partly as follows:

"From the Berg camp we ascended the scree and talus slopes to a ridge above the camp at about 6,500 feet. We were wary for many reports of a small grizzly had been given by the drillers. On this first day a cairn was found which had been lost for some years so the trip was worthwhile. On the second day of climbing we descended the other side of the ridge by glissading down to about the 5,000-foot level searching for the second of the lost cairns. Nothing was found in the valley so we ascended a prominent ridge to the north of the camp. The ridge was quite rotten, like the rest of the rock in the area and care was needed to prevent dislodging large rocks. The descent from the summit of the ridge at just over 7,000 feet to the camp at 5,000 feet was at an angle of about 40 degrees and very enjoyable as we covered the entire distance in three long glissades."

Rob's reference to the cairn having been 'lost' for some years reflected the historical problem of locating cairns in the area. Depending on the amount of snowfall during previous winters and the summer climate, the rubbly ground surface did not become snow-free until early in the fall on some occasions.

Larry Candido worked at an unrelenting pace. After our regular workday we frequently met at night at his office to review progress and discuss strategy. These sessions frequently lasted to near midnight. Just prior to the start of the trial, he contracted some form of tendonitis in his lower legs which was extremely painful and for many months afterward, including the trial period, he walked on tip-toe to avoid bearing weight on his heels.

Larry's dedication to every detail of the case as he prepared for the trial was most impressive. There was no abatement in his drive throughout the lengthy proceedings. At the end of each court day Larry would return to the offices of Lawson, Lundell and discuss progress with a group of other trial lawyers who could advise on strategy. He always had readily available precedents at the trial to support his particular case where the other three lawyers didn't. He would arrive with all of these neatly stacked in order for Justice Dryer's attention.

One of the most important outcomes of our sessions occurred one evening when we were discussing the Tustin plan. As noted earlier, this was a critical piece of evidence that Kennco became aware of when Charlie Ney and Jack Gower, General Manager of Kennco Exploration (Western) were shown a copy of the map prepared by John Tustin on behalf of the Lead Empire Syndicate in 1966. Larry was dissatisfied with the available copy which was not complete and consequently lacked impact as evidence even though it tended to support Kennco's belief that the posts lay farther east than claimed by Sanft and Patmore.

I queried Larry: "How much effort did Kennco spend in the 1966-7 period in trying to determine the whereabouts of Tustin?"

"Well," he replied, "I don't know the actual details, but they apparently satisfied themselves that he couldn't be located. Also, you remember that in Patmore's recent discovery he was asked about Tustin, having worked with him at the property in 1951". Larry well recalled his rejoinder, "Oh ... Tustin was an old man then ... he would be dead by now."

The more I cogitated, it seemed that a further effort should be made to at least confirm whether he was alive or dead and Larry agreed. I said, "It shouldn't be all that difficult - Tustin is an uncommon name. As a start, I'll check the principal telephone directories in Canada, at the B.C. Telephone office, starting tomorrow".

The next day I located the phone numbers of 84 Tustins living between Vancouver and Halifax in principal cities across Canada. There were a few major cities in the prairies with no Tustins, the principal concentration being in Ontario. My heart raced, when on the fourth try I found myself speaking to John Tustin's _father_! He was somewhat guarded, so I provided him with the bare but essential details of my enquiry and after some hesitation he gave me his son's telephone number in Ontario where he was still enjoying a career as a surveyor. He explained his reticence in providing information at the outset; John had been required to give evidence at several hearings and trials in the past and he didn't want to unnecessarily burden him with additional problems.

I phoned John Tustin's home immediately and was fortunate to find him at home. I explained the background of my call. "You know Mr. Barr, this is quite a coincidence. I was cleaning out my basement only three weeks ago and threw out a lot of old records. I came across my old survey notes and the original tracing of the work I did at the Lead Empire property in 1951." My heart just about stopped, but he continued, "I was about to throw it out after all these years, but I didn't." He chuckled.

He not only agreed to provide us with all his original data, he was also willing to come out to Vancouver and be available to describe his recollection of the work he had done and most importantly where the cairns for the Lost claims were located. It worked out to our mutual benefit as he had the opportunity to visit his parents. We spent several evenings with him. A colourful man, with a remarkable memory, he spoke quite loudly because of a hearing problem. This condition occasionally worked to his benefit as in his later court appearance as a witness for Kennco, he took time, if needed, to consider his response to questions. The trial Judge, Justice Dryer, would try to interrupt him on occasions, and John would simply carry on, whether he had heard the judge or not, necessitating Justice Dryer to bellow out "Mr. Tustin, just a minute now!"

As we had not advertised Tustin's availability prior to the trial, the defendants were shocked when he was called as a witness. Larry had anticipated the effect of this announcement with considerable relish and said to me "You just watch their faces when I call John Tustin." As I was prepared, I was watching two of the principal defendants at the time and the effect was quite dramatic - one's face went pale and the other's distinctly flushed, as if witnessing someone returning from the grave!

Another outcome of our reviews was my suggestion, following my visit to the Berg property and after having access to the Tustin map, that we should have a map prepared by an independent survey company from existing air photos post-dating the 1951 period which might reveal enough pits, trenches, and possibly cairns and which with contouring could be used to compare the position of the Tustin plan with topographic features and existing surveys of the Berg claims. Although I felt the pits and trenches might be detectable by a professional mapper, I was not so sure of the cairns. My testimony at the trial expressed my concerns:

"There were not as many loose rocks in the area of the Leggett cairn as in the area of the Parnell cairn. It was difficult to locate the Parnell cairn since it blended in with the rest of the territory. I located the two western side cairns first, that was not too difficult, and then I travelled to the east. I should have gone further down the slope; I was about 40 feet from it when I first spotted it."

Preparation of a map was commissioned to Photographic Surveys Limited of Vancouver. The completed map showed most of the pits and trenches and compared very favourably with the Tustin Plan when oriented to fit the plan. At the trial Mr. Rolfe, an experienced photogrammetrist gave evidence on the preparation of the plan and the relative elevations of the topography in the vicinity of the Parnell cairn which he claimed to be accurate to within a few feet.

At the trial I was subsequently to provide a graphic illustration to the Jury of the effect of rotating the Tustin Plan about the centrally located No. 1 posts, depending on whether it was centred on the Leggett or Parnell Surveys. In the former case three mineralized drill holes completed by Kennco in the northeastern part of the Berg deposit could be included on the Lost claims, contingent on the orientation of the claims.

The trial commenced on April 9, 1973 and continued intermittently for 24 court days, ending with the Jury's verdict on June 15. Judgement was subsequently delayed by further arguments not requiring a jury, which were provided by counsel on October 11 and 12, 1973.

The case was tried before the Honourable Mr. Justice Victor Dryer, with a jury of the County of Vancouver in the presence of Larry M. Candido and David L. Rice, counsel for the Plaintiff and Arthur W. Johnson counsel for the Defendant, Sierra Empire Mines Ltd., and Hugh J. McGivern, Q.C. counsel for the Defendant, William Henry Patmore and A. Boyd Ferris, Q.C. counsel for the defendant, Godfrey Frederick Sanft.

Following impanelling of the Jury, Justice Dryer explained their duties to them, carefully noting that they were to be judges of the facts and to determine the weight to be given to the evidence provided, whereas he was to be the judge of the law and that they would be required to follow his directions on the law. He was again to emphasize this later at the conclusion of the trial in his charge to the Jury.

After opening arguments were provided by respective counsel, witnesses for the Plaintiff were called, the first being Jack Parnell, the partner and surveyor for Underhill & Underhill whose testimony in locating the No. 1 post for the lost 1 - 4 claims on July 8, 1966 has been partly quoted previously. His total time on the witness stand took part of two days as did testimony provided by several later witnesses including the late Charles Ney, John Tustin, myself, William Patmore, and S.R. Leggett.

Charles Ney had participated in the original staking of the Berg claims, first visiting the Lost claims in September, 1963 as noted in part of his previously recorded testimony, where he examined the area "very generally". It was Ney and George Stewart who in 1965 searched unsuccessfully for the posts based on the verbal report from Sanft that the No. 1 posts were on the ridge above Bergland Creek, rather than where they were located by Parnell about 650 feet to the east.

John Tustins' evidence was to be particularly compelling because of his memory for names and events backed up by his data which included original field notes and an original copy of the 1951 survey of the Lost claims,; and traced copies from the original notes and metal tags located in the No. 1 cairn of the Lost I - 4 claims. He provided much information, some of it considered extraneous but lending conviction to his testimony. Selected portions of his testimony follow:

"I left Prince Rupert in 1951 and went to Vancouver to the office of Wood & McClay and met J.E.R. Wood, who seemed to be in charge of Lead Empire Syndicate. Met Sanft and Patmore and another man who was apparently Wood. I was given instructions which are Exhibit 150 under the word 'plan'. Left Vancouver the beginning of the third week in July, met Patmore. Found him in a beer parlour with a man he said he had hired named Don McNeil."

Justice Dryer was to note the witness who was called at the trial was Clifford McNeil (nicknamed "Cap").

"I had a man named Roy Manhard coming up, who had been hired as a labourer by the three principals, Patmore, Sanft and Wood. We -- Manhard and I --found two small cairns about a hundred feet apart right on top of the crown. We examined them and stopped and had lunch. We were looking for the initial post for Lost I and 4. We continued and about an hour later we found it. It was on the right-hand side, on the northern side, about 650 feet from the ridge at the south side. It seemed fairly new -- two or three years old. It looked fresh, three or four feet high. It might have been higher because the rocks were spread out. We dug down into the centre of the cairn and we found no can, no tags. We went back to camp.

"The next day Manhard and I went back. Lloyd dug down into the big cairn. I went searching around on the south side, the Kid Price Creek side, looking for anything that would interest geologists. Lloyd hollered to me from the top of the crown."

Not included in Victor Dryer's comments at this time was the ensuing comment that Lloyd Manhard reportedly hollered "John, I found the tags! " For effect Mr. Tustin turned to the jury and "hollered" the statement with his hand raised to his mouth. It was most dramatic and unexpected.

There were many humorous moments during the trial and I only wish that I had written some of these down at the time. I have talked to some of my acquaintances with Kennco at the time and some of the witnesses for their recollections and have added them to my own.

Because of its proximity to the Berg claims for access purposes between 1948 and the early 1960's, prior to the building of the road from Houston, three of the fifteen witnesses who testified at the trial were either born there or had lived there and had visited the area of the Lost claims. During the trial, Hugh McGivern was to observe:

"Everyone from Burns Lake is a liar - I should know because I came from there."

Inasmuch as all the witnesses from Burns Lake were representing the Defendants the purpose of the comment was not clear, but it did produce an appropriate response in court.

One of the witnesses for the Plaintiff was George Stewart whose principal testimony concerned his dealings with Fred Sanft in trying to locate claim posts for various Sierra Empire claims with which Mr. Sanft was supposed to be familiar.

On one occasion during cross-examination by Mr. Farris, counsel for Mr. Sanft, George was asked how he felt about Mr. Sanft. He responded quite abruptly "I could have killed him!" Mr. Ferris made some comment about life being so cheap and later offered in his defence of Mr. Sanft a quote from Othello "You can kill me if you will, but you cannot rob me of my good name." I recall being most impressed with the ease with which he was able to remember such a quotation. Twenty-two years later, recognizing that I was not sure of the wording as quoted I visited the West Vancouver library and spent almost an hour scanning Shakespeare's tragedy before discovering the license taken by Mr. Ferris in his recollection. For Iago's comment to Othello in Act 3, Scene 3 was as follows:

"Who steals my purse steals trash; 'tis something, nothing;
'Twas mine, 'tis his, and has been slave to thousands.
But he that filches from me my good name
Robs me of that which not enriches his
And makes me poor indeed."

On another occasion, Larry Candido held a copy of Tustin's map supported on a rail in front of the Jury while illustrating a particular point that he wished to emphasize. Mr. McGivern, lawyer for Mr. Patmore, was not in very good health and he toddled over beside Larry and supported himself on the rail, placing his hand over the portion of the map critical to Larry's point. Larry had to move Mr. McGivern's hand.

Justice Dryer's charge to the Jury commenced at 10:30 a.m. on Thursday June 14, 1973 and ended for the day at 4:30 p.m., to be continued on the following day at 10:00 a.m. and continuing following a break for lunch until 2:12 p.m. The Jury returned shortly before 6:00 p.m. for further instructions and following their dinner they spent another two hours before returning with their verdict at 9:30 p.m.

The charge to the Jury fills 164 pages and can be broken down as follows:

Pages

2 - 9   Respective duties of the Jury and the Trial Judge.

10 - 34   Review of sections of the B.C. Mineral Act as applied to question to be put to the Jury.

34 - 44   Claims against the corporate defendant and Dr. Patmore as applied to questions to be put to the Jury.

44 - 58   Validity of the Berg claims as applied to questions to be put to the Jury.

58 - 97   Justice Dryer's summary of the principal testimony provided by the 15 witnesses.

98 - 104 Further comments by counsel to clarify concerns expressed by Mr. Justice Dryer on the map (Exhibit 251) discussed in the testimony provided by Mr. Steiner.

105 - 118   Point-by-point comments on Question 17 related to the Berg claims.

112 - 121   Point-by-point comments on Question 18 related to the Berg claims.

122 - 131   Summary of Mr. Candido's arguments.

131 - 135   Summary of Mr. Johnson's arguments.

135 - 137   Summary of Mr. McGivern's arguments

137 - 138   Summary of Mr. Candido's reply.

139   Final instruction to Jury which included the following exchange:

THE FOREMAN: My Lord, do we take these?
THE COURT: Oh yes. Take your questions with you.
MR. MCGIVERN: And the exhibits.
THE COURT: You can take any of the exhibits with you.
MR. MCGIVERN: If we can find a wheelbarrow.

As noted previously there were at least 254 Exhibits accepted at the trial. Based on the 164 pages of the Charge to the Jury during a two day period and the 24 court days of the total trial, the testimony and arguments probably filled about 2,000 pages.

Pages 140 - 155 included further comments by Mr. Candido and Mr. Johnson to the Court in the absence of the Jury which was later recalled by Justice Dryer to consider these proceedings as discussed on pages 156 - 164. The court also returned to request further instruction on the curative clauses of the Mineral Act as it applied to the Berg claims which was discussed on pages 161 - 163. The curative clauses were remedial provisions for unintentional errors or omissions in staking and recording procedures which provide for validating claims.

The following questions put to the jury illustrate the amount of technical information that they were expected to have absorbed during the trial, and the degree of judgement they were expected to exercise over the charges of fraud and conspiracy.

Berg Trial Questions

Dispute between plaintiff and corporate defendant

1. What was the position of the number 1 post (or posts) of the Lost 1, 2, 3 and 4 mineral claims when originally staked in 1948?

    (a) Was it the centre point shown as the number I post (posts) by J.M. Parnell in his survey map, Exhibit 132?

    (b) If not, was it the centre point shown by S.R. Leggett on his survey map, Exhibit 131?

    (c) If the answer to both (a) and (b) is 'No', has any other position been established by the evidence to be the position of the number 1 post of the Lost 1, 2, 3 and 4 mineral claims when originally staked in 1948?

If the answer to (c) is 'Yes', what was the position of that post or posts?

2. Regardless of the location of the initial post (or posts) what was the orientation of the location line of the Lost 1 to 4 mineral claims?

    (a) Was it north 07 degrees, 00 minutes, 00 seconds east, as shown by J.M. Parnell on Exhibit 132?

    (b) Was it south 65 degrees, 55 minutes, 05 seconds east as shown by S.R. Leggett on Exhibit 131?

    (c) If the answer to both (a) and (b) is 'No', has any other location line been established by the evidence as being the location line for the Lost 21, 2, 3 and 4 mineral claims as originally staked fin 1948?

    (d) If the answer to (c) is 'Yes', what was it? Give its bearing if possible.

3. If the correct location lines of the Lost 1, 2, 3 and 4 mineral claims are shown by S.R. Leggett on Exhibit 13 1, did the original locator of the Lost 1, 2, 3 and 4 mineral claims in 1948 fail to write at the number 1 post the approximately compass bearing of the number 2 posts of the Lost 1, 2, 3 and 4 mineral claims?

4. If the answer to question 3 is 'Yes', was there a bona fide attempt on the part of the locator to provide the approximate compass bearing?

5. If the answer to question 3 is 'Yes', was the said failure to write the approximate compass bearing of a character calculated to mislead other persons desiring to locate claims in the vicinity.

6. Is the evidence of location or record on the ground of the Lost mineral claims 1, 2, 3 and 4 difficult of ascertainment through the act or default of any person other than the recorded owner thereof or his agent duly authorized?

Claims against corporate defendant and Dr. Patmore

7. Did the defendant William Henry Patmore maliciously provide false statements as to the position of the Lost 1, 2, 3 and 4 mineral claims to S.R. Leggett?

8. If the answer to question 7 is 'Yes', did the making of those false statements cause any damage to the plaintiff?

9. If the answer to question 8 is 'Yes', what amount of damage was so caused to the plaintiff?
10. Did Sierra Empire Mines maliciously provide false statements as to the position of the Lost 1, 2, 3 and 4 mineral claims to S.R. Leggett?
11. If the answer to question 10 is 'Yes', did the making of those false statements cause any damage to the plaintiff?
12. If the answer to question 11 is 'Yes', what amount of damage was so caused to the plaintiff?
13. Did the defendants Sierra and Patmore conspire together to cause the plaintiff damage by maliciously and falsely altering the position of the original location posts and orientation of the location lines of the Lost mineral claims so as to cause the Lost claims to encroach on the plaintiff's claim Berg 18?
14. Deleted.
15. If Sierra and Patmore did so conspire, did that conspiracy cause damage to the plaintiff.?
16. If the answer to question 15 is 'Yes', what amount of damage was so caused to the plaintiff.?
Contents between plaintiff and corporate defendant
17. Subject to whatever rights there may be or have been in the Lost 1, 2, 3 and 4 mineral claims, did the plaintiff Kennco Explorations (Westem) Limited duly locate and record the following mineral claims? And they are listed.

(a) Answer yes or no.

(b) If the answer is no, what was defective? And so on for each of those claims.

18. Is the plaintiff Kennco Explorations (Westem) Limited the present holder of the said Berg mineral claims subject to whatever rights there may be or have been in the Lost 1, 2, 3 and 4 mineral claims?

In charging the Jury, Justice Dryer provided instructions which appear to apply in general to any jury.

In explaining a conspiracy he was to provide an example and cite another reference:

. . . this was used, as a matter of fact, by John Diefenbaker. Dealing with conspiracy he said that if four men meet on the corner, and one man produces a bottle of rye, and another man produces a bottle of scotch and some glasses, and the four of them then have a drink, if you prove *that* (emphasis mine) you have not proven a conspiracy; but if you prove that four men meet on the corner, and one man produces a bottle of gin, the second man produces a bottle of vermouth, the third man produces some glasses, and the fourth man produces a shaker and ice, then you have proven a conspiracy, because you can argue, with some sense, from those acts, that they must have had a previous agreement.

I am now going to read to you what was said by another judge, more formally, but it puts the same idea across:

"A conclusion of that kind--"

That is, a conclusion from overt acts that there is a conspiracy.

"--is not to be arrived at by a light conjecture. It must be plainly established. It may, like other conclusions, be established as a matter of inference from proved facts; but the point is not whether you can draw that particular inference, but whether the facts are such that they cannot fairly admit of any other inference being drawn from them."

Now, here again, if the conspiracy is proven, the plaintiff must prove that it suffered damage from the alleged conspiracy; that is, the onus is on the plaintiff to prove the damage.

What damage has been proven? The only evidence I can recall is that which I mentioned in respect to paragraphs 9 and 12; that is, the time of Mr. Barr and Mr. Winter, and the secretaries and the draftsmen. The costs of the lawsuit are taken care of in another way, they are not a concern of yours here.

A most important part of the Charge to the Jury was its final portion, when Justice Dryer summarized the positions of the three counsels: Mr. Candido, Mr. Johnson and Mr. McGivern. As noted earlier this occupied 16 of the 164 pages and although an excellent summary in assisting the Jury in its deliberation, it is unfortunately too lengthy to quote verbatim. In addition, many references are made to specific exhibits with which the Jury were expected to be familiar but which could not be expected of the reader.

In his discussion Justice Dryer made specific reference to some provisions of the law concerning the surveying of mineral claims as provided in Section 33 of the 1960 Act and in regulation 5.06 of the Surveyor-General's regulations.

Section 33 of the 1960 Act reads as follows:

"When a mineral claim is being surveyed, the surveyor shall be guided by the records of the claim, the sketch plan on the back of the declaration made by the owner when the claim was recorded, posts numbers 1 and 2, and the notice on post number 1, the initial post, and any regulation made by the Surveyor General."

Regulation 5.06 of the Surveyor-General's regulations reads as follows:

"Should one or more of the location posts be obliterated or lost, the evidence used for restoring or re-establishing the said post or posts may be in the form of a statutory declaration by the locator of the claim, or some other person who was present at the time of location, or has definite knowledge of the location. This shall not relieve the surveyor from the obligation of procuring any further evidence corroborative, or otherwise, as can be obtained, and of reaching a conclusion from all the evidence available."

Justice Dryer indicated that, "Mr. Leggett said that he relied on regulation 5.06 of the Surveyor-General's regulations," possibly intimating that Mr. Leggett had not given enough consideration to Section 33 of the 1960 Act.

Justice Dryer made the above references after commenting on Mr. Candido's lengthy discussion concerning the procedures used by Parnell and Leggett in determining the position of the No. I posts of the Lost 1-4 claims. Justice Dryer obviously thought the interaction of his references important for he uncharacteristically repeated both provisions twice, one immediately after the other.

Our friends Dick and Jessie Bowles from Toronto were visiting Vancouver for the final day of the trial and we had arranged to meet them for dinner. When we met we emphasized that it should be a leisurely dinner as we were awaiting a verdict which we hoped would come later that evening. Larry Candido had the telephone number of the restaurant and I was called to the phone at about 10:00 p.m. He said that we had won on most counts. I remember calling from the phone, "We won!" Other patrons must have thought it was a game - and as it turned out, "It ain't over 'til it's over."

The June 16, 1973 issue of The Vancouver Sun carried an eye-catching article under a headline spanning six columns which read, "Kennco wins long legal battle over claims." After providing some background the article stated that, "the jury decided that the

Kennco claims had been properly staked and were valid, something which Sierra Empire had disputed. . . but that no fraud or conspiracy had been committed" as Kennco had charged and that "the jury failed to deliver a verdict that indicated Sierra Empire and Dr. William Patmore of Vancouver had conspired in order to deprive Kennco of its claims."

After the first flush of victory, we were dismayed to learn that the positive outcome was significantly diluted because the judge proceeded to disallow most of the jury's verdicts regarding the validity of Kennco's claims!

Justice Dryer delivered his Reasons for Judgement on January 8, 1974, summarized as follows:

1. The claim against the Defendant Godfrey Frederick Sanft is abandoned without costs.

2. The claims against the Defendant William Henry Patmore are dismissed with costs to the Defendant on the basis they would be payable as though the amount involved exceeded $30,000.

3. The claim against the Defendants Sierra Empire Mines Ltd. and William Henry Patmore for damages in respect of their alleged fraud in altering the positions of the original claim posts and orientation of the location lines of the Lost Claims and providing fraudulent and false information to the surveyor S.R. Leggett, upon which he acted in conducting a survey of the Lost Mineral Claims and preparing the Leggett Plan, is dismissed.

4. The claim against the same parties for damages for allegedly conspiring together to cause the Plaintiff damage and injury by falsely altering the positions of the claim posts and orientation of the location lines of the Lost Claims so as to encroach on the Plaintiff's said mineral claims Berg No's. 14, 15, 18, 64, 66, 76, 78 and 265 is dismissed.

5. The claim by the Plaintiff against the Defendant Sierra Empire Mines for a declaration that the four Lost Claims are void or invalid is dismissed.

6. The court declared that the true boundaries and orientation of the location lines of the Lost Claims are as shown on the Parnell Plan.

7. The court declared that the Plaintiff's claim for a declaration that the fourteen claims known as the Berg Nos. 14, 15, 64, 65, 66, 76, 77, 78, 83, 84, 85, 86, 265, 266 are valid, is dismissed, except with respect to that portion of Berg No. 15 which has been taken to lease.

8. The court declared that the Berg No. 18 claim is valid insofar as it does not cover ground occupied by the Lost No. 4 and Lost No. 2 claims.

9. The court declared the Leggett Plan invalid.

10. The court ordered that the counterclaim of the Defendant Patmore was abandoned at the commencement of the trial and the counterclaim of the Defendant Sierra Empire was abandoned during the trial.

11. The court ordered that the Plaintiff and the Defendant Sierra Empire Mines Ltd. bear their own costs save that the Plaintiff will reimburse Sierra Empire to the extent of one-third of the out-of-pocket expenses paid by the Defendant to the Sheriff on account of the Jury.

The most disappointing result of this decision was Justice Dryer's reversal of the Jury's decision concerning the validity of thirteen of the fifteen Berg claims. The reason for Berg No. 18 being declared valid, was that of all the claims being contested, it was the only one which had been staked with two posts, which Justice Dryer considered necessary. Berg No. 15 was declared valid, within the limitation described, because of the curative provision applied to leased claims, i.e. they could only be declared invalid by evidence of fraudulent procedures.

Backed by the conviction of Larry Candido, Kennco appealed the verdict which was heard by the Honourable Mr. Justice Bull, Mr. Justice McFarlane and Mr. Justice McIntyre in the B.C. Court of Appeal. Counsels for the appellant were Larry M. Candido and David L. Rice, and counsel for the respondent was Arthur W. Johnson.

The Reasons for Judgement of the Honourable Mr. Justice Bull were delivered on March 18, 1975. The preamble describes the background of the case. The remaining 16 pages provide a succinct description of the reasoning applied by the judges in reaching their verdict and is a remarkably lucid treatise to someone like myself who has not been a frequent reader of similar documents. Some examples of the observations and decisions made by Justice Bull in his Reasons for Judgement are provided in the following direct quotations:

The jury's favourable verdict for the appellant was rejected by the Judge because, inter alia, two posts were not used except in the case of Berg 18. The Judge also properly told the jury that the application of the curative provisions of section 39 (supra) was for its determination. It is not known, of course, whether the jury found the staking and recording on the evidence to be in accordance with the statute as interpreted by the Judge, or whether it, following the Judge's charge, applied the curative provisions of section 39. . . .

In my opinion the trial Judge erred in his conclusion that two posts were always needed it in order to properly locate a claim in compliance with section 27 of the Mineral Act. There is no doubt that claims must be staked with two posts (which by definition includes rock cairns as were used in the claims in issue) not more than 1500 feet apart, and the location line of the claim is the direct the straight line between them, which must be marked by some means such as blazes in timbered country or rocks or mounds in open country. I agree that there cannot be a validly located claim without those essential requirements being met. But it is clear to me that the statute provides that in certain circumstances those two locating points and the location line between them can be notionally placed without the actual marking posts ("No. I Post" or "Initial Post" and "No. 2 Post" or "Final Post") being physically placed at all. In such cases witness posts can be used. They are legal but they do not themselves mark the location line of the claim or its ends where the No. I and No. 2 posts must go. Such witness posts can be, and often are, completely outside, and sometimes a distance from, the claim itself. The witness posts are merely directional signs carrying all the requisite information and actual positions of the No. I and No. 2 posts and of the location line between them. . .

I have concluded, after a careful review of the evidence and exhibits, including the many photographs of the country and the locations of the ice fields and slide areas shown on many of the sketches forming part of the affidavits of location filed with the Mining Recorder, that there was sufficient evidence upon which the jury could find that the witness posts used were justified in view of the terrain.

As a result of the Court of Appeal's deliberations, Mr. Justice Bull allowed Kennco's appeal to the extent of declaring the validity and good standing of all twelve Berg claims under appeal under the terms requested, i.e. insofar as they did not cover ground occupied by Lost 1-4 mineral claims as shown on the Parnell Survey. Kennco was entitled to its costs of the appeal and a cross-appeal by Sierra Empire by the Judgement.

The final outcome of the claim dispute was quite bizarre. Although Sierra Empire attempted to have the Court of Appeal Judgement reviewed by the Supreme Court of Canada, there was apparently insufficient basis for it being heard. Sierra Empire subsequently developed financial problems and was unable to settle some of its outstanding accounts. Included was Mr. Leggett's survey firm who arranged for the Crown to seize title to the Lost claims.

On May 27, 1977 the Lost 1 to 4 claims were auctioned off in Prince Rupert, B.C. with little advance publicity. Present to bid at the auction was Mr. Vic Ryback-Hardy on behalf of Mr. Leggett. Coincidentally Mr. Ryback-Hardy was later to become an employee of Kennco. Although Kennco had been alerted to the forthcoming auction, it was decided to inform CanexPlacer who at that time still had the Berg claims under option. Mr. David Howard, Assistant Exploration Manager for Canex-Placer, who had been supervising the Berg drilling program when Rob and I visited the property in 1972, represented Canex-Placer at the auction. When the bidding started Vic waited until the bid had reached $6,000 and then asked Dave Howard, "Do you really want them?" to which Dave responded in the affirmative. Vic then said, "We didn't want the claims to go for $1,000 as we needed at least $5,000 to get our money back". David Howard told me recently that as a successful bid at the auction had to be paid in cash, he was armed with cheques in increments of $1,000 and $2,000. Both he and Vic, who had not met previously, had an enjoyable dinner together and split a bottle of wine in celebration. More than ten years later the Berg property and the Lost claims are held by Placer Dome following take-over of the assets of Canex-Placer, a wholly-owned subsidiary by Placer Developments Ltd., in January 1978 and by amalgamation of Dome Mines Ltd., Campbell Red Lake Mines Ltd. and Placer Developments Ltd. in August 1987. The last major program at the property was by Placer Developments Ltd. in 1984. Work being planned in 1989 consisted of a $180,000 program which was to include environmental base-line surveys, an access road evaluation and a study of possible area drainage problems in a future open pit. The 1989 program would permit the company to vest its 51% interest in the agreement with Kennco. By the end of 1989 actual expenditures in the property by Kennco, Placer Dome and its predecessor companies were about $3 million. Ed Kimura, Exploration Manager, Western Canada for Placer Dome, reported to me in September 1989 that, "the Berg is not viable at present as a stand-alone operation and is on the shelf as potential future reserves. Little work has been done on its gold content and as assessment work is required on certain of the claims by 1990, the requirement will probably be satisfied by additional gold studies".

At the outset of this saga, I mentioned my emotional involvement with the Berg property during a relative short period and the ultimate consequences of the Berg trial. I will list them, not necessarily in order of significance with the exception that the first was the only real tragedy.

A former acquaintance who had been employed as a geologist with Kennecott Copper Corporation in the 1960's became associated with a brokerage firm in Vancouver in the early 1970's. As I understand, he was responsible for arranging considerable financing for Sierra Empire, partly from European sources, being convinced that Sierra Empire had a good chance of winning the lawsuit. Sierra Empire's stock was listed for trading on the Vancouver Stock Exchange in 1971 and its B.C. Charter was cancelled in December 1978. The following is a record of the trading range of the stock:

Year	High	Low
1971	$4.30	$1.00
1972	$2.10	$0.75
1973	$4.00	$0.60
1974	$1.30	$0.47
1975	$0.30	$0.08

1976	$0.15	$0.06
1977	–	–
1978	–	–

Severe stress created by Sierra Empire's loss of the lawsuit reportedly contributed directly to his death in 1974 or 1975.

2. As a direct consequence of the Berg trial, Subsection (4) of Section 85 of the B.C. Mineral Act 1960, which provided for trial by jury, was later repealed. I only learned about this several years ago and wondered at the reasoning applied by the bureaucrat responsible for its repeal if it was truly based on the Berg trial. It seemed to me that if this was the case, the reasoning was clearly wrong in that the Court of Appeal supported the jury's decision.

3. Forty years after its discovery and the expenditure of over $3 million in actual dollars and close to $6 million in current dollars, the deposit is still undeveloped but remains of potential interest. The Berg property is currently (2003) controlled 51% by Placer Dome Inc. There are many such deposits which, during their early exploration, were regarded as almost certain producers by their enthusiastic explorationists - and others.

Those investors in Sierra Empire stock in 1971 and again in 1973 when the stock sold at $4.00 or more per share with 1.2 million shares issued, placed a value of $4.8 million not on the lead-zinc-silver deposits on the Lost claims, but on the gamble that they would be acquiring a significant portion of the Berg deposit.

The 1973-74 Canadian Mines Handbook in its description of the properties held by Sierra Empire Mines Limited, presumably based on information provided by the company, makes no specific reference to the silver prospect covered by the Lost 1-4 mineral claims, which was the basis of the formation of the company in 1968. Its principal property asset was the Berg deposit, described as follows:

(l) Copper-silver-molybdenum pros, 47 cls (4 cls in dispute), 65 mi S of Smithers, Omineca div, BC; 1969-72, surf explor, trenching & d d & road bldg, cost approx $240,000; indicated ore res 15,000,000 tons aver 0.57% copper, 0.05% MoS_2, & 0.1 oz silver per ton.

The correctly placed Lost claims sold for $6,000 in 1977 - that may even have been too much. However, part of the Lost Claims may be mined in the future should a production decision occur at the Berg property – but the product would probably only be considered as waste rock as the mining proceeds outward from a future open pit operation at the Berg property.

African Safari, 1969

By 1968 Kennco had elected to focus on the base metal and related mineral potential of Canadian carbonatite and related alkali-syenite occurrences as one of its generative projects. This led to an active study of layered intrusions: a diversified series of mineral and lithological complexes containing significant mineral deposits on a global scale and also to associated deposits related to mafic igneous rocks. Included are the copper-nickel-platinum mines associated with ultramafic to mafic rocks at the rim of the Sudbury Irruptive in Ontario; the recently discovered Voisey Bay copper-nickel-cobalt deposit in Labrador; the world-famous Bushveld igneous complex in South Africa with its remarkably persistent Merensky Reef hosting major platinum mines, plus chromite, iron and tin mines each associated with different lithologic units within the complex; and the Palabora carbonatite complex with economic concentrations of copper, phosphate, vermiculite associated with a north-east trending structure containing more than a dozen other carbonatite deposits and extending for over 3000 kilometres through southern Africa.

Kennecott's first involvement in Canadian minerals was its interest in the discovery of a significant economic source of titanium dioxide, based on the recognition of ilmenite occurrences in the Havre St. Pierre or Allard Lake region of Quebec in 1941 by the late J.A. Retty, then with the Quebec Department of Mines. Following prospecting, geological exploration, aeromagnetic surveys and diamond drilling, eight separate ilmenite occurrences were discovered. One of these was the Lac Tio deposit, the world's largest ilmenite reserve. Metallurgical expertise for treating the ore was provided by the New Jersey Zinc Company with which Kennecott formed Quebec Iron & Titanium Limited. A plant was constructed at Sorel, Quebec in 1950 following definition of initial reserves of over 100 million tons of ilmenite averaging over 82% combined iron and titanium oxides. Still operating in the 1990s, it has the capacity to produce about one million tons of titanium slag, 600,000 tons of high grade iron and 300,000 tons of steel per year.

In the 1950s, Kennco also had extensively explored the Oka carbonatite complex in southern Quebec, with its associated niobium oxide and rare earth minerals in a project largely supervised by the late Jack Gower.

Although many layered intrusions and related occurrences had been documented and explored in Canada by 1968, many others remained to be discovered and even known occurrences to be re-evaluated.

At the time, Gordon Lister, a senior geologist, was on Kennco's eastern staff and had recently supervised a successful deep drilling program on the Lac Tio deposit. Although the deposit's pre-production reserve had been established by diamond drilling to a depth of about 100 metres, its potential size was based on a gravity survey completed by the late Hal Fleming, Kennco's chief geophysicist. The program completed by Gordon Lister verified Fleming's remarkably accurate projection of the form of the deposit, with its ideal stripping ratio conforming to the reduction of the surface area of the deposit with depth. Following 20 years of production, the remaining reserves at Lac Tio were estimated at 184 million tons grading 35% titanium dioxide and 40% iron!

I had heartily recommended that Gordon should become Kennco's expert on layered intrusion deposits in Canada. In support of his activities, Kennco's president John

Sullivan suggested that Gordon should be given an opportunity to visit the Palabora deposit. Quite unexpectedly, while planning for his visit was in progress in April 1969, Kennecott's exploration arm in the U.S. requisitioned his transfer to the U.S.

Arrangements had been made for me to attend the IX Commonwealth Mining Congress in London, England on May 5-9, followed by a short vacation, so on very short notice, my plans were somewhat modified to provide for a visit to Palabora. It seemed to me like too great an opportunity to travel all that distance to visit only one mine, so I arranged an itinerary which provided for visits to the properties listed belowin Central and South Africa between April 15 - 30. I remain indebted for the support provided by John Sullivan, Lowell Moon and others in approving and partly arranging my itinerary.

I was provided with excellent maps, samples and up-to-date references by the company geologists during my visits. In addition, I took copious notes and photographs wherever I went. Almost every evening I recorded my principal impressions, both in a diary and in a draft of a proposed report on my visit. My final report, written in summary style, emphasizing principal impressions of geology, ore controls, statistical and production data, included about 65 pages of text, 25 maps and sections and 55 photos. Copies of the report were subsequently reproduced for circulation by Kennecott's Exploration Services Group. I have excerpted portions of these reports later in this section

The two-week period in Zambia and South Africa were to provide among the most interesting and vivid memories of my career. I was exposed with little to no introduction to over 60 individuals, almost all involved in the mineral industry, some of whom invited me to their homes, and all of whom made me feel most welcome. I recorded their names every evening in my diary and corresponded with several, including Tom Molyneux, our last contact being only two years ago.

In addition, I was exposed during my short visit to superlative scenery, geology and mineralogy, exploration methods and unfamiliar animal life, such as I could never have envisaged and which were to leave an indelible impression with me. It was so impressive that I vowed that some day both Rene and I would revisit both Central and South Africa - which we were able to do for four weeks in October-November, 1980.

I left Toronto for Africa on the evening of April 12 via BOAC, arriving at Heathrow airport in London the following morning where I was booked into a nearby hotel to await an evening flight to Nairobi. Following departure from London, I dozed for only a couple of hours and at dawn eagerly started to catch views of the relatively barren landscape in Egypt near a tributary of the Nile. Enormous clouds rose to altitudes of 15,000 metres above the desert which supported isolated patches of scrub, gradually giving way to more verdant growth as we flew south toward Nairobi.

Date	Itinerary
April 15 – 17	Zambian copper-cobalt deposits: visits to Rhokana and Nchanga open pit and underground mines, Bancroft underground mine; exploration sites; travel to Nairobi.
April 18	Nairobi - Johannesburg; rental car drive to Santa Barbara Lodge near Pretoria.
April 19 – 20	Meetings with University of Pretoria geological personnel - (arranged for tour of Eastern Bushveld complex with Tom Molyneux, Ph.D. candidate), drive 500 km to guest house, Palabora Mine.

	Itinerary
April 21 – 23	Visit Palabora Mine; drive back to Pretoria via Kruger National Park.
April 24	Visit Premier Diamond Mine at Cullinan and drive on to Carltonsville.
April 25	Visit Western Deep Levels gold mine in Witwatersrand.
April 26 – 28	Visit numerous sites throughout Eastern Bushveld with Tom Molyneux, including Mapochs iron mine, Onverwacht platinum-bearing pipe, chromite mine, Apiesdoringdrai magnesite mines and another iron mine at Kennedy's Vale, celebrated chromitite seams at Dwars River and exposures of the Merensky Reef.
April 29	Visit Rustenburg Platinum mine and mill in the Western Bushveld.

On arrival at Nairobi, following clearance through Kenyan customs, I was again provided with a hotel room at the Stanley Hotel by BOAC to await a connecting flight to Ndola in Zambia. I took the opportunity to visit the nearby Nairobi Game Reserve situated on the outskirts of Nairobi. The two-hour drive through the game park was an unbelievable transition from the crowded city and afforded abundant views of nearby game and a welcome introduction to a part of Africa's wildlife that I had longed to see in its natural surroundings. The cost of entrance fee, car and driver was the equivalent of $15 and a lunch which included a quart of beer and a chicken salad $1.50!

In the evening I flew westerly to Ndola for my visit to mines in the Zambian Copperbelt. Permission to visit Anglo-American corporation's principal mines in the Copperbelt had previously been obtained for me from Dr. Arnold E. Waters Jr., Vice President of Anglo's Canadian subsidiary. The excellent itinerary provided was due to his personal interest and efforts.

At the time of my visit to the Copperbelt in 1969, the Republic of Zambia was less than five years old, having been formed on October 24, 1964, following the declaration of independence of Northern Rhodesia. Virtually every social gathering I was to attend in my brief visit revealed the problems and anxieties faced by the largely European population during a period of transition still in active progress. Since 90-95 percent of the country's gross national product was derived from mining, Zambianization affected the future of Anglo-American Corporation and Rhodesian Selection Trust, who controlled all 19 developed ore bodies in the Zambian Copperbelt, which collectively produced over 600,000 long tons of copper annually, equivalent to 15 percent of the world's annual production. By 1990 this had dropped to 6.5 percent. Many of the mineral industry personnel I was to meet were obviously concerned about their future in Zambia. At the time, all labour up to and including shift bosses were Zambian, which included two mine captains at Nchanga. Geological personnel I met were all whites. At the time, senior Anglo-American geologists working in the Copperbelt earned about $10,000 annually and fringe benefits included good housing at an average cost of about $20 - $30 per month including utilities with about $2 per month for all medical benefits, plus 6 statutory holidays and 55 days paid vacation per year. By contrast, the average for a Zambian labourer was about $2.50 per day. The staple diet for a Zambian labourer and his family consisted of maize and a month's supply cost about $8.00. All Copperbelt personnel's housing was subsidized and the fringe benefits were considered generous.

The Zambian Copperbelt extends northwesterly through Zambia into Zaire (formerly the Congo Republic). The Copperbelt and its mines lie within an area about 450 kilometres long and about 80 kilometres wide.

The copper deposits of the Copperbelt lie within sedimentary rock assemblages named the Mine series of the Katanga system of late Precambrian age. In Zambia the ore deposits lie within the Lower Roan Group of the Mine series, whereas in Zaire they lie within the Upper Roan. [xvi]

By 1969, most of the deposits had been mined by both open-pit and underground workings, e.g. Rhokana, and Nchanga, although Bancroft's two active ore bodies, both lying on the nose of domal structures, were being mined only by underground methods.

Bancroft was noted for its successful operation under extremely wet conditions, being the wettest mine on the Copperbelt and perhaps the world at the time. It had to pump an average daily volume of 65 million gallons of water from the mine workings, the water being derived from stratigraphically controlled acquifers. One of the most notable features of the Copperbelt mines was their copper contents in ore. Following the start of production at the Roan Antelope Mine in 1931, the Zambian mines had produced almost 10 million long tons of copper by 1969 and reserves totalled about 800 million tons averaging 3.2% copper in oxide and sulphide form. By contrast, Zaire reserves were in excess of one billion tons averaging 5.2% copper principally in oxide form.

By comparison, in the late 1960s, Canada had no significant copper production or reserves related to sedimentary copper deposits, the bulk of its production being derived from massive sulfide deposits in the Precambrian Shield. This was to change dramatically with the discovery of the Bethlehem Copper deposit in the Highland Valley area of British Columbia in 1954 which ushered in the porphyry copper and molybdenum boom in Western Canada. The porphyry coppers in British Columbia had an average grade in the late 1960s of about 0.6% copper characterized by the presence of only sulfide copper mineralization with rare surficial oxides and by being mineable, like some of the African deposits, by low-cost open-pit methods.

In Western Canada by the late 1960s, there were several significant operating open-pit copper mines; however, there was no comparison with the size of the open pits I was to visit in Zambia, nor for that matter the more closely related world-class porphyry-copper deposits in the Western U.S.

Following my arrival at the Ndola airport in Zambia on April 14, I was driven to Anglo-American's Jacaranda Lodge in nearby Kitwe with Bill Vance, an Australian, who was also an Anglo-American company guest. We arrived in casual clothing to find ourselves in the middle of a welcoming reception for both of us. The reception was followed by a sumptuous dinner and further discussions with our hosts.

I awoke next morning in my lavish room to a gentle knock on the door followed by the entry of the guest-house servant who handed me a morning paper and informed me that breakfast would be served in my room. Following a bath in a bathtub that would have accommodated a sumo wrestler, I searched for my shoes only to discover that they had been picked up for polishing during the night and were outside the door in the hallway.

I was driven to Rhokana where I was introduced to the mine and geology staff. I was then provided with information on the geology of the mine area by John Coles, Chief Geologist, followed by an underground visit to the Central Shaft area with an opportunity to photograph many of the mine's interesting features, including intense drag folding within the South ore body and the abrupt footwall contact of overlying copper-bearing shale with footwall conglomerate.

Following a lunch with mine staff, I was provided a memorable field trip accompanied by C.J. Borgheutz, a field geologist with Zambian Anglo Mines Services Ltd., to a portion of the property lying west of Rhokana and under exploration. Emphasis was placed on soil geochemical surveys on the extensions of known mine areas and on stream surveys in regional work.

Areas of interest like the one we visited were being followed up by test pitting. As I was to learn, this is a far different procedure than that employed in Canada where pits or trenches were excavated either by hand or with mechanized equipment, generally to shallow depths.

On the Zambian Copperbelt anomalous soil geochemical areas were defined as having values in excess of 200 parts per million (ppm) total copper in arenaceous[17] environments and 500 ppm copper in argillaceous[18] environments. Samples were collected at a depth of 0.4 metres at 30 metre intervals on lines spaced 300 metres apart.

Areas of interest based on soil sampling were then tested in detail by hand excavation of circular pits through the lateritic soils to average depths of 12 metres. The pits had an average diameter of 0.8 metres and were dug by native labourers. Two labourers could average 8 - 10 m per day and the cost of labour averaged about 90 cents per m. Total cost, including sampling and mapping by geologists was about $6.00 per m. By 1969 a total of 40 km of headings had been excavated on the Copperbelt since 1925.

During digging, one of the labourers is progressively lowered by his partner by a winch suspended over the hole. Excavated material is raised to surface. On reaching bedrock at the base of the laterite, with additional sampling a geologist later maps the profile of the pit. I recorded the following observations in my report:

I was given the opportunity of descending into one of the headings and using a geologist's log of the pit, I found geologic data accurate to the nearest foot against a tape suspended down the pit wall. Mapped features include type of residual rock or soil, attitudes of veins and residual bedding.

Native workers slice channel samples down the face at intervals of about 5 feet and these are analyzed principally for total copper and cobalt contents. When anomalous values are obtained, i.e. generally over 1000 ppm copper, cross-cuts may be driven between headings for more detailed information.

I simply could not envisage digging a 'cross-cut' by hand from the base of one of the pits for 30 metres to the base of the next pit on a survey line. Apparently the only guidance is a pole laid across the top of the first pit directed to a picket at the second pit.

The maps produced from headings are extremely detailed and reflect a complete range of lithologies, with structural and facies changes. Based on interpretation, diamond drilling at an average cost of $20.00 per foot is carried out.

Airborne geophysical methods, including magnetic, electro-magnetic and scintillometer surveys, have been used in the past in promising areas prior to completion of geochemical studies.

I was informed that on occasion, labourers and geologists have encountered snakes at the base of pits upon being lowered down. One of the geologists later mentioned one incident when a geologist reaching the base of a pit suddenly realized that a snake had reached his head and was advancing up the rope. When he hollered to the wincher, the

---

[17] arenaceous: sandy

[18] argillaceous: composed of clay

snake had reached the top of the hole and was struck by the wincher. It fell back down the hole hitting the geologist, but not harming him.

Following my field trip, I returned to Jacaranda Lodge suffering increasing stomach cramps. On the advice of the manageress, well acquainted with these types of maladies which were frequently suffered by recently-arrived overseas visitors, she gave me some brandy and advised against eating dinner. I managed to remain sufficiently sociable not to retire until about 11 pm.

I spent an uncomfortable night and still had cramps the next morning. April 16 was to be a very full day. I was driven about 30 miles northwest to Chingola and to the nearby Nchanga mining operation where I met Roland Lawton, Assistant to the General Manager. He introduced me to Pete Freeman, Chief Geologist and a former roommate of Joe Brummer who I worked with at Kennco for several years. Pete was one of the most accommodating individuals I have ever met. He provided me with a most informative description of the local geology and ore controls and then arranged for a tour of the extensive Nchanga and River Lode open pits accompanied by Alistair Walker, a geologist.

The Nchanga deposit, the largest in the Copperbelt, was discovered in 1924, but production did not commence until 1939. At the time of my visit current production totalled about 830,000 tons of ore per month of which 570,000 tons was mined in open pits and the balance from underground. Reserves quoted in 1968 were 25 million tons grading 3.95% copper.

In 1969 the Nchanga open-pit dimensions were about 1500 x 600 metres with a maximum depth of 200 metres and a waste to ore ratio of 14:1 - an unbelievable ratio compared to any North American open pit mine at the time. Overlying laterite was stripped off at a rate of about 3,000 tons per hour by huge bucket wheel excavators. The labour-intensive nature of the operation, like most South African mines, is indicated by the payroll which included 9770 employees for the year ending March 31, 1968.

By contrast in 1994, the number of employees engaged on a full-time equivalent basis in **all** the mining operations in British Columbia including Cominco's smelter at Trail, totalled 9,280. Their **average** annual salary was $55,700.

During lunch at Nchanga Lodge, I met Mr. Peterson, the first Mine Manager at the Roan Antelope mine in 1928, also Sandy Legatt, Senior Geologist, with whom I later toured the Chingola open pit along with Pete Freeman. We returned to Pete's office where we discussed Zambianization and were later interviewed by Michael Kongode, Editor of *The Miner in Zambia*, an Anglo-American publication. I later received a copy dated May 2 containing a photo of Pete, Sandy and myself under the caption "Canadian Sees Mine" and providing quotes about the differences in Canadian and Zambian exploration attributed to myself.

That evening I met Pete's wife Theresa (Terry) from Quebec and later had dinner at Nchanga Lodge with Bill Vance, Pete and Terry. The manageress provided me with the keys to the liquor cabinet after dinner, and entrusted me to lock up and keep the keys until morning. It was another very pleasant evening on the Copperbelt.

Once again I was awakened by the delivery of the morning paper and coffee. I was driven about 6 km north to Bancroft for a visit to this underground mine, famed as described earlier for its ability to carry on operations under severely wet conditions. Following a brief meeting with K. Opperman, Assistant to the General Manager, and Ed Dawson, Mine Superintendent, I was provided with a description of the local geology and toured underground workings from the No.3 Shaft with John Clutten, Assistant Geologist. Underground, I also met Ed Sulkowski, a Canadian, who was the Diamond Drill Foreman.

In the afternoon we drove back to Nchanga Lodge and then left for Ndola via Mufulira, a longer route, necessitated by the closure of the road to Kitwe by a blast of 24 tons of high explosives planned that afternoon. The dynamite was detonated by army and mines experts after a train carrying the explosives was derailed at a siding near Kitwe. The aftermath of the explosion was a crater 80 metres in diameter and 15 metres deep tarnished with twisted metal.

I was dropped off at the airport by my driver for a return flight to Nairobi. I had an interesting time at customs at the Ndola airport. The customs officer casually asked if I had any Zambian currency. I had about 10 Kwacha. He politely asked "May I see it please?" I handed the bills over and watched him writing on a form. When completed, he handed it to me, but retained the 10 Kwacha - certainly not a fortune- but enough to purchase two pair of Bata leather shoes (8 Kwacha) as advertised in *The Miner of Zambia*. The form was a conditional receipt stating that if the recipient returned to Zambia within a year, it could be exchanged for the equivalent amount, or alternatively, if the recipient could find just cause, he/she could write the Zambian Internal Revenue department and try to justify why the monies should be returned. It was a form of extortion, for there was no foreign exchange office in Zambia, nor could Zambian currency be purchased outside of the country. Later I thought "I wonder what the name Kwacha means?" The best guess I came up with is that it rhymes with "gotcha"!

The flight to Nairobi was uneventful. The following morning I flew to Johannesburg, enjoying excellent views of Mt. Kilimanjaro, the Zambesi River and the changing African scene. I was met at the airport following a very quick customs and immigration check by a Hertz agent for a pre-arranged car rental. While at the airport I was also fortunate in contacting Mr. North, Chief Metallurgist for Johannesburg Consolidated Investment Co. Ltd. (J.C.I.), concerning a visit to the Rustenburg Platinum Mines Ltd. at Blaskop in the Western Bushveld. A previous contact had been made through Dr. D.J. Malan, Consulting Engineer for J.C.I. who was absent. The visit was confirmed for April 29, the day before my scheduled departure for Edinburgh.

The travel agent in Toronto had made arrangements for me to spend a couple of nights at the Santa Barbara Lodge situated about 35 km west of Pretoria, the capital of South Africa and about 100 km by road from Johannesburg. I had enough problems as it was reaching Pretoria, having to accommodate to a stick shift after several years of automatic, in addition to driving on the wrong (left) side of the road! Leaving Pretoria, I found that I didn't have adequate information on the route to the lodge, and went far out of my way before locating the dirt road leading to the lodge from the main highway.

After checking in at the lodge, I soon befriended Wilhelm Olivier, the manager and realized that the lodge was really an isolated resort. It was well constructed and ornately designed and furnished. I was to later learn that this was the work of Italian prisoners of war during World War II. The only other occupants were a South African film group producing a film called 'Petticoat Safari'. Wilhelm introduced me to Cliff Butchett, the Assistant Director, and I subsequently met all the cast including Memsie, Elaine, Jimmy, Hal and John as principals. They were all genial South Africans and I was soon invited to have dinner with them. Following dinner, we viewed one of their recently filmed movies "Majumba" which focused on the 1881 war between the Boers and the British. An informal dance followed and later I was invited to attend a swim at the resort - quite casually being informed it was a 'skinny dip'. I thanked them and made my apologies and went back to my lodging to work on my report for a few hours.

The next morning, I drove into Pretoria where I had previously arranged to visit the geological department of the University of Pretoria. I met Gero von Gruenewaldt, a staff member and Ph.D. candidate in Geology. By good fortune he introduced me to Tom

Molyneux, another Ph.D. candidate at the university, who had been completing a study of the magnetite seams in the Eastern Bushveld complex as his thesis. He was anxious to return to the area to collect samples for age datings and he readily agreed to accompany me to places of geologic interest in return for the opportunity to obtain his samples. We agreed to meet on April 26 at the university.

We talked until after noon about geology, when I was invited by Gero to have lunch with him. We drove to his home where I met his wife Judy who had spent her early life in the Zambian Copperbelt and was teaching English while Gero was completing his Ph.D. requirements. In the afternoon Gero and Judy very kindly drove me around Pretoria and I returned to the lodge later that evening.

The next morning, April 20, I checked out of the lodge and drove into Pretoria and then headed north on the highway to my next objective - Palabora, a distance of about 500 km. This was a lonely trip, but the first occasion that I had to get significant impressions of the rural part of that portion of South Africa. I later noted in my diary, somewhat cryptically:

South Africans are terrible drivers - beetle along the road at 80 mph and take terrible chances - considerable glass on the highway. Cops under shade trees. Numerous little rest stops on roads. Country all very open with the exception of 20 mile wide strip in Drakensburg Mountains which shows gradation from west to east from organ pipe and Mickey Mouse cactus (Saguaro) to planted areas of pine, tea plantations and jungle-type vegetation for a few miles. To the north it changes to stunted scrub with ubiquitous termite mounds. Natives are walking or cycling throughout the entire northern section of the highway.

Near Palabora my diary records the rather pathetic cry, "Give me something, master," presumably when I had slowed down enough to hear this plaintive request.

I checked in on arrival at the Palabora Guest House - a single men's unit. My final entry for the day was: "Far cry from Copperbelt but most welcome. Didn't meet a soul."

On April 21, I noted that I woke up at 6:00 a.m., had breakfast at 6:30 a.m. and arrived at the mine at 7:30. Arrangements for my visit to Palabora had been made through Dr. J.A. Sadler, President Rio Tinto Canadian Exploration Ltd.

On arrival I met Dr. J.J. Schoeman, Chief Mine Engineer; Nick Steenkamp, Chief Geologist; Peter Murray, Geologist; Peter Dewaal, Mineralogist; Win van Kralinger, Geologist and H.J. Scheepers, Supt. of Vermiculite Division. All of these individuals provided considerable time and information in describing the Palabora deposit.

I was to devote 15 pages of text in describing the history of early development of these deposits and their later development in my report to Kennecott which included radiocarbon datings in the vicinity of old native workings indicating mining activity as early as 700 A.D.

As early as 1912, vermiculite, copper and phosphate mineralization had been noted at Palabora by Dr. Hans Merensky. A short period of apatite production occurred between 1932 - 1934, but was terminated by high operating costs. Re-investigation of the deposits by Dr. Merensky in 1946 led to the discovery of higher grade phosphate associated with copper, magnetite and radioactive uranothorite. Vermiculite production commenced in 1946.

Another 20 years were to pass before the full potential of Palabora was recognized and adequate infrastructure was installed to place Palabora into production, which provided for a tax-free period of about three years for payback of about $75 million.

As indicated earlier, Palabora is a remarkable carbonatite deposit with layering controls on mineralization in a pipe-like structure and with remarkable vertical continuity to a depth of over 1000 metres tested by diamond drilling. In describing the significant features of the deposit, I ventured that, "depending on future economics, deep drilling has, in effect, established reserves which probably total in excess of one billion tons averaging about 0.7% copper to a depth of 3,500 feet."

I spent parts of April 21 and 22 examining the pit area at Palabora and the local geology in the company of Bill (Win) van Kralinger. I also examined part of the vermiculite pit and was indebted to Mr. Scheepers who gave me some superb specimens of vermiculite and phlogopite which I still retain and admire on occasion.

In the afternoon I left Palabora and drove through the northern part of the Kruger Park to Oliphants Rest Camp on my return to Pretoria. It was an interesting drive as one dropped in elevation from the veldt into a more humid, almost jungle environment. I stopped for the night at Oliphants in a small kopi containing three beds within its circular area with screened netting over windows protected by louvered shutters. The rental for the night was only 3 Rand ($4.50); however, it was compensated somewhat by the 95º Fahrenheit temperature which I recorded as I was writing at 8:00 p.m.

During the night I was awakened by the opened window shutters banging against the kopi in strong winds. Without lights, as the power plant was turned off at 11:00 p.m., I walked quickly to close the shutters, stepping on a sharp object. The next morning I checked for that sharp object intending to remove it for future visitors. Underneath my bunk I spotted a small black creature, which on closer inspection revealed itself as a black scorpion. I encouraged it to enter one of my small plastic sampling bags and stuffed it into a briefcase intending to release it later.

In my diary for April 23 I noted how impressed I was with Kruger Park: "Saw most common animals except things like leopard, cheetah, and rhino." In addition, I had driven, "almost continuously from 6:20 a.m. to 8:00 p.m.," to reach Witbank where I checked in to the Boulevard Hotel, recommended by dinner companions the previous evening. It was great advice and I enjoyed giant HongKong prawns in the dining room.

Tea was served in my room at 6:00 a.m. By now I had learned that it was a customary service prior to breakfast. However, as I had arranged to arrive at the Premier mine at Cullinan by 7:30 a.m., I had to leave before breakfast.

Although Kennco and Kennecott had no interest in diamond exploration or mining at the time, I could not pass up the opportunity to visit a diamond mine, particularly such a famous one and my visit had been pre-arranged through Dr. Arnold Waters Jr. of Anglo-American mentioned earlier, consulting engineers to the DeBeers group.

En route to Cullinan I remembered that I had a small traveling companion with me who would probably appreciate some fresh air. I stopped on the side of the road beside a cultivated field far from any residences and carefully grasped the plastic bag containing the black scorpion which I had retrieved from under my bunk at Oliphants, some 600 km to the north in a far different environment. I dropped it on the ground where it immediately jabbed at the blades of grass with its small tail. It didn't even bother to wave good-bye!

I arrived at the Premier mine at about 7:30 a.m. where I met Mr. E.J.B. Sewell, General Manager, Mr. Nathan, his assistant and Peter van Blommestein, Public Relations Officer. The Premier mine had no geologic staff and relied on a geological consultant for periodic guidance, about three months every three years.

Peter accompanied me on an underground tour and a visit to the plant which included an inspection of part of the recent day's diamond production, in addition to some typical coloured diamonds from the mine.

Rather than attempt to summarize my impressions at the time of my visit, I have elected to excerpt portions of my report to Kennecott Copper:

The discovery of diamonds in South Africa ushered in the period of great mineral discoveries on which the prosperity and industrial development of the Republic has largely been based.

The development of the Premier mine was preceded by a colourful prospecting period during which Percival White Tracey, a prospector who had worked claims on the original De Beers mine in Kimberley, is said to have traced diamondiferous stream gravels in 1902 to the vicinity of a farm owned by Joachim Prinsloo, a man who had a profound distrust of fortune-seekers. Eventually Tracey, a building contractor named Cullinan and a third man named Jerome formed a syndicate to purchase Prinsloo's farm for a price of $100,000 - truly a small fortune at that time.

Open cast mining began in April 1903, the diamond-bearing kimberlite being raised by rope haulage. Washing pans were used to separate the diamonds from the sludge. In 1904 a pulsator plant was added to speed up operations. In January 1905, the famed 3,024-3/4 carat Cullinan diamond, the world's largest, was discovered.

The mine was closed down for about one year at the outbreak of World War I and again from 1932 to 1944 because of the depressed state of the diamond industry.

In 1944, a decision to resume mining was made following appointment of Anglo American Corporation consulting engineers to the De Beers group. Ten months were required to pump out about 900 million gallons of water that had accumulated in the open pit following removal of 107 million tons of rock. The pit at this time was at a depth of 610 feet. A total of 133,796,542 loads[1] had been washed for a recovery of 19,199,795 carats, an overall average of about 22 carats per 100 loads.

By the end of 1949, three shafts had been sunk and a heavy media separation pilot plant erected to determine whether this method would improve on jigging as a means of diamond recovery. Tests proved satisfactory and the first unit was brought into operation in February 1950, followed at intervals of one month, by three additional units.

In 1960, following sampling over an 11-year period, a decision was made to build a plant for the retreatment of over 100 million tons of material dumped as waste in earlier mining operations.

Tests had indicated the presence of large quantities of small diamonds, mostly of industrial quality, in the tailings dumps. A new plant, capable of handling 500,000 tons a month went on-stream in October 1962 and increased the mine's total treatment capacity to 900,000 tons per month.

Although there are several kimberlite pipes in the Pretoria district, the Premier is the only known diamond-bearing pipe. It is also unusual in that kimberlite pipes of much the same character occur throughout Africa which are reportedly of late Cretaceous age, whereas the Premier pipe is almost certainly Precambrian in age, i.e. indicated age of 1290 million years.

The kimberlite pipes, which are volcanic craters of explosive origin, number many hundreds in the Republic of South Africa and occur in clusters, probably related to rift structures. Only a very small number of them contain diamonds in sufficient quantity to be workable and of these, five belong to the DeBeers group.

---

[1] A load is equal to 16 cubic feet of broken rock and weighs about 1600 lbs. or 0.80 of a short ton; 100 loads, the usual unit or reference is thus equal to 80 short tons.

Kimberlite, which fills the pipes, is a porphyritic ultrabasic lava with an unusual diversity of accessory minerals. Among the more important constituents is olivine, partly or wholly altered to serpentine, with the pyroxenes enstatite and diopside. Phlogopite mica and garnets are commonly present and ilmenite is frequently an important constituent.

Inclusions or xenoliths comprise fragments of wall rocks traversed and occasionally of rocks which formerly lay far above the present land surface which have since been removed by erosion. At Premier the pipe is divided into two sections by the inclusion of a large block of Waterberg quartzite. This 'float' or 'floating reef', as it is called is presumed to have fallen into the pipe when it was molten.

A different class of xenoliths are the 'Cognate' variety which are considered to have been formed at great depth in the earth's crust.

At surface kimberlite weathers to a soft yellowish or reddish speckled clay known as 'yellow-ground'; in depth yellow-ground changes progressively to a harder, less weathered rock known as 'blue ground'. At greater depths the blue ground becomes harder and does not disintegrate readily on weathering.

Through 1970, five great pipe areas had been discovered in South Africa and included the Kimberley group of pipes, the first of which was discovered in 1870, Premier in 1902, Williamson in the 1940s, Finsch, which commenced production in 1966 and in the late 1960s a cluster of five or more pipes discovered at Lesotho, Central Botswana by Anglo American, the largest of which reportedly measured 3000 x 4000 feet.

The Premier pipe covers 79 acres, measures 2900 x 900 feet in area, elongated northerly and intrudes felsitic rock. Yellow-ground occurred to a depth of 35 feet and blue-ground below. The blue-ground is composed of a variety of kimberlite tuffs representing at least six different eruptions, arranged in a crudely concentric pattern about the vertically dipping pipe.

Apart from kimberlite with abundant mica, basaltic kimberlite deficient in phlogopite occurs. Fragments incorporated in tuff and breccia include large boulders of Waterberg sandstone, smaller fragments of shale, quartzite, felsite granite and limestone, representing various wall-rocks intruded by the pipe.

The various kimberlite tuffs are recognized primarily by colour differences. At the time of my visit there was no published or recordable information on variations in grade in the different tuffs. Apparently certain sections of the pipe are noted for particular types of the highly prized coloured diamonds, for which Premier mine is famous. Especially characteristic of the Premier mine are steel-blue diamonds of great brilliance and stones with a peculiar blue-green opalescence, known as Premier 'oilies'.

At a depth of about 1700 feet below surface, the pipe is cut by a flat-lying tabular body of gabbro about 200 feet thick. The marginal portions of the kimberlite ore baked for several feet. Steeply dipping earlier dykes of carbonatite up to 20 feet long and 8 feet wide cut the kimberlite.

Diamonds produced at Premier are about 85% industrial and 15% gem quality. Part of one day's production which I viewed included seven gem stones averaging about 20 carats each and several hundred stones of smaller size. At the time, the seven large stones were valued at $100,000.

About 600 tons of concentrates are produced from 21,000 tons per day (tpd), mined and consist of ilmenite (88%), pyrite (10%), minor magnetite and diamonds.

The Premier mine was mined by open cast methods to a depth of 910 feet. Underground development through 1969 had proceeded to a depth of about 1700 feet.

During mining a 45-foot slot, 200 feet in depth is cut through the centre of the long axis of the pipe. The walls of the slot are stepped in benches at vertical intervals of 50 feet to form a slope of 65 degrees. Lines of 68-foot square cones are excavated throughout the length of the bottom of the slot. Kimberlite blasted from the benches collects in the cones and gravitates to the level below through grizzlies to the haulageway. Automatic hoisting using 12-ton skips is used to convey ore and waste to surface.

At the time of my visit production included 80% kimberlite and 20% waste composed of Waterberg quartzite. As indicated, the mining rate was about 21,000 tpd. An additional 6000 - 9000 tpd of former waste present in old dumps was also being treated.

About 4,000 individuals were employed at the Premier mine in 1969. About 700 were Europeans and 3,300 Africans. Underground employees totalled about 2,000 Africans and 330 Europeans. Underground operations were based on three shifts per day and a six-day week.

Diamond recovery is a moderately simple operation. Ore is crushed to minus 1-1/4 inches and is separated into plus and minus 7 mesh fractions. The plus 7 mesh product is treated in a sink and float plant and the float (waste) is recrushed to minus 3/8 inch and retreated. The minus 7 mesh ore is treated separately in jigs, the jig tailings being discarded with the minus 3/8 inch float tailings. The sink product and the jig concentrates pass over vibrating grease tables to which only the diamonds, which are wettable, adhere. The diamonds are recovered following scraping of the greased surfaces and boiling of the scrapings.

A summary of operations for the year ending December 31, 1967 follows:

Loads treated	9,785,769
Carats recovered	2,376,879
Carats per 100 loads	24.29
Cost per load treated	R 0.66 (U.S. $0.93)
Cost per carat recovered	R 2.73 (U.S. $3.85)

My visit to the Premier mine was to be recalled in later years on more than one occasion. Between 1986 - 1988 I represented CSA Management Limited as Exploration Manager of the First Exploration Fund Limited Partnership 1986, 1987 and 1988. These were three flow-through funds which raised $104,925,000 for exploration purposes for the junior mining sector in Canada. (The 'Juniors' are recognized as having no producing mines.)

One of the companies funded was Dia Met Minerals Ltd. whose president was C.E. "Chuck" Fipke, a geologist/geochemist who had taken an early interest in exploring for diamonds in British Columbia and later in the Northwest Territories. Chuck was to gain an international reputation and well-deserved acclaim when he discovered the Lac de Gras diamondiferous pipes in the Northwest Territories in 1990. Chuck achieved this over several years by patiently tracing indicator minerals related to diamond-bearing pipes, in glacial till for 200 miles easterly from the Mackenzie River area. The discovery resulted in the greatest staking rush in Canada's history and occurred at a time when mineral exploration was in one of its doldrums.

In 1990 I was surprised to learn that Kennecott had once again elected to become active in Canadian exploration. Although it was still interested in base metal exploration it was to become a key player in the diamond boom. By 1995 it had spent over $10 million in diamond exploration and its 1995 budget included $23 million in a major underground development program. In Canada its reactivation was a pleasant surprise and led to my having an opportunity to represent its interests in the Stikine River area and the Galore Creek deposit through my participation as a representative of the B.C. and Yukon

Chamber of Mines on the Lower Stikine Management Advisory Committee. I was also to enjoy an association with Mark Fields, a Kennecott Canada Inc. geologist who shared an interest in distance running. In addition, Eric Finlayson, Western District Manager in Vancouver, very generously allowed me to access Kennco's old files on several occasions in order to provide accurate data on some of my recollections. Included was a history of Kennecott's exploration in Canada from 1946 - 1974 prepared on behalf of Kennecott Canada Inc. in 1994.

Following my visit to the Premier mine I drove on to Carltonville for a visit to the nearby Western Deep Levels gold mine on the following day. En route I made note of the following road signs in Afrikans/English:

    Witwatersrand / Ridge of the White Waters

    Hou Links / Keep left

    Skool Pad / School Road

    Och! / Robot

Much of my drive was in a total downpour and eventually my car sputtered and I managed to pull it off the road before it came to a complete stop. I had the hood up trying to determine the problem when a vehicle drew up behind me. A South African army officer came alongside, volunteering to help. He rapidly detected a crack in the distributor cap and wiped off the water, cheerfully sending me on my way.

Once again I was indebted to Dr. Waters Jr. who obtained permission for me to visit the world-class Western Deep Levels Limited operation, also controlled by Anglo American.

Having worked at McIntyre Gold Mine in Timmins, Ontario and Wright Hargreaves gold mine in Kirkland Lake, Ontario as a student and having visited many gold producers in Canada and the U.S., I looked forward to visiting this major gold producer in one of the largest and most unusual gold reservoirs in the world - the Witwatersrand system and its basin, extending for 180 miles southwest of Johannesburg. Again, I provide the following excerpt from my report to Kennco:

My visit consisted of an underground tour of one of the stopes[19] on the Carbon Leader Reef at a depth of about 10,000 feet and an examination of several workings exposing portions of the hanging-wall and foot-wall. Several surface sink-holes caused by the removal of water in the course of mining operations were also examined.

My visit coincided with that of Jock Smith, geologist in charge of Mount Isa's exploration work in south Australia and Jacques Bauer, a French geologist employed by Aquitain in oil exploration in Mozambique. We were introduced to A. Tennant, General Manager, and our tour was conducted by Franz Wager, Chief Geologist, Alan Turner and Robin Messenger, Geologists at the mine. Mr. Van Lionden, Underground Manager, Carbon Leader Reef section, provided me with some pertinent facts on production statistics.

Gold was discovered near Johannesburg in 1886 within conglomerate beds in a well stratified succession of sediments, subsequently to become known as the Witwatersrand system. The beds occur in a basin of Late Precambrian rocks comprising a lower division about 15,000 feet thick, composed of interbedded shales and quartzites, the former predominating, and an upper division, about 9000 feet thick, which is dominantly quartzite.

Within the Transvaal and Orange Free State goldfields, outcrops of Witwatersrand rocks cover relatively small areas, aggregating not more than 1500 square miles. These outcrops occur at the rim of a major synclinal basin oriented northeasterly, with

---

[19] Stope: undergound mine working from which ore is extracted

diameter of about 180 x 90 miles. Most of the basin is buried under younger formations and exploratory drilling indicates that the known ore-bearing horizons underlie an area of at least 10,000 square miles.

The gold-bearing horizons are highly indurated conglomerates, or groups of conglomerates of great persistence in the upper division of the Witwatersrand system. Uranium is an important by-product at many of the mines.

The ore-bearing horizons or reefs are payable in different districts and local variations between economic and sub-marginal deposits also occur. The ore-bearing conglomerates occur throughout the upper division in various goldfields. Most of the mining through 1969 during my visit had taken place in shallower portions of the Witwatersrand basin. Large barren gaps also existed particularly in the southeastern portion of the basin and around the rim of a positive area in the central part of the basin. Portions with complex faulting had also proven uneconomic to mine.

In 1961 working revenue for mines in the Witwatersrand basin was $793.7 million, working costs averaged $7.06 per ton and working profits $4.43 per ton milled. The average gold content was 6.142 dwt per ton, i.e. 0.307 oz per ton. In 1961, 73.1 million tons were mined containing 22.5 million ounces of gold compared to 30.0 million ounces recovered in 1967 at an average cost of $21.00 per ounce.

Uranium has been an important by-product of the Witwatersrand goldfields, production having commenced during the period 1954 to 1956.

The average uranium content for the goldfields through 1969 was about 0.5 lbs per ton U_3O_8, although the range at the different goldfields is great, i.e. about 0.4 lbs per ton at East Rand, 1.1 lbs per ton at West Rand, 0.35 lbs per ton at Far West Rand, 0.65 lbs per ton at Klerksdorp, and 0.35 lbs per ton at Orange Free State.

Witwatersrand production in 1960 was 12.4 million pounds U_3O_8 from 24.1 million tons treated.

Ore reserves declared at various mines were normally about a two-year feed for the mill. Effects of rising costs on reserves were considered particularly significant, i.e. increase in 18 years from $3.70 to $8.40 per ton for an average of 4.0% per annum by 1967, but with no real prospect held in doubling of the price of gold in an equivalent period in the future! (By 1980 it had reached an unbelievable US$900/oz.) As indicated in the preceding section, uranium is an important by-product of the Witwatersrand goldfields. Depending on future demand and declining gold reserves in many of the goldfields, the value of uranium production was considered to significantly prolong the life of many of the gold mines.

By 1943 exploration and negotiation on land rights covering the present Western Deep Levels holdings had begun. The lease occupies about 16 square miles and contains a reef strike length of seven miles, one of the longest on the Witwatersrand.

Of ten deep boreholes in the early exploration stage, nine were completed, and of these, five intersected both the Ventersdorp Contact Reef (V.C.R.) and the richer Carbon Leader Reef (C.L.R.) which lies about 2500 feet vertically below. Two boreholes intersected the V.C.R. only and two the C.L.R. only.

Estimates indicated that 104 million tons of payable ore would be available for milling on the C.L.R. horizon and at least 11 million tons of payable ore on the V.C.R. horizon. In 1956 a production decision was reached based on the establishment of a mine with an eventual capacity of 200,000 tons a month.

Initial planning was based on the expectation that mining would ultimately reach a final depth of about 13,000 feet, the deepest gold mining operation in the world. Actual production commenced in March 1962 from the V.C.R. horizon.

A significant feature of the operation is that Western Deep Levels became the first deep level mine to receive direct government incentives through a special tax concession. The total redeemable capital expenditure incurred by the company in excess of the aggregate profits earned was to be treated as an assessed loss for taxation purposes, and no taxation or state's share of profits would be payable by the company until this assessed loss has been absorbed by profits. The assessed loss for taxation purposes at December 31, 1968 was calculated at R 58 million[2] , down some R 15 million from 1967. The company paid out R 13.75 million in dividends in 1968.

A typical geological column in the Western Deep Levels area contains two ore horizons, the V.C.R. and the C.L.R. which have distinctly different characteristics.

The V.C.R. which lies between depths of 5,000 - 11,610 feet within the property consists of a bed of quartz pebble conglomerate which tends to be intermittent but is up to 6 feet thick. It contains coarse pebbles, i.e. about 2-3 inch diameter, and evenly distributed gold values. Average stoping width is reportedly 45 inches. Average grade is about 0.45 oz gold per ton.

The C.L.R. is a finer grained quartz pebble conglomerate, i.e. pebbles less than one inch in diameter and frequently less than 1/2 inch in diameter in a horizon which is generally 1-4 inches in thickness and which is very persistent. It is highly mineralized with interstitial pyrite and pyrrhotite. Its most distinctive characteristic is its richness in carbon which generally occurs as a seam averaging about 1/4 inch in thickness at the base of the reef. This carbon-rich footwall portion is generally highly mineralized in gold. The C.L.R. horizon lies between depths of 7,500 - 13,000 feet on the property. Average stoping width is about 39 inches. Average grade is about 13 dwt per ton in gold, i.e. 0.65 oz gold per ton.

The reefs dip at about 21° south in the property area and are separated from each other by a thick assemblage of continental quartzites, grits and conglomerates comprising the Main Bird and Kimberley-Elsburg series. In the mine workings excellent cross-beds, mud-cracks and ripple marks occur. The V.C.R. lies immediately on the old Witwatersrand land surface on which the overlying Ventersdorp lavas were extruded. The gold content of the V.C.R. was derived from the weathering of older gold-bearing conglomerates of the Witwatersrand system.

Most of the uranium content, present as detrital grains of uraninite occurs within the C.L.R. horizon and averages about 0.45 lbs per ton. A uranium plant was being built at Western Deep Levels and was scheduled to start treating up-graded ore from the C.L.R. early in 1970 at a rate of 70,000 tons per month.

At the time of my visit Western Deep Levels contained two twin-shaft systems, No. 2 and No. 3 which lay 8,500 feet apart in the northern part of the property. The 20-foot diameter upcast ventilation shafts that form one half of the twin systems were each about 10,000 feet deep. The two 26-foot diameter main shafts were sunk to about 6,500 feet for V.C.R. exploitation and from that depth two sub-vertical shafts with their own hoisting facilities were sunk to 10,000 feet to exploit C.L.R. ore.

The total hoisting capacities of the two shaft systems, including ore and waste, was about 360,000 tons per month, based on hoisting for 20 hours per day. Personnel transportation was a problem, as 5,500 men had to be moved by each shaft system daily. This was performed by double-deck cages which conveyed 80 men per trip, 3,000 feet per minute, and by a smaller service hoist hauling a double-deck cage with a carrying capacity of 20 persons per trip at 2,000 feet per minute.

---

[2] Rand (R) was equivalent to $1.50 (U.S.)

Mining was by long-wall method and as indicated previously stoping widths varied from about 39-45 inches. Because of rock pressures, rock-burst incidence was frequent and considerable support was required to maintain essential mine openings. In practice, in the working area, no pillars were left and closure of footwall and hanging wall occurred in about 6 months.

Water control was an essential factor because of the high water content of overlying formations. Each shaft system had an initial pumping capacity of 7 million gallons per day, and this could be increased to 30 million gallons per day, if required. In 1968 water pumped to surface averaged 2 million gallons per day. Removal of water from near-surface water-bearing dolomites during the course of mining operations caused soil collapse producing sink-holes with diameters of several hundred feet and up to 30 feet deep. A gravity survey indicated trends of critical areas and led to demolition of some 50 homes in the nearby residential area.

In preparation for deep mining, data on temperatures of deep boreholes were gathered and together with data from nearby shallower operations were used to predict probable thermal gradients. The data indicated temperatures of 93°F at 6,000 feet increasing to 122° at 11,000 feet.

In addition to conventional ventilation systems, underground cooling plants were used to improve working conditions. During our visit to a stope area, in which temperatures were reportedly above 90°F, working conditions seemed difficult because of the combination of high temperature, high humidity and low stope height. Apparently Africans and whites were climatized to the working conditions over a period of several weeks during which time little to no work output was expected."

I recall emerging with others from our cage at a depth of about 10,000 feet. As we walked away from the station, we passed by several benches filled with newly hired workers, all of whom were Africans. They were sitting there talking and apparently quite content. This was part of the climatization process.

As we walked down the drift, I noted that interspersed along the drift were huge I-beams which had been formed into oval shapes to provide support to the drift from the pressure of the overlying rock settling against the floor in nearby worked-out stopes. I was shown an area beside the drift which I had thought was unmined. The geologist pointed out a fracture near the floor of the drift which paralleled the drift and was in reality the location of the collapsed roof against the floor of an adjacent work area.

I had to squat entering the relatively flat-lying stope which was inclined at about 15 degrees. Interspersed within the working area at intervals of about 25 feet were criss-crossed piles of large timbers, about 5 feet in length extending from floor to roof as support during mining. As mining proceeds at the working face, the roof was gradually collapsing about 150 - 200 feet from the face.

I took my camera out of my pack to take some photos of the miners operating drills while reclining, as working space from floor to roof was only about 30 - 35 inches. All I could see through the lens was fog. Because of the high humidity, I should have had the lens cap off on leaving surface in the cage.

It was most impressive to have the opportunity to actually see the thin band of 'ore' that constituted the Carbon Leader Reef, realizing that it could be mined economically even though the associated dilution rate with waste rock removed, was over 100 times the actual width of the CLR. Under normal underground mining conditions, where veins of ore are involved, the allowable dilution rate is commonly 10 - 25 percent of the vein width.

I was unable to retrieve a sample of the ore in the stope; however on reaching surface, one of the geologists very kindly provided me with a fine specimen which I still treasure. It includes a portion of the CLR about 2 inches square by 1/4 inch thick with about 2 inches of each of the quartz pebble conglomerate footwall and hanging wall. With the naked eye, one can see minute rods of gold intermeshed within the carbon extending vertically from footwall to hanging wall. Based on the average gold content of about 0.65 oz of gold per ton and the average stoping width of about 39 inches, the gold content in the carbon seam is about <u>75 oz</u> per ton!

At December 31, 1968 there were a total of 15,209 personnel employed at Western Deep Levels Limited. 13,731 were Africans, with 11,761 or 85% working underground, compared with 1478 'Europeans' of whom 1,010 or 68% worked underground in supervisory roles.

At the time of my visit, the mine operation was among the most profitable in the Witwatersrand, as reflected in the summary of 1968 operations:

Tons of ore hoisted	3,667,000
Tons of waste sorted	454,000
Percentage of waste sorted	11.85
Tons milled	3,384,000
Yield - oz / ton	0.554
Ounces of gold produced	1,875,435
Residue value - oz / ton	0.125
Working revenue	R47,710,000
Working costs	R21,854,000
Working profit	R25,856,000
Working revenue per ton milled	R14.10
Working costs per ton milled	R 6.46
Working profit per ton milled	R 7.64
Cost per ounce	R11.65

The proportion of tonnage milled from the C.L.R. horizon averaged 78 percent.

In 1969 gold was valued at $35.00 U.S per ounce. Following our re-visit to South Africa in 1980, several mines on the Witwatersrand were re-milling their tailings, which at 0.125 oz gold per ton was economic with the gold price reaching record levels of over $800 per oz.

As at December 31, 1968, ore reserves at Western Deep Levels were estimated as follows:

	Tons	Stope Width (inches)	Gold Value oz / ton	U_3O_8 Value lb / ton
V.C.R.	2,292,000	53.8	0.433	-
C.L.R.	2,623,000	40.0	0.845	0.45
Total*	4,915,000	45.43	0.653	-

* C.L.R. U_3O_8 reserves kept separate.

It's interesting to note that at the rate of mining in 1968, the reserves would be mined in about 15 months! Yet in 1991 the estimated production was 1,435,000 ounces of gold and reserves were 2.9 million ounces compared with 3.2 million ounces in 1968 and the production of about 35 million ounces in the 23-year interval!

After leaving the mine, I drove Jacques Bauer to his hotel in Johannesburg and continued on to Pretoria.

The next morning I met Tom Molyneux, as arranged, at about 8:00 a.m. at the University of Pretoria. Tom was to accompany me on a tour of the Eastern Bushveld to examine various sites providing exposures of parts of the Bushveld Intrusive Complex (BIC).

The BIC is a unique layered intrusion, dwarfing all other known complexes. Occupying an area of 67,340 square km, the next largest established economic complexes were the great Dike, Zimbabwe (3,265 sq. km) and the Sudbury Complex, Canada (1,342 sq. km). Again, I have elected to provide direct excerpts from my report:

The BIC is composed of a variety of igneous rocks noted for their layered nature balance and for the persistence of lithologic and mineralogic units over vast areas. It occupies an easterly oriented, roughly oval area 260 miles long and about 100 miles wide with its south central edge centered on Pretoria.

The most commonly accepted origin for the complex is one of successive eruptions of the various components into the Pretoria Series and consolidation as a sequence of layered intrusions.

The BIC reaches a maximum thickness of about 30,000 feet. Its basin-like configuration is somewhat reminiscent of the setting of the Sudbury Complex, particularly in the western section which is characterized by a moderately flat present-day internal surface rimmed by inward dipping Pretoria quartzites.

The age of the Bushveld granite was determined several years ago as 1950 ± 50 million years. Previously this younger part of the Complex had been regarded as Late Precambrian in age, and differentiated from the same magma that produces the BIC phases. It is now known to be intrusive into the layered mafic rocks - however, it retains its economic importance by hosting the tin deposits of the region.

After leaving Pretoria, we drove east to Witbank, about 100 km from Pretoria and then took a side road through the eastern part of the Bushveld mostly within the Hi-veld (high field). It was beautiful open country which provided many stops of interest at well-exposed outcrops of the Bushveld Complex in the Steelpoort River region. The Steelport River drains northeasterly through the Complex, its waters eventually reaching the Indian Ocean about 550 km northeast of Pretoria in Mozambique via the Oliphants and Limpopo rivers.

At one of our stops we encountered two young African girls attired in brightly coloured native costumes, including leggings. In spite of Tom's polite request, they would not permit us to take photographs - and left us as they went on their way giggling merrily. At one particularly scenic spot we had a picnic lunch of beer and sandwiches. Our final geologic stop was completed in fading light at 6:00 pm. We drove on to Fortrecher Bat, a small community which claimed to contain a resort hotel. Although we had no trouble locating the facility, it was obviously very run down. We were fortunate to locate the eccessively hospitable owner at a nearby party. Having anticipated a somewhat memorable dinner, we purchased cold canned corn, some sardines and crackers from the owner and following our let-down were in bed by 9:00 pm.

Our breakfast the next morning included porridge with thick cream, the porridge derived from corn. Again we visited many sites of geologic interest within the Eastern Bushveld Complex which included several former or active small mining operations. Among them were the former Onverwacht platinum pipe west of Burgersfort; Apiesdoring magnesite mine near Burgersfort; Winterveld chrome mine; the magnificent exposures of the Dwars River chromitite seams and the Kennedy's Vale iron and vanadium mine.

During our travels on April 27 we were invited to have tea at the home of a mechanical engineer at Kennedy's Vale, an extremely welcome respite as we were both quite dehydrated!

That evening we sought accommodation at another small community near the Mapochs Iron mine, about 30 km northwest of Dullstroom on the Steelpoort River. We stayed that evening at a very fine hotel, whose owner, Col Minaar, was most hospitable. Tom had friends in the area from his previous trips within the region and we were invited to the home of Phil and Joyce Liford. Again, this afforded an opportunity for me to learn much about the concerns of the white population in South Africa as much of the evening's discussion focused on apartheid and the uncertainty of the immediate and near-term future for whites. During the evening we were invited to enjoy some of Phil's home brewed beer, which most unfortunately disagreed with my unacclimatized stomach and I spent a somewhat wakeful night with about three hours of sleep.

The next morning we set off at about 7:30 a.m. to visit the Mapochs iron mine which lies within the Upper Zone of more acidic rocks, including diorite and granodiorite in addition to gabbro, in the Main Plutonic Phase of the BIC. It includes 21 magnetite seams based on studies by Tom Molyneux, which lie within 1000 m from the base. The uppermost seam is the thickest (10 m) averaging about 50% iron, 18% titanium oxide and 0.2% vanadium oxide. The magnetite seams increase in vanadium oxide content and decrease in titanium oxide content toward the base as determined by Tom.

At Mapochs Mine the main magnetite seam is about 8 feet thick and averages 55% Fe, 14-16% TiO_2, 1.6% V_2O_5. At the base of the seam a chrome layer 2 inches thick is underlain by disseminated sulfides in a 2-inch thick section. The sulphide layer averages about 0.5% copper-nickel (45% chalcopyrite, 50% pyrrhotite, 5% pentlandite) and contains up to 2 dwt per ton platinum and a trace of sphalerite.

The Main Zone, directly underlying the Upper Zone, includes the lowermost three magnetite seams of the BIC. The Upper Zone composed principally of norite, gabbro and anorthosite has a maximum thickness of about 5,000 m, its base being placed at the top of the Merensky Reef.

The Critical Zone is defined as the layered succession between the Merensky Reef and the Main Chromitite seam. It generally includes at the base the pyroxenite which overlies the Main Chromitite seam and which merges into feldspathic pyroxenite, norite, anorthosite and pyroxenite, which are often alternating and banded rocks.

The maximum thickness of the Critical Zone in the Eastern Transvaal is about 3500 feet, but in faulted and folded sections in the Steelpoort River area it may only be 500 feet thick. Southeast of Rustenburg in the Western Bushveld the zone varies from 110-2200 feet thickness.

At Mapochs mine we met J.J. Pieterse, Manager, who accompanied us on a tour of the Mapochs mine operation.

We then drove westerly to visit the new Loskop dam site where Tom wanted to obtain fresh samples of Bushveld granite for age dating purposes. We continued on to Pretoria where I dropped Tom off at his flat before checking in at a motel.

Although I had not received any confirmation during my absence from Mr. North, Chief Metallurgist of Johannesburg Consolidated Investment Co. Ltd. (JCI) concerning my pre-arranged visit to the Rustenburg platinum mine operation northwest of Pretoria, I left the following morning at about 7:30 a.m.

This was a memorable morning drive through rolling country to the rim of the BIC and through a pass descending from the rim into the basin. Just below the height of land, descending from the higher hills, I applied my brakes as a very large baboon suddenly loped across the road glaring at me with upraised arms. I have never forgotten this encounter with a probable split from a long distant ancestor; the shoe might have been on the other foot!

At Rustenburg, I explained my lack of contact with Mr. North, but I found sufficient preparations had been made. J.C.I. is a financial and investment company which owned a controlling and management interest in Rustenburg Platinum Mines Limited.

I found that a considerable amount of information on Rustenburg's operations was classified, some of it I felt unreasonably, for instance the number of employees.

My visit included an underground tour of Rustenburg's B-5 operation in the company of Dave de Wet, Captain, and a tour of the existing plant and a new addition scheduled for completion in August, 1969. The latter included an electric blast furnace and provisions for expanding the facility in the future. The plant visits were in the company of Mr. Bob Wallace, Asst. Metallurgist, J.C.I. Platinum Complex, who was largely responsible for the design of all plants. We had a typically enormous lunch at the nearby Safari Inn.

In the mid 1920's a farmer-prospector discovered platinum in a dry river bed in the Lydenburg area of the Eastern Transvaal. Samples sent to the late Dr. Hans Merensky led to his taking an active interest in discovering the source of the platinum and the eventual recognition of its association with a member of the Bushveld Complex which was named the Merensky Reef.

A platinum boom commenced and when depressed prices followed the 1929 crash, only two companies survived. These were Potgietersruist Platinums Limited and Waterval (Rustenburg) Platinum Mining Company which merged in 1932 to form Rustenburg Platinum Mines Limited. In 1947, Union Platinum was formed to mine the platinum reef lying about 60 miles north of Rustenburg. In 1949 it was amalgamated with the Rustenburg company.

In 1969, Union Corporation Limited had been developing the Impala mines, lying 10 miles northwest of Rustenburg on the Merensky Reef and this caused concern to J.C.I. as it represented its first serious competition in the Bushveld Complex. Output from Union Corporation's mine was planned at about 100,000 ounces platinum per year.

J.C.I. and others carried out exploration of the Merensky Reef over a period of years and many prospect pits and trenches attest to the vigor of their exploration efforts. It did not appear, however, that much drilling had been done on the down-dip projection of the Merensky Reef prior to my visit in 1969.

The Merensky Reef varies from about 1 to 4 feet in thickness and consists of pyroxene gabbro. It forms one of the important layers of the Bushveld Complex and lies at the top of the Critical Zone, in the central part of the Complex at the top of the belt of chromitite seams, as described earlier.

The platinoid minerals within the Merensky Reef include sperrylite ($PtAs_2$), cooperite (PtS), laurite (RuS_2) and stibiopalladinite ($Pd_3 Sb$). Much of the platinum mineralization is in solid solution in nikeliferous pyrite, pyrrhotite and pentlandite. Metals recovered from the ore are platinum, palladium, rhodium, ruthenium, iridium, osmium, gold, copper, nickel.

The geology is generally so regular, with an average dip of 9° in the Rustenburg Section and about 20° in the Union Section, that only one geologist and two helpers were employed by the mine. They worked only on problem areas. Faulting appears to be minor and offsets are only a few feet.

The Rustenburg Section is currently developed over a length of about 10 miles and to a maximum depth of 1500 feet.

The Rustenburg mine is developed by a series of surface winzes collared at the surface trace of the ore horizon and dipping in the plane of the ore. The ore is generally oxidized to a vertical depth of about 100 feet, but ground is solid below this depth and requires little to no support during mining.

Surface winzes are spaced about 500 feet apart and ore is mined in panels which advance along sub-drifts spaced at about 40-foot intervals and extending out for 250 feet from the winze. All ore is removed, no pillars being left, and to date closure of walls in the 28-30 inch stopes does not exceed 6 inches. Apparently minor rock bursts have been encountered at depth (1500 feet).

Every fifth winze is opened as an incline shaft dipping at 14° in order to get box capacity below the ore horizon at depth. The second and fourth levels at incline distances of 1000 feet and 2000 feet from surface are opened as haulage ways. Muck is conveyed by small handcars to the winzes where it is slushed to the haulages and loaded on locomotive-hauled cars and transferred through grizzlies into side-discharge hoppers on the incline shafts. Two hoppers, carrying a total of about 12 tons per trip, haul the ore to surface Hoisting capacity by this method is reportedly 800 - 1200 tons per day per shaft.

Rustenburg is reportedly administered by three separate mining developments with separate managers, each consisting of about 6 sections, there being 17 in all, with a mining captain having overall responsibility for each section.

Under the mining captain there are 6-7 shift bosses with 3-4 Europeans and 100-200 natives per shift boss. The total mining complement at Rustenburg is probably about 2400 on this basis. Apparently mining is carried out by one shift operating six days per week, while milling is continuous with three shifts operating seven days per week. By inference on milling and smelting statistics, the current daily mining rate is probably about 10,000 tons, including the Union Section.

The feed to the mill probably averages about 0.2 oz. platinoid metals per ton of ore. The ore is initially crushed to minus 6" plus 1-1/4" and passed along two conveyor belts which move at 50 feet per minute through a hand-picking team of about 50 Africans. The 'waste', consisting of light coloured footwall anorthositic gabbro is removed, the average culling amounting to about 20 percent. Mr. Wallace feels that the practice is not satisfactory because of potential losses of ore in the immediate footwall. After further crushing the ore passes through ball mills where it is reduced to 65% minus 200 mesh.

Gravity concentration using corduroy tables recovers free platinum which is sealed in cans and flown to Johnson, Matthey & Co. Ltd. In Great Britain for refining.

Flotation recovers a copper-nickel concentrate which is pelletized without additives (except water) and fed into blast furnaces with coke and limestone. No silica is required. Converter matte is recovered which contains an average of 48% Ni, 28% Cu, 22% S and 50 oz. per ton platinoids. Matte production is reportedly 60-70 tpd.

These latter data suggest an error in platinum reporting as total platinum production at Rustenburg is currently reported to be about 700,000 ounces per year, and most of the recovery reportedly occurs in the gravity concentrate.

Construction is proceeding on a separate plant due to go on stream in August 1969. The plant is several miles from the current plant and will feature a large electric blast furnace with provision for expansion of smelting facilities. Production capacity of new mines should reach about 850,000 ounces per year by the end of 1969. Production rate will have been increased 250 percent over a period of seven years. Capital expenditure in the five-year period 1967-71 will probably be in the order of $50 million.

Significant features of the Rustenburg Platinum Mines operation based on my limited visit and as reported to Kennecott were:

1. The topographic setting of both the western portion of the Bushveld Complex and the Sudbury Basin are similar, with flat internal portions surrounded by inward dipping rims of basal rocks - actually norites at Sudbury, and quartzites of the Pretoria series (Archean basement rocks) in the Bushveld. The similarity of metal production from both areas is well known, although only annual production of platinoid metals are comparable, i.e. about 400,000 ounces at Sudbury and 700,000 ounces at Rustenburg.

2. The Merensky Reef horizon with which platinum is associated at Rustenburg, like most layers of the Bushveld Complex, is remarkably continuous. The consistency of the geology at two exposures visited, i.e. that at Rustenburg and another in the Steelpoort River area of the Eastern Transvaal, 225 miles to the east, is truly remarkable.

3. Similarities between both the Merensky Reef and portions of the Witwatersrand Goldfields were noted. The most obvious similarity is the association of most of the gold values in the part of the Western Deep Levels mine with the Carbon Leader Reef. - a carbonaceous pebble conglomerate horizon several inches thick and the occurrence of most of the platinoid mineralization at Rustenburg with a chrome seam 1/4 to 1 inch thick. The mining column, about 28-44 inches in thickness is similar at both localities, as is the ache in one's legs after visiting stopes.

4. Because of its persistence, vast extent, and important platinum content, the Merensky Reef is the world's greatest known source of platinum. Although prospected largely by J.C.I., most exploration has tested only the near-surface portions of the reef. The deepest workings at Rustenburg are currently about 1500 feet deep and most workings are much shallower. Considering the depth at which mining is currently being practiced at the Witwatersrand, i.e. 10,000 feet, the future outlook for increased platinum production from the Bushveld Complex rocks is most attractive.

5. J.C.I. reportedly hold grants covering most of the known exposures of the Merensky Reef, but I am not sure of the down-dip extent of these grants.

6. The association of base-metal mineralization, i.e. copper-nickel with the platinoid-bearing horizon may be significant as a prospecting guide. It is doubtful whether other layered intrusions elsewhere in the world have been investigated as intensively as the Bushveld Complex for platinoids.

Bob Wallace generously provided me with considerable verbal information on the operation both during our tour and later while having lunch at the nearby Safari Inn. I never could get used to the enormous servings provided at South African hotels and restaurants. It tended to reflect in the average size of the patrons.

After a two-hour drive I checked in at the Kevin Grove Hotel in the Saxonwold district of Johannesburg. After dinner, I worked on a draft of my report on the Rustenburg mine visit until midnight.

The next day I shipped about 30 pounds of mineral specimens to Canada at a cost of $75.00 (R53). On returning my rented car to Hertz, the total rental for 2000 miles in 12 days with a 20 percent corporate discount allowed was about $200 or 10 cents per mile, without gas.

Labrador: Kiglapait, Harp Lake and Voisey Bay

My memorable trip to South Africa in 1969 was principally designed to provide on-site familiarization with the Palabora mine, a world-class carbonatite[20] intrusion and also the Bushveld layered intrusion and its diverse assemblage of economic mineral deposits. The visit led to a renewed focus by Kennco on layered intrusions in Canada which could warrant further exploration.

By coincidence, my first assignment following graduation from the University of Toronto in 1950 was a temporary job with Kennco in the Thunder Bay district of northwestern Ontario. I supervised a small field party searching for copper-nickel occurrences associated with the basal portions of the Crystal Lake gabbro[21] and affiliated sill-like intrusions. Planning for the project had not been particularly thorough, to say the least, as when we assembled in Port Arthur, it was delayed for two days while I learned to drive, as none of the party had a driver's licence! Very fortunately, my examiner overlooked a few minor deficiencies in my brief experience and gave me a passing grade.

We spent considerable time climbing up the steep slopes to the base of the relatively flat intrusions with their near-vertical cliff-like exposures in order to examine, map and sample mineral occurrences. In several areas it was particularly arduous work as a thick growth of alder almost obliterated the base of the intrusions.

The most encouraging mineral occurrences were in the Pigeon River area. In a report which I prepared for Kennco following the completion of the program, I recommended diamond drilling of three separate occurrences. One of these was to become the Great Lakes Nickel Corporation Ltd. deposit.[xvii]

In 1952 Kennecott Copper Corp. were to become interested in the copper-nickel potential of the nearby Duluth Gabbro in northern Minnesota, in a similar geologic setting which is probably genetically related to the intrusions in the Thunder Bay district.

Research in late 1969 and early 1970 by Kennco indicated three layered anorthositic[22] gabbro complexes lying to the east of the Ungava trough in Labrador. All three bodies, named the Kiglapait, Harp Lake and Michikamau intrusions, appeared to be of similar age, composition and probable lopolithic[23] form. Little information was available on their exploration potential; however, it was decided that the Kiglapait intrusion, the smallest and most northerly, with a surface area of about 500 square kilometres, should receive an initial reconnaissance survey.

---

[20] Carbonatite: high-carbonate intrusive rock
[21] Gabbro: granular igneous rock, composed of calcic plagioclase, a ferromagnesian mineral, and accessory minerals
[22] anorthositic: composed mostly of anorthosite, a plutonic rock high in plagioclase
[23] lopolithic: large lenticular intrusion

Kiglapait lies within a moderately rugged sub-arctic region on the Labrador coast about 50 kilometres north of Nain, the nearest community. Lying north of the tree-line, it affords a high degree of exposure with only a thin veneer of soil cover and shrubs.

During our planning for the project, we learned that a small shack situated on the coast on the east side of the Kiglapait intrusion was available for our use. The location was known as Village Bay, which belied its name, as the area was completely uninhabited. We also decided to rely on Nain as a supply point for trans-shipping aviation gas which would be required for a G-2 float-equipped helicopter that we decided to charter from Universal Helicopters Limited of Gander.

Our field party consisted of myself, David McAuslan, senior geologist; Ray Goldie, temporary geologist and Tom Webster, field scout. We flew from Toronto to Goose Bay on August 1, 1970, arriving at Nain via an Eastern Provincial Otter on August 2.

It was my first visit to Labrador. Our flight in the Otter revealed a significant transition from the timbered, relatively subdued landscape in the south to the increasingly rugged and barren coastal region as we approached Nain.

The village of Nain is the northernmost municipality on the Labrador Coast, about 370 kilometres north of Goose Bay. The population in 1970 was about 800, but by 1995 it had reached 1000, mostly of Inuit and English derivation. The settlement lies within a bay at the east end of a peninsula, well sheltered by islands from the Labrador Sea.

Nain has been an important settlement for the fishing industry for over 200 years. The oldest building is a church, established in 1771 by the Moravian Missionaries, which lies near the foot of a long pier extending into the bay. The dominantly white buildings with their red, blue and occasional green roofs all lie within several hundred metres of the shore of the bay, the settlement's appearance being somewhat softened by the surrounding spruce trees at low elevations.

During our brief stay at Nain, we walked through the settlement. I climbed a 200 m hill above the community to photograph the surroundings and on the way back through the settlement I also photographed half-a-dozen sled dogs tethered to posts waiting for the summer to end. I was to see similar clusters of sled dogs throughout the Arctic during visits to various exploration sites - they always seemed quite at ease and were never threatening.

We also contacted Henry Webb, a long-time resident at Nain, who had agreed to transport our barrels of aviation gas up to Village Bay in his dory. We flew on to Village Bay and settled in at the shack after unloading our supplies from the Otter and setting up two sleeping tents.

Our surroundings were quite spectacular, with a small iceberg floating within the bay and the central part of the Kiglapait intrusion rising to an elevation of about 1000 metres immediately to the west. The camp site itself was a broad grass-covered area, partly surrounded by boulders amidst which grew a profusion of wild flowers. Broad patches of snow remained along raised beaches in one section and at scattered intervals throughout the mountains to the west.

The next morning Henry Webb arrived with his load of fuel, which was unloaded directly to the shore. As our helicopter had not yet arrived, I decided somewhat extemporarily, but perhaps influenced by the many occasions when I had waited for helicopters, to make use of Henry and his dory on his return journey to Nain.

He agreed to take me back along the shoreline to a point about 10 kilometres south of our camp near the southern extremity of the Kiglapait Complex. We stayed close to shore as I wanted to ensure that if I had to follow it back to Village Bay, I would have no problem returning to camp.

I recall that Henry was somewhat apprehensive about dropping me off. "Are you sure you'll be all right?" he asked. I assured him, appreciating his concern, but assumed that it was a natural reaction in a seafaring man considering the prospect of a landlubber contending with the vagaries of terra infirma. I set off equipped with my usual field gear, which included aerial photos and clear film overlays on which to plot observations, or sample locations as required.

After a most absorbing day, I was within one kilometre of our camp, looking forward to arriving in time for dinner. As I walked on with a steep mountain rising above me, I was eventually forced to traverse immediately above the shoreline until I reached a cliff face dropping into the sea. Checking the aerial photo, and recalling my observation of the shoreline from the dory, I realized that there was a 50 - 100 metre wide cliff spur separating me from what appeared to be accessible ground beyond.

I couldn't believe the reality of the situation, considering the opportunity that I had to observe the shoreline while in the dory. Finally it dawned on me that I was looking at high tide; it had been low tide when we passed earlier. I considered my options. I could wait for low tide when it would be dark. I had no flashlight. I could walk back south to a point where I could ascend the steep mountain above me, and make a descent to camp, without any guidance as to the distances and time involved. I finally gambled on the data in the aerial photo augmented by my visual recollections from the dory.

After placing my camera with aerial photos, compass and a few minor items in my pack, I lowered myself into the sea, which was fortunately quite tranquil and set off holding my packsack above my head and paddling vigorously with my other arm. On reaching what I believed to be the half-way point, it was so cold I thought, "My God, I'm not going to make it." The alternative must have seemed worse, for I continued on and by good fortune reached a pull-out point.

Again I was most fortunate, for it was still sunny. I stripped quickly and was still wringing out clothing when I heard a sudden "whissh!" to seaward. I turned around and unbelievably saw the spray from the blowhole of a whale, not more than 70 metres offshore. After donning my clothing and gear again, I continued on without incident to camp, arriving to a concerned group. The whale was to remain at Village Bay for much of the next day, along with a sizeable iceberg, before continuing on its journey.

Following the arrival of Fred Wagner with his helicopter, we completed our reconnaissance surveys within six days. Interestingly enough, although we covered the entire Kiglapait Complex in reconnaissance fashion, the only wildlife we saw were several caribou well within the complex. By 1990, the western part of Labrador contained more than 500,000 caribou, reportedly the largest contained caribou population in the world.

I considered our relatively brief reconnaissance of the Kiglapait Complex a very successful venture. Although no significant sulfide mineralization had been previously reported, we discovered copper-nickel concentrations partly concentrated by 'layering' in two portions of the intrusion about 1600 and 4800 metres above its base. The best grades encountered from surface sampling consisted of 0.1-0.2% copper and 0.04% nickel across thicknesses of 8 - 15 metres with no associated precious metal values. This was clearly uneconomic and our attention was immediately focused on the Harp Lake Intrusion, situated some 150 km to the south.

Prior to leaving Village Bay, Fred Wagner, an ardent fisher, had several opportunities to cast his line into the sea from a rocky prominence near camp. The product of this effort and skill was that Fred caught several Arctic char, one of which I photographed, weighing nine pounds. Fred kindly donated one of these to me, caught the day before we left for Goose Bay and our home in Toronto. I was able to have the fish frozen overnight in Goose

Bay and with it securely wrapped the next morning, I rolled it up in my sleeping bag to provide additional insulation.

Arriving in Toronto, Tom Webster and I claimed our incredible amount of baggage comprising camping equipment and our sleeping bags. We managed to stuff all of this equipment into the rear of Tom's station wagon and drove off on the 401 to drop me off at my home in Etobicoke. Along Highway 401, both Tom and I became aware of the highly noxious odour of decaying fish. We immediately attributed it to the Arctic char wrapped in my sleeping bag, but before we committed ourselves to some possibly disastrous action, the truck which we had been following and which was evidently the real cause of this malevolent odour turned off, enabling us to save my char for a particularly memorable dinner!

Having got my feet wet, so to speak, in the Kiglapait venture, we next focused on the Harp Lake layered intrusion, some 150 kilometres south of Nain and broadly centred on the southwest end of Harp Lake, extending about 50 kilometres northeasterly and drained via a river 50 kilometres easterly into Ugtoktok Bay, south of Hopedale on the Labrador Coast.

In spite of our limited success at Kiglapait, we were enthusiastic about the economic potential of the Harp Lake Intrusive Complex, covering an area of about 10,000 square kilometres, or two times that of the Duluth Gabbro. David McAuslan was assigned overall responsibility for planning a helicopter-supported reconnaissance of one month's duration to determine whether more detailed investigation was warranted.[xviii]

Kennco concluded that there was a definite potential at the Harp Lake Intrusion for economic concentrations of nickel, copper and platinum metals, "such as are known to occur in the Bushveld Complex, the Duluth Gabbro and others."

In late August and September, 1971, Kennco carried out a one-month program including stream sediment coverage over 45 percent of the area of the intrusion, geological mapping, prospecting and sampling showings detected initially as rusty (colour anomalies) from aerial reconnaissance. Of 37 colour anomalies detected, 13 contained mineralization of further interest with 15 remaining to be investigated.[xix]

The results were considered encouraging and the Newfoundland Government was approached with the intent of obtaining a concession over the Harp Lake Intrusion.

On June 3, 1972 the Newfoundland Government approved Kennco's application over a reserve area of 12,160 square kilometres for a four-year period with total exploration expenditures of $250,000 required, and a minimum of $50,000 in each of the first and second years. During the 4-year exploration period, Kennco could select an area or areas not exceeding 78 square kilometres in aggregate for which the government would issue 5-year development licences, through payments of annual rentals of $1.30 per hectare. The company was then entitled to apply for a mining lease, to be issued provided all obligations had been met.

Kennco's budget proposal for the 1972 Harp Lake program was not approved. I arranged to have a joint-venture proposal offered to 17 companies. Chevron Oil and Selco Mining Corp. Ltd., each agreed to provide $50,000 to earn a 25% interest in the concession within a two-year period ending December 31,1973. Their offers were accepted.

Helicopter-supported programs, each of about two months' duration, were completed in 1972 and 1973 from a base camp established at Dave's Pond in the east-central part of the Intrusion. An average of 10 men were employed.

The program included additional airborne reconnaissance, prospecting, detailed geological mapping in the vicinity of showings, induced polarization and ground magnetometer surveys which successfully traced mineralized trends and trenching and sampling. Aerial reconnaissance in the three-year period 1971-3 detected a total of 70 sulfide gossans in addition to over 100 "colour" anomalies, mostly caused by rusty weathering barren rocks.

I visited the projects in 1972 and 1973, being accompanied in August 1972 by Jim Finley and Gerry Pollock of Chevron Resources and Selco Mining Corp., our joint-venture partners. Other familiar individuals on the Harp Lake project included Tom Webster, Al McOnie, a geologist from New Zealand and Rick Sebastian. In 1972, Rick completed a B.Sc. thesis on some aspects of silicate and sulphide mineralogy of samples collected from most of the principal showings.

I found the region somewhat similar to Kiglapait; however, it was not nearly as austere, with far more vegetation and small ponds scattered throughout the area in addition to several large lakes, including Harp Lake with its impressive cliffs rising steeply from the shore.

On one occasion, Dave McAuslan and I visited our two prospectors who were working in a partly timbered area. We sat with them chatting around a fire at their camp, when Dave commented on their wood supply, noting the size of the blocks they had cut from a spruce they had felled and the extremely close spacing of the annual growth rings. Trees being quite scarce, we were surprised at the 30 centimetre (one foot) diameter width as the area was near the northern limit of the tree line. We cut several slabs off one of the blocks and I later sanded and polished one of them to accentuate the growth rings. The tree was over 400 years old! The prospectors had also made several picket lines for reference purposes from small trees with a diameter of about 2.5 cm (one inch). We usually considered these as 'young' trees; however, a cut and polished fragment I retained was about 150 years old, the tree rings being so closely spaced, they could only be differentiated with a magnifying glass! We cautioned the two prospectors about over-cutting for firewood.

Dave and I were most impressed with the conviviality of three Newfoundlanders who had been hired as helpers on the project. They shared a tent and could be heard in the late evening laughing over various events they had experienced. Nothing ever seemed to distress or perturb them and they always looked on the bright side, taking every opportunity to effectively ridicule the mainlanders' smug views of their off-shore cousins. One of our most ardent recollections of this fine group, was the closing statement supposedly directed to a mainlander by a Newfoundlander who said (phonetically) "Yes boy'oh, we'se suave (rhymes with wave) and de-bonner (rhymes with honour)." They were quieter that evening as they heard our supportive laughter.

I left Kennco to join DuPont Canada Inc. and form DuPont of Canada Explorations Limited in January 1974. I never learned at what stage the joint-venture agreement on the Harp Lake project was terminated; however, it was probably in 1974-5.

In 1993, some 20 or more years later, two struggling prospectors - Albert Chislett and Chris Verbiski from St. John's, Newfoundland were grubstaked by Diamond Fields Resources to look for diamonds, as well as gold and base metals in Eastern Labrador. The general area lay about midway between the Kiglapait and Harp Lake layered intrusions and about 50 km south of the community of Nain.

In September, Chislett and Verbiski were prospecting on the northwest side of Voisey Bay about 35 km southwest of Nain, when they were attracted by a prominent gossan. Within 15 minutes of reaching the gossan, the two knew they had made a potentially significant discovery.

Chislett recalled the discovery nine months later: "We broke some fresh rock and could see the stringers of chalcopyrite shooting through the gabbro and knew we had about 1-2% copper over an area 500 metres long and between 40 and 80 metres wide."

Following their discovery the pair researched government literature and maps while planning for staking a land package, based on favourable geology, within the region. They learned that between 1984-7, the Geological Survey of the Newfoundland Department of Natural Resources had carried out a mapping program in a 60-kilometre wide area westerly from Voisey Bay to the Quebec border - a distance of about 130 kilometres. The survey indicated the presence of a previously unrecognized layered and massive mafic intrusion named the "*Reid Brook*" intrusion.

Verbiski and Chislett learned that their "discovery" had been previously mapped by government geologist Bruce Ryan as a pyritic gossan, presumably as no base metal mineralization had been observed because of the leached and weathered nature of the gossan.

The land package staked was to total about 1800 square kilometres and included the Kiglapait and Newark Island intrusions. The staking activity which commenced in January 1995, soon attracted attention based on early reports of the discovery and was to extend south and west to include the Harp Lake intrusion.

By mid-May 1995, diamond drilling of the discovery had included 81 vertical holes, covering a strike length of 600 metres of the westerly oriented deposit, mostly on a 50-metre spacing. Although still open on strike it was estimated that the portion of the deposit covered by drilling probably contained 20-25 million tonnes grading 3.71% nickel, 2.14% copper and 0.15% cobalt in an area 500 metres long and about 275 metres wide.

Although much more exploration will be required, the deposit appears to be one of the richest nickel-copper occurrences discovered to date in Canada.

Considering the recent depressed economic conditions in Labrador and Newfoundland with severe reductions of allowable catch imposed by government on the fishing industry, the discovery could not have come at a better time. It should certainly have a beneficial impact on the residents of Nain.

In 1995, an issue of The Northern Miner made a passing reference to Kennco's early exploration in Labrador in the Kiglapait and Harp Lake areas. I telephoned Eric Finlayson, Western Manager for Kennecott Canada Inc. in Vancouver and asked whether he was familiar with the post-1973 work in Labrador. As he had only joined the company a few years earlier he was unable to comment. However, he mentioned that he had heard several unflattering comments about Kennco "missing Voisey Bay". I decided to write a letter to the editor of The Northern Miner to describe Kennco's work completed 20 - 25 years before the Voisey Bay discovery. It was published in the September 25, 1995 issue.

As I had mentioned Dave McAuslan's significant contributions in the exploration work which were culled from my article I contacted Dave at his home in Janetville, Ontario and forwarded a copy of both The Northern Miner article and my submitted draft. Since leaving Kennco in 1974 and Shell Oil's Canadian exploration subsidiary in 1982 Dave has been an instructor at nearby Sir Sandford Fleming College's Earth Sciences Department.

I have excerpted the following paragraphs from Dave's reply to my letter in November 12, 1995, reflecting a few of his own recollections:

"Ed Butler from St. John's was the cook at Dave's Pond in 1973. I hired him on the recommendation of Canada Manpower in St. John's He arrived with two very heavy suitcases which I lugged to the cooktent. He was an alcoholic and the suitcases were full of Screech. He did his drinking at first but was eventually found out by Réjean Dallaire, the helicopter mechanic. Réjean was in camp during the day and found Ed incapable to make lunch. Réjean somehow sobered Ed up whereupon Ed asked him for help. Ed told Réjean that he was terrified of the bush and if the booze was put somewhere out in the woods he would not touch it. Réjean hid the screech a mere 50 feet from the tent and Ed never went near it for the rest of the summer. He was thereafter an acceptable cook with an amazing collection of jokes and an incredibly crooked cribbage player (things like "15-2, 15-4 and a pair make 8"... but so quick you would never notice). Ed once in a while after dinner would say "party tonight Dave?" and if it seemed appropriate a bottle would be fetched.

"Ron Emslie, the Geological Survey of Canada geologist who was mapping the Harp Lake intrusion while Kennco was active in the area named several lakes. The names appear on the map accompanying his summary report which was published in 1980. Dave's Pond is one of these names. As I recollect Ron did whatever was necessary to make these names official (I guess they would be in the Gazetteer of Place Names). Dave's Pond was named by him in part to acknowledge the helicopter support made available to him.

"The Voisey Bay discovery more than any other important find in the last 10-15 years has made me reflect about the teaching I have been doing in contrast with mineral exploration as a career choice. A strong interest in Cu-Ni mineralization from INCO roots and the Labrador experience I guess. I have been doing watercolours for a few years now and as a way to address some feelings triggered by the Voisey discovery I have painted a series of pictures relating to my time at Kiglapait and Harp Lake. One of them (the photo is of it) is a landscape of Port Manvers Run from Village Bay; I guess your cliff is just around that first point."

Dupont of Canada Exploration

DuPont of Canada Limited (DOC) came into existence on July 1, 1954 following segregation of its manufacturing operations and products from Canadian Industries Limited (CIL). However its origin can be traced back to the founding of E.I. du Pont de Nemours and Company Inc. in 1802 when Eleuthère Irénée du Pont de Nemours, a native of France founded a factory near Wilmington, Delaware to make gunpowder.

Throughout the 1900s the DuPont name has universally been recognized as a prominent leader in the chemical industry. Only an aged sector of the public would be aware of its emergence from an explosive manufacturing background, through diversification and innovative research in the 1920s and earlier to its present position.

For DOC, the 1960s was a decade of record growth; net income rose from $6.6 million to $16.2 million between 1960 and 1969 and sales from $99.9 million to $228.5 million. However, the company's performance in 1970 was seriously affected by a variety of factors, including a general reduction in economic activity, increases in competitive imports, reduced tariffs and an increased value to the Canadian dollar, all of which combined to focus mostly on the fibres business.

To remediate, DOC imposed several austerity measures and in addition sought to develop new and more profitable approaches to the marketplace. DOC had shown its ability to diversify since 1954; however diversification into explosives, finishes and polyethylene were all within fields in which expertise had been acquired in the pre-segregation period. In addition the company was well supported by the expertise and extensive research organization within its U.S. parent.

The explosives divisions of both DOC and DuPont U.S. afforded the opportunity for strong contacts with the mining sector. Art Baker, a senior executive in DOC's Explosives group, and President of DOC's exploration subsidiary and Vice President, Operating Services in 1979, was highly supportive of diversifying into the mineral resource sector initially through an affiliation with an established organization.

The diversification proposal into 'mining' was approved at management level and subsequently by the DOC Board in early 1970 in the belief that it offered a superior long-term opportunity. As a result, in 1970 DOC entered into a joint mining / exploration venture with Lacanex Mining Company Limited, which had mineral properties in Canada, the U.S. and Mexico. The company was headed by W.H. (Bill) Gross who had been one of my professors at the University of Toronto in 1948. For exploration purposes, Ducanex Resources Limited was formed in which equity was shared on a 50:50 basis by DOC and Lacanex, and funding 70% by DOC and 30% by Lacanex up to $750,000 in any single project and on a 50:50 basis thereafter. This "inequitable" funding arrangement was eventually to prove unacceptable to DOC, fair as it seemed as an entry requirement.

In August 1970, the DOC Board also authorized investment in the capital stock of Lacanex and through Ducanex, in Pure Silver Mines (30%) and Tormex Mining Developers Limited (27%), companies developing gold-silver mines in Mexico. It was envisaged that cash flows derived from these future operations would enable Ducanex to be self-supporting by 1973.

Differences between the objectives of the two partners and problems in the Mexican operations resulted in a review of DOC's position and objectives which included majority control, substantial projects and a Canadian focus.

In late 1973 the DOC Board approved the formation of DuPont of Canada Exploration Limited (DOX), as a wholly-owned exploration subsidiary. The option to renew the shareholders' agreement between Lacanex and DuPont, which included Canadian exploration, was subsequently terminated in 1974.

In late 1973, although I was aware of Ducanex's Canadian exploration, I had no specific knowledge about its background. At the time, I was approaching completion of my third year as Vice-President and General Manager of Kennco. I was quite apprehensive about the Company's future having witnessed the Toronto office closure in 1971 where 14 of 40 staff were laid off prior to my appointment. A head office was subsequently established in Vancouver. In 1972 six additional staff had to be laid off.

After some soul-searching I then advocated a policy of farming-out custodial properties in order to increase the odds of some return on the Company's investments, which included many significant prospects. By mid-1972 agreements were at advanced stages of negotiation on the Berg, Huckleberry, Whiting, Carmi, Chappelle and Sam Goosly properties in British Columbia and Harp Lake in Labrador. Although this policy was well received by Kennco's U.S. parent, I found further budget cuts and the consequent reduction of exploration activity demoralizing.

Throughout my 22 years of experience with Kennco in Canada, our principal goal, which was generally shared by Kennecott's international exploration group, was the discovery and development of porphyry-copper/molybdenum deposits, similar to the major deposits mined in the Southwestern U.S. by Kennecott, including:

Mine	State	Production Period	1958 Ore Reserves	
			Tons (millions)	% Cu
Chino	New Mexico	1912 -	125.0	1.2
Ely	Nevada	1908 -	25.0	0.8
Ray	Arizona	1911 -	265.6	0.93
Bingham Canyon	Utah	1905 -	700.0	1.0

The 1958 references are shown for two reason: (1) convenience, as I have data on the reserves at that time (Joklik, 1960), and (2) the ore reserves shown were those of operations which had produced for about 50 years, the reserves exceeding any known in porphyry-type deposits in Western Canada at that time. The closest was the Bethlehem Copper operation, which produced 440,000 tons of copper between 1962-82 with initial reserves of 60 million tons grading 0.63% copper. Bethlehem Copper was to herald the start of the porphyry-copper molybdenum era in B.C. Although Kennco either staked or had options on many of the deposits which eventually reached production, none matched the corporate parameters for capital investment which included a minimum five-year payback, 10% annual contribution to parents' earnings and a minimum 20-year life. The only exception, Kennco's decision to place the B.C. Molybdenum property at Alice Arm, B.C. into production, spurred by a political/emotional decision in the wake of Bethlehem

Copper's mine opening, ended in less than five years (1967-72) with operating and capital losses.

In retrospect, most of the many properties in which Kennco once held an interest such as Brenda, Endako, Lornex, Equity Silver (Sam Goosly), became highly profitable or were expected to be as late as 1996, based on recent production decisions (Huckleberry, Kemess).

One evening in mid-November, 1973 I received an unexpected phone call from Art Baker, Director of the Chemicals Group for DOC, whom I had never met. He was staying at the Bayshore Inn in Vancouver. Following his introduction and apology for his "cold contact" as he called it, he came right to the point – would I be interested in considering a change of position involving the establishment and management of an exploration organization in Canada on behalf of DOC? It didn't take long for me to reply, as the opportunity to be involved in developing a new exploration team with a company committed to diversification into the resource sector was most appealing. We arranged to meet the following day.

Our meeting lasted less than an hour. I was immediately impressed with his direct approach and congenial nature. He was obviously enthusiastic about the diversification by DOC, and anxious to have an exploration organization established. As I warmed to his proposal, my only concern was his expectation that the exploration office would be based in Eastern Canada. Having only moved back to Vancouver from Toronto in 1971, a return to the East would have been too disruptive to our family. I immediately said that the position appeared most attractive, providing it was based in Vancouver. Art cogitated only briefly and said, "That wouldn't be a problem. DuPont has a Vancouver office and is planning a move to a new location. Your group could be accommodated with them." We then agreed that with appropriate notice, plus some time off, my employment with DOX would commence on February 1, 1974, almost 23 years following my start of 'full-time' employment with Kennco on April 1, 1951.

I notified Lowell Moon, Director of Exploration for Kennecott Copper Corp. in New York of my resignation on November 19, 1973 to be effective December 31, 1973 and followed up the telephone call with a letter expressing appreciation for my association with Kennco and its affiliated companies, both personally and professionally. I received a very commendable letter from Lowell in response, the last line of which revealed his careful interpretation of words and his sense of humour:

I wish you all the success in the world* in your new undertaking.

*Unless you make the competition too tough

Lowell loved poetry and dabbled at it himself on occasion. In 1971, Kennco and several other exploration companies were attracted by Dr. Harry Warren, the noted Canadian geochemist, to experiment with the use of dogs in sniffing out the presence of sulphide-bearing boulders buried to shallow depths in till. The use of dogs in exploration for this purpose had been successful in the discovery of significant mineral deposits in Finland.

Early in 1972 I was to report to Lowell on progress in our 'dog-sniffing' exploration. Lowell wrote back, noting that he had "reflected upon some of the various methods of prospecting which have been used over the years" and that he had "made a few notes, a copy of which is attached". The copy is reproduced in its entirety, as follows:

>PROSPECTING IS GOING TO THE DOGS
>– or –
>I really think

> That sulfides stink.
> **1910** Rugged man; pick, pan and shovel
> Living in the crudest hovel.
> **1930** Rugged man; canoe and plane
> Looking for bonanza vein.
> **1950** Black box in plane with trailing bird.
> The latest wonder – had you heard?
> **1960** Trace elements in soil and tree.
> The only way as all can see.
> **1970** Rugged man and canine true.
> Sniff out sulfides is what they do."
> L.B. Moon
> January 20, 1972

I responded with the following telex:

> Enjoyed your historical review on prospecting in era 1910 to 1970. It has prompted this message:
> Hark ye now, we would not tarry
> Budgets are firming and snows are soon gone
> If Huckleberry is this year to marry
> Time for courting is all but done.

At the time we were in the midst of the annual budgeting process for the 1972 season. Harry Burgess, Kennecott's V.P. Exploration in this era, used to insist that this burdensome task absorbed about six months of his time every year. I was obviously anxious to obtain approval for a joint venture on Huckleberry with The Granby Mining Co. Ltd. following the policy which I had proposed in 1971.

Huckleberry, a promising porphyry-copper-molybdenum property near Houston, B.C. was discovered through geochemical exploration by Kennco in 1960, staked in 1962 and explored at an overall cost of $338,500 through 1971. Drill-indicated reserves were estimated at 67 million tons (0.25% cut-off) grading 0.39% copper and 0.022% molybdenite at a stripping ratio of 1.73:1 in a potential open-pit to a depth of about 500 feet. In addition two shallow holes were drilled in a quartz-sericite alteration zone one mile to the east.

I subsequently negotiated an agreement with Granby which provided that it could acquire a 50% interest in the claims for $1.5 million in expenditures with a minimum of $500,000 to be spent by the end of 1974. Granby drilled 16,190 metres in 65 holes on the Main Zone in the next two years but failed to earn an interest. The property remained idle until 1988 when Noranda completed additional geochemical surveys over the entire property before dropping its option with Kennco.

In 1992 New Canamin Resources Ltd. optioned the property, initially concentrating its effort on the Main Zone in 1992-93. However a water quality monitoring well drilled in February 1993, 1200 metres east of the proposed Main Zone open pit for potential tailings disposal purposes in the general area of the two shallow holes drilled over 20 years earlier by Kennco intersected 0.905% copper over 8 metres! The New Canamin agreement with Kennecott Canada was remarkably similar to the earlier agreement between Kennco and Granby. It provided for an option to acquire a 100% interest in the property by spending $1.5 million over five years, with Kennecott retaining a right to buy in for 60% at a production decision.

Following a major drilling program in 1993-5 focused on the East Zone during which New Canamin merged with Princeton Mining Corporation, diluted minable ore reserves were 91 million tonnes in the two zones grading 0.517% copper, 0.064 grams gold per tonne and 0.014% molybdenum. A positive feasibility study in 1995 outlined a 13,500 tonne/day operation for 18 years at an initial capital cost of $137 million.

In March, 1994 just prior to the New Canamin-Princeton Mining merger, Kennecott Canada Inc. relinquished all its rights in the Huckleberry property by accepting a $100,000 cash payment and 2.5% of New Canamin's issued capital totalling 214,000 shares valued at a deemed price of $1.50 per share. The right to re-acquire a 60% interest in Huckleberry had been "retained against the possibility of discovery of a world-class deposit, i.e. greater than 400 million tonnes ore reserve (of an unstated grade!). New Canamin's Huckleberry development is toward a smaller, 9000 tonne per day, twenty-year operation, which does not fit with Kennecott's targeted size requirements".[1] The quotation was reminiscent of my much earlier understanding of Kennecott's target objectives for a viable porphyry-copper/molybdenum deposit as described in the early part of this chapter.

By March 1996, Princeton had received both federal and provincial government approval to proceed with development of the Huckleberry mine, overcoming late land claims filed by a local native band on the Sierra Legal Defense Fund. Subsequently, in January 1998, Honorable Chief Justice Williams ruled against a petition filed in late 1995 by the Sierra Legal Defense Fund on behalf of the Council of Cheslatta Carrier Nation aginst the B.C. government alleging lack of due process in the environmental assessment of Huckleberry.

The Huckleberry Mine was acquired by Imperial Metals Corporation in April 1998 as a 50% owned subsidiary. Commercial production began in October 1997 following investment of capital costs of $142 million.

The Huckleberry saga is a lengthy diversion from my recollections of Lowell Moon. It reflects the spasmodic nature of most mine developments and the changing role of actors involved between the initial hopes and aspirations of success and eventual realization of the initial dream.

Another interesting facet of Lowell's character was his shrewdness. Shortly after my appointment as Vice-President of Exploration for Kennco, he asked me to visit him for a day or so to discuss our Canadian exploration program. He wasn't specific about the nature of the visit. Following my arrival in New York, we spent part of a day on general discussion concerning Kennco's exploration strategy at the time and its fit with Kennecott's corporate objectives, plus the need to consider personal reductions to meet a specific budget level.

At this time I learned that staff salaries were based on the Hay Point System. Lowell's level at the time was 1164 points as Director of Exploration whereas a Clerk/Typist in Kennco's Vancouver office had a level of 85 points. The system is, of course, widely used in industry even at this time. However, Kennco did not employ the annual type of performance review that I was to later adopt for DOX in order to comply with DOC's practice. Lowell confided that he personally rated all of Kennecott's 150 - 200 exploration personnel who operated on a world-wide basis. He did this, not for salary decision purposes, but in order to rate the Corporation's annual budget submissions on a project-by-project basis, based on his own assessment of the relative confidence he placed on each geologist's report and promotional abilities. He used a point system of 1- 9 in developing factors reflecting those with a pessimistic, conservative or cautious approach

---

[1] New Canamin Resources Ltd. News Release, March 31, 1994.

to assessing the potential viability of a project from others at the other end of the spectrum who tended to be optimistic, progressive or rash. His use of these factors was intended to provide a levelling effect on individual assessments.

Just prior to breaking off for the day, he surprised me by requesting a budget summary by the next morning for discussion purposes, which would reflect proposed expenditures and personnel requirements for the following year, tied to a fixed budget allocation. Needless to say, I saw nothing of New York that evening, working through the night until about 6:00 a.m. the following morning in order to comply with his request.

Following a fairly hectic year-end period in 1973, providing for my termination responsibilities with Kennco and official farewells to staff, our family left for a welcome three-week vacation, which included visits to Oahu, Maui and Hawaii.

Art Baker had not expected that any significant exploration programs would be completed in 1974, rather that it would be a period of acquiring staff and developing exploration strategies. Accordingly he was quite surprised when I forwarded him a copy of a letter of intent dated February 8, 1974 providing terms of an agreement to be negotiated with Kennco on the Chappelle gold-silver property in B.C. The formal agreement was not to be approved for almost two years; however, exploration by DOX was initiated in 1974. In addition I was most fortunate in assembling most of our initial staff within a matter of months. Key personnel were David McAuslan, Senior Geologist who had worked with me in Eastern Canada prior to our joint move to Vancouver in 1971 and Annikki Puusaari. From a reasonably lengthy association, I regarded Dave as one of the most capable and energetic geologists that I had been privileged to work with and I was somewhat overwhelmed by his decision to join our new enterprise.

Annikki was more than familiar with a mining background having grown up in Mount Isa, Australia. She had emigrated from Australia a couple of years earlier and was employed in a secretarial capacity with Lornex in Vancouver. She made it quite clear during our initial meeting that she sought advancements to more senior positions through performance and that her tenure would probably not be lengthy. I was impressed with her from the start and ensured her that I would remain aware of her objectives. Starting off as secretary in our temporary office with only a couple of staff, she was to remain until our mutual departure in 1984 with the well-deserved title of Office Manager – a position description not otherwise assigned to female personnel in DOC's organization at the time.

Other original DOX members hired in 1974 included Gerald Harron with experience in Eastern Canada, Marshall Smith, a geologist with a strong background in geochemistry, Tom Drown, a recent geology graduate and Keith Jones our Chief Draftsman, all of whom were to remain with DOX until 1984.

Sometime later we were fortunate in attracting Christopher B. Gunn with his considerable exploration experience in Eastern Canada and the Northwest Territories. Chris had been recently employed with Derry, Michener and Booth, Geological Consultants with an international reputation affording Chris an opportunity to obtain quite diversified experience.

Although we were to add additional staff in subsequent years, particularly in 1981, including John Korenic, John Kowalchuk and Gordon McCreary, I had resolved to maintain only a core group of permanent personnel and to add only temporary personnel or consultants on peak demand, in order to avoid the necessity of implementing staff reductions in low budget years which tend to be a cyclical phenomenon with demoralizing impact.

In my 23 year tenure with Kennco I had witnessed this cyclicality on three separate occasions, apart from the much shorter commitment to exploration in Canada by many companies for a variety of reasons. However, to be fair, Kennecott's commitment to exploration in Canada has been one of the longest, if not the longest of foreign-based corporations. In 1996 it is particularly active, with a head office for Canadian exploration headed by John Stephenson in Toronto and a very active exploration group based in Vancouver. Thus its term of activity in Canadian-based exploration has reached the half-century mark.

In any event, during the decade of DOX's existence which was to follow, it was only necessary to lay off one individual during periods of corporate austerity. The high level of morale enjoyed by the exploration team can be reflected by the subsequent attendance at annual and bi-annual reunions in Vancouver, through my writing in 1996, which normally attract 8 - 10 individuals. Even with a mine in production during part of the period, we had only a maximum of 12 permanent personnel based in Vancouver of which 10 were exploration-related. After a few months in temporary office space in the downtown area of Vancouver, our emerging exploration group moved to a new office space shared with DOC's Vancouver personnel. Our combined operations required the complete ground floor of the recently constructed Sandwell building on Alberni Street. During our initial tenancy the DOC group were headed by C.R. Dick Asher, a long-term DOC employee. He was to be succeeded in the mid-70s by Dick Hermon. The DOX group always appeared to appreciate its association with the DOC representatives, who were generally bullish about the corporation's aspirations in its entry into resource sector exploration activities, even if some questioned the diversification effort.

As can be imagined, a tremendous challenge existed for DOX to justify its creation as a mineral sector subsidiary of a corporation with a recognized prominent national and international reputation as a chemical company with an extremely diversified range of products. There were obviously many basic differences and aspirations, based on experience, between the two. DOC was a well-established entity, oriented to growth, with a record of achieving increased cash flows and profits through innovative research and a dedicated and motivated organization.

From its roots as a producer of explosives in the 1860s, with little concern for safety, DOC had established an enviable safety record, a century later, officially initiated in 1931 through its predecessor organization when a "No-accident record plan" was adopted. In 1928, the first year Canadian Industries Limited, a DOC predecessor, maintained accident frequency statistics, the rate stood at 24.73 lost-time accidents per million person-hours worked. By 1954 the rate had dropped dramatically to 2.2 and by 1970, four years before DOX was formed, DOC's lost-time accident rate was 0.23 injuries per million person-hours worked, a new record for the company.

In 1974 there were no comparable statistics for explorationists in Canada. Nonetheless, it was generally accepted within the mineral industry that mineral exploration, like all exploration activities to that time, was an activity subject to a much higher level of occupational hazard than that which would be experienced in a plant-controlled environment.

I can confidently confirm that the safety aspects of DOC's decision to embark into a diversification tied initially to the exploration sector of the Canadian mineral industry was apparently never seriously considered, as I certainly would have been aware of this objective. At that time, DOC had no personnel with experience in mineral exploration and its principal focus was in attracting a suitable individual to form and direct an exploration subsidiary.

An additional concern which emerged <u>after</u> DOC's commitment to form a mineral exploration subsidiary was the question of a time-frame and cost in which it could be anticipated that acceptable economic returns could be anticipated following the original investment in the exploration sector. DOC did not provide any specific guidelines concerning preference of commodities to be sought or whether the subsidiary's activities were to be confined to Canada, North America or elsewhere, although it was understood that the principal activity would be in Canada.

Other key personnel to join DOX in the 1970s included Keith A. (Sandy) MacLean, whom I had previously attempted to attract to a position with Kennco while I was District Manager for Eastern Canada in Toronto. Sandy was to be a significant contributor to DOX's efforts in British Columbia and Yukon. Of particular note were his contributions at the Chappelle property, later to become the Baker Mine. Through his personal associations, Sandy was to attract Martin Kierans, a consulting geologist, to provide an independent and as it developed a supportive reserve estimate for the Chappelle property in 1979.

Louise Eccles, a graduate in geology from UBC was also to become a welcome addition to DOX's exploration team commencing in 1976. Louise, at six feet tall, commanded an unusual presence in this era. By nature, sensitive, good-natured and physically extremely well-developed, she easily competed with her male counterparts in any particularly strenuous or demanding endeavours. I was initially to be informed of her cooperative nature by Marshall Smith while he was involved with a property examination in Southern B.C. for DOX which was supported by several trailer units, one being an office-sleeping kitchen unit, partly occupied by Louise as the sole permanent resident. The portion of the trailer unit devoted to dining contained storage lockers for supplies unreachable by all but the loftier humans. These, inexplicably, were used to store such essentials as breakfast cereals. Louise cheerfully brought these items down and returned them to their designated location without the need of a stool or chair, which would have been the requirement for all the other (male) members of the party.

DOX Strategy & Growth

By 1977, three years after its formation, DOX had considered 400 property submissions and joint venture proposals. Although DOX's annual exploration expenditures by this time averaged about $1.5 million, total exploration expenditures on all projects in which DOX was involved, including its joint ventures, was about $3 million. By that time DOX was solely funding all exploration costs on the Chappelle gold-silver property, optioned from Kennco, on which some encouraging results had been obtained from surface diamond-drilling.

By then DOX had evolved an exploration strategy, which had been expressed in several white papers to senior DOC management. In 1979, I received an unexpected phone call from Herbert Lank, who had retired as Chief Executive Officer of DOC in 1965. Following his retirement he had tried unsuccessfully to persuade his successors to have some professional writer undertake the task of recording the history of DOC and its predecessor companies. Finally in April 1979, he was told that he was the logical person to supervise such an assignment. His call was specific – would I be willing to provide him with a chapter on DOX. I suspect that he knew the answer before asking the question, because all I asked for was how detailed and by when.

The DuPont Canada History by H.H. Lank and E.L. Williams was published by DOC in 1982. My contribution, published almost verbatim, occupied about half the chapter entitled "Digging for Profits". Keith Jones, our draftsperson, prepared the half-page diagram consisting of a simplified geologic map of Canada, entitled "Exploration Activity 1974 - 1980" which showed the location of our principal projects.

Our strategy and growth occupied less than two pages of the volume as the succeeding section was devoted to a description of our principal projects at that time: (1) Great Slave Reef lead-zinc, Pine Point area, NWT, (2) Chappelle gold-silver property B.C. – to become Baker Mine, (3) our emerging interest in the Bell Creek project, Timmins area, Ontario with Canamax Resources Inc. – to become the Bell Creek Mine in 1987, was not referenced.

In the interest of representing my best recollections of our strategy and growth philosophy as written at that time I have reproduced the pertinent portion of Herbert Lank's history as follows:

DuPont of Canada Exploration Limited (DOX) was incorporated by February 8, 1974 and field activities were initiated that year. The objective was the developing of opportunities in the mineral resource area which would compete for capital on a risk / return basis with projects developed in DuPont Canada's other business areas, thus increasing the number of investment opportunities open to the company. The business of the mineral venture includes the acquisition (through discovery or purchase), extraction, processing and marketing of metallic and non-metallic minerals, principally in Canada.

DOX consists of a group of eight people (five of whom are geologists) and up to 25 temporary employees operating out of Vancouver, B.C.* Since incorporation, the emphasis has been placed on the search for, development and/or acquisition of, tangible mineral prospects containing readily marketable commodities, excepting oil and natural gas. The exploration strategy is based on selecting target areas where evidence exists of economic grades of mineralization but where past exploration has not defined sufficient tonnages for a production decision. This strategy can be adapted to a wide range of exploration ventures, from grass-roots investigations to relatively well-explored deposits.

Growth through discovery rather than acquisition has been the primary objective. Annual exploration expenditures have grown from $1,000,000 in 1975, the first full year of exploration, to $1,800,000 in 1979 and were forecast to increase each year over the next five years, to maintain this level in 1979 dollars.

In the period since its inception, DOX has emerged as a viable exploration organization. Its principal tangible assets are its interest in the Great Slave (Northwest Territories) joint venture and the Chappelle (British Columbia) project.

The organization has followed an exploration strategy based on the following guidelines:

1. Expanding exploration in environments containing mineral deposits of economic grade, as opposed to riskier grass-roots reconnaissance projects.

2. Exploration in accessible locations, with the objective of generating early cash flows. Over 90 per cent of the exploration expenditures have been in this category.

3. Joint venturing with other companies in order to broaden exposure, increase the possibility of discovery per DOX dollar expended and provide skills not present in DOX by associating with companies who have expertise in a particular aspect of the business. Since inception, DOX has been involved in 28 projects, of which 15 have been joint ventures.

4. DOX to act as Operator in a majority of joint ventures in order to maximize use of the expertise of its exploration people and to exert control over the exploration strategy. DOX has been designated Operator in nine of the 15 joint ventures in which it has participated.

5. Concentrate on exploration for marketable commodities. DOX has maintained a relatively heavy emphasis on the exploration for base and precious metals, principally zinc, lead, copper, silver and gold. More recently uranium, coal, molybdenum, tungsten and tin.

Since its formation in early 1974, DOX had processed 650 exploration proposals and property submissions. Its principal assets are situated in Western Canada.

Although DOX proposes pursuing its current course of growth, consideration will also be given to growth by acquisition of interests in more established properties. Once a discovery of economic significance has been made, the exploration company has several alternatives:

(a) Sell all or part of its interest.
(b) Elect not to participate in further development and allow its interest to be diluted.
(c) Maintain its interest and participate in the mining venture as the Operator or as a non-Operator.

*Four employees and the pilot were killed in the crash of a helicopter near Stewart, B.C. July 3, 1980. The four: Chris Gunn, Sandy MacLean, Ruth Nussbaumer and Ian Shaw, son of Tom Shaw of DuPont Canada's Explosives Division.

Our actual strategy and growth did not diverge significantly from this. Events referred to in the footnote, the terrible tragedy of July 3, 1980, which took the lives of four DOX 'permanent' and 'temporary personnel' are described elsewhere (Darkest Days).

Mexican Odyssey

I have mentioned DOC's involvement with Lacana Mining Corporation which was to lead to the formation of DOX. In mid-1977, Art Baker asked me to prepare a comprehensive valuation of Lacana's assets by the end of 1977 as a basis for determining Lacana's present value in support of a proposed sale of its interest in Lacana. Earlier valuations had been made by Graham Farquharson of Strathcona Mineral Services in April 1974 and by William Hill in November 1976 with an up-date in May, 1977.

I relished the assignment. Although other requirements prevented me from devoting a full-time commitment to its completion I was able to visit eight of the eleven Lacana properties between July 19 - October 12, 1977. Although the three others were not examined, I was able to review pertinent data with exploration personnel familiar with the properties. Apart from examinations completed on my own with appropriate Lacana personnel, four were in the company of Art Baker in Mexico and Guatemala, two with Gerry Harron in Quebec and British Columbia and two others with Lacana personnel in Mexico and the U.S.

Although I could provide considerable information – mostly of a technical nature of questionable interest to readers – only three of the examinations appear to be particularly noteworthy: Guanajuato, Encantada and Santo Tomas are in Mexico which collectively accounted for over 90% of the total discounted value of Lacana's assets based on Hill's and my own 1977 reports.

Guanajuato is of particular interest with its colourful mining heritage, topographic setting and surviving colonial structures, which make the city a living monument to a prosperous and turbulent past.

Guanajuato lies about 275 km northwest of Mexico City at a mean altitude of 2000 metres within a relatively rugged mountain range with a maximum relief of 900 m. Spanish prospectors discovered rich silver and gold-bearing veins at La Luz in 1548 and nearby Rayas, which lies within the present city boundaries in 1550. The major production of the district did not start until the discovery of the extremely rich La Valenciana Mine, 5 km. northwest of Guanajuato in 1558. For 250 years the mines produced 20% of the world's silver, creating wealthy colonial barons who lived opulent lives at the expense of Indians who worked the mines and gave their allegiance to the Jesuits.

In 1765 King Charles III of Spain imposed severe taxation on the barons and further alienated them and the Indian miners with his decree in 1767 banishing the Jesuits from the Spanish dominions. In 1810 Miguel Hidalgo, a priest and rebel leader in nearby Dolores, led an independence movement resulting in the capture of Guanajuato. When the Spaniards retook the city they retaliated with the infamous 'lottery of death' in which Guanajuato citizens were drawn at random to be tortured and hanged. Independence finally resulted in 1821, which led to the construction of mansions, churches and theatres which have survived the period and are among Guanajuato's many attractions.

While in Mexico City with Art Baker in 1977, we walked through Sullivan Park where we admired the many paintings displayed by local artists. Art had previously bought several paintings there – this was his 20th visit to Mexico in about five years and he admitted to being somewhat travel-worn. I bought an oil painting during our visit which still hangs in our hallway. It depicts old homes in Guanajuato astride a sub-level roadway interconnected by stone archways above the road, draped with floral arrangements and Spanish lamps. The sub-level 'road' was constructed centuries ago by the Spaniards as a canal to control overflow from the nearby river during periods of flood. When modern flood-control was established, the road resulted.

I still recall my astonishment on learning about the early history of mining in the district, which commenced a bare half-century after Columbus 'discovered' America. It is unusual in mining districts for operations to continue for 450 years as is the case with Guanajuato, where tourists in the 1990s could still visit the La Valenciana Mine which re-opened in 1968 following shutdown after the 1911 Mexican Revolution.

I visited Guanajuato in early October, 1977 with Art Baker to inspect the five mines of the Torres Mining Complex which consisted of Las Torres, Pedros, Cebada, Bolanitos and Peregruia. All the gold-silver orebodies occurred in vein systems traced for up to four miles in length and extending through a maximum vertical range of 800 m, bottoming at an elevation of about 1700 m.

Prior to 1968 the Guanajuato mining camp was generally considered to have been worked out. In 1964 the potential of the camp was re-evaluated by Antunez, a Mexican geologist who recommended a deep drilling program of 30 holes to test unexplored portions of the most productive structure along a 30 km. strike length. The project was abandoned after the first eight holes failed to intersect ore-grade material. In 1967 Bill Gross and Jarier Moreno reviewed Antunez's work and recommended further diamond drilling which resulted in new discoveries.

At the time of my visit a flotation plant was still processing ores at a rate of 300 tons per day recovered from the ancient Valenciana zone at depths of 500 m below surface, using the original shaft constructed in the 1770s.

Art Baker and I visited the Encantada mining operation with Antonio Alvarez, Lacana's Chief Geologist - Latin America on October 10 - 11, 1977. The property consisted of two adjoining silver-lead mines, the Encantada and Los Angeles, located at the base of a mountain slope in the semi-desert region of the State of Coahuila in northeastern Mexico. Although accessible by road we arrived by a chartered fixed-wing aircraft, landing on a 1200 m gravel airstrip situated near the Encantada townsite. A dirt road links Encantada to a paved highway at La Cuesta, 100 km. to the north. Concentrates produced at the mine were trucked to a smelter at Torreon, about 700 km. to the south at a cost of 356 pesos ($16.00) per tonne.

The Encantada discoveries are relatively recent. A vein-like replacement of massive hematite (iron-oxide) with rare lead oxide was discovered by a shepherd in a surface outcrop lying about 50 m. above the top of the major La Prieta orebody in 1954. About 1.3 million tonnes of silver-lead ore grading 22% lead and 900 gms silver per tonne (26 oz. per ton) were mined between 1956-9 from the pipe-like orebody between elevations of 1690 - 1790 metres before operations were suspended. Antonio Alvarez descended 100 m. down the old shaft in a bucket and was impressed with mineralization in the walls. Lacana became interested and exploration and development led to renewal of production in 1973. Commencing in 1977, about 560,000 tonnes of ore had been produced grading 16% lead and 11.5 oz. silver per ton. Proven and probable ore reserves developed in five ore bodies totalled 2.9 million tonnes grading 7.6% lead and 13.1 oz. silver per ton.

One evening, Art Baker and I were invited for dinner with Octavio Alvidiez, the mine manager and his wife. Although the main meal of the day in Mexico is commonly held at noon and I always felt, quite logically, complemented the need for a siesta in the early afternoon, our meal was sumptuous and the description of life in the remote camp was most interesting.

Following our visit, arrangements had been made for Antonio and I to fly to the Santo Tomas porphyry copper deposits, a dormant exploration project owned by Industria Penoles (55%) and Lacana (45%). The property lies in rugged mountainous terrain on the west side of Sierra Madre Occidental in the most northerly part of the State of Sinaloa. Rio Fuerte (Strong River) and its tributaries cut through the Sierra Madre Occidental and the main river encircles part of the property as it flows southerly to its mouth on the Gulf of California near Los Mochis about, 140 km. to the southwest.

The examination of the Santo Tomas property was to be one of my most memorable trips. En route to the property we flew over the area and I took several photos of the rugged terrain, jeep roads and bulldozer trenches. I was impressed with the size of the river which averaged over 100 metres in width. Antonio had provided me with the barest of information on our proposed examination. I knew that we were to be met by a Lacana employee at an airstrip accessible to light aircraft near the Village of Choix where we would be driven about 40 km. to the property area. The driver was to arrive by road with our baggage from Los Mochis in order to cross Rio Fuerte and drive up a road about 120 km. on the west side of the river to Reforma mine, owned by Penoles, which lay on the north side of the river from the Santo Tomas deposits. I asked Antonio how we were going to cross the river. "No problem", he said, "there is a shallow section where we can step across". I was to learn that Antonio could be very unreliable with such details, apart from being quite devious.

On approaching the airstrip at Choix, I saw the remnants of several aircraft lying on the side of the short runway. On landing I was surprised to see about half a dozen armed militia rapidly approach our aircraft who urged us to leave the plane without our baggage. We watched as they made a thorough search of the plane and our bags, before waving us on to our waiting vehicle. It turned out that the wrecked aircraft included some they had fired upon while trying to take off, as this part of Mexico was an area where marihuana and other drugs were grown. The pilot of our aircraft appeared quite unconcerned by the commotion. He had obviously experienced many similar incidents in the past.

On reaching the San Tomas property area the driver walked along with us for a short distance from an impassable portion of the road before returning to the truck for his 250 km. trip to reach a point about 10 km. distant on the other side of the river! It was a particularly hot day for a gringo such as I in mid-October. The ridge that we traversed while heading toward the exploration workings was bare of any trees. Although I retained my yellow, short-sleeved shirt, Antonio stripped to reveal a well-tanned torso. About a half-hour later, as we neared the exploration cuts, several enormous green hornets hovered above me and descended in a slow but purposeful fashion. They were not interested in Antonio – only myself. I waved them off as best I could but they were persistent. Antonio yelled, "Don't let them sting you!" I had no intention of providing them with an opportunity – however, they were relatively easy to fend off because of their slow approaches. It suddenly dawned on me that as they had not been at all interested in Antonio, who was bare-chested, they might have regarded my yellow shirt as some sort of enticing flower. I pulled it off and thankfully they lost further interest.

After several hours of examining the extensive showings and collecting a few samples for reference, we descended some 500 metres to the river valley, traversing toward the region where Antonio felt we could cross the river by "stepping across". It was so hot that we both tried to take shelter amidst some relatively bare shrubs. Eventually we reached the river near the point where we would have to cross to gain access to the end of a road on the other side of the river, above the valley, to access the Reforma mine.

Quite suddenly and unexpectedly, near the river, we reached a small adobe building which looked uninhabited. Strewn about the building were many artifacts which were easily identified by the eyes of a geologist, including relatively crude mortars and pestles, and rocks used to grind maze by hand, called 'manas'. It was so unexpected in the midst of this barren region that I felt that I had stumbled on a treasure trove.

Much to our surprise a lean middle-aged individual appeared around the edge of the building. Antonio explained our presence. Mr. Hernandez was apparently a rancher, eking out a living to support his two teen-aged daughters. He had noted our interest in the artifacts and, in addition, that I was equipped with a camera. He walked back to the building with Antonio, to re-appear shortly after offering me a magnificent small axe-head carved from porphyritic volcanic rock. In return he asked if I would take some photos of him and his daughters. I was stunned – as I would have willingly done so without expecting a 'reward'. I took several photos and obtained his address, which was care of the Reforma Mine across the river.

Some time later, Antonio emerged from the building with Mr. Hernandez and waved at me to follow as they trudged off toward the river. Mr. Hernandez carried several small boards with him not more than a metre in length resembling miniature surf boards. As we reached the river I was handed one of these with a length of cord. Both Mr. Hernandez and Antonio began to disrobe and I followed suit as our method of crossing of the Rio Fuerte became evident.

As we waded out into the river, which was remarkably refreshing, we reached a point where we had to start swimming, pushing the boards with our packs and clothing ahead of us. Soon the current could be felt as we neared the central part of the river. Eventually we emerged on the other side some 400 metres below our starting point. After putting our clothes back on, we walked up the other side for about 800 metres. Without any urging from Antonio I felt that Mr. Hernandez's guiding service required adequate recognition and offered him a reasonable fee which appeared more than adequate. We shook hands and watched him re-enter the river before starting up the steep slope toward the road. Within a few minutes we were completely dry and I was conscious of the weight of my back-pack filled with the samples I had collected on the Santo Tomas property. We eventually reached the road above and walked several kilometres to the Reforma mine to be greeted by our driver.

As I off-loaded my back-pack, Antonio reached down and pulled out a large rock, grinning as he showed me a mana (grinding stone) which he had retrieved near the rancher's adobe. I should have suspected the possibility I suppose, since he had stuffed an underground battery, belt and lamp into my back-pack following our underground visit at the Encantada mine, only a day earlier! He had called at my room as I was cleaning up after returning to surface, having left my lamp, belt and battery at the mine change house. I had not opened my pack since returning to surface and had assumed that it contained only the samples collected underground. I was speechless when he calmly opened the pack retrieving the items before returning to his room.

I never did learn whether the rancher received the photos I sent to him. I often gaze with wonder at the stone axe-head he so generously gave me, realizing that Stone-Age Indians lived in the area until only recently.

I completed my report on the Lacana evaluation on December 31, 1977. My assessment indicated a value of $23.4 million in Canadian funds discounted at 15% for Guanajuato, Encantada, Blizzard, Scott, Minorsa and Pinson and by 20% for less tested assets. This compared with a value of $27.7 million in Canadian funds for Hill's earlier estimate, the difference being attributable to slightly higher values by Hill for Guanajuato and Encantada and the lack of any assessment for the potential of the newly discovered Blizzard uranium deposit in B.C.

Within a year (October 1978) I provided an update in which the present value for Lacana had soared to $38.9 million Canadian, principally due to recent drilling results obtained on the Blizzard uranium deposit as reported by Gerry Harron. The vagaries of the exploration business, influenced by so many parameters, are reflected in the subsequent totally unexpected six-year government-mandated moratorium on uranium exploration projects in B.C. which virtually eliminated any future value of such deposits. The moratorium period coincided with significant uranium discoveries in Saskatchewan which were of a much higher grade and which would have eliminated any competitive opportunity for financial success at Blizzard.

Lest any knowledgeable readers question the valuation ascribed to the Blizzard property at the time, it should be noted that in 1978 Ontario Hydro purchased a 10.5% net interest in the property for Cdn $5 million, which was the basis for the inference that Lacana's 30% interest in the property corresponded to $14.28 million.

Late in 1980 I received a suitably inscribed memento indicating that on October 27, 1980, DuPont Canada Inc. had sold 1,147,809 shares without par value of Lacana Mining Corporation to Western Mines Limited via Wood Gundy Limited as financial advisor to DOC. Of course, I had to make reference elsewhere to what exactly that entailed. My reference (The DuPont Canada History) indicated that DOC sold its investment in Lacana for $12,626,000 with a gain on disposal of $7,916,000. In the same year DOC acquired 100% interest of Ducanex Resources Limited when Ducanex purchased and cancelled all of its issued shares not owned by DOC.

For several years after my first visit to Mexico with Art Baker, DOX maintained an active interest in potential mineral opportunities in Mexico. This without any objections being raised by DOC or DuPont U.S., recognizing that our principal area of interest remained Canada.

I was to examine several other interesting mineral prospects in Mexico following my 1977 trip with Art Baker. A number of these were prompted by a relationship which developed between DOX and Terra Mining & Exploration Ltd. in the mid-1970s. One of DOX's exploration ventures was a reconnaissance program in the Great Bear Lake area, N.W.T. directed by Marshall Smith. Terra's high grade silver-copper producer, the Silver Bear mine which operated intermittently between 1969 - 1985, was then in operation and was visited by DOX personnel, including myself, on several occasions.

Alvin Harter of Edmonton was Terra's president at the time and he approached DOX as to its possible interest in examining several properties in the Alamo silver belt in Sonora, Mexico in which Terra had obtained a 49% interest from the owner of Compania de Minas Nuevas S.A. One of these was the La Cumbra tungsten prospect in the Alamos district, which was considered to have an open-pit potential based on the widespread dispersion of values. I flew to Hermosillo in the early part of 1978 through arrangements made with Terra, travelling on the same flight as Archie McCutcheon who was visiting the Terra properties with the possibility of securing a drilling contract. On arrival we were fortunately met by a Terra geologist who was fluent in Spanish and knowledgeable about such matters as entry requirements to Mexico. I had obtained a business visa, but Archie had not had time to have one issued. An imposing Mexican immigration official checked my passport and visa as I stood nearby waiting for Archie to join me. He insisted that he was travelling for other than business purposes, but the official leafing through his brief case enquired "Senor, if you are here for a vacation why do you need all these reports?" The Terra geologist quickly approached the two, greeting the official warmly with an arm on his shoulder and a firm hand-shake. Within a few minutes, the official turned to Archie and wished him a pleasant stay, suggesting that in future he should avail himself of a business visa if there was any possibility of mixing business with pleasure.

As we left the terminal building, we asked the Terra geologist what had been said to avert the problems. "Oh nothing much", he replied, "I just gave him $50."

When we met Alvin at Alamos I was somewhat surprised to be introduced to Dr. Wolfgang Lehmann, a geologist representing a German company also interested in examining some of Terra's mineral interests in Mexico. He was an engaging individual, sturdily built and fluent in English.

The next day we were both driven out to the La Cumbra prospect in a dry desert environment with typical, widely scattered scrub and cactus growth. The workings consisted of numerous trenches and pits scattered over the area with no apparent geologic trend to the workings. After a brief introduction to the various locations Wolfgang and I went our separate ways, to meet for lunch around noon.

Throughout the cloudless, hot day we could hear the sound of four miners working under a lease arrangement to recover tungsten by a primitive labour-intensive method which apparently obtained about 50% of the available tungsten from gravels. The 'ore' was shovelled into an inclined box by one miner and fed to a slatted board also on an incline. Air was forced upward beneath the slatted board by a foot-operated pump which propelled the feed upward to drop below its starting point on the inclined plank. The tungsten 'ore' in the form of fragments of schellite, about three times heavier by volume than the other particles became crudely separated from the lighter fragments and was screened off by hand. The 'thumping' sound produced by the dispelled air carried for some distance and during the seven or eight hours on the property there was rarely a lull from this eerie reverberation.

At the end of the day, after having produced a map of the workings and collected numerous samples for assay, I decided to return to the first pit I had mapped to check some details. It was a large circular pit about 20 feet in diameter and more than 6 feet deep with the material that had been shovelled out on its rim. Feeling quite weary I clambered on to the rim of the pit and before looking down, Wolfgang's voice exploded from the bottom of the pit, "Take one more step and you're a dead man!" I gasped before both of us burst out laughing at Wolfgang's apt association of the pit with a World War II shell hole.

My mapping and assay results revealed that the trenches and pits were roughly aligned to expose a parallel system of structures containing the tungsten values and that they were too narrow to be of economic interest separately and too widely dispersed for bulk mining.

Another interesting property examination in Mexico was completed in the company of Chris Gunn. The property was the former La Fortuna mine in the state of Durango in the western foothills of the Sierra Madres, 45 miles northeast of Culiacan, the state capital of Sinaloa. Richer portions of the gold-silver-copper deposit had been mined in the 1920s by underground methods and the upper portions of these workings were accessible by adits. As early as 1919, E.H. Cook, a mining consultant had defined a proven reserve of 2.9 million tons grading 0.3 ounce gold per ton with a further 1.3 million tons of an undetermined grade in the probable category.

Early in 1979 DOX expressed interest in examining the property through contacts made with a young Mexican government state geologist who had recognized the potential for a bulk-minable operation. The Mexican government owned the deposit and was most interested in labour-intensive underground mining operations.

Chris and I flew to Culiacan in March, 1979 and met the geologist who had made all the necessary travel arrangements for our examination. He had informed us that apart from our personal gear and examination equipment we would also require light sleeping bags because of the limited accommodation at the small village of El Portezuelo on the south bank of the Humaya River, some 5 miles by an old road from the mine.

In order for all three of us to travel into the property in a single flight and for overall safety reasons we flew to El Portezuelo the following morning arriving there at about 7:00 a.m. in order to take advantage of the colder air. On approaching the landing site we could appreciate the pilots' concern as the landing strip was only about 1000 feet long and lay on the flat north bank of the Humaya River. It included three old concrete building foundations interspersed with bare earth and was approachable for landing and take-off in only one direction because of a steep curve in the bank of the river.

We were met by two caballeros with one spare saddle-horse, the assumption appearing to have been made that neither Chris nor I could ride and would walk to the property. After stowing our limited personal gear we walked behind the three riders who included the Mexican geologist who had arranged to return that evening to Culiacan, our return being planned on the morning of the fourth day.

It was incredibly hot and humid and the others soon left Chris and me behind. As we trudged on, approaching the mine, Chris became dehydrated and we were fortunate to locate an old explosives storage area. After he cooled down, we continued on, reaching the workings where we rapidly gained access to one of the old adits. The walls of the old adits were covered by a thick veneer of oxidized rock and it was difficult to obtain fresh samples. We spent the day there examining most of the workings and allowed for an additional day to complete the examination. While we were in one of the adits we disturbed several bats which flew off toward the entrance, startling both of us. Fortunately we didn't encounter any threatening insects.

Following our return to El Portezuelo we were both accommodated in the home of the residents who provided us with breakfast and dinner, our fare consisting principally of tortillas, rice and eggs. We slept on the bare compacted earth floor, sharing our accommodation with a wandering goat and several chickens.

The next day was much like the first, with one important exception. We insisted on having horses for both of us, a concession which we were granted and which provided a welcome diversion from walking. On the following day we made a reconnaissance traverse of the surrounding area. Along the shallow muddy embankment of the Humaya River we noted neatly aligned marihuana plants – a common sight in some of the more remote interior regions of Mexico, which wary geologists avoid like the plague so that their presence is not misrepresented as a possible official interest. At another location in a dry hilly region containing scattered desert vegetation, we came upon a long line of army ants crossing the sandy path we were following. They were 'coming and going' along the same route, the difference being that one line was bearing tiny fragments of leaves obviously borne from some distance.

Returning to El Portezuelo late that afternoon we saw a truck laden with cases of Coca-Cola driving up the river bed. We later learned that, although very poor, the villagers loved Coca-Cola and that trucks from Culiacan used parts of the Humaya River at low water to access the remote villages in the district.

The following morning our aircraft returned on schedule to return us to Culiacan. The flight off the 'airstrip' was made without incident, except that both Chris and I and, I suspect, the pilot also, were greatly relieved once we were airborne as we appeared to require the full length of the strip and a steep turn to follow the river while gaining altitude.

I often wondered during the passing years whether La Fortuna ever reached production. Chris Gunn, Sandy MacLean and I all recommended interest in the property by DuPont, although senior management decided against supporting another Mexican venture. However, on reading the July 22, 1996 issue of The Northern Miner, I was surprised to note an article entitled 'Alamos Minerals to gain Fortuna'. There are several 'Fortunas' in Mexico. I made contacts with representatives of Alamos Minerals and a map was kindly provided by David Duval, formerly Western Editor for The Northern Miner, which revealed that the two were one and the same. The present owner is San Fernando Mining which has agreed to sell the property to Alamos Minerals for U.S. $50,000 plus 2.3 million common shares (current value Cdn. $1.10 / share). The reserve estimate by Fluor Daniel Wright in 1994 was 2.7 million tonnes grading 5.5 g gold (0.161 oz), 36 g silver (1.05 oz), 0.24% copper with recoveries of 92% for gold, 55% for silver and 70% for copper. Alamos proposes a feasibility study which will include a 50,000 tonne bulk leach test.

Apparently, the Mexican government installed an 80 ton-per-day flotation mill on the site in 1987 and processed sulphide ore intermittently until 1990. The concentrate was shipped to a Mexican government smelter.

Many new mines have developed in Mexico during the past decade with a more favourable climate for foreign investors. Perhaps La Fortuna may yet produce as a bulk minable deposit!

Darkest Days

In the late 1970s we gave high priority in Western Canada to exploration for precious metals in geologically prospective areas in B.C. and Yukon Territory. The approach was patterned after my previous experience with Kennco in grass-roots programs in Western Canada guided initially by silt sampling for base metals, principally copper. However, our emphasis was to be initially focused on the collection of heavy mineral samples from stream sediments, followed by the staking of areas considered as probable source regions prior to detailed discovery surveys.

By 1980 we had obtained considerable coverage over areas in north-western B.C., much of it in helicopter-supported programs in remote areas. This had led to claim staking over numerous promising target areas. Early follow-up surveys on the most anomalous targets for gold-silver with extraordinarily high values had proved disappointing due to nugget effects in the screened samples. Our emphasis shifted to the Iskut River / Unuk River region in what was to develop during the late 1980s as one of the most prolific regions for newly discovered precious metal deposits in B.C.

At a relatively early stage of this program we had retained the services of several experienced claimstaking individuals and groups. One of these was Hi-Tec Resource Management Ltd., through Malcolm Bell the personable owner-operator, to complete virtually all our staking requirements over target areas we proposed to evaluate in the Iskut River area. Through Malcolm we also obtained expert climbers, several with geological backgrounds, who completed sampling of several showings for us on highly exposed cliffs on the south side of Iskut River.

Our principal objective for 1980 was to obtain preliminary geological and more detailed geochemical coverage for these targets. In addition we planned to evaluate the "DOC" claims, an old gold prospect near the head of the South Unuk River which we had optioned from Tom McQuillan, a veteran prospector in northwestern British Columbia. I had approached Tom with a proposed deal on the DOC claims in early 1980. He was receptive to our offer which led to our option agreement with work required in 1980. I have called Tom a legendary prospector which indeed he was. I have no doubt that my personal recollections concerning him, which span such a brief period of this literature, can only serve to reflect his extraordinary character and generosity.

In the early 1930s he back-packed supplies and equipment into the upper part of the Unuk River area to exploration camps in this rugged and relatively inaccessible region. The only route used in this era was by launch from Ketchikan, Alaska to the mouth of the Unuk River - still in Alaska at Matney's ranch, a distance of 120 kilometres. For navigation up the Unuk River for 26 kilometres, specially constructed flat-bottomed shovel-nose river boats powered by outboard motors were required.

Transportation conditions into the Sulphurets Creek region at the head of the Unuk River in 1935 are vividly described in the following excerpt from the B.C. Ministry of Mines Report, 1935:

"The river is navigable to the first canyon, a distance of about 16 miles from Matney's ranch at the mouth. This stretch can be covered in one day. At favourable stages of water the First canyon is also navigable to near its head, where a sharp bend in the upper river-channel produces an extremely dangerous overfall and whirlpool. Navigation is at its best when the ice goes out in the early spring, usually about the beginning of May. At about the middle of November navigation begins to be impeded by ice. During part of the winter, dog-team transportation over the frozen river may be possible. Except under very favourable conditions, the stretch of the river between the First canyon and the Boundary, a distance of about 7 miles, is not navigable and can only be negotiated by means of continuous and arduous "lining".

Starting at the International boundary, a trail extends for a distance of approximately 17 miles along the west bank of the river to a point about 1½ miles above Sulphurets creek. Along this trail convenient cable crossings have been constructed across Harrymel (North Fork) creek, across the Fourth canyon to the south side of sulphurets creek, and across Ketchum creek 1½ miles above this creek.

Accommodation at the mouth of the river can be arranged with Messrs. McQuillan and King and farm produce may be procured from the ranch of Harvey Matney.

Arrangements for transportation up the river can be made with Messrs. McQuillan and King or Bruce and Jack Johnstone, of Ketchikan, who are familiar with the intricate, swift-water navigation of the stream. Quoted rates covering people and freight are 9 cents per pound to the Boundary and 16 cents per pound from the Boundary to Sulphurets creek. With the completion of trail facilities and the planned introduction of pack-horses, rates from the Boundary on may be proportionately reduced."

Not mentioned in the above report is the fact that supplies, i.e. freight from the Boundary to Sulphurets creek at the quoted price of 16 cents per pound were back-packed. However, in the Depression period one cent per pound per mile of back-packing was obviously considered acceptable, notwithstanding the need to travel back on the difficult route on most occasions without pay.

Tom's early role in this emerging well-mineralized but still remote region is recognized by McQuillan Ridge, extending for 12 kilometres on the east side of the Unuk River to Sulphurets Creek and by the 'McQuillan Vein' on the north side of Sulphurets glacier in the heart of the mineralized district.

I first met Tom while examining a mining property in the Stewart area in northwestern B.C. in 1960. The circumstances are worthy of record for they provide just one more example of the unpredictable importance of good luck, of being in the right place at the right time which is a feature of many significant mineral discoveries. In the summer of 1959 one of Tom's prospecting partners was walking along the road by the old Premier gold mine, controlled at the time by Silbak Premier Mines Limited. The property, staked originally in 1910, became one of British Columbia's principal gold producers with continuous production from 1918 - 53 and intermittent production through 1967, reopening again in the late 1980s. By 1967 it had produced 1.8 million ounces of gold, 41 million ounces of silver and important lead-zinc by-products from 4.8 million tons of ore mined. It was closed in the summer of 1959.

As he walked past an old glory hole following a period of heavy rain, Tom's partner noted that a wide quartz vein had been exposed from waste rock which had sloughed into the glory hole. On examination he saw what appeared to be very high-grade mineralization within the vein. After some thought he cut down a number of branches from nearby fir trees and covered over the exposure.

Tom and his two partners in Bermah Mines Ltd. quickly approached Silbak Premier Mines management and were successful in securing a one-year lease on the property from the company to terminate on September 23, 1960. The lease was quite specific as to length and the number of men that could be employed at the mine.

The vein discovered in 1959 was parallel to and 5 metres on the footwall side of the main vein that had been stoped in the glory hole. During the latter part of 1959, Tom and his partners shipped 62 tons of high-grade ore sorted from the slough in the bottom of the glory hole. The shipment contained a total of 650 ounces of gold and 16,829 ounces of silver.

The B.C. Ministry of Mines report, 1960 contains the following information on the 1960 production period:

In 1960 a short sublevel drift 30 feet long was driven along the new vein and a raise put through to surface in ore. High-grade ore as a shoot about 35 feet long, 4 feet wide, and 100 feet down dip was benched down through the raise and drawn off through the sublevel.

On the termination of the lease on September 23rd, the company bought the lessees' equipment and, with a crew of twelve men, continued to mine high-grade ore until November 1st, 1960, when operations ceased.

Production during 1960 amounted to 1,282 tons of high grade ore, 1,239 tons being mined by Bermah Mines Ltd.

The smelter returns from the ore shipments to Tom and his two partners reportedly provided a net profit of about $300,000 including the cost of mining.

I have also mentioned Tom's role with his prospecting partners in unselfishly sharing their meagre supplies with a starving field crew supervised by Al Lonergan in 1956 in the headwaters of Snippaker Creek, a south tributary of Iskut River, some 45 air kilometres north-west of the DOC claims (see "Searches").

I considered our proposed program in the Iskut River / Unuk River area as one of our most significant undertakings in 1980 and one which would require experienced early evaluation of the precious-metal potential of the numerous claims staked by DOX in the region. Sandy McLean had been actively involved during the previous two years in the evaluation of our Chappelle gold-silver property which advanced to the pre-production stage following the DOC Board decision in early 1980. Chris Gunn, another Senior Geologist, well-experienced in gold property evaluations was also available. Both Sandy and Chris worked well together, sharing several common interests, including the intricacies involved in evaluating placer gold deposits and sailing, not that either were considered prerequisites for the 1980 program! Incidentally, Chris was a popular instructor at the B.C. & Yukon Chamber of Mines Placer Mining School.

In order to provide both of them with the opportunity to focus their energies on the challenging property evaluations, I also enlisted the assistance of Bill Smitheringale as an on-site consultant to supervise the proposed program on the DOC claims. I had last worked with Bill in 1954 when as a Senior Geologist I supervised a far-ranging program of property examinations and general reconnaissance in British Columbia. One of our more memorable experiences was the examination of the Grace Claims lying about 12 kilometres south-east of the DOC claims on the east side of the South Unuk River. (See "Introduction to babies, helicopters, perennial snow and ice.")

In support of our 1980 program we established a base camp on the south Unuk River bank below the DOC claims and arranged to share one of two helicopters chartered for the season by Esso Minerals Canada to assist in an exploration program based at Mitchell Creek about 24 kilometres to the northeast.

The helicopters were both Bell 206 Bs (Jet Rangers) owned and operated by Vancouver Island Helicopters Ltd. (VIH) a subsidiary of Vancouver Island Airlines Ltd. Apart from use by Esso and DOX, the helicopters had also been used during the summer by Hi Tec on their staking programs for DOX.

In mid-June I visited the exploration crew and had an opportunity to examine several of the claim areas. I spent one day with Sandy examining mineral showings located by the crew on the Warrior claims on the south side of the Porcupine Glacier, about 65 kilometres north-west of the DOC claims, and was much impressed with the apparent mineral potential in this remote area.

Shortly after my return to Vancouver, I was alone as Rene left for a stay with her great friend Helen Langford, a classmate from her School of Nursing days at the University of Toronto. Helen had been a bridesmaid at our wedding and the two had kept in touch over the years. We were to spend many enjoyable short summer visits with Helen and her husband Walton at their cottage on an island in the Lake of Bays in Northern Ontario and this was where Rene was staying in early July, 1980.

At about 1:00 am on July 4, 1980 I was awakened by a telephone call. It was Bill Smitheringale. I'll never forget his words, "Dave, I'm afraid I've got some really bad news. There's been a helicopter accident up here". I believe I interrupted to ask if there had been any injuries. "It looks like at least four people have been killed, although identifications aren't possible at this time", he said. On questioning he named four DOX individuals plus the pilot of the helicopter who were believed to be involved, although only four bodies had been located during the late evening search and it was hoped that one other individual might still be alive.

I couldn't believe what I was hearing. I didn't know the four individuals. After recovering slightly from my shock, I realized that I should report the information to senior management in Toronto. Accordingly, I telephoned Gordon Whitman, awakening him with the tragic information.

Conscious of what I was about to face and lacking direct experience with any sudden death, let alone of this magnitude, with identification uncertain, at that early hour of July 4 I took the easiest preparation I could think of, swallowing a couple of sleeping pills that happened to be available.

An on-site investigation on the morning of July 4 by RCMP and others at the crash site indicated that all five individuals had been killed, the information being passed to me by Bill Smitheringale. We discussed the need to close down the camp temporarily and return to Vancouver with Cynthia Hamilton and Jacques Dupas, the only two remaining DOX temporary employees on site.

Those killed in the tragedy were: Robert Cecil Clarke, pilot of the helicopter, aged 26 with 1485 hours of helicopter flying experience and DOX employees, Christopher Bruce Gunn, aged 40; Keith Alexander (Sandy) MacLean, aged 40; Ruth Anne Nussbaumer, aged 23 and Ian Ross Shaw, aged 20, both university students. Ian was the son of Tom Shaw, Western District manager of DOC's explosives division in Vancouver.

A review of the events leading to the accident and the initial investigation formed part of a report prepared by a DOC/DOX Investigating Committee dated August 22, 1980, which I chaired.

Between 8:00 - 9:00 a.m., Vancouver Island Helicopter Bell 206B, C-6SHE piloted by Robert Clarke arrived at the DOX base camp. It took Cynthia, Jacques and Bill to a site on the DOC claims immediately above the base camp, returning within 10 minutes to pick up the remaining DOX employees. Sandy and Ruth were dropped off on the Warrior Claims on the north side of Iskut River about 55 kilometres north-west of the DOX base camp; Chris and Ian being dropped off about 12 kilometres to the south of the Warrior claims on a tributary of Snippaker Creek. Both parties were to carry out geological traverses, collecting rock samples.

The aircraft returned to work for Esso Minerals and as prearranged the pilot picked up the party working on the DOC claims at about 6:00 pm, returning them to base camp before leaving to pick up the other two parties. He picked up Sandy and Ruth first and continued on to a designated pick-up site in the Snippaker Creek area. The time for the second pick-up was estimated at about 8:00 pm with the return flight to base camp estimated at about 30 minutes.

The actual accident is described in the August 22 DOX / DOC report in the following selected excerpts:

2.5   The helicopter crashed shortly after takeoff at a point estimated to be about 2000' from the last pick-up point which was identified by footprints. Evidence indicates the tail skid gear hit first contacting the rock surface. The machine probably pitched forward, the tail boom separated and it landed inverted and fire ensued. The wreckage was distributed over an elliptical area about 200 feet long. There was considerable damage and much fragmentation. The crash site was approximately 100' lower in elevation than the pick-up point. Appendix 3 contains photographs of the helicopter involved and the terrain in the vicinity of the crash-site.

2.6   The weather was clear and the temperature is estimated at 15 - 20' C. The area of the crash was close to the toe of a glacier. It is common in these areas to have winds coming down the face of the glacier and the valley below. Typically at the time of evening of the crash, these winds are diminishing. Overall, the winds and weather were discounted as factors by VIH following their investigation.

2.7   The base camp became increasingly concerned when the helicopter did not return as planned and tried to contact the VIH base at Stewart in order to initiate a search at about 21:30 hours. Contact was finally made at 22:00 hours. The search by four helicopters commenced about 22:30 hours and the wreck was discovered about 23:30 hours. D.A. Barr was notified in Vancouver about 00:50 hours 1980 July 04 from Stewart, B. C.

2.8   On the morning of 1980 July 04, the R.C.M.P.; the Chief Pilot and Chief Engineer of VIH, together with the DOX consultant visited the site. The bodies of the DOX employees and the pilot were removed late that day under the direction of the R.C.M.P. to Prince Rupert for autopsies.

The portion of the report covering the investigation noted that the aircraft involved was manufactured in 1977 and had recorded about 1200 total operating hours. There were over 1000 helicopters operating in Canada in 1980, the Jet Ranger being the most common type. Neither DOX nor VIH knew of any prior problems or deficiencies concerning the helicopter on the day of the accident.

The load at the time of the accident was estimated to be 1230 lbs. assuming fuel tanks were 1/2 full.

This was within authorized limits for the conditions encountered.

VIH had about 38 employees, 18 helicopters including 10 Bell 206s and operated from four bases on the Pacific Coast of B.C., the nearest being at Stewart, about 100 kilometres to the south. VIH had been in the helicopter business for 25 years and had a good reputation. The crash was VIH's first fatal passenger incident. Based on a visit by the six-person investigating committee to VIH's main base in Victoria which included DOC safety and aircraft division personnel from Canada and the USA, VIH management was considered effective, the pilot was properly trained and experienced and maintenance was acceptable. Bob Clarke had been employed with VIH for 18 months after employment with another helicopter firm. He had flown 1485 hours in helicopters with flying times in 1980 of 50 hours in April, 42 hours in May and 150 hours in June. He had had one previous accident in September 1979 which was not similar to the fatal accident. On July 3 his flying time estimated at 5-1/2 hours and his duty load at about 12 hours was not considered excessive by industry standards.

Transport Canada investigators visited the crash site on July 4 and subsequently removed significant portions of the wreckage for testing in Victoria and by the National Research Council in Ottawa.

A study was completed of recent helicopter accidents and the results were summarized as follows:

3.11   Helicopter accident and fatality experience was obtained from Transport Canada. Basically since 1970 the analysis indicates that 83% of the accidents can be attributed to the pilot. This also holds true for fixed wing planes. During the years 1974 to 1978 inclusive, there were 225 accidents with 22 fatalities. The average is 15 accidents and 1.5 deaths per 100,000 hours flown. An analysis of DOX helicopter usage indicated approximately 800 people hours in the year ending 1979 August 31. On this basis DOX would have sustained one death every 80 years. In fixed wing exposure, the DOX rate would have been one death every 800 years.

The Investigating Committee concluded that:

The basic cause of the accident was unknown. Probable causes were:

1. Equipment failure;
2. Pilot error or pilot incapacitation.

The Committee were unable to identify evidence to establish any of these as basic cause. The final Transport Canada report might contain new information.

I prepared an addendum on 'Helicopter Safety in DOX' to the Investigating Committee's report in the form of a memorandum to senior management with copies to the Investigating Committee. In the memo I noted that although there were no deficiencies in helicopter operation, practice or procedure uncovered by the Committee which were considered to be contributing factors to the accident, the safety of DOX's operations could be substantially improved by eliminating any unnecessary risks. My concluding statement was the following:

The tragic accident, which took the lives of five persons, like most accidents, was unexpected and deeply shocking. At this stage of investigations into the cause, both by Transport Canada and by representatives of DOC and DOX, there is no known explanation. Hopefully, a plausible cause will result, only because it could help to prevent a recurrence of such a tragedy to DuPont, or to others relying on helicopter transportation in the future; however, the possibilities of detecting a positive cause appear remote at this time.

DuPont has responded positively and without delay in its independent enquiry into the accident, making available its qualified and highly concerned safety personnel, all of whom have actively contributed to the investigation. Considerable data have been obtained by the Investigating Committee, which have led to the consensus that much can be done to reduce helicopter accidents and fatalities on an industry-wide basis by promoting a programme of safe procedures and incident reporting designed to eliminate or substantially reduce unnecessary risks in helicopter operations.

During the formative years of DOX, the Investigating Committee Chairman expressed to senior DOX management his appreciation of the emphasis on safety by DuPont in all its operations, noting that such an emphasis had no parallel, to his knowledge in mineral exploration organizations. This led to the development of a safety awareness policy by DOX which culminated recently in a Vice Presidents' Award, of which all DOX personnel can be justifiably proud. All four DOX employees involved in the helicopter accident were attracted to their vocation by an appreciation of the wilderness, its demands and rewards, and all were staunch devotees of safe working practices.

Although the potential hazards of mineral exploration have been enunciated in the past by DOX, those individuals not directly involved in exploration cannot be expected to appreciate the diversity and extent of the potential hazards, encompassing as they do, most of the hazards encountered in the transportation and mining industries and those experienced by personnel working in all types of terrains in remote localities. The recent tragedy has focused attention on only one of these potential hazards, i.e., helicopter transportation of personnel in exploration work. Although there were no deficiencies in helicopter operation, practice or procedure uncovered by DuPont's investigating committee, it was concluded that the safety of our operations can be substantially improved by eliminating areas of unnecessary risk. It is proposed that a similar committee meet in the near future to consider other areas of DOX's operation with the same objective.

The most proactive result of this proposal was my recommendation for the formation of a Safety Committee by the B.C. & Yukon Chamber of Mines in September 1980. 1 have chaired this active committee since its formation; however, that activity is covered in a later chapter.

My principal concern on the morning of July 4 was the realization that I had the responsibility of informing the wives and parents of the four DOX victims of their sudden calamitous deaths and that there was virtually no time available because of the imminent public announcement. The aftermath of dealing with the tragedy was to require virtually all of my attention for several months. Not only was it required, I was so deeply shocked and affected that my sole objective was to provide whatever assistance I could to those who I felt needed it most - particularly Nina Gunn and Alison MacLean. In return, I was to receive reciprocal support from all four families, many other friends and acquaintances and DuPont's management group.

On July 4 1 informed Tom Shaw of the loss of his son at our office in Vancouver; we shared the same space in the Sandwell Building on Alberni Street. Shortly after I telephoned Alison MacLean at her home in West Vancouver , followed by a call to Nina Gunn at her home in Deep Cove in North Vancouver. Alison had no advance information, and nor had Tom Shaw. Naturally, both were deeply shocked by the tragic news. Nina had received a phone call from a friend who had overheard a radio report concerning a fatal helicopter accident in the Stewart area. Her friend was with her when I phoned, but the confirmation was obviously a terrible shock. Ruth Nussbaumer's parents were travelling and could not be reached on July 4.

Both Alison and Nina had the heart-breaking task of explaining the loss of their father to young children. Sandy was survived by twin daughters Andrea and Kirsten and Chris left his son Ian and daughter Erica. The ordeal was equally difficult for Tom and Rosemarie Shaw and Peter and Margaret Nussbaumer, both of whom had a surviving son and daughter.

Several days after the accident I met Sandy's sister Bonnie and husband Bill Beveridge who had arrived from Ottawa to be of assistance to Alison. Nina Gunn was fortunate to have the nearby company of John and Maris Ratel, two very close friends.

During the period prior to funerals for all four DOX victims and for some time later, I maintained a running memorandum with the help of Annikki Puusaari, on the numerous arrangements required on behalf of the four families. Included were assistance in the preparation of obituaries for Alison and Nina, contacts concerning funeral arrangements, disposal of personal effects including the contents of the offices of Sandy and Ian, contacts with RCMP, coroner's office, life insurance companies, legal representatives and DOC's personnel involved in the provision of benefits to the estates of the victims.

The cooperation I received in attending to these matters was almost overwhelming as so much had to be arranged in so little time. The area which caused me most concern was the time that would be required for the completion of autopsies and the issuance of death certificates. The completion of autopsies affected planning for funeral services at a most sensitive period whereas a delay in the issuance of death certificates was considered a potential major inconvenience.

On July 6 1 learned that the autopsies were being performed by at least two pathologists at the Prince Rupert Hospital, one being Dr. Charles Leake, coroner from McBride, because of his expertise with helicopter accidents. The remains of the victims were to be released to Ferguson's Funeral Home in Prince Rupert for furtherance to Vancouver.

Death certificates were eventually made available at isolated Dease Lake, not Prince Rupert and I made a special trip there to pick them up and return them to Vancouver for the signatures of W.S. Carpenter, Coroner on August 7, 1980.

If the shock of Bill Smitheringale's phone call on July 4, informing of the tragic accident heralded the worst day of my life, attending four funerals / memorial services on July 10 and 11 was a close second.

Within a week I returned with a new crew supervised by Louise Eccles which included individuals who had been working with Marshall Smith in southern Yukon. I stayed with them at the camp on the DOC claims for about one week. While there I had the opportunity to fly into the Snippaker Creek area where I visited the site of the accident. I had the helicopter pilot leave me alone for a couple of hours while he flew off to service other party members. It was a brilliant, cloudless day with no indication except on the barely visible crash site that this was the scene of the worst helicopter accident in British Columbia to that time and possibly to the present. The site was less than 500 metres from the scree-covered toe of the Snippaker Creek glacier within its broad, scoured valley which rises steeply on either side to craggy snow-covered ridge crests, culminating at the head of the valley in two ice-falls surrounding a nunatak, reminiscent of Galore Creek. Where I stood amidst coarse fragments of scree which partly supported a light growth of brush and fireweed, the vista and setting was similar to dozens of other sub-alpine valleys in the coastal mountains. Why and how could this terrible loss of life have occurred?

Over a couple of hours I built the largest cairn that I have ever constructed or for that matter seen; at least seven feet high with a similar diameter at the base.

After my return to Vancouver I visited W.R. Chandler Memorials Ltd. on Fraser Street in Vancouver and arranged for them to prepare a bronze plaque attached to a grey base and engraved with the following dedication:

IN LOVING MEMORY OF
ROBERT CLARKE
CHRISTOPHER BRUCE GUNN
KEITH ALEXANDER MACLEAN
RUTH ANNE NUSSBAUMER
IAN ROSS SHAW
PILOT AND EXPLORERS. KILLED IN A HELICOPTER ACCIDENT.
JULY 3, 1980.
GONE BUT NOT FORGOTTEN.

I arranged to have it crated for shipment and picked it up in mid-September and shipped it to Vancouver Island Helicopters at the Stewart base to be held until further notice.

On October 3, 1980 I flew to Stewart with Marshall Smith with the object of examining the basal area of cliffs on the south side of Iskut River near Snippaker Mountain where we had recently staked mineral claims to cover the assumed source of anomalous gold-silver values obtained in several creeks draining the region. We hoped to obtain some visual evidence of mineralization in the creek beds or below the cliffs for the values. The crated plaque, some sand, cement and a small shovel and pail formed part of our cargo.

Before returning to Stewart we visited the crash site on Snippaker Creek and embedded the plaque in concrete at the base of the memorial cairn. This was followed by a moment of silence while the three of us stood beside the cairn. Unlike my previous visit, it was a cloudy day with light rain. I took several photos of the cairn and the plaque, copies of which I later forwarded to the surviving wives and parents. A close-up of the plaque was most dramatic, the rain drops having puddled on the base, appropriately resembling giant tears.

On returning to Stewart we intercepted a radio message from the helicopter base, asking us to land at a mining camp on Bear Pass near the road linking the Cassiar-Stewart highway with Stewart, as a Hughes 500 helicopter with three miners aboard was overdue. As we flew toward Bear Pass we encountered an increasingly low ceiling, but we had no trouble locating the camp.

A small group of men were clustered together near the landing area and we spoke with the manager who provided more details concerning the overdue helicopter and its crew. Apparently the aircraft had taken off that morning at about 9:30 am for the mine site which was located several miles from the camp in a mountainous section. The manager stated that the pilot had waited for some time before departing for improved visibility and had eventually taken off with the three miners through a hole in the clouds but the pilot had failed to return from his short flight. We were immediately alarmed by this information. The manager showed us the location of the mine site on a map and we headed off in the general direction, encountering impassable cloudy sections en route. Within 10 minutes we spotted several bright white objects about 2000 feet above Bear Pass on a steep mountain slope just above the timber line. A closer inspection showed that they were parts of the helicopter tail assembly and window frames, distributed among other parts of the helicopter in an area with a diameter of about 50 feet containing a blackened core area. There was no evidence of movement from several visible bodies. Thinking that there might be some chance that one or more of the men might be alive but too injured to move, and more for the presence of someone nearby for possible comfort, I asked the pilot if he could get close enough to drop me off with the helicopter's medical kit. He said that it was too risky, but I believe that he had made his mind up that there was no possibility of a survivor. Based on the catastrophic nature of the impact that must have occurred, it was hard to argue. I took several photos of the crash site. It was about 5:30 pm but the lighting conditions were still satisfactory. Two of the photos were later used by Transport Canada as support material in its accident report.

On April 7, 1981 1 received a letter from the Vancouver Coroner's Service with a copy of the coroner's Report of Inquiry into the DOX helicopter accident which included the accident investigating report. The conclusions of the report are excerpted as follows:

3. CONCLUSIONS

3.1 Findings

1. The pilot was properly licenced.
   The pilot's medical category was valid.
   The pilot had received regular flight checks and was qualified on type.
4. The aircraft was properly certified and maintained.
5. Weight and balance limitations were not exceeded.
6. There was no evidence of aircraft failure or malfunction prior to the crash.
7. The aircraft was destroyed in the crash.
8. The pilot and four passengers received fatal injuries during the crash.
9. This was a non-survivable accident.
10. The deaths of all on board were accidental and unnatural.
3.2 Cause of the Accident
   The cause of this accident remains obscure.

4. SAFETY RECOMMENDATIONS

Since the cause of this accident is obscure it is not possible to make recommendations for corrective action.

Among the many individuals who wrote to or telephoned me to express sympathy at the time of the tragedy was my good friend Bob Cathro who reported that his company had experienced a helicopter accident at an exploration site on July 4, 1980 when an employee walked into a tail rotor of a helicopter during loading and was instantly killed. The pilot was badly burned in the ensuing crash.

The Snippaker Creek and Bear Pass crash sites are only about 65 miles apart. Research into the incidence of fatal non-scheduled rotary and fixed wing aircraft accidents in Western Canada has shown that by far the greatest number have occurred in this part of British Columbia. The accepted reason is that the conditions associated with the Coast Mountains produce the most unstable weather patterns in the province, with a combination of heavy precipitation, high winds and rugged topography. That is probably only part of the reason. The region is also one of the most highly mineralized portions of the Coast Range mountains with the most difficult and restricted access, there being virtually no roads within the area until only recently. Those working in the area or visiting it have relied, to a great degree, on both fixed wing and rotary wing aircraft to gain access. Almost all the ensuing accidents to aircraft were related to support of mineral exploration and mining projects in the area and I am aware of a dozen or more.

Baker Mine (Chappelle Property), Toodoggone River Area, B.C.

The Chappelle property, so-named after a cook on an early survey through the Toodoggone River area in the Omineca Mountains of northern B.C., was to absorb a significant part of my time with Kennco and later with DuPont. Its serendipitous discovery by Gordon Davies, a Kennco prospector in 1969 and its subsequent exploration and development into the first totally aircraft-supported mine in B.C. were a couple of its trademarks. Its related activity focusing on the gold-silver potential of the district led to a major staking program in the late 1970s, culminating in the development of two other mines (Cheni and Shasta) and the extension of the Omineca Road into the region. Expenditures attributed to this relatively modest period of exploration, development and mining in the area between 1970 and 1990 exceeded $150 million in actual dollars.

I was author and co-author of two papers on the property, both published in The Canadian Mining and Metallurgical Bulletins. In the first 'Chappelle Gold-Silver Deposit, British Columbia', February 1978, I described the history of exploration, geologic setting, geochemical and geophysical investigations, genesis and economic considerations. A production decision was anticipated at the time, although it was not to be approved until February 1980. The influence of environmental considerations related to a mining production decision was evident even in the late 1970s as evidenced by my final paragraph:

"From the outset, exploration, development and financial planning for eventual production at Chappelle has been directed toward the objective of a neat and tidy mining community with high standards of industrial health and safety, and with minimum disturbance or pollution of the surroundings. To this end, chemical and biological studies of water resources in the immediate property area were completed in 1976 to establish acceptable base-line information on the aquatic environment. On completion of any mining operation, all disturbed and unseeded areas affected by such an operation would be re-vegetated. Thus, although the property lies near the eastern boundaries of Tatlatui

Provincial Park and Spatsizi Plateau Wilderness Park, it will be no more of an environmental hazard than the public roads providing access to the region."

The second paper, 'The Baker Mine Operation', published in September 1986, was a joint effort with contributions from those involved in the mining operation, including Tom Drown – exploration and assistant mine geologist; Terry Law, a DOC employee seconded to the Baker operation as manager of mining operations; Gordon McCreary, who joined DOX from DOC as a mining engineer; Bill Muir, mill superintendent from late 1981 to 1983; Jim Paxton, mine geologist; Bob Roscoe, mine manager; plus myself as general manager. It was unique as a paper reviewing the complete cycle of a small mining operation in a remote setting, which showed that industry and wildlife can co-exist in harmony. It also compared the feasibility assumptions with actual mining experience.

The earliest recorded prospecting activity in the Toodoggone River, an easterly flowing tributary of Finlay River, occurred about 6 km. north of the Chappelle property in the early 1930s. Lead-zinc mineralization in skarn about 1500 m south-west of the property was also discovered, staked and explored by Cominco Ltd. during this period.

In 1968 Kennco carried out geochemical reconnaissance surveys in the search for porphyry copper deposits in portions of the Hogem and Cassiar batholiths within the Cassiar-Omineca mountains. Several base metal anomalies were defined which were considered of further interest. One of these was a relatively weak copper-in-stream sediment value which led Gordon Davies to the base of a talus[24]-covered slope above the timber line where he located fragments of pyritic quartz float. His curiosity was aroused by the precious metal possibilities, so he collected a few samples for assay. Gordon and the rest of us at Kennco were astounded when one of the samples assayed 85.7 grams gold / tonne (2.5 oz/ton) and 2229.0 grams silver/tonne (65 oz./ton).

Following claim staking, exploration between 1970-72 consisting of prospecting and trenching resulted in the discovery of quartz veins near the mineralized talus and at six other locations in the surrounding area.

The only significant discovery occurred in Vein A, near the site of the original discovery. Detailed surface sampling indicated an average content of 34.3 grams gold/tonne (1.0 oz/ton) and 617.4 grams silver/tonne (18 oz/ton) for a length of 230 m and across an average width of 3.0 m. Two X-ray drill holes showed persistence of mineralization to a depth of at least 20 m.

In 1973 Kennco decided that the property did not meet its target criteria and an agreement was negotiated with Conwest Exploration Ltd. During the year, Conwest built an 850 metre airstrip at nearby Black Lake and a 3 km access road from the airstrip to the property. An adit was driven for 160 m to intersect Vein A at a depth of 40 m below surface and 50 m of drifting were completed on the vein. Only two of eleven underground drill holes intersected high-grade mineralization and Conwest terminated its option in late 1973.

As mentioned in an earlier section, I negotiated an option on the Chappelle claims with Bob Stevenson of Kennco about one month after joining DOX in January 1974. Over the next five years DOX completed over 9,000 m of diamond drilling in 96 holes and underground development work including 100 m of cross-cuts and 400 m of drifting and raising. By 1979, following further drilling and underground work, a drill-indicated reserve was estimated by Martin Kierans based on 28 drill holes at 93,700 tonnes averaging 0.99 oz gold/ton, and 19.84 oz silver/ton at a $125.00/ton cut-off.

---

[24] Talus: broken rock fragments

During the exploration stage I had an opportunity to provide my son Rob with a short summer job at the property. He had spent an earlier period that summer as an assistant in an exploration program supervised by Dave McAuslan in the remote Back River area of the Northwest Territories. The change from the barren land environment with its infestation of black flies and mosquitoes was a welcome one; however, the experience convinced him that although he relished the outdoors and particularly a mountainous environment, mineral exploration was not his bag.

In the summer of 1979 our younger son David spent about three weeks based at a lodge at Moosevale, then the end of the Omineca Road, which had been extended by the government as a potential road to resources for about 240 km north of Germansen Landing on the Omineca River over a 25-year period. He worked as a rod-man during the construction of an airstrip at Sturdee River about 40 km north-west of Moosevale and 10 km by a proposed road linking the airstrip to the Chappelle property. The best part of the job as far as David was concerned was the daily flight by helicopter from Moosevale to the airstrip and return – particularly the latter!

For many years I have joined others in the mining sector emphasizing the need for road access to most mining properties in order to support a viable mining operation. Early exploration and even development can often be supported by aircraft, as shown by the emerging use of float-equipped fixed-wing aircraft commencing in the 1930s in many parts of Canada and by helicopters in the more mountainous regions in western Canada starting in the mid-1950s. Although the obvious need of access roads in most cases was recognized by the public and bureaucrats until only two decades ago, the growing preservationist movement has successfully provided this same influential sector with sufficient misinformation to convince it otherwise.

Road construction was suddenly stopped at Moosevale in 1978, reportedly for budgetary reasons but almost certainly in response to a growing environmental pressure citing potential adverse impacts on wildlife, particularly caribou in nearby Spatsizi Wilderness Park. This was a severe disappointment to DOX and other exploration companies, which were relying on continued construction to provide road access to their properties for mining purposes, following any production decision.

Not generally recognized is that the Omineca Road construction had been partly supported by aircraft, requiring airstrips at intervals of about 70-80 kilometres along the road route. In 1978 I approached the Ministry of Energy, Mines and Petroleum Resources for possible support with an airstrip at Sturdee Valley after it had been recognized by Bob Stevenson as an obvious and beneficial site for our Chappelle operation. I received immediate support which led to Nielsen's contract for the 1000 metre long airstrip in 1979.

Following completion of DOX's final feasibility study in 1979 a production decision totalling $12.5 million was approved by the DOC Board of Directors in February, 1980. The decision was backed up by Kilborn Engineering (B.C.) Ltd.'s feasibility results which concluded that the project would be financially viable under nine separate cash flow projections using various gold and silver prices plus capital and operating cost assumptions. The DCF after tax rates of return calculated by Kilborn ranged between 45-65%. Indicated payback for the capital costs ranged from 15 to 21 months and the defined reserves provided an operating period of about 2 times the payback period. Undoubtedly an unstated encouragement to the DOC board in reaching the production decision is that it coincided with the highest recorded market prices ever attained by gold and silver being in the US $900 and $30 per ounce range, considerably higher than the selected feasibility values of US $350 gold/oz and US $11.50 silver/oz.

By 1979 total exploration and development costs – principally in diamond drilling and underground development were $3.3 million of which DOX's portion was $2.2 million. In the latter part of the summer period of 1979 DOX contracted with Bud Nielsen for lengthening of the Sturdee Valley airstrip to 1600 metres in anticipation of committing to a totally air-supported operation with a PWA C-130 Hercules aircraft.

Sandy MacLean and I had spent a long day in August 1978 flagging out a road route from the Sturdee Valley airstrip to link up with the old road from the Black Lake airstrip to the Chappelle property. This route was roughed out in 1979 by Bud Nielsen. When the production decision appeared imminent we arranged for Bud to upgrade the road in late 1979 and early 1980 in preparation for the delivery of 4000 tons of ATCO trailers and construction equipment, fuel and supplies to the property. However, we were to encounter an unexpected and unappreciated predicament when the overall contract for the construction of the mill and camp buildings, tailings impoundment and other components of the operation was let to Dillingham Construction of Vancouver, a large union operation. Bud Nielsen was a small, non-union contractor and, of course, the two could not work on the same job site according to union regulations.

Bud did the best he could to complete the upgraded road prior to the deadline imposed by Dillingham in order to meet their own scheduled completion date of late December, 1980. Finally, Sandy and I had the unpleasant task of having to inform Bud that the completion deadline had to be enforced. Sometime later he sued DOX on the grounds of a cancelled contract. The case was heard in Prince George and I was required to accompany DOX's lawyer from Vancouver. As expected, my meeting with Bud accompanied by his wife was brief and uncomfortable for both parties. The case was dismissed after a very brief hearing. I was to meet Bud once again in 1980. He had travelled from Fort St. James to attend the funeral service for Sandy MacLean and had also sent me a personal note of sympathy over Sandy's death.

The construction period for the Baker mine, which I had named after Art Baker who had died suddenly in 1979, required 213 flights with the Hercules. In addition 250 return flights were made with a company-based Navajo Chieftain for transportation of personnel and small freight. In late 1980 DOX purchased a Twin Otter aircraft for $2.8 million that was used almost exclusively for support of the Baker mine operation. The aircraft was based at a company-owned hangar in Smithers and was maintained by a 5-man crew contracted from Innotech Aviation. It was to log 1325 return flights to Sturdee Valley during which 1645 tons of freight and personnel were carried. In addition it flew the output of the mine in the form of 60-70 lb. dore bars at intervals to Vancouver for refining.

The construction phase of the Baker Mine presented many challenges typical of remote mine developments. Within 400 days of the production decision, the first dore bar was poured at Baker Mine after airlifting over 10 million pounds of freight and transporting over 1300 passengers between Smithers and the Sturdee airstrip.

In celebration of the first dore[25] pour, I sent the following telegram to the DOC head office:

"Baker Mine gave birth on April 2, 1981 to quintuplets with reported aggregate weight of 2623.3 ounces. Offspring resting comfortably in vault. Details to follow. Doctor reports further deliveries anticipated within week."

On average the silver to gold ratio throughout the 30-month operation was 20.6 to 1.0, thus each dore bar was about 95% silver and 5% gold. A total of 37,558 oz. gold and

---

[25] dore: gold and silver bullion

742,198 oz. silver were produced from 87,740 tons milled prior to closure with recoveries of 94% for gold and 84% for silver when calculated heads from mill feed reached designed levels.

As indicated in considerable detail in the paper on 'The Baker Mine Operation', the mining operation compared poorly with the feasibility assumptions provided by Kilborn. The milling operation was initially plagued by the unconventional use of a single-stage dewatering section utilizing a belt filter which could not adequately process the finely-ground ore. A drum filter was eventually added and with it process development throughputs reached design levels in late 1981. Tonnage dilution had been estimated by Kilborn at 10-20% with waste material containing no gold or silver values. In practice ore dilution averaged about 65%, because of the extensive faulting, incompetent wall rock, and difficulties in grade control. Eventually the upper portions of the deposit below original shallow open-pitting had to be mined by open pit with unplanned dilution. Although cut-off grades sere 0.3 oz/ton gold equivalent, dilution eventually forced by-passing of lower grade ore blocks with removal of 40% of the original reserve tonnage and 34% of the assumed value content.

Pre-production capital costs were $14.4 million compared to original project estimate of $12.5 million ±20%, or 15.2% higher. In addition, $2.8 million was added in the first year for purchase of a Twin Otter aircraft to provide more reliable transportation than would have been possible with charter aircraft. A hangar was also constructed at Smithers for adequate inspection and maintenance.

The underground mining method employed was cut-and-fill. The milling operation employed conventional cyanide-leach process with cyanide destruction initially followed by alkali chlorination and subsequently by the newly developed Inco SO_2 / air process.

Despite the demoralizing problems encountered and mostly resolved, the Baker Mine met its operating costs on a monthly basis with only one exception during the 30-month operation. However, it failed to repay the full capital investment by a significant margin.

On a more positive note, an excellent working and living climate was developed at the mine through the close relationships maintained among all personnel. The initial rotation allowed for a 21-day-in and 7-day-out schedule. It was modified in 1982 to a 14-day-in and 7-day-out basis, primarily at the suggestion of employees. Efficiency was not affected, productivity was improved by about 12% and employee satisfaction was solidified.

Because of DOC's emphasis on safety in all its operations, safety was a prime concern from the outset of the operation. In 1982 and 1983, Baker Mine was awarded the British Columbia Small Mines Safety Award, presented annually by the West Kootenay Mine Safety Association to the safest small mine operation in B.C. In addition, the Baker operation recorded an accident-free air operation service provided by Innotech over the mine life of three years.

Recreation on the site centred around a satellite receiving dish that brought in four TV channels and one radio station. A VCR complemented the system that was distributed throughout the bunkhouses to each individual room.

Bob Roscoe, our mine manager, was an avid cross-country skier and covered many hundreds of kilometres throughout the surrounding area on his frequent outings in the long winter months. I had a pair of mountaineering skis which I had stored in my room on site. However, I made little use of them as I found the skiing conditions in the wind-swept mountainous areas quite treacherous. Thin crusts of snow frequently covered 'pot holes' which could cause abrupt falls without any warning during down-hill runs.

As previously mentioned, the surrounding area contains a diverse wildlife population, noted in particular as an important habitat for Osborn caribou, with herds of up to 300 animals having been observed within several kilometres of the mine site during rutting season, normally late September and early October. During the exploration period we had reports of caribou, always curious, close to drill sites. Later they were occasionally spotted around the mine workings. I have a photo taken by our caretaker in July 1988 showing three caribou grazing on fresh grass seeded on the tailings pond below the mill site in 1984-5.

During our periodic fuel transportation campaigns with the Hercules aircraft, the plane would, on rare occasions, have to make more than one pass for a proposed landing in order to allow time for caribou on the airstrip to be encouraged to leave!

There were two prominent wolf packs within the area during the operation. Each contained 6-8 animals. I only saw them on one occasion while driving along the road from the airstrip to the mine in winter, when we followed one of the packs for almost a kilometre as they loped ahead of us before veering off the road.

Moose frequent the lower areas below timberline and could frequently be spotted in the open swales and around several small ponds near Black Lake. Goats were also visible on some high mountain crags from a rough road extending from the property to the nearby Lawyers property. Our caretaker reported seeing a large grizzly bear near the camp from a window of an ATCO building and running for his rifle in the event of an unwelcome encounter.

Smaller fur-bearing animals spotted near the mine included wolverine, marten, fisher, beaver and hoary marmot. Bob Roscoe reported being followed by a wolverine on one of his cross-country ski trips. Our explosives man loved animals. He was an early riser and following breakfast he would begin his chores driving up to a small hut near the main adit where he had sundry supplies. Included was some feed for several of the small fur-bearing animals that frequented the area, including hoary marmot. He mentioned that on one occasion he had fed three different species at the same time during these early morning sessions.

Considering the isolation and the variety of individuals employed at Baker Mine, it was surprising that there were hardly any serious brawls among the men in a camp containing up to 44 company and 3 contract personnel. The most serious damage to any of the buildings by company personnel was by a burly bulldozer operator who became somewhat unhinged and wrecked part of the entrance to the cookhouse with the blade of the bulldozer before he was subdued by four or five of his fellows.

The greatest potential threat to the camp buildings occurred in January 1985 after the mine had shut down and was in a care and maintenance mode. It was a bizarre episode and warrants separate treatment.

A Burning Desire

Following the suspension of mining activities at the Baker mine in late November, 1983 the property was placed on a care and maintenance basis. During the early part of this period a considerable amount of equipment and supplies were removed from the property and sold. However, the mill building and the ATCO trailer complex, which accommodated personnel, were left virtually intact.

Following mine closure, I had the good fortune to make the acquaintance of Don Davidson, then a self-employed geological engineer living near Hudson Bay Mountain, just west of Smithers. Don had worked for many years as an exploration geologist with Climax-Molybdenum Co. Ltd. and affiliated companies and was retained on a part-time basis after the company phased out of Canadian activities. The nearby Hudson Bay Mountain (Glacier Gulch) molybdenum project with its office and core storage facility was one of its principal exploration assets in B.C. in the early 1980s.

Don also provided consulting and caretaking services to the mining industry through his company, North Central Services Ltd. I arranged to retain Don's caretaking services for the Baker mine property commencing on June 1, 1984. Don selected Denis Barnett of Smithers as a replacement for the initial on-site caretaker on December 1, 1984. A caretaking stint of about 5-6 weeks was envisaged. Denis, then in his mid-20s, had worked at the Baker mine as a heavy equipment operator during the summer of 1984 and was known to be a very reliable and competent individual.

A similar caretaking arrangement had been provided by Serem Inc. for the nearby Lawyers gold-silver property located about seven miles by rough road north-west of the Baker mine. The Lawyers prospect, another Kennco discovery of the early 1970s, was then in an advanced stage of exploration with a drill-indicated reserve of 1,249,070 tons averaging 0.182 oz. gold/ton and 6.96 oz. silver per ton. Its on-site facilities included a large bunkhouse and a separate kitchen/dining building.

Like DuPont, Serem was concerned enough about its camp security to maintain caretaking services at this very remote location in the winter period. Serem had retained a young couple for this purpose and both had been on-site for their second winter, since late September when Denis arrived at the Baker camp. In late December they were joined by two of their sons, aged 16 and 20, over the Christmas, New Year period. John and Jane[1] had spent six months at the Serem camp during the previous winter, and were considered totally dependable as caretakers. Used to long periods of isolation, both were avid artists and spent considerable time painting, John actually mixing his own paints.

Both the Lawyers and Baker camps were equipped with fairly reliable radio-telephone communication through the B.C. Telephone operator in Smithers and maintained a pre-arranged schedule. As early as Christmas eve, Denis sensed that Jane was having some problem when he overhead her break down and sob during a conversation. However, he could not have imagined the bizarre event which would involve both only ten days later.

About 1:30 pm on January 3, 1985 John, Jane and their two sons arrived unexpectedly at the Baker camp by snowmobile and skis – the sons being hauled behind by an attached rope. They explained that they were en route to the Sturdee airstrip – about 10 kilometres distant in the Sturdee River valley where they had arranged to be picked up for a flight to Smithers. They appeared tired and all except Jane lay down on beds in the Atco trailers for a rest. Denis noted that Jane appeared restless and moved in and out of the kitchen area several times.

In the late afternoon Denis smelled smoke and located the source at the main electrical junction box in the kitchen complex. He quickly extinguished the fire which had burned part of the wall and ceiling. He was surprised to see that a book which he had previously loaned to Jane had been used to start the fire.

Recognizing that he was facing a potentially lethal situation, Denis rapidly accumulated various personal items, including an unloaded company rifle, and ammunition stored in one of his packs. He secreted this equipment outside the trailer complex.

Shortly after he was approached by both Jane and John and coerced into assisting them light another fire in the kitchen area. Apparently he had sufficient time to soak an appropriate part of the selected area with a fire-extinguisher. Appearing to submit to

---

[1] Names changed from author's original report to DuPont to eliminate identities in what could have been a life-threatening event.

Jane's demands he assisted her in lighting the fire and then quickly closing a door behind them, said, "There, it's lit!"

At this stage Jane and John told him that they had burned down the kitchen/dining building at the Lawyers property and intended to do the same at the Baker camp. Denis emphasized to me later that neither of the sons shared in any of this activity or supported it, on the contrary they criticized their parents for their actions. The reason provided by the couple for their irrational behaviour appeared to be part of a preconceived mission against society and industry.

Denis was forced to accompany the couple with their sons to the Sturdee airstrip that evening. He realized how confused and irrational both parents were by that time. John was obviously suspicious of any action that Denis might take in interfering with the 'mission.' He had said to him, "I'll kill you if you break the link." He also wondered whether arrangements had actually been made for a flight to Smithers. The parents had informed him that the aircraft would be a Beaver piloted by Mel Melissen, an owner/operator of former Smithers Air Services Ltd. Mel had been reported missing with 7-8 passengers in a flight from Vancouver Island to Smithers in late 1983 which never reached its destination and was never located. The company no longer existed.

During the trip that evening to the Sturdee airstrip, the snowmobile hit a snowbank and came to an abrupt halt. Denis described the difficulty that he had had on previous occasions in extricating the snowmobile in similar situations. He said, "John seemed to have 'super-human strength' and easily pulled the machine back on to the hardened track by himself."

When the group arrived at Sturdee airstrip, Denis immediately attempted to start a fire in the oil stove of one of the buildings on site which contained several bunks. John physically resisted his effort without any explanation. At this stage, Denis was totally concerned with his personal safety and he resolved to risk taking very positive action. At the first opportunity he left with the snowmobile, returning to Baker mine without incident late that night.

On arrival, he checked and extinguished a minor fire in the kitchen area. After failing to contact anyone on the radio-telephone, he spent a restless night anticipating the sudden appearance of John and Jane, intent on fulfilling their mission. He kept himself occupied by completing an inventory of the groceries and other supplies remaining on site, securing the camp/mill complex buildings and turning off water and power to the buildings, anticipating an imminent departure.

He was able to contact Glacier Helicopters Ltd. in Smithers the next morning and without divulging any of the events of the preceding day, he requested a helicopter at the Baker camp as soon as possible. The helicopter arrived in the early afternoon.

En route to Smithers, Denis checked for the whereabouts of the couple and, sure enough, located them walking up the road to the mine at a point about 2 - 3 miles from the airstrip. They appeared oblivious to the presence of the helicopter. The pilot landed the machine on the road in front of them intending to try to pick them up. Realizing that they appeared to be determined "to walk through the helicopter" he lifted off to avoid a possible accident because of the low rotor blade clearance with the ground.

On arrival at Smithers, at about 2:30 pm, the incident was reported by Denis to the R.C.M.P. who immediately dispatched two officers by helicopter to Sturdee Valley. Tom Britton of Central Mountain Air Services also left for the airstrip with a Cessna 195 aircraft equipped with ski-wheels.

The R.C.M.P. located the couple within one mile of the Baker Mine site shortly before dark. They had walked the 8 miles along the partly packed snowmobile trail in about 6 hours and had discarded mittens, headgear and some outer garments during the walk.

Corporal May later informed me that the officers had no doubt that the pair were determined to burn the Baker complex, as part of their "mission". They were loaded on to the helicopter, a brief scuffle with John ensuing in the process.

John and Jane's two sons were returned to Smithers by Cessna late that afternoon. The couple were handcuffed to a bed and chair respectively and spent the night in the small building at Sturdee Valley with the helicopter pilot, two R.C.M.P. officers and Tom Britton.

They were taken to Smithers hospital on arrival in Smithers the next morning. According to Corp. May they later agreed to allow themselves to be committed after the R.C.M.P. had determined that they appeared to be in need of psychiatric help. They were later transferred to a hospital in Terrace.

I was notified of the incident by Don Davidson, who had been contacted by Denis on Friday evening, January 4. Later that evening I spoke with Denis, the R.C.M.P. and Tom Britton by phone. After learning that the only description of the fire damage at Baker Mine was based on information provided by Denis, as the site had not been visited by the R.C.M.P., I decided to fly to Smithers on Monday and check the damage and security with Don Davidson for insurance purposes. After an aborted attempt to reach Baker Mine by helicopter on Tuesday, we arrived at the mine at about noon, Wednesday January 9.

The principal fire damage was as described by Denis and the repairs were estimated to be in the range of $600 - $800. I took several photos of the damage for reference purposes. An inspection of all rooms in the camp complex showed no other fire damage or attempt to start a fire. Prior to returning to Smithers we flew over the Serem mine site and I took several photos of the gutted remains of the kitchen / dining building where damage was later estimated at about $70,000.

While in Smithers on January 7, I obtained a verbal statement from Denis which I scrawled hastily on several pages and which formed the basis of the above description together with data provided by Corporal May on January 8 and later by Peter Tegart, then Western Canada Manager for Serem Inc.

I also learned that John and Jane's pet cat, tragically recognized by the pair as either a Satan cat or a God cat had been thrown onto the Serem camp fire as an offering. Peter Tegart suggested that paint mixing and fumes in a poorly ventilated room, such as would exist at the Serem Camp in winter could have contributed to the problem. In addition, John was suffering from hepatitis and jaundice at the time.

No charges were ever laid against the unfortunate pair in this bizarre incident and it appears that both have fully recovered from their affliction and its uncertain cause. However, inhalation of fumes from paint and additives over a prolonged period in a relatively unventilated atmosphere would appear to be a most probable cause.

Great Slave Reef Joint Venture

One of DOX's most promising and unusual prospects was named The Great Slave Reef or GSR property. Its name reflected its location on the south side of Great Slave Lake in the Northwest Territories covering potential lead-zinc deposits within an ancient barrier reef. Geologically, the occurrences are classified as carbonate-hosted lead-zinc deposits of the Mississippi Valley-type and are one of a spectrum of mineral deposits formed in sedimentary rocks sometime during the evolution of sedimentary basins.

Most persons interested in geography are familiar with the location of The Great Barrier Reef, the world's largest active reef which stretches for 2000 km along the coast of Queensland in Australia. It encompasses about 2500 individual reefs, more than 600 islands in a 50 km wide belt up to 300 km from the mainland and is the largest structure created by living organisms - principally corals, of which over 350 specimens have been recognized in the reef.

Drilling on the Great Barrier Reef has shown that it can be over 500 metres in thickness. Coral is formed by a marine polyp, closely related to sea anemones and jellyfish which secretes warm water, i.e. above $17.5°$ C and it will not develop below a depth of 30 metres as sunlight does not provide sufficient penetration. Its development to thicknesses of up to 500 metres reflects conditions of continuing submergence of the seabed on which the reef lies, with new polyps growing on their dead predecessors. Its accretion is thought to have taken some 600 million years.

As a geologist accustomed to exploration for minerals contained within non-life supporting environments, the exposure to the GSR project was to provide a welcome diversion.

The history of discovery of lead-zinc occurrences near Pine Point on the south shore of Great Slave Lake is interesting. The earliest recorded report is linked to sourdoughs travelling through the area in 1898 to seek their fortunes in the Klondike gold fields. A particularly observant prospector noted that Indians near Hay River on Great Slave Lake were using fragments of galena – a lead sulfide, to weight their fishing nets. Indians led the prospector to the location of the galena which lay scattered with fragments of limestone in a clearing on a ridge crest which formed a pathway for animals. No outcrop was exposed anywhere in the area. The prospector collected several fragments which were assayed for silver, as there was no interest at the time in lead at that location. As there was no significant silver present in the galena, the prospector headed on for the Klondike.

Ed Nagle, a fur trader, reportedly staked the first claims in the area in 1898 and deposits were noted by Dr. Robert Bell of the Geological Survey of Canada in 1899. Following trenching in the area, an examination of the exposed occurrence was made by by Ted Nagle, son of Ed Nagle, and a partner in 1927. Nagle secured an option on 16 claims for Cominco. In 1929 a consortium of Cominco and other claim holders outlined about 500,000 tons of 15% combined lead-zinc mineralization on the property. Despite extensive surface work no further occurrences could be located. However, Cominco maintained 104 mineral claims at Pine Point until 1948 when large concessions surrounding the known area were acquired which were shared by Cominco with two other companies.

Pine Point Mines Limited (PPM), controlled by Cominco Ltd., was formed in 1951 to finance a drilling program on the property. By 1955 a total of 5 million tons of ore averaging 11% lead-zinc combined had been outlined in several deposits. At this stage exploration work was terminated and PPM then focused on the feasibility of production, recognizing the need for cheaper transportation of concentrates from mined ores to a smelter than available by existing roads in order to improve project economics.

Key factors in the production decision were to be timing and good luck in addition to the confidence of Cominco geologists that the mineral belt could contain a major lead-zinc mineral resource. The 1950s coincided with a period of enormous optimism by Canadians on the potential economic growth of the country through resource development. The focus was on northern development. In 1955, Gordon Robertson, then Deputy Minister of Northern Affairs and Commissioner of the Northwest Territories, recommended to the Gordon Royal Commission construction of a railway from Grimshaw, Alberta for 650 km north to Hay River and beyond to Pine Point as a project of national interest under the Roads to Resources program.

In 1961 an agreement was finally reached between the Government of Canada, the Canadian National Railway and PPM for construction of the railway. Cominco as manager and holder of a 78% interest in PPM, undertook to bring the property into production. In return PPM committed to shipments of 215,000 tons of concentrate per year for ten years and freight charges of $20 million.

The Government of Canada also directed the Northern Canada Power Commission to build a 25 megawatt hydroelectric plant 250 km east of Pine Point on the Taltson River to supply power to Pine Point, Fort Smith and to meet other future needs. Cost of the Taltson plant was underwritten by PPM. Between 1962 and late 1965, the Pine Point development was completed and included construction of an access road, mine development, concentrator construction, a townsite which eventually accommodated 2000 people, the railroad, Taltson power plant and transmission line.

Further exploration drilling continued during the 1962-5 period but by early 1964 only 1.9 million tons of 4% lead-zinc combined reserves were delineated – considered uneconomic at the time. Although not optimistic, the exploration staff tested known geophysical surveys over a known orebody in 1963. Included were magnetic, electromagnetic, gravity and Induced Polarization (IP) surveys. The IP survey provided a strong response and supported further survey recommendations over surrounding areas with several anomalies resulting. In 1964-5 an additional 13.1 million tons grading 13.7% lead-zinc combined were discovered by diamond drilling – quadrupling the pre-production reserve!

The high-grade nature of the new discoveries enabled PPM to ship untreated ore to Trail, Kimberley and into Montana for milling and direct smelting during mill construction at Pine Point. From late 1964-70 a total of 1.4 million tons grading 46.0% lead-zinc combined was shipped. These early ore shipments combined with the three-year federal income tax exemption then available for new mines, had a major positive impact on the rate of return from the project.

As if Lady Luck had not smiled enough on the PPM operation, another unexpected bonus evolved. Pyramid Mining Company Ltd., one of the companies involved in a late 1965 staking rush in the area, drilled IP anomalies on their property and discovered two ore bodies, the largest (Pyramid) being 19.3 million tons grading 8.7% lead-zinc combined. The property was sold to PPM in June 1966 for 526,400 shares of PPM, then valued at about $40 million. The Pyramid ore body was to be the largest of the 70.8 million tons to be mined by PPM between 1964-87 from 48 deposits ranging in size from 20,000 to 19.3 million tons. The two Pyramid ore bodies supplied one third of the total ore mined at Pine Point.

The total cost of the Pine Point mine development and supportive infrastructure was $125.1 million of which the Government of Canada provided $88.1 million and PPM $37.0 million. These early investments were to lead to allegations by some of the public and others that the Government of Canada or the local population were not adequate beneficiaries of the project, or alternately that the shareholders of PPM received too great a benefit.

As noted by D.L. Johnston, PPM's president in 1988, "It is undeniable that the Pine Point development was an unqualified success as a mining investment. The shareholders benefited directly through a generous dividend policy ($339 million). They were not the only beneficiaries however". The Government of Canada received $98 million in corporate income taxes, $60 million in personal income taxes, $11 million in NWT royalties and an estimated $600 million in foreign exchange earnings. The municipality of Pine Point received $7.5 million in local taxes. The CNR received an estimated $400 million in freight charges – based on 11.4 million tons of ore and concentrate shipped on the line compared with the 2.15 million tons PPM guaranteed. A freight survey in the mid-1970s showed that the Pine Point freight was less than 40% of that carried by the system. The Northern Canada Power Commission derived $100 million in charges above the cost and amortization of the Taltson power system and in power charges. Employees were paid $246 million in wages and salaries and the Canadian economy benefited by an estimated $500 million in materials, supplies and services purchased.

Thus, after repayment of its initial investment plus later capital costs, the PPM development returned 14.3% to its shareholders and 85.7% to other beneficiaries in its 25 year period.

Exploration work and production criteria completed by PPM provided considerable data concerning geological and mineralogical controls of the lead-zinc mineralization in the PPM deposits. The deposits occur in a Middle Devonian carbonate barrier reef sequence composed of the Pine Point and Sulphur Point formations which were built on a marine platform sequence. At its time of formation, the reef lay on the margin of an evaporate basin to the south and a deep water shale-hosted environment to the north, presently covered by Great Slave Lake. The reef, which is about 10 - 15 km wide and up to 200 m thick, barely outcrops beneath shallow till cover on the east and central part of the PPM property. The reef and its underlying and overlying formations dip gently to the west and parallel the south shore of Great Slave Lake.

The lead-zinc deposits are controlled by three mineralized trends which parallel the northeast trending barrier reef and, in addition, by northwest trending cross-structures with which solution channelling developed karst cavities. The structures created by solution channelling localized at the intersection of both vertical and lateral structures provided favourable depositional foci for 'prismatic' and 'tabular' lead-zinc deposits.

In 1975 the PPM property covered a 60 km length of the barrier reef and an area of about 1000 sq km. Extending southwest to a point west of Buffalo River where the base of lead-zinc deposits lay at a maximum depth of about 150 m, <u>all</u> deposits were considered mineable, if large enough, by open pit methods.

At this time and only one year after DOX's formation as an active exploration company, I was approached by Bruce Spencer, then a senior geologist with Western Mines – the predecessor of Westmin Resources Limited. Bruce was aware of DOX's interest in joint venturing and proposed that DOX consider participating with Western Mines in exploration of their recently staked property covering an area of 622 sq. km. on the western extension of the PPM property. The property extended for 43 km west to Hay River and was accessed by both a road and the CN railway from Hay River to Pine Point. The proposal had considerable appeal as there was little doubt that the same potential existed for new discoveries of the type encountered at Pine Point Mines.

We soon reached agreement on a 50-50 joint venture with Westmin Resources acting as operator. From the outset we envisaged that any mining operations which resulted would probably require mining by underground methods because of the anticipated depth of burial of any deposits discovered and that extensive de-watering of the highly porous limestone formations would be required.

Hugh Snyder, an affable transplanted mining engineer from Britain, was president of Western Mines and Art Soregaroli, a prominent geologist, became VP Exploration in 1976. I was to enjoy a relationship with both for many years. Art was, and still is, one of those rare individuals in the mining sector who has held senior positions as a geologist in academe, government and industry. His sudden move from Vancouver to a new position in Ontario in 1971 corresponded with our move back to Vancouver and had vaulted me – most apprehensively – into accepting the position of president of the Geological Association of Canada – Cordilleran Section, as his replacement. In 1992, quite by coincidence ,Hugh and I were both to be elected directors of Dickenson Mines.

In March 1981, Western Mines changed its name to Westmin Resources Limited and the company sold a 10% interest in the joint venture to Philipp Brothers (Canada) Ltd., but retained its position as manager. Throughout the active life of the project, Alf Randall was to act as a most responsible and capable project geologist, with his wife, Marie, also a geologist, acting as his assistant in the early phase of the venture. Peter Mason also spent nearly as much time as Alf on the project acting as surveyor and geologist.

The till cover on the GSR ground was considered too conductive and thick to provide detection of lead-zinc deposits by IP methods. Accordingly, the extensive exploration experience and geological controls developed by PPM formed the basis for initial diamond drilling, which commenced in the fall of 1975.

The principal guide for early exploration on the GSR ground was the recognition that ore deposits on the PPM property occur along mineralized 'trends'. Originally it was believed that the major northeast-southwest trending MacDonald fault system underlying the area created structural offsets to which barrier development and orebody deposition were related. Later gentle folding along these axes and local to regional northwest trending cross-structures were considered to be foci for karst development. In any event, three main 'trends' were recognized within a centrally located 10 km width of the 25 km wide reef. Most of the deposits occur on the more centrally located 'Main Trend'. Virtually all of the GSR exploration was focused on a 15 km strike length of the Main Trend, immediately to the west of the GSR property.

In February, 1976 Bruce and I arranged a visit to the property – it was my first exposure to the area. We arrived by aircraft at Hay River, experiencing the usual shock of an abrupt change from a balmy 10º - 15º C at Vancouver to minus 25º C at Hay River. We traveled by U-Drive rental on a brilliant sunny day along the road to Pine Point, turning off on a drill road to Alf Randall's trailer camp. We spent the rest of the day with Alf, examining various maps and cores and visiting the drill site.

Although we had obtained some encouraging signs of mineralization in our drilling program, no significant deposit had been discovered by then – our 25th drill hole. All holes drilled on our program and at Pine Point were vertical and as both vertically-oriented 'prismatic' and horizontal 'tabular' deposits have been mined at PPM, of the deposits associated with the Main Trend, about 70% are prismatic. These are obviously not ideal targets for discovery with only geological guides available rather than the more specific targets obtained from geophysical surveys.

Bruce and I spent that evening at the Pine Point motel where, after a pleasant dinner, we engaged in a wide-ranging discussion concerning our project and related considerations until the small hours of the morning. As I recollect, we consumed a generous amount of a particularly satisfying product recommended by Bruce and available at the local liquor store. Prior to turning in, we decided to go for a stroll around the Pine Point residential area with its neatly arranged homes, an enervating experience at a temperature of about minus 30º C.

The next morning, en route back to the airport, on a whim we decided to briefly re-visit the core shack to see whether any noteworthy developments had occurred. Alf spotted or heard our car as it approached the trailer. He opened the door and with a delighted yell, held up a one-metre length of massive sulfide core. Subsequent assays were 22.5% lead-zinc combined over 5.2 m (17 ft.). Additional drilling outlined the X-25 deposit which contained 3.8 million tonnes grading 10.2% combined lead-zinc on a 60 m (200 ft) grid spacing of 59 drill holes using a 2% lead-zinc cut-off in calculating the reserve. Of interest, a closer spaced 15 m (50 ft.) grid spacing with 109 drill holes and the same grade cut-off indicated 3.2 million tonnes grading 9.7% lead-zinc combined.

Considering that the entire PPM production decision had initially been based on 5 million tons of 11% lead-zinc combined in several deposits, we were naturally elated, to put it mildly.

In June 1979 I was requested to prepare a Mining White Paper on the status of DOX's activities. Part of the exercise included an estimated value of DuPont's interest in the GSR and Chappelle properties. At the time, DOX's share of exploration costs since 1975 at the GSR project totalled about $2.1 million and five deposits (including the X-25) had been discovered. Based on a conservative high and low price consideration for lead and zinc, I estimated that DuPont might expect to receive between $10 - 20 million for its equity in the GSR project.

In August 1979 the DOC Board arranged for a meeting in Edmonton, Alberta at which I was asked to provide a presentation on the GSR joint venture. I prepared a handout and provided a slide show. In addition, I had one-half inch slabs of Pine Point Mines 'ore' prepared and polished on one side from a large piece of high grade lead-zinc mineralization which I had obtained during a visit to the M-40 deposit – the only one mined by underground methods as a test case in a de-watered area by PPM. After my presentation I invited the directors to select one of the slabs as a memento of a forthcoming visit which had been arranged at the Pine Point operation following the meeting. I cautioned the directors to handle and pack the slabs carefully as the specimens were brittle and would easily break if struck. The directors moved swiftly to the table containing the slabs exhibiting a commendably competitive spirit. Shortly after, I was approached by a senior director from E.I. DuPont de Nemours in New York, who held a broken slab in his hands. "You were right Dave – it is brittle – I was able to break it with my bare hands!". I'm not sure whether he had resolved to test his strength or by nature he had learned to question everything he heard – a not uncommon trait of many successful executives.

Our following trip to Pine Point was very successful, despite an unforeseen delay in one vital part of my planned itinerary. I recall gazing down at the Hay River airport as our aircraft circled for a landing. I could not see any sign of a bus which had been chartered to take our party of about 12 - 15 individuals to Pine Point. On landing I learned that the bus had been seconded by an emergency and that another would be at the airport in about a half-hour. It arrived as re-scheduled, but it was a dilapidated vehicle with grimy torn seats and it emitted a somewhat nauseous odour inside from years of smoke and cargoes of uncertain nature and origin. I was very disappointed having envisaged a luxury vehicle for our distinguished guests and considerably relieved when it arrived at Pine Point.

As was DuPont's custom during rare field trips for its directors, a two-day retreat had been arranged for the party at nearby Jerry Bricker's Fishing Lodge at Snowdrift near the east end of Great Slave Lake. I had been invited to attend and enjoyed two days of fishing for lake trout. Our party was segregated into two fishers per boat, each boat being operated by a staff member who assisted in preparing fried fish for lunch at a convenient site. Quite appropriately the largest catch – a 20 pounder – was taken by Bob Richardson, DuPont's President, while the rest of us all had catches in the 10 - 15 pound range which were frozen and boxed for transportation on our departure.

Publicity resulting from PPM's excellent performance and DOX's exploration results led to an additional staking spree in 1975. Involved was a joint venture between Cominco, Pine Point Mines, Aquitaine Company of Canada Ltd., Gulf Resources Canada Ltd. and Samim. The area staked was equivalent to the combined holdings of PPM and GSR and extended for an additional 50 km to the west of the GSR ground. If nothing else, it helped to convince senior management in DOC that our expectation of underground mining of high-grade lead-zinc discoveries in the area had validity.

By September 1983, the GSR joint venture geological reserves at a 2% lead-zinc cut-off grade were 7.31 million tons grading 3.57% lead and 7.26% zinc in seven deposits. Cost of discovering the reserves was essentially identical to the historic experience at PPM, being about $0.013/kg lead-zinc combined ($1983). In the previous decade PPM had drilled an average of 52,000 m annually in its exploration programs, compared with an average of 18,000 m annually at the GSR joint venture since inception of activities in 1975. Most of the GSR drilling was completed on the eastern 11.5 km portion of the Main Trend.

DOC announced its decision to withdraw from the mineral business in late 1983. At the time, an economic study was in progress by the GSR Joint Venture which was focused on the high grade R-190 deposit (573,000 tonnes grading 9.84% lead and 18.34% zinc). The study was a two-phase evaluation to provide for revision of mining methods and costs considered in previous economic studies and a hydrological study with grouting tests with the object of determining the estimated cost of de-watering.

More than a decade later, in 1997, no further exploration or development activity has occurred because of the high pumping costs envisaged from tests completed by the Joint Venture.

From its outset, the GSR Joint Venture was an exciting project which yielded encouraging results and further demonstrated the remarkably high lead-zinc content of the Great Slave Reef, its persistence and the relatively low cost of discovering reserves. Oil well exploration drilling over 200 km west of the Pine Point district has encountered the same structure at a depth of about 5,000 m with similar lead-zinc values. The reef remains a future major lead-zinc resource which could be viable with appropriate grouting techniques. A similar technological break-through to that which provided for the economic development of the Saskatchewan potash reserves would be required.

As has been my practice in recording some of the more interesting and memorable events I have experienced, I have requested and received the following additional comments from Alf Randall on his recollections.

On our West Reef project we also intersected some gas in our drilling and had a near incident when one hole caught fire and was saved by the drillers quickly pulling the rig off the hole. We had numerous incidents of H_2S gas in our drilling elsewhere, but mostly on the West Reef part of the property.

We were involved in a number of staking rushes in competition with Cominco and Pine Point in which we generally came out ahead using fewer people. These included the original staking of the immediate Pine Point extension when we had 17 natives from LaRonge running around the swamps. Pine Point management were surprised and dismayed when they discovered us staking in their backyard. They made this discovery when they came into our camp unannounced and found us organizing tags for 1700 claims on a 4 x 8 sheet of plywood. Staking rushes occurred on two other occasions when we extended the property to the north and south to cover the expected North and South Trend axes. We were able on both occasions to beat out the competition for the favoured ground.

Cominco were at first very reluctant (as might have been expected) to share information with us but with time, relations improved and we generally had good rapport and were able to gain much from our discussions with them on the specifics of the geology and they in turn were able to gain some information about the regional extent of controlling geological features and distribution of mineralization. They also probably became more accommodating as it became apparent that we would probably end up shipping out ore to Pine Point Mines for milling.

Upon our discovery of the X25 deposit the Cominco crew, when they first heard about this, thought we had a real bonanza as their largest deposit had been designated X15 indicating 15 million tons discovered on their X-line. However, our designation was indicating the discovery hole #25 on our X-line.

We also had another interesting incident at the time of the discovery of the R190 deposit. As a result of this incident I had to call Bruce Spencer so I told him that I had some <u>real good</u> news, some <u>bad</u> news and some <u>real bad</u> news. The real good news was that we had intersected some 100 feet of massive sulphide in hole R190, the bad news was that Cominco were attempting to stake up our north boundary and the real bad news was that I had sunk one of the drilling contractor's $90,000 Nodwell track carriers in a creek some 20 miles back in the bush during our efforts to beat Pine Point at staking and we might have to pay for it if it couldn't be recovered.

Bell Creek Gold Mine, Porcupine District, Ontario

Within DOX's organization, Gerald Harron, Senior Geologist was our principal source of base and precious metal submissions from Northern Ontario and Quebec. Prior to joining DOX, Gerry had spent about five years working for Placer Dome, the Geological Survey of Canada and the Quebec Ministry of Natural Resources in exploration and metallogenic studies of mineral deposits in the Appalachians and Canadian Shield. Many of the property submissions originated from Bob Middleton, a long-time friend and acquaintance of Gerry's who was vice-president exploration for Rosario Resources in the late 1970s.

By mid-June 1978, DOX was participating in options and joint ventures with 22 mining and oil companies in 26 projects in Canada, of which 5 were from Ontario and Quebec. Quite remarkably, three of these were destined to reach production: the Chappelle gold-silver property as the Baker mine in 1981; the Scott zinc-copper-silver mine in the Chibougamau area, Quebec in 1986 and the Musselwhite gold mine at Opapimiskan Lake in north-western Ontario in 1997. Two were joint venture interests held by Ducanex Resources - the exploration subsidiary comprised of DuPont and Lacana Mining Corp.

In the latter part of 1978, Gerry and I discussed a possible joint venture with Rosario on a property in Murphy and Hoyle townships in the eastern part of the Porcupine Gold Belt, located about 8 km north-east of Timmins, Ontario. The nearby Owl Creek gold deposit owned by Kidd Creek Mines Ltd., which was discovered in 1969 and placed into production as an open pit mine in 1978, lay less than one kilometre from the eastern boundary of Rosario's property.

The joint venture proposal was attractive considering its excellent land position covering about 25 sq. km in Canada's most important gold mining district, which had a 60-year record of continuous operations at the time, from 20 current and former producers, which had yielded over 50 million ounces of gold.

We entered into a joint venture with Rosario on a 50-50 basis in 1979 with Rosario acting as operator. Although several minor gold showings had been discovered on part of the joint venture property, the general area is covered by a monotonous mantle of muskeg and drift and is essentially devoid of outcrops. The earliest exploration work carried out by the joint venture included ground geophysical surveys consisting of detailed magnetometer, horizontal loop electromagnetic and induced polarization programs over selected areas. This was augmented by overburden drilling to obtain basal till samples for lithologic and mineralogical determinations. Gold anomalies derived from the analysis of heavy mineral contents in the till samples combined with results of geophysical surveys provided targets for the early drilling programs. The geological data base which helped to guide subsequent exploration programs was developed from a combination of geophysical surveys and diamond drilling programs.

By the end of 1982 exploration by the Murphy-Hoyle joint venture had led to the discovery by diamond drilling of three distinct zones lying within an auriferous belt extending westerly from the Owl Creek zone.

These discoveries were named the North, Bell Creek and Marlhill zones. The most promising was the North Zone, discovered in 1982, which had been outlined to a vertical depth of 250 m, across an average true width of 4 m and for a length of at least 350 m at a depth of 200-250 m, with the deposit open at depth. The North Zone actually consisted of two parallel auriferous formations lying 20-50 m apart. The 'upper horizon' was the more continuous, higher grade and thicker of the two zones.

Although the Bell Creek Zone, lying some 250-300 m to the south, had a similar geologic environment it appeared to be of much lower grade at 3.3 grams gold per metric ton compared to 6.2 grams gold per metric ton for the upper horizon of the North Zone. The Marlhill Zone lying 400 m northeast of the North Zone and traced for a length of 300 m by shallow drilling across widths of 1-7 m appeared to have a much more variable content, averaging about 5.1 grams gold per metric ton.

As the exploration programs increased in intensity in 1983 and through the pre-production stage, I visited the property more frequently. All the on-site exploration proposals and programs were managed by Canamax Resources Inc., an offspring of Rosario's, and my involvement was initially with Gerry Harron on joint venture management meetings in Toronto and later as a joint representative for DOX and CSA Management Limited in 1984-5. Canamax provided excellent supervision through Fred Johnston and were fortunate to have Al Philipp, a geologist, as a dedicated on-site representative throughout the exploration and development stage.

Unlike the Baker mine, the Timmins district deposits contained visible gold – always an attractive phenomenon for the geologist. At Bell Creek it was so consistent that Al was able to quite accurately predict the grade of auriferous drill cores. His ability was uncanny and we joked, "Why do we have to have these drill cores assayed, when Al can do it so accurately!"

At this stage of my narrative on Bell Creek, it is probably appropriate to review the varied corporate interests that were to be involved during the exploration, development and production at the project in a relatively short time frame.

The early part of my career was carried out within a fairly structured and stable corporate environment, albeit for the occasional periods of staff reductions. My six-year involvement with the Bell Creek project and its aftermath was to witness the more volatile side of exploration's corporate structure, dominated by abrupt changes in ownership and policy.

Rosario Resources, like Amax of Canada Ltd., were both wholly-owned subsidiaries of Amax Inc., a major U.S. corporation engaged in exploration, mining, smelting and refining. In 1981 Rosario sold its interests in its Timmins properties to Amax of Canada Ltd. and the Bell Creek project then came under the direct management of Amax

Minerals Exploration headed by John Hansuld, vice president of exploration, and a long-time acquaintance of mine. A further and somewhat novel modification of our joint venture management occurred in December 1982 when Canamax Resources Inc. was formed to acquire the mineral rights and other assets of Amax Minerals. Amax Inc. advanced $22.4 million to make expenditures on behalf of Canamax in exchange for shares in Canamax. In July, 1983, in its initial offering, Canamax sold 2.2 million shares and 5,180 units for $29 million. Bell Creek was its principal project.

On November 11, 1983, a day remembered annually for the end of hostilities in two World Wars and other conflicts, DOC announced that its exploration activities would be phased out by the end of 1984. The carefully worded announcement which follows reflected months of preparation, which included senior management's respectful consideration for the DOX personnel that would be directly affected.

"We went into the mineral venture to determine whether it represented a suitable means of diversification. We are now convinced that to make mining a significant part of the Company, we would have to greatly increase our investment. This would mean foregoing other opportunities. After carefully weighing the alternatives, we have decided to withdraw from our mineral venture in favour of business more closely related to our strengths in manufacturing, marketing and research.

"There are 12 employees involved. Several will be offered other positions. It is unlikely that we have suitable jobs for the others who are specialists in geology."

This latter development was to have little immediate impact on the joint venture, for it would be business as usual through 1984. During that transition period I determined, with DOC's support, to attempt to sell DOX as a business, with its employees if possible, to a third party. We produced a comprehensive corporate profile with many coloured plates which described the business of DOX, its major and minor projects, staff profiles and our Vancouver office assets. Copies were forwarded to about 50 Canadian mining organizations, commencing in January 1984. At the time, DOX had evaluated 1,150 submitted and self-generated proposals across Canada and Mexico and its office contained 12 people, including seven professionals. It had spent $20 million on exploration and development activities, placed the Baker gold-silver mine into production and had a 50% working interest in the Bell Creek project, which was to achieve production at a level of 180 tons per day on a custom-milling basis on January 1, 1987 and at 385 tons per day with its on-site mill in September, 1987.

Our brochure attracted the particular attention of CSA Management Ltd. of Toronto, a firm controlled by Donald McEwen, which managed $217 million in two gold-oriented mutual funds (Goldfund and Goldtrust) and a closed-end gold investment company (Goldcorp Investments Limited). Don saw an opportunity to enter the mining and exploration business with a going concern and was quoted in the June 6, 1984 issue of the Globe and Mail as commenting, "This deal is just what we were looking for. It has catapulted us five years ahead."

CSA agreed to spend $10 million on development and exploration to earn a 68% interest in the DOX assets. Although the agreement was to be modified within a short time frame it was to directly affect my future plans which include initiating a consulting practice under Barrda Minerals Inc. in 1984. However, I was to be closely tied to the McEwen interests for the next 10 years, as described in the next section.

The eventual decision to proceed with production plans at Bell Creek on an initial custom milling basis using the nearby Pamour Porcupine Mine's gold-mill attracted the attention of this long-term entity in the Timmins district. Don McEwen had formed CSA Minerals Corp. to hold the assets acquired from DOX plus other mineral interests in 1985. This company was eventually rolled into Consolidated CSA Minerals Inc. and the name was changed to Pamorex Minerals Inc. in April 1987 to reflect Pamour's control following acquisition of an interest in the original DOX assets by Pamour in 1987.

As indicated earlier we enjoyed a good working relationship with Canamax in the Murphy-Hoyle joint venture through the exploration stage. However, by 1984 Canamax was anxious to proceed with accelerated programs at four different projects in Yukon, B.C. and Ontario, all in the underground development stage. As early as late 1982 DOX had exerted an influence at the Murphy-Hoyle project which was geared to approval of budgets on a staged basis justified by results. Rather than approving a $1.4 million budget in November proposed by Canamax as a two-stage program for 1983 it recommended approval of a $650 thousand diamond drilling budget and suggested tabling of a supplementary budget by Canamax in mid-1983. Again in mid-1983 DOX questioned the assumption of a mineable reserve at almost double the gold content of that actually indicated by diamond drilling, to partly support a road construction proposal without a firm budget. The road was eventually built through the muskeg, using DuPont Typlar, a geotextile to support the surface road material.

The above illustrations reflect only a minor divergence of opinion on the overall program which was subsequently followed. Prior to a production decision and commencing in 1984 a shaft was sunk on the North Zone to a depth of 280 m with drifts on four levels. The results of underground sampling provided an excellent comparison with that indicated from the diamond drilling program. By December 31, 1986 proven and probable reserves on the North Zone to the 240 m level were 526,000 tons averaging 0.177 oz gold per ton versus a drill-indicated mineable reserve of 557,000 tons averaging 0.175 oz gold per ton in late 1983.

By December 31, 1986 over $13 million had been spent on the joint venture and the cost of a 400-ton-per-day mill and associated structures was estimated at about $8 million. Some delay occurred in the subsequent production plan for Bell Creek when the shaft and change-house burnt down in 1987. However, actual production commenced in 1987 with ore initially shipped to the nearby Pamour mine mill. The 385-ton-per-day on-site mill at Bell Creek was in operation in mid-September 1987 using the Merrill Crowe process. By December 31, a total of 50,163 tonnes had been milled to produce 9,558 ounces of gold in the 50-50 Pamorex-Canamax joint venture.

Proven and probable reserves in the North and Marlhill zones at December 31, 1987 totalled 866,160 tonnes grading 0.189 or gold/tonne. In addition, drill indicated reserves for the North Zone below the 240 metre level were 451,400 tonnes averaging 0.138 oz gold/tonne and 299,700 tonnes average 0.12 oz gold/tonne at the nearby Bell Creek Zone.

On August 31, 1988 Pamorex Minerals sold its 50% interest in Bell Creek to Canamax for $16.5 million plus certain net smelter return interests.

Results of underground exploration on the four Marlhill Zone veins explored by drilling and from an underground ramp in the pre-1989 period supported a production decision. Between 1989 and 1991 Marlhill produced 25,400 oz of gold from 141,470 tonnes. Three of the four veins were mined to maximum depth of 149 metres with the MI vein mined for a strike length of 228 metres.

Bell Creek operated from September, 1987 to December 31, 1991, producing 121,240 oz gold from 569,600 tonnes milled. In December, 1991, Bell Creek mine and its related assets were sold by Canamax to Falconbridge Gold Corporation for $3.5 million. Proceeds, net of amounts of mine and mill shutdown were used to eliminate debt. An additional 6,015 oz gold were produced in 1992-3.

Like many other mining ventures, the Bell Creek story did not end here – it merely entered a hiatus – a common feature in the sporadic lives of mining properties. In 1995 Kinross Gold Corporation acquired a 100% interest in the Hoyle Pond Gold mine east of Bell Creek which commenced production as an open pit in 1985. It also acquired a 100% interest in the Bell Creek Gold Mine which had been maintained on a standby basis through 1995. Kinross expanded the gold reserves in the Hoyle deposit and also increased the capacity of the Bell Creek mill to 1,000 tonnes per day.

With its interests in the Hoyle Pond and nearby Owl Creek gold deposits, Kinross sold extensive claim holdings, including all of the former Bell Creek property except the mill assets to Pentland Firth Ventures in 1994 for $3.4 million, retaining a 50% back-in right plus the right to be operator of any deposit found, upon completion of a positive feasibility study. Pentland Firth's extensive drilling programs in 1995 in the vicinity of the Marlhill Mine, Allerston and Schumacher III properties, all of which were part of the original Bell Creek holdings, outlined geological resources of 221,700 oz gold. Additional success in 1996 leading to pre-feasibility work was in progress in 1997.

CSA Management Limited & Related Companies

My association with Don McEwen and, following his untimely death in 1987, with his son Robert was a most welcome one and provided a degree of continuity with progress in the DOX assets initially acquired by CSA Management through CSA Minerals Corp. and subsequently consolidated CSA Minerals Inc. in January, 1986. Of the 21 original property interests transferred by DOC only Bell Creek eventually reached production, as described.

In 1986, through Don McEwen's contacts, CSA Management became a participant in a new tax-directed exploration fund designed specifically for the junior mining sector in Canada. Named First Exploration Fund 1986 and Company Limited Partnership it also had the backing of Merrill Lynch and Dominion Securities Pitfield, acting as broker agents. Heavily involved in the organization of the fund was Peter L. Bradshaw, then a vice-president of First Capital Corp. which controlled the general partner in the fund.

Because of public concerns about expenditure of funds allocated for specific projects, an audit procedure was implemented involving CSA Management as auditor of exploration programs and the release of expenditures to approved budgets. I was designated as Exploration Manager and Len Bednarz was Finance Manager. The flow-through share funding procedure as a tax incentive met with good response and the 1986, 1987 and 1988 First Exploration Funds raised $109,540,000 for over 87 companies involving about 100 mineral properties across Canada.

An innovative approval procedure for applicants was developed which included review of the applying company's financial status, legal assessment and a geological assessment completed by a consulting board. For all three years, Alastair Sinclair, a geological professor at the University of British Columbia was chair and the review board was composed of geological consultants, including myself, Bob Stevenson, Ed Thompson, Marcel Vallee and Ralph Westervelt; and as First Exploration Fund representatives: Gerry Carlson and Mark Fields.

The geological review board met frequently in Vancouver and would commonly approve $5 million or more in budgets ranging from as low as $125,000 to in excess of $1,000,000. Although I examined a few of the properties in B.C. for audit purposes while exploration programs were in progress, during the three years I contacted over 50 geological consultants in Canada for this purpose. A standard agreement form provided for the procedure, remuneration and reporting requirements. The audits were highly successful.

An interesting benefactor, albeit indirectly and at a very preliminary stage was Diamet Minerals which had been involved in diamond exploration in south-eastern B.C. and was also approved for a very modest exploration program in the Northwest Territories. While the latter program was in progress I received a telephone call from Charles Fipke, President of Dia Met Minerals Ltd., asking whether his company could carry out exploration off the property, while in the area. I indicated that this would be acceptable providing no related costs for such work were charged against the approved budget. Later I was to learn that the work done off the property consisted of helicopter-borne prospecting of glacially deposited gravel trains derived from far to the east in the Northwest Territories. The gravels of interest contained indicator minerals associated with diamond deposits, in this case scoured from source and transported by glaciers for hundreds of miles toward the Mackenzie River. Later exploration programs were to lead to staking of claim blocks over 300 kilometres to the northeast of the Mackenzie by Chuck Fipke, aided by Stuart Blusson, his partner and helicopter pilot, who was also a geologist previously employed with the Geological Survey of Canada. Dave Mackenzie, a director of Dia Met and a long-time pilot, was also involved in the field program as a Super-Cub pilot and sample collector. The discovery of diamond deposits in the Lac de Gras area by Dia Met was to be heralded as one of the most sensational historic mineral exploration successes in Canada, and led to Canada's position as the third largest diamond producer in the world by 2002.

My involvements with CSA Management led to my appointment as Vice President – Exploration for CSA Management in 1988, a position I held until 1994. Rob McEwen very generously included me at most board meetings of related companies. I also had the opportunity to visit many of the exploration ventures being carried out by these companies, which included former properties owned by DOX.

For a period I was also a Director of Dickenson Mines Limited and Goldcorp Inc. and had several opportunities to visit the Dickenson Mine – also known as the Arthur White Mine, one of the two principal gold producers in the Red Lake area of northwestern Ontario.

Claims covering the Dickenson Mine and the adjoining Campbell Red Lake Mine to the west, were staked by Arthur White in the early 1940s. He split the property in half, selling the Campbell Red Lake and retaining the property which became known as the Arthur White or Dickenson Mine. Production commenced at Campbell Red Lake in a 300 ton per day mill in June 1949 and at Dickenson in a 150 ton mill in December 1948. On my first visit to Dickenson Mine in the early 1990s, Dutch Van Tassell, Dickenson Mines' Vice President Exploration, provided me with a summary of the mine geology and the results of past diamond drilling programs exploring for new gold deposits. With the aid of drill sections he pointed out various drill hole intercepts containing higher grade intercepts which had never been followed-up with more detailed drilling.

The historic lower grade gold ores at Dickenson, averaging in the 0.3 – 0.5 oz/ton grade over a 40 year period, being only about 50% of those at the adjacent Campbell Red Lake Mine, had always been an exploration and production concern. At an early Board of Directors meeting, I recommended that selected areas of higher grade diamond drill intercepts should be followed up with more detailed definition drilling to try and boost the mine production grade. This was approved at the $150,000 level, but the program was never carried out because of higher priority given to necessary development drilling. I remember being disappointed, but realized that funds for exploration were tight at the time.

Dutch had an imaginative sense of humour, and I always enjoyed his company. On leaving the Dickenson Mine on that first visit, we were walking away from the mine office to catch our flight back to Toronto, watched by some of the staff. Suddenly Dutch stopped and said to me, "Dave, do you know what those guys are saying?" Of course I had no idea. He continued, "They are saying, 'there go the seagulls.' They fly in, eat up all our grub, crap all over us and then fly out again!'" I couldn't stop laughing – ruefully mindful of the gulf that sometimes exists between brass and staff.

In 1994, I had briefly discussed with Rob McEwen the possibility of higher grade ore at depth in the mine based on a possible overall easterly plunge of ore-bearing structures from the area of the higher grade Campbell deposits. In two subsequent communications in late December and early January 1994 I provided data based on available references and an opportunity to examine Campbell Red Lake reports on file at Placer Dome's offices in Vancouver.

I never learned whether these aided in any way in supporting a $10 million exploration and development program at the mine announced in February 1995, which I understand was recommended by Dutch Van Tassell. In any event the success of the program was a tribute to the measures and policies introduced by Rob McEwen following a lengthy period of difficult operating problems encountered at the mine. Success in exploration work, apart from recognition of favourable environments for viable deposits, requires adequate funding, optimism, and persistence and also a measure of good luck. The dramatic change for Dickenson evident in the late 1990s is a stunning and well-deserved example.

With my mother, Cali, Colombia, September, 1931

With my mother and father, San Salvador, 1950

Finishing mile run (Magann Cup), followed by Pete Bremmer
Sports day, Upper Canada College, 1945

Senior Intercollegiate Harrier Championships, 1949-50, University of Toronto
Back Row: Barry Lowes, Staff Member, Pete Niblock, Manager
Front Row: George McMullen, George Doull, Dave Barr, George Webster
Absent: Dunc. Green

At Mile 5, Aiken Lake Trail, near
Germanson landing, BC, Sept, 1951

Dorothy-Elizabeth property, Omenica, BC July, 1953
l. to r. John Anderson, myself, Eric Persson

End of season at Dorthy-Elizabeth property, August, 1953
Rear, l. to r. Wilf Christian, Curly Okabe, John Anderson, Sandy Dean, Bob Bradfield
Front, l. to r. Gord Davies, Ed Jones (cook), Eric Persson, myself

Our wedding day – Swastika, Ontario, December 29, 1953

Fly camp on Scar Creek, 6 miles south of Mt. Waddington, BC
Sandy Dean and myself, July, 1954

Bill Smitheringdale and myself on Leduc Glacier, en route to visit Wendell Dawson's "Lucky Lady" copper prospect near the future Granduc Mine, July, 1954

We flew singly with the pilot in the "Copper Queen," a home-made aircraft chartered in Alaska. It crashed without serious injury on the return flight after Bill and I flew out, July, 1954

Herman Petersen, pilot from Atlin, BC, myself and Charlie Ney unloading supplies from Beaver aircraft at Chutine Lake BC, January, 1959

The family at my parents' home, Duncan BC, Christmas, 1960

David, Rene, Rob and Susan at Niagara Falls, August, 1966

Our 25th wedding anniversary, Vancouver BC, December 29, 1978

Family in Vancouver, 2000

Northern part of the Great Wall, China, near Badaling. September 1, 1990

Trail up Dudh Kosi River, north of Namche Bazaar, Nepal, October 18, 1986 Mt. Ama Dablam (22,493')
on right; distant view of Mt. Everest (29,028') to left of Mt. Lhotse (27,923')

At the pyramids near Cairo, Egypt, May 9, 1985

Rene at the Amber Palace, Jaipur, India, October 7, 1986

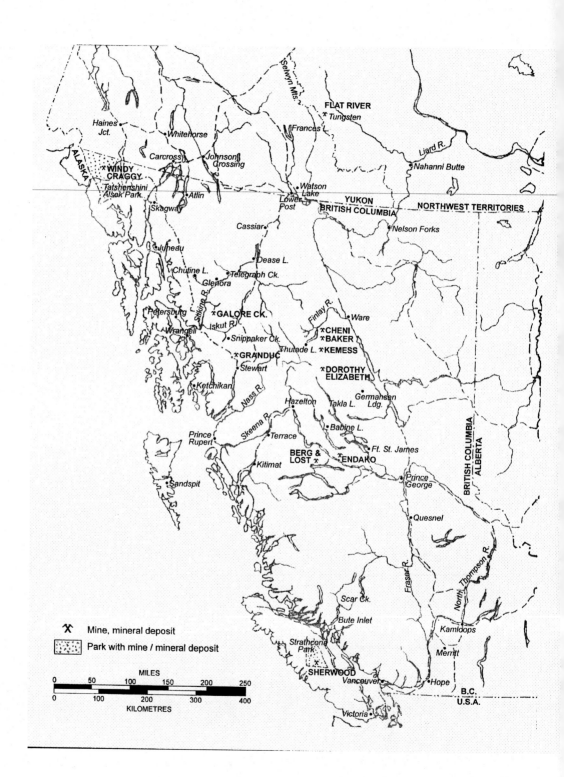

Aerial view of Galore Creek camp, Stikine River area, BC. Snow being dug out to an average depth of 10 feet for tents. May 14, 1962

Galore Creek camp, July, 1963

River boats "Judith Ann" and "Totem" leaving Wrangell, Alaska with 80 tons of supplies for Galore Creek summer exploration project May 8, 1962

Exploration personnel leaving Galore Creek following completion of summer project. l. to r. Jack Bush, Ross Blusson, myself, Don McKinnon, John Anderson, John Nuppunen and Bela Gorgenyi. September, 1963

Myself, David McAuslan, Senior Geologist Kennco Exploration Canada Ltd., and pilot in Kiglapait area, Labrador, August, 1970

Charlie Ney, pilot and myself (photographer) following crash of Bell 47D1 at Johansson Lake, near the future Baker Mine, BC, June, 1971

Production commenced at Du Pont of Canada Exploration Ltd.'s gold-silver Baker Mine, February 1981 at a milling rate of 100 tons per day. Photo July 1981

The Baker Mine was supported by a Twin Otter aircraft C-GDOX piloted by Innotech Aviation personnel. Shown above in April 1981 are, l. to r. Bob Roscoe, Mine Manager, Tony Kremenchuks, Du Pont of Canada Explosives Division, and myself, DOX General Manager and VP Exploration.

I photographed Du Pont of Canada Exploration Ltd.'s crew at a party on Argo claims, Snippaker Creek, BC, August, 1981

BC & Yukon Chamber of Mines Safety Award presentations for 1998. l. to r. Carl Edmonds, Homestake Canada Inc., Annual Safety Award; Peter Fox, Fox Geological Services Inc., Ten Yaer Safety Award; Dave Barr, Safety Committee chair; David Terry holding Five Year Safety Award on behalf of North American Metals Corp.; Bruce McKnight, Boliden Western Canada Ltd., Ten Year Safety Award

Brent Hawkins, Rob and myself on the summit of Mt. Whitney (14,494') California, September, 1983

David and Rene with new Kawasaki, Vancouver, 1998

Rob, Dave McAuslan, Jack Whitlock on our climb of Mt. Rainier, Washington (14,410'), June, 1973

At the finish of the London Marathon on Westminster Bridge, April 12, 1992

With Susan, Stanley Park, Vancouver, after winning our age groups in national Duathlon Championships, September 23, 1990

Start of the World Triathlon Championships, Montreal, September 11, 1999

Start of the 2003 Triathlon Canada National Age Group Championships for 40-plus men's age group, myself on far right (75-9), my friends Mike Ellis (65-9) second from my right and Mike Stokotelny (70-4) to his right. Edmonton, Alberta, July 12, 2003

Cycling portion of above

Rene and myself among other Canadian triathletes in "Nations Parade" at World Triathlon Championships in Cancun, Mexico, November, 2002

Finish at Queenstown ITU Triathlon World Championships, for Silver Medal, 75-9 Age Group, December 6, 2003

Early Characters

Donald W. Coates & Robert B. Fahrig

Don Coates and I share somewhat similar backgrounds, like several of my other friends and acquaintances. We both graduated from the University of Toronto in 1950 - Don in Mining Engineering and I in Mining Geology. Both courses were in the School of Practical Science and the content in the first and second years was identical, with a greater degree of specialization occurring in the final two years. Consequently, we had opportunities to intermingle - particularly at a three-week survey camp at Minden, Ontario following completion of our third year.

Apart from a commitment to acquiring a degree in mineral industry related studies, which incidentally required evidence of practical experience obtained in at least one of our summers between academic years, both of us were active athletically - Don in hockey, where he played for the senior faculty team and I mostly in track and field and harrier. We also had a common acquaintance - Bob Fahrig, a classmate of Don's with whom I had driven out west in 1949 and Doug Williams, another classmate, en route to summer jobs. Our transportation was a vehicle advertised as being required in Vancouver; drivers were offered free transportation in exchange for getting it there. Bob and Doug both had guaranteed summer jobs in the west whereas I was seeking one in the oil exploration sector, and as agreed, they dropped me off near Calgary with a very small bag containing my worldly goods but flush with $10.00. Although it is not relevant to this story, I can report that I arrived in Calgary on a Friday and actually landed a possibility of a field job with a seismic crew to be confirmed over the weekend and a drafting job with an oil company which saw me actually working Friday afternoon!

I met Bob Fahrig again in the winter of 1951, when he was passing through Vancouver and I was a new employee with Northwest Explorations Limited, the predecessor to Kennco Explorations Western Limited, based in Vancouver. We spent an entertaining evening together which included a visit to a bootlegging establishment in the downtown area which Fahrig ferreted out in the wee hours. Nothing was impossible for Bob.

In May 1952, 1 was driving through Burns Lake en route to a property examination, when I met Don, quite by accident, on a muddy street corner. I learned that he had been approached by Bob with an offer of a job as Mine Superintendent at the Emerald Glacier mine, a small silver-lead-zinc operation located in steep mountainous terrain 100 kilometres south of Houston in the Tahtsa Lake area. Bob was the mine manager and the operation, in the development stage, employed about 30 men.

Years later, Don was to recall his brief experience at the property, which was highlighted by several incidents in which both he and Bob were actively involved. Before recording Don's recollections of this period, obtained in February 1994 - some 42 years later - let me describe a brief, but somewhat related incident involving Bob Fahrig while we were at Ajax, the University of Toronto's temporary overflow facility which accommodated first- and second-year engineering students commencing in the fall of 1945, following World War II.

Bob returned late one evening from a visit with friends to one of the pubs at nearby Whitby, determined to complete some homework starting just before most of his unit's residents were preparing for bed. The noise level caused by individuals tracking back and forth to the washrooms evidently disturbed his concentration to the point where he yelled down the corridor for all and sundry to keep quiet. There was an inadequate reaction to his request so Bob returned to his room and appeared shortly after with a loaded Mauser, which he had obtained from a German officer during the war. He fired one or two shots down the hallway above the heads of several individuals, which caused bedlam, but fortunately no other harm. Bob was later reprimanded by the 'house representative', and there were no further repercussions.

Don recalled his arrival at the Emerald Glacier property quite vividly:

"It was May 1952 when I got up there and we had the camp at about 6000 feet. The first level was at about 6400 feet and the second at 6800 feet. There was a bunkhouse, cookery and showers. Fahrig had a cabin with dimensions of 12 x 16 feet that he lived in. It had a bed on each side with a night table in the middle, a stove near one end, a drafting table at the other end, with a Spilsbury radiotelephone for communication with Vancouver.

"It was pretty windy up there, as we were above tree-line. The cabin was cabled down so that it wouldn't fly away. When I got there, there was still a considerable amount of snow around. It was the first time I had ever been in the mountains in my life.

"I remember that one night during my first week at the property, I was lying in my bunk almost asleep and Fahrig was also in his bunk. Suddenly I hear this rattle-rattle-rattle sound and I listen for a while and I hear it again! Fahrig does a little cursing and sits up in bed. So I look across and Fahrig is sitting up in bed with a flashlight in one hand and this Mauser in the other. He's pointing his flashlight down on part of the floor and this little mouse has jumped into the waste-paper basket and is standing there in the middle of the light. So Fahrig takes a shot at him... BANG! The mouse jumped away - moving about 18 inches. Fahrig followed him with his flashlight trying to get another shot at him while cursing away.

"The thing that bothered me the most was that underneath my bed is where we kept our blasting caps - a nice dry place as the blasting caps have to be good and dry before you put them on the lead. I figured if the mouse ran under my bed, Fahrig would follow him around and just as likely hit the blasting caps and that would be the end of our day!

"This went on for two or three nights. The little guy would do a lot of rattling.... on would go Fahrig's flashlight.... he'd get the gun and take more shots at him.

"I soon asked Fahrig where he got the gun and he told me about getting it from a German officer. I started talking more about the gun, telling him how much I admired it and eventually I offered to buy it from him. We dickered back and forth and finally agreed on a price. In addition to the gun, he sold me the inside holster, which he had made up when he came back to Canada; a holster that sat on the hip and a box full of shells. I immediately took it down to the bottom of the hill where I kept my car and put the gun, holsters and shells into the trunk.

"Next night the little mouse comes out again and Fahrig yells at me 'where's that gun.' So I said to him that I had been practising shooting and I'd taken it down to the car. He knows how I feel about the gun, so he doesn't say much, but he's acting kind of funny. About three weeks later, he receives this large trunk from Ottawa which his brother had been storing for him. He was really in a cheerful mood when it arrived. He opened it up, reached in and pulled out this sailor's cap - put it on his head and marched up and down the room. (Fahrig had been a submarine officer during World War II). Then he put on his old uniform. Finally, he reached down into the trunk and pulled out a Thompson sub-machine gun! He takes it and places it over the top of his bed. That night he sits there on his bed with the Thompson cradled in his arms, waiting for the mouse to come out.

"When it finally appeared, he let a burst off, but still missed the mouse. Although he put a bunch of big holes through the floor, it was hunky dory with me, as long as he didn't shoot at me, because by then I had removed the blasting caps from underneath my bed.

"Some days later I came back down to the cabin, having been up on the hill in the afternoon and Fahrig is sitting on his bunk pondering. Fahrig suddenly said 'I know how to do it.' I questioned him and he added 'I know how to get rid of the mouse.' We used to debate a lot so I questioned him further and he said 'I'm going to take an electric blasting cap and cover it with cheese. I'm going to stretch the lead out to here, then wait for the mouse to come out and then I'm going to blast him.' I said that it wouldn't work. He said, 'Why not?' I said, 'Well I don't think you have enough amps and voltage to make that blasting cap go off. Anyway, how are you going to make the blasting cap go off?' He said, 'With a flashlight battery.' So then I said again that it wouldn't work. I didn't really know, but we always argued a lot. He said, 'Come on I'll show you.' He hands me the battery, and then he goes out the door. He was holding a battery with one lead and I was holding the battery with the other and the blasting cap was at my feet. He then pointed to the blasting cap and said, 'Here, touch it.' Well, I did and it went off right at my feet, with bits of the cap splattering me!

"Of course you weren't supposed to be in possession of a working machine gun. Fahrig would loan it to me once in a while and I would take it out on the hill and practise shooting at a bunch of beer bottles. I remember that the bullets cost fifty or seventy-five cents each and the drum held about twenty bullets. Man, it was expensive!

"The men at the camp knew that we had the machine gun. One guy that we fired went into town and told the RCMP about it. The RCMP drove out to the camp to confiscate the gun. They got to the bottom of the hill where we had a loading ramp to load ore on to the trucks. We were warned by an operator at the ramp by radiotelephone that the RCMP were on their way up the hill and had been asking about the gun. Fahrig and I walked to the edge of the hill and you could look right down and see the flash of light as the sun hit the windshield with the vehicle turning up through the switchbacks. Fahrig just stood there and threw his arms in the air and yelled out: 'Here, they've got me. They're coming.' Fahrig was very courageous, and totally fearless. I said that he should hide the gun. He looked at me and said 'Ah! Don't worry, don't worry, they won't make it. At the third switchback they'll boil over.' And that's exactly what happened. They were looking up at us, and we were looking down at them and they never got in. Eventually Fahrig took the gun home with him. Soon after he married a girl in Duncan and left the machine gun with his father-in-law when they went to Colombia. Later somebody broke into his father-in-law's home and stole it."

I forgot to ask Don how a smart individual such as Fahrig never bothered to buy a mouse trap.

It's just as well; this story would never have developed!

Don recalled another incident at the Emerald Glacier camp that I had not heard previously, so

I also recorded it.

"The Emerald Glacier property had at one time been explored by Cominco. They had an old powder magazine on the way up to the upper level. The powder in there must have been a zillion years old, right? Every morning we picked up a couple of cases of this old wet powder. Now this stuff was really wet. Nitro was pouring out of these old wooden boxes. We were taking the crew up on those old 4 x 4 jeeps, so we would throw a couple of boxes in with them plus a couple of good cases when we were drifting because we didn't want to blow out the drift. We were trying to use up this old powder which we usually used in the stopes. We were drilling up holes and then we'd blast out the vein and then we'd blast in the wall for fill. It really didn't matter if you had a misfire. You would just load it up and blast it off again. So every morning we'd go up and we'd tear open these small cases of wet powder and then we'd haul them into the stope and blast with it. Eventually it got to be a bit of a problem because the guys didn't like travelling with it. So I decided to empty the old powder from the magazine and take it into the back of the upper level of the mine where we were stoping. I planned to just put it in the back of one of the old drifts. So we hauled it all up and I think there were 160 cases. You know, you were supposed to have about 24 hours' supply at the most in the powder magazine. Well, we no sooner got it all stashed back there when the Mines Inspector shows up at the mine. Incidentally, the Mines Inspector was in your class, or a year ahead, but I can't think of his name. (John W. Patterson, B.A.Sc., Toronto 1949 appointed Mines Inspector, Prince Rupert, April 1952) Anyway, up comes this guy - we recognize him . . . and we know we're in deep trouble. He came up in the middle of the afternoon so we sort of entertained him until the evening and then it was really too late for him to go up. So we told all the guys to go up and get the powder out of the back of the drift and take it up over the back of the hill and leave it there. And also that we wanted it done right away. In addition we asked them to take the old powder magazine underground and clean it up and make a rim mallet as if we know what we're doing and a new wedge so that it looks as if we're opening up the boxes the proper way. They were supposed to do that in the night shift.

"They came down in the morning and the shifter says, 'we had problems.' So I said 'what's your problem?' And he says 'the truck broke down.' So I say 'Jesus'. So we immediately organized everybody and told them all to get up the hill real quick. The Mines Inspector then came out and said 'Well, I'm ready. " And I said 'Oh, I forgot you. " Now everybody is up there and you can see these figures running back and forth at the top of the hill. We delayed him as best we could until the job was done. Eventually he got up there, but he must have known something was up, because the hammer and the mallet were both brand-new and unused and the powder was only sufficient for a 12-hour supply. Everything looked really beautiful! And so he made only a few comments about our stopes and then he took off.

"Later Fahrig and I took a jeep up to where the old powder was now stashed to decide what we were going to do with it. Again we had a debate. Here was this great big mass of wet powder in boxes, about four tons of it. So Fahrig said, 'I'm going to burn it.' I said, 'If it was me, I wouldn't burn it, I'd blow it up, that's easiest.' Fahrig said 'No, no if you blow it up, it might not all blow and there would be pieces around and some bear might eat it or some damn thing. No, we're going to burn it.' So he reaches into the jeep and he gets an axe and he breaks open a box, just like that. And you know that thing can go! Right? So I say to myself, 'God if it blows, I'll be a dead duck anyway, so I might as well be stupid and I'll blow up with him.' So I got an axe and we bust open all the cases and we pile it all up in one pile. So then we try to get it to burn and it wouldn't burn because it's so damned wet. So Fahrig jumps into the jeep and goes down to the camp and gets some diesel and brings it up and he lights a fire with some newspaper and gets some dry wood from the powder boxes and eventually gets a fire going. The powder starts to burn and I remember the formula: $PV = NRT$. 'You're going to raise the temperature buddy' I yell to Fahrig and then I take off. I start running over the hill, probably a quarter of a mile as fast as my poor old legs would go. I turn around to see if Fahrig is right behind me, and I can see him, he's still back at the fire! I see him take a piece of wood and he rams it into the fire. I couldn't believe it . . . and that's Fahrig, standing there warming his hands at the fire!"

Finally, I asked Don to recall for me his experience of drowning which I had heard from him years earlier.

"It was February 1971, and I had been on my own for about two years after leaving Midwest Diamond Drilling in January 1969 and forming my own company. Things were really tough. I had a part-time job and I was working hard to make payments because I'd already bought a couple of drill rigs on time. If I got a job I would ask people for an advance.

"Anyway, I heard that there was this nice job out in Tasu, (owned by Wesfrob Mines Limited, wholly-owned subsidiary of Falconbridge Nickel Mines Ltd., producing iron and copper concentrates on the west coast of Moresby Island). It was underground and surface and I know where I could borrow or rent an underground drill, and I had a surface rig of course. So I phoned them up on a Tuesday and I said that I was a drill contractor and that I understood they had some work at Tasu and could I have an opportunity to tender on it. The guy I spoke to hemmed and hawed and finally said O.K. I told him, 'Don't worry about it, I'll do a good job.' So he said, 'But you know, if you want to bid, you have to have your bid in by Friday morning! So I said, 'All right, all right leave it with me!

"So I figured if I got up there Tuesday, I could see the geologist Tuesday night; see the property Wednesday morning; come out Wednesday night; write my bid up Thursday and have it down there Friday morning. I phoned him back and said 'I can make it!' I got the reservations and came up to Sandspit on CP Air. I had a connecting flight on a Trans Provincial Beaver on floats right through to Tasu. I landed at Sandspit and it was a miserable, crummy day. I was reading a book called *Khrushchev Remembers* - my wife gave it to me for Valentine's Day. I get off the plane at Sandspit and I walk to the float base and I say to the agent, 'I'm on the flight to Tasu. What time does it leave?' And he says, 'Well, there's two flights to Tasu as there's so many people ... you're on the second flight.' Instead of getting there at about four o'clock in time to see the geologist, I wouldn't get in until seven o'clock, and I won't get a chance to see him. His name was Hermitson and I start saying, 'I've got to get in there' and I'm making a bit of noise, and he says, 'I'm sorry I can't help you ... go and sit down!' So I sit down.

"A little later he calls the first flight and I'm reading my book. Everyone files out - passengers and the pilot. The last guy going out the door turns to me and says, "Did you really want to go? I said, 'Yes'. He said, 'Take my place,' and I thank him. Right? I jump on the plane and I'm sitting in the seat right behind the pilot. A great big fat guy is sitting next to the pilot. He works for the company doing the catering at the mine. I'm next to the door. There were two guys sitting in those little swing seats in the back. Plus a little bit of luggage.

"Off we go, and I'm reading my book because I'm really intrigued with Khrushchev . . . and I've flown quite a bit in my life. Every once in a while I look out to see what the country is like. We come to Tasu and we fly around the open pit and I look down at that and then we come up to the far end and we come around to land in the bay. We land and I go back to reading my book. We slow up awful fast. You don't have to be a rocket scientist to say, 'There's something funny,.' So I look over the pilot's shoulder to see what's going on and there's a wave of water right up to the top coming at us. And I say to myself, "Jesus we're going to go down!' No, that's not the first thing I say to myself ... I say to myself 'We've hit a deadman you know? That's what we did we hit a deadman.'

And I was in a plane where the pilot did that before and he gunned it and lifted it up on to the steps and he ran it into shore. Well, anyway, he's pulling back, he can't pull it up, and he's gunning it. I tell you he's gunning it and he's pulling for his life and it ain't lifting! All I know is we're in trouble. I'm watching this and I say 'I'm getting the hell out of here!' So I took off my belt and I started to open the door which I had trouble doing because it's opening against the wind. I can't get it open. And it's a good thing too, because I can't swim!

"Seconds later the plane nose-dives and it goes down and I go flying over the pilot's shoulder into the cockpit area and bang my head and end up in the bottom all crunched up. The plane is upside down. The next thing I know I come to, the pilot has scrambled up by the door. The big fat guy is on top of me and I can't move. The guys are scrambling all over the place and the water is coming in. That's kind of an amazing thing, because at the time I'm a little confused, and I don't realize that I'm upside down and I'm not exactly sure what the story is. So when I see the water coming in I say to the guy, 'You'd better hurry up because I'm going to go under pretty soon.' and he says, 'Well, I'm having problems opening the door.' And I say, 'Well O.K., but you know I'm going down pretty soon and he says, 'Well, I'm doing my best.' This is the pilot - a young guy (not the big fat guy).

"Eventually I'm under water. I pull up, take a big gulp of air, and down I go. The big guy then moves out, but by then I've got very little energy left and I move up to the nearest door and it's not open and I can't understand that. I move over to the other door and it's not open either and then I go, 'I'm finished.' And I remember that it's nice and warm, very pleasant and I kept talking to myself, saying, 'They're going to get me, don't worry, it's nice and warm.' I didn't panic. I held my breath and I pushed it out.

"On surface, they started counting heads and they realize that there is one missing. The pilot dives down and pulled on the door and got it open and just before he ran out of air he reached inside and that's where I was lying. With extra exertion he grabbed me by the collar and pulls me to surface.

"By then a boat had come out from the village and they had said, 'Take this guy first because he's not with us anymore.' Apparently as they were pulling me up into the boat they rocked me back and forth trying to get me into the boat and I started breathing. And then I can remember for sure the noise I made trying to get fresh air into my lungs. Everything was like a dream. I was nice and warm and I look around and there's the plane, but upside down, held up by the pontoons.

"They put me into the hospital and the doctor checked me over and pronounces me great. I had to bum some money off him to get a pair of pants because everything was lost. I went and saw the geologist. He saw me at night, which was pretty good of him. And the next morning I got up and went and visited all the different drill sites. I went back around noon to catch the plane to return to Sandspit. They wouldn't let me on because I didn't have a ticket, because I'd lost it on the plane! Eventually I got on and they took me back to Sandspit where I had to go through the same damned exercise because I've got no ticket. I had no baggage, because everything I'd had was ruined.

"I go home, bid the job. I got down Friday morning. I give it to him. I don't get the job! And I thought that was pretty crummy! I lost about 300 dollars worth of material, you know, my hard hat, my underground clothes and so on. So I thought, 'What would be fair?' So I took 60% of it and charged them 220 dollars. Trans Provincial wouldn't pay me. I had to go to a lawyer to collect my money. And do you know, no one picked up the phone to say, 'Are you living, are you dead?' Nothing, and I thought that was pretty crummy. Not only that, I didn't get the job.

"I think I survived because I'm a student of Maxwell Maltz' psycho cybernetics, where you do affirmations by talking to yourself. And the moment I was going under, I started to talk to myself. And I talked and I talked and I talked. And it was all positive. Like, 'I am calm, I know I'm going to get out of here, take it easy it's not a problem,' that kind of stuff.

"It turned out that the problem on landing at Tasu was that the pilot had forgotten to lift the wheels." (the Beaver was equipped with retractable wheels housed in the pontoons and used for carrying the aircraft up or down ramps at the air bases.)

At the end of our interview, Don said:

"I got busy last night as I went to play hockey; and meant to bring a picture of the plane which one of the guys at Tasu sent to me."

Don has always been an avid hockey player and at 65 he still plays three times a week in Vancouver: two times with two different 'old-timer' teams and once each week with the 'young guys.' He has been a regular member of the Prospectors All-Stars who annually play a team from Teck Corporation at the Prospectors and Developers Association of Canada convention, held in Toronto in March. I can personally attest to his remarkable condition and ability, as I played on a couple of occasions with the Prospectors All-Stars at the PDA convention while I was based in Toronto in the late 1960s. Don has also played with several old-timer teams from Canada at various world competitions held in the US and he has been on one or more age-group championship teams at these events.

I was so impressed by Don's survival episode that I decided to follow it up with the hope of identifying the pilot who was responsible for saving his life. Incredibly, I located him within a few hours with several phone calls. The first was to Ken Mair, a pilot with Trans Provincial Airways (TPA) from 1968 - 1993, whom I had flown with in the past when the company was based out of Terrace. Unfortunately TPA incurred debts of about $6 million and went into receivership in March 1993, its assets partly being acquired by Harbour Air in Prince Rupert. Ken had previously helped me locate the whereabouts of Doug Chappell, another pilot who I flew with on many occasions and whom I have written about in my recollections. Ken recalled the incident at Tasu and identified the pilot - Reg Young. He knew that Reg was currently working as a pilot flying Martin Mars, on Vancouver Island, although he didn't know his actual whereabouts. I had seen two of these remarkable aircraft at Sproat Lake on Vancouver Island where they are based, when we landed by helicopter at the air base in 1991. At the time I was en route to make an examination of the Sherwood Mine in nearby Strathcona Park. The Mars, the largest float planes built prior to Howard Hughes' Spruce Goose with its 219' wing span were apparently designed for trans-oceanic passenger service. They were used during World War II partly for troop and cargo transport by the US Navy, as they were the only flying boats that could fly between the west coast and Hawaii in a head wind. One such flight took 16 hours. Four Martin Mars were purchased in 1959 by the Forest Industry through Forest Industry Flying Tankers (FIFT), a company currently owned by MacMillan Bloedel Ltd. (51%) and Pacific Forest Products and TimberWest Limited (49%). As I write this, only two Martin Mars survive. The Mars have a wing span of 200 feet and the tip of the tail stands 50 feet above the ground. The modified aircraft used for fire-fighting can pick up 30 tons of water skimming above lakes at 70 mph, taking 22 seconds to fill its water tank.

The company's office in Port Alberni agreed to provide me with Reg Young's home phone number when I described my quest. On contacting him, I learned that he has been living at Port Alberni for 13 years and I believe he joined FIFT after leaving TPA, which would be about 1980. Reg is a native Canadian, having been born and brought up in the Queen

Charlotte Islands. He recalled the incident and he essentially confirmed Don's description, with only minor modifications. He was concerned about my possibly publishing a description of the event because of his responsibility for the aircraft and the passengers as a pilot. I assured him that my focus was on Don's survival as one of the more momentous recollections of events in his life, and the fact that Don did not know the name of the individual responsible. Reg impressed me with his candour, modesty and obvious courage and focus in a potentially tragic situation.

He recalled that after getting the door open and reaching surface, he asked one of the floating passengers to count heads. He immediately realized that one passenger was missing and dove down, located a door, pulled it open and fortunately Don was right there. He pulled him out and brought him to surface.

Reg has flown extensively on the West Coast. A new map of the Queen Charlotte Islands recognizes his relationship to the Islands with Reg Young Lake.

Bob Chaplin, Frank Cooke, Spud Huestis

I spent several weeks in the summer of 1957 with Bob Chaplin completing geological reconnaissance surveys for Kennco in the portion of the Guichon Creek Batholith lying south of the Highland Valley area in southern B.C. The period coincided with the exploration activity on Bethlehem Copper Corporation's properties, then being directed by American Smelting and Refining Company (ASARCO) and its team of geologists, including Bill Stevenson and several others on contract, namely Bill White and Bob Thompson, both professors at the University of British Columbia. It was the start of the porphyry copper/molybdenum era which was to have such a significant impact on the B.C. exploration sector in the next two decades and on the economy for an additional three decades.

We camped at various localities accessible by our 4-wheel drive vehicle and I learned to appreciate Bob's company. He was, and remains, a physically tough and imposing individual, and a dedicated worker with an ebullient attitude. We enjoyed the brief periods of relaxation in the evening, after the day's work had been adequately recorded. We secreted bottles of beer in the cool streams nearby during the day, revealing an ingrained attitude common to prospectors toward protection of valuable resources until fully claimed. These greeted us both before and during our simply prepared dinners which all too often consisted of T-bone steaks.

Bob spent the following summer in the Highland Valley area working under Joe Brummer who was supervising a comprehensive program of geological and geochemical exploration over the Guichon Creek batholith. My principal job in 1958, commencing in May and extending through October was supervising an exploration program on the Krain property in the north central part of the batholith. The program was initiated in the previous fall. As I realized that it would be continued the following year I arranged to store all the moveable equipment for safekeeping at a warehouse in Ashcroft.

Prior to our departure from Vancouver in May to commence our 1958 program, Joe Brummer had asked me if I could spare any equipment for his own program. I mentioned a few items including a couple of space heaters and some cots. Joe arranged to have Bob meet us at Ashcroft with a truck to pick up the items. We arrived at Ashcroft just before Bob, and we were starting lunch at a restaurant when he appeared. I gave him the keys to the warehouse as he was obviously anxious to pick up the gear and leave. We were leaving the restaurant when Bob reappeared with the keys. I looked at the 3/4 ton truck which was well packed with a tarpaulin covering the contents. "Good God Bob", I said, "that looks like a lot more than a couple of space heaters and some cots. What's going on?" I walked over and drew back part of the tarpaulin revealing a significant portion of our equipment. Bob wore a sheepish grin as he reached into his pocket and handed me a note he had received from Joe. It read, "Pick up two space-heaters and x-number of cots, plus anything else you have room for. " Such was the cooperative attitude among some explorationists in that hectic period that we did help Bob unload all but the space heaters and cots onto our trucks.

In the summer of 1954 Bob had worked as a student assistant on an exploration crew on a lead-zinc property in the Vangorda Creek area, Yukon, where he helped in the drill-core logging program. The property had been discovered in 1953 by Al Kulan, who was working as a prospector for Prospectors' Airways. Kulan had been led to the general area by two local natives who were aware of the presence of several rusty outcrops.

Bob did not return to the Vangorda Creek area until 1964 when he was working with Kerr Addison Mines Limited, having joined the company in the fall of 1963, about one month after Al Kulan had decided on completing a program in the Stikine River area. Kerr Addison also had extensive holdings in the Vangorda Creek area. They proposed to work initially in the Stikine and then move up to the Vangorda Creek area later in the season. Meanwhile, a small company named Dynasty Explorations Ltd. was formed which included Al Kulan, Gordon Davis (who had worked with me as a student assistant in the Stikine area in 1960-61), Aaro Aho, R. Markham, John Brock and several others. At the time of Al Kulan's departure from Kerr Addison, it was understood that he would receive a 5% interest in any mineral discovery in which he had been involved while he was employed by the company. Immediately prior to Kulan's departure, Kerr Addison had just completed an airborne magnetometer survey over parts of the Vangorda Creek belt. However, Kulan learned that Kerr Addison did not intend to honour the 5% interest arrangement. Bob Chaplin acted as intermediary in subsequent discussions between Kerr Addison and Kulan, as Kulan refused to talk to Kerr's representatives directly. One evening Al said to Bob, "Bob, you had better tell Kerr to go in and stake those magnetic anomalies within the next week or two weeks, or I'm going to go in and stake them myself."

Kerr immediately rounded up the necessary personnel, which included Bob, and completed the claim staking in the area of interest in the spring of 1964. The date coincided with the announcement of the major base metal discovery by Texas Gulf Sulphur Co. Inc. in the Timmins area of northern Ontario, which was placed into production as the Kidd Creek mine in 1967.

Bob then worked with Kerr in the Stikine River area and visited me at the Galore Creek property when I was supervising the Stikine copper program in its fifth consecutive year. Bob recalls that when he phoned me about making a visit to the property, I had said, "Yes Bob, you can come in, but I want it strictly understood that this is to be a social visit." When Bob arrived at the property I wasn't there to greet him and he made his way to the office where we had an up to date model of the deposit based on drilling information which was covered with a cloth when not in use. There was no one else in the office. Shortly afterwards, I arrived back from a visit to a drill-site and Bob, hearing my voice outside the door, locked it! I banged on the door and Bob said, "I'm sorry, Dave, you can come in and look at whatever you want, but I want you to understand it's strictly a social trip." Apparently I was then allowed in, and Bob recalls that I showed him everything of interest anyway! On being reminded of this incident, I recalled to Bob that one of the many Japanese visitors to Galore Creek in its active exploration stage had been caught by John Anderson completing a tracing on rice paper of one of our maps. It was confiscated with profuse apologies from the unlucky visitor.

After having some disagreements with Kerr Addison supervisory personnel in the summer of 1964, Bob quit. He got a ride to Telegraph Creek in a helicopter piloted by Leo Lannin of Okanagan Helicopters, whom Rene and I had flown with into Harrison Lake in the spring of 1954 where I examined an old gold property. Bob hitchhiked from Telegraph Creek to Cassiar where he spent a few days. He then went to Watson Lake by bus. He had not made up his mind on a destination, but he had $500 on arrival at Watson Lake. He elected to take the bus to Whitehorse where he met Gordon Davis and Al Kulan who had just completed staking ground over a discovery they had made in the Swim Lake area, near Vangorda Creek. Bob decided to tie on some claims on his own with their concurrence. Following his staking they made him a proposition whereby Dynasty would issue him shares of stock for his claims and he would join the company. Bob agreed. He received 27,000 vendor shares and he also bought some additional stock. He went to a bank to obtain a loan to purchase an additional 8,000 shares. Fortune smiles, as the bank manager was a Dynasty shareholder. He gave Bob a loan of $3,200 to buy the shares. Bob's collateral was $400! Eventually he had a total of about 40,000 shares.

He left Dynasty after the 1964 field season and in 1965 formed a syndicate funded by Kennco Explorations, Cominco Limited and Asarco to explore in the Terrace area. During that program he broke his arm and on a trip to Vancouver in June to have his cast changed he met Gordon Davis who informed him of the major lead-zinc discovery by Dynasty which subsequently developed into the Anvil deposit. He made a quick trip to the area and then returned to Terrace to complete the summer exploration program. Dynasty stock was trading at $7 at that time. He was tied up in the bush until October and had no occasions to check on the price of the stock which was to reach a high of $16 in 1965, peaking at $23 in 1966.

In retrospect, Bob's decision to go to Whitehorse versus Vancouver, when in Watson Lake in 1964, was to have a dramatic impact on his future and provides an excellent example of being at the right place at the right time.

In 1974 Bob had travelled to Godlin Lakes in the Northwest Territories, to determine whether he and his partner, Paul White, should stake ground tying on to a recent lead-zinc discovery by Welcome North Mines. There was a large tent camp at the property and many of the explorationists, including Bob, carried firearms for protection against the bears. One evening several of the party decided to test their markspersonship (assuming some ladies were present) using tin cans as targets. Bob carried a 357 magnum in a holster in his packsack as it had fallen out of the holster on one occasion. On the same day he suddenly had an opportunity to return to Whitehorse on short notice by helicopter, a flight of 300-400 miles. On approaching the Whitehorse airport, Bob noted that the Canadian Pacific flight for Vancouver was at the terminal building preparing for departure. He jumped out of the helicopter and was met by Paul White who had Bob's large duffle bag with him containing Bob's return ticket to Vancouver. Bob had placed the day pack on top of the duffle bag, and in his haste to retrieve the ticket, he dropped the day pack on the ground forgetting that it contained his firearm which unfortunately was loaded. The gun discharged. Bob thought it was a car backfiring. Paul said, "Are you okay Bob?" Bob wondered what he was talking about, then he smelled the powder and looking down saw a slit in his pants. The bullet had ricocheted off the pavement, taking a slice about 1/8 inch deep off his leg. He gave the gun to Paul for safekeeping and walked into the airport to check in. After boarding the aircraft he checked his leg and found that the wound was only bleeding slightly. He later went into the washroom and stuffed toilet paper against the wound, secured by his work socks. Another old friend, Robert Gale, was on the same flight. On arriving in Vancouver Bob was welcomed by his wife. While driving away from the airport Bob said, "Well, maybe we'd better go to the hospital."

"What for?"

"I've just shot myself in the foot."

On arriving at the hospital, Bob was attended by a doctor who noted the powder burns.

"Is this a gun shot wound?"

"Yes it is." Bob replied.

"Well, did you report it?"

"Well no", Bob said, "it happened at the airport in Yukon Territory, and that sort of thing happens all the time up there and people don't bother reporting it." Typical Bob comment!

Later Bob received a small package in the mail. It contained a tape-deck which included a song entitled, "Dumb-Dumb Elmer, the 76th worst gun in the west", who also shot himself. I confessed to being somewhat disappointed with Bob's version of this story when he recalled it for me in March 1990. I had heard a highly embellished version of the same event years earlier. My informant claimed that a stewardess, walking down the aisle of the plane en route to Vancouver, looked down to see a stream of blood trickling down the aisle which she traced to his seat. On asking Bob if he was hurt, he was supposed to have replied, "Oh no, it's nothing serious."

In September 1977 I completed an examination for DuPont of a recent tungsten discovery, made by Welcome North Mines Ltd., in the Ross River area which had attracted considerable attention from potential joint venture participants. In our small party was Al Kulan who had been involved in the discovery and who acted as our guide. A week after returning to Vancouver I was stunned to learn from an acquaintance that he had been fatally shot by John Rolls while drinking beer in the Ross River Hotel bar.

Al and John Rolls had been partners in the construction of the Ross River Hotel. Al eventually became owner of the hotel. Subsequently, the original partners had a serious falling-out and the two had not spoken to each other for three years before the murder. The cause of the tragedy was attributed partly to John's jealousy of Kulan's success and partly to a fantasy he developed that Kulan was plotting his financial downfall.

John Rolls was tried before Mr. Justice Harry C.B. Maddison and jury in a case heard March 28 to April 4, 1978. He was convicted on April 4, 1978 of first-degree murder and sentenced to imprisonment in the British Columbia Penitentiary at New Westminster for a term of life without eligibility of parole for 25 years.[26]

Some years later, his son Jack, who had provided helicopter support in a program in the Texas Creek area, near Ross River, in which I had been involved with DuPont, Bob Chaplin and Teck Explorations, was killed in a helicopter logging accident in the Pitt Lake area of B.C.

Al Kulan's death was preceded on May 27, 1977 by the equally tragic and untimely death of Aaro Aho on his farm near Ladysmith, B.C. when a tractor that he was driving rolled over, pinning him beneath. Shortly after, a story began circulating in western mining circles concerning an elderly native in the Vangorda Creek area who had been greatly disturbed by the exploration activity and the subsequent mining which he considered an unwelcome intrusion within his hereditary lands; he had purportedly placed a curse on those directly involved.

Of course this involved Gord Davis and John Brock, among others, in addition to Aro and Al, and I always felt sympathy for these explorationists, whether they were superstitious

---

[26] "Cashing In" by Jane Gaffin, 1980 (p 210)

or not. Their subsequent success and continuing longevity, if it can be so called thirty years later, thankfully belies the story.

The exploration sector has always attracted those with a sense of adventure who have frequently found themselves in life-threatening situations in the wilderness environment. Although the tragic deaths of Aaro Aho and Al Kulan were not exploration-related, the relatively high incidence of fatalities recorded in the exploration sector is well substantiated.

Among the many mutual acquaintances of both Bob Chaplin and myself were Frank Cooke and Spud Huestis. The common time frame was the Highland Valley exploration activity in 1957, although I had heard of Frank Cooke much earlier. Frank was one of the more colourful individuals that I have had the good fortune to meet, although our association was all too brief. As the following intertwined relationships reveal, individuals like Frank delight in living on the edge and appear to gravitate naturally to permissive situations and environments that satisfy this yen.

One of Bob's recollections concerning Frank Cooke was very similar to one of my own. He had worked with Spud Huestis and Frank for several years, partly in the Highland Valley and also on mining property examinations. On one trip Bob and Frank were together examining the Blue Grotto prospect near Pitman in the Terrace area of west-central B.C. To access the property they had to cross Hardscrabble Creek; adequately named, as events were to prove.

As is, or was the case, when attempting crossings of turbulent waterways in such remote areas, large trees have frequently fallen and successfully spanned the creek or river, to be limbed by those seeking to cross. Frank nimbly made such a crossing uneventfully, while Bob, being less experienced, cut a pole to steady himself. Half way across he dug the pole into the river bed, and lost his balance as the current caught and swept the pole downstream. As he fell he managed to reach up and grab a limb, preventing himself from being swept under the log. About 100 metres downstream was a logjam and he clung on to the limb for dear life.

Frank looked back and immediately jumped onto the log, swiftly reaching Bob. Such was Frank's strength and balance that he reached down, grabbing Bob by the shoulder with one hand and hauled him back on to the log. Bob would have weighed 180 pounds or more at the time.

In his earlier days, Frank Cooke was a logger on Vancouver Island. In those days, jobs were hard to get. In spite of that, loggers would work for a while, then leave. Frank used to say, "There were three crews - the guys working, the guys coming and the guys going." Apparently some of the guys would steal stuff when they left. They favoured those old grey Hudson Bay blankets. The logging company that Frank was working for could not keep a supply of blankets on hand. One day, in order to eliminate this problem, the logging operator went into the bunkhouses and took out all of the blankets. He had a blanket made which was mounted on a roller. When all the loggers got into bed, they had to pass this blanket over themselves.

Eventually, Frank recounted this story to his son who listened with rapt attention. Finally, his son asked, "How thick was the blanket?" Frank never hesitated in his reply: "Two inches." And then he smiled.

Bob Chaplin and Frank Cooke, while working for Spud Huestis in the Pitman area near Terrace in 1962, drove into Prince Rupert one weekend. It was Bob's first visit to the city. Bob recollects that they were walking along the main street. "All of a sudden we heard a crash as this window broke on the second storey of a building above us. Suddenly this body came hurtling out of the window. There were two loggers standing on the street immediately below, and a drunken Indian woman landed on the two loggers, knocking them flying. She got up, thinking that they had attacked her and she starts kicking them. She then stomped off across the street into a bar."

After recovering from this introduction to Prince Rupert, Bob and Frank continued their walk. Bob recalled needing a restroom and entering a beer parlour. He looked over the crowd trying to locate the restroom. His gaze was distracted momentarily by a large native sitting alone at a table. They looked at each other and suddenly the native said "Hi!". Bob returned the greeting. Bob said that the native suddenly jumped off his chair, advanced on Bob and said, "What did you say?" in a threatening tone. Bob placated him with some difficulty as he retreated out of the beer parlour.

In 1957 while working at the Krain copper property in the Highland Valley area, I arranged to drive down to Vancouver one weekend with Frank Cooke. Although I had met Frank on several other occasions, this was the first, and only time that we were together for a prolonged period, our drive to Vancouver taking almost seven hours. My only regret was that in those days I had no access to a tape recorder, as Frank provided a lengthy and thoroughly entertaining description of some of the more colourful events in his life as a logger, member of the Royal Northwest Mounted Police (predecessor to the Royal Canadian Mounted Police), horse wrangler, and miner.

One of Frank's recollections of his experiences as a member of the Northwest Mounted Police stationed at Prince George was one of his more difficult apprehensions of a logger, who was unwilling to submit to arrest for a misdemeanour. A fist fight broke out between the two which attracted a considerable following. Frank and the logger sustained the altercation for several blocks as Frank steered the logger toward the station, finally subduing him after a running fist fight lasting almost half an hour.

On that sunny day in August we drove out of Highland Valley along the road to Ashcroft. Coincidentally, we were stopped briefly at a logging site where, on an earlier occasion, I and most of our exploration crew, driving to Vancouver for the weekend, had spent several hours fire-fighting in our go-to-town clothes. Not this time: while the logging operator was trying to recruit fire-fighters, including us, Frank was pleasantly and equally emphatic about our driving on.

Bob Chaplin recalled similar drives with Frank as they were regular events while he was working for Spud Huestis. Bob and Frank used to take a bottle of whisky with them on their drive to Vancouver which they would purchase at Ashcroft. They would stop periodically, as Frank knew all the little springs en route. They would consume the whole bottle on their drive and would repeat this process on their return from Vancouver. Bob was careful, when recalling this, not to identify the driver.

Years ago Frank and Spud Huestis had taken a trip to Victoria to obtain some information from the Minister of Mines concerning Bethlehem Copper Corporation's proposed mining operation in the Highland Valley. Spud and Frank had spent the evening prior to their meeting with the Minister in Victoria where they had driven to a restaurant for dinner. As they were parking their car they got into an altercation with an individual who had come out of the restaurant. The individual came over and poked his head into the open window next to Spud who was driving. Spud punched him and drove off. As they left, Frank said, "Drive around the block, Spud, and we'll hit him again." The incident was apparently reported in the morning paper under the heading 'Local Man Attacked by Two Unknown Assailants.' The Minister, who knew Spud and Frank quite well, was reading the paper when they arrived for their meeting. He jokingly said, "I guess neither of you would know anything about this?"

Spud is still remembered for his life-long career in mining from prospector through to Chief Executive of Bethlehem Copper. He and his close associate, Pat Reynolds, were responsible for attracting the financing from Japanese banking institutions which was required to place the Bethlehem Copper Mine into production as British Columbia's first porphyry copper operation in the early 1960s. Canadian financiers were skeptical about the viability of such an operation - which was to herald in the porphyry copper era which provided such an impetus to the mining industry throughout the 1960s and 1970s.

In his later years, Spud was prominent as a fundraiser. Bob Chaplin recalls being invited to lunch one day with a group of others by Spud at the University Club where Spud was a member. The only item on the menu that Bob recalls eating was turtle soup. After finishing his soup, Spud asked Bob whether he enjoyed it. Bob replied, "Oh, it was great!" Spud said, "Well, that's good because it's going to cost you three hundred bucks, as I'm raising money for the B.C. & Yukon Chamber of Mines Building Fund." Bob laughingly recalls that he had his name on a small plaque commemorating donors as evidence of his expensive meal.

One of Frank's more ribald experiences occurred in Ottawa during his service with the Royal Northwest Mounted Police. Frank was instructed to deliver a bouquet of flowers to the residence of one of the Ambassadors of a Latin American country. Frank undoubtedly would have cut a magnificent figure in his uniform with his impressive physique and handsome features. At the door he was met by the Ambassador's wife who smilingly reached for the flowers with one hand, the other grasping his privates as she announced, "Oh, you are ze nice boy!"

Somewhat disjointedly, I remember two other much earlier incidents involving Frank Cooke which bear mention, as they partly reflect the attitudes and survival episodes of the times.

As will be evident, the first, which I remember as so humorous 40 years ago, is so dated that it would raise even my eyebrows. Frank had been a cohort of "Skook" Davidson, an almost legendary 'outfitter' in northeastern B.C. with whom I had had very early dealings in arranging packhorse support in the 1950s from his base at Terminus Mountain, south of Watson Lake. This was before Frank became a member of the RNWMP.

On one occasion in the late, late fall, Skook, Frank and another individual working for Skook lost virtually all their packhorses, supplies, etc., while on their way back to their base at Terminus Mountain. Seeking only to survive, they decided to head for Ft. Ware. Half starved and within a few days of reaching Ft. Ware, they stumbled on a trapper's cabin in late evening. Frank recollected that although Skook was by far the oldest, he successfully fended the other two off as he tested the trapper's meagre assortment of wares. Stunned, Frank looked over to see Skook gobbling down the upturned contents of a can, which included the tail end of a mouse.

"Jesus, Skook," he yelled, "look at what you're eating!" Skook looked, ran outside the cabin and vomited.

They were so tired that sleep soon overcame them. Frank recalled that the next morning Skook could not find his false teeth. Frank said, "Check outside, Skook," and sure enough his teeth were securely implanted in the dead mouse.

If you think that humour is dated consider this incident that I thought so humorous in the early 1950's - also dealing with Frank and Skook.

In the same period when Frank was working for Skook, they had one of their frequent binges at Manson Creek, about 150 kilometres north of Ft. St. James. Four were actually involved in the evening spree of which only one was an 'outsider.' The following morning, the unfortunate outsider pointed to the other three, gleefully observing that he was the only one without a black-eye. Apparently Frank looked at Skook, who winked, and within the twinkle of an eyelash, all four were so encumbered, thanks to Frank.

Pilots

Many of the early Canadian bush pilots, who played such an important role in supporting mineral exploration, mining and other ventures in the North are legend. The exploits of the more well-known pilots have been well documented, in autobiographies, biographies and other references. Following World War II, pilots were so numerous and although the services required of bush pilots by their diversified clientele were no less demanding, their exploits have not attracted the attention and awe that was focused on their predecessors.

The 1950s were to provide a dramatic stimulus to exploration in the more mountainous, remote regions of Canada lacking the profusion of lakes, which afforded such reasonable access by float-equipped aircraft within the Canadian shield. Here, of course, support was provided by helicopters. The pilots of rotary-wing aircraft were a similar breed, mostly very competent, devoted to their companies and considerate of the demands of their clients, some of which were totally unreasonable. In order to survive in a dangerous occupation, they had to know the limitations of their machines, their own abilities and the unpredictable vagaries of weather conditions. They were, and remain, a courageous group of individuals.

I was privileged to accumulate over 1400 hours of helicopter-time, principally in the 1954 – 1965 period, and to a lesser extent, from 1971 to the present, with 30 – 40 pilots. In addition, I have flown throughout Canada with as many fixed-wing pilots commencing in 1946. Many of our operations in Western Canada were totally dependent on helicopters for transportation of all supplies from the nearest airstrip or lake accessible by fixed-wing aircraft and for the daily movement of geological personnel or diamond drillers to their working area. As described in another chapter, we relied on Hercules C-46s to transport most of the equipment and supplies to construct the Baker Mine gold-silver operation in the Toodoggone River area north of Smithers, B.C. in 1980 and Twin Otter aircraft to maintain it with periodic Hercules campaigns during the three-year production period.

In the late 1950s and early 1960s, ground-to-air and air-to-ground communications involving helicopter-supported exploration programs, or others, were in their infancy. In the earlier part of the period, radio communications were non-existent and the normal retrieval of a dropped-off explorationist was dependent on a reasonably accurate description of his drop-off and/or pick-up location at base camp, in the event of a helicopter malfunction.

Most of us carried a one square yard fragment of brilliant red nylon to attract a helicopter. We also relied on reflections directed to the helicopter from a compass glass in sunny weather and we became quite adept at homing in on a helicopter by sticking our arm out as a guide to the reflection point from the 'mirror' to home in on our target. Later we advanced to 'pencil-type' emitted flares as described under "Survival Adventures."

No one who has spent an arduous day in the field, be it mountainous or otherwise, and who in that era was not equipped with any emergency gear, can imagine the calming effect after a long, hard day, of hearing the distant drone of an approaching helicopter. Our pick-up system worked in most cases, although there were exceptions. In my narrative I have mentioned those in which weather, helicopter mishaps, sudden emergencies demanding helicopter use, and others interfered with the system.

I recall one which did not involve the 'hard-rock' sector and more importantly, had a 'happy' ending. An oil-exploration party working in the foothills of Alberta in the 1950s had an arrangement, as did I, while working in 1949 with a seismic crew in Alberta, whereby over-time days worked during the month were taken off normally as a 2 – 4 day liberty at month-end. A helicopter pilot of one of the crews relaxing at a bar in Edmonton on such a furlough with several of his crew happened to mention a geologist who had been dropped off in the mountains two days earlier. Fortunately, considerable anxiety as to his whereabouts developed as the group realized that he had not been seen in the interval. The pilot immediately returned to base camp and retrieved the unfortunate individual.

I really wish to stress the courage exhibited by most of the helicopter pilots that I had the good fortune to fly with. As I have mentioned elsewhere, possibly ad nauseam, they were not only adept pilots, but most had the bush mentality of their fixed-wing predecessors. In this I wish to stress their compatibility to their somewhat lonely environment, shared by us geologists, which made them excellent companions. Most of them were family types, as were many of us explorationists, caught up in an era where, as for most of our predecessors, the love of adventure, the opportunity to visit places not previously recorded, soon become secondary to the pull of family life. Yet, as in the past, circumstances one way or another compelled many to continue this diverse relationship. As in other fields, most were caught up with the adventuresome aspects as single individuals prior to being exposed to the compelling ties of family life. For some it was a difficult, if not impossible, situation to abandon.

I have made frequent references to the fixed-wing and rotary-wing pilots that I flew with. On only one occasion was I in a position to provide an adequate recollection of one pilot – this being Herman Peterson, whom I greatly admired both as a pilot and an individual. I had the opportunity to record an important part of his life, which I have provided in the following section, together with briefer recollections of Hughie Hughes and Bill McLeod.

Bill McLeod

When I first met Bill McLeod in the late 1950s, he was already an experienced fixed-wing and helicopter pilot. My most vivid memory of Bill, a good story-teller, was one of his more incredible experiences on the Alcan project in the 1950s.

Bill was one of several helicopter pilots supporting crews working on the construction of a major tunnel designed to divert the easterly flow of waters in the Nechako River system from its headwaters into the Kemano River to the west, for power generation purposes in the aluminum smelter built at Kemano. The region is one of heavy precipitation and the river is subject to flash floods in several of its tributaries. On one of these occasions, a bulldozer with operator and passenger became stalled while trying to re-cross one of the flooded creeks near base camp. The water level rose at an alarming rate and the two men elected to climb onto the roof of the cab.

Another worker spotted their predicament and ran to base camp for help. Following a short discussion with base camp personnel, it was decided that Bill should complete a quick reconnaissance of the site by helicopter and report back. In the meantime, a pick-up truck had arrived at the site. With one of the two occupants already preparing to jump into the water and try to reach shore, the driver of the pick-up successfully threw a length of rope to the top of the cab, attaching one end to a nearby tree. The panic-stricken individual on the top of the cab secured the other end around his waist with the intention of jumping into the water and having the flow pull him to shore.

At this moment Bill arrived on site and hovering over the top of the cab, saw that prompt action was required. He decided to make a slow approach toward the cab in the restricted valley to see if it was practical to extricate the pair. As Bill recalled, "Before I knew what had happened, I felt this jar, the passenger door flew open, this individual was sitting beside me and slammed the door shut. I applied full power, but as the helicopter rose upward I suddenly lost control, with the helicopter arching off to the shore and dropping. By the grace of God there were no trees in the way, just large alders and I crashed into these without too much damage to the aircraft and none to ourselves."

Of course, Bill was not aware that his passenger was attached by a rope to a tree, and by a miracle the knot was not a slip knot! The end of the story is anti-climatic. Within minutes a large powered shovel with a sizeable bucket appeared on the scene and lifted the other occupant off the cat roof. Within half an hour, the water level had dropped dramatically and the bulldozer had been retrieved.

Bill was to support our small crew at Galore Creek in 1962, where we had a very eventful, satisfactory and trouble-free operation as noted elsewhere.

Hughie Hughes

As described in another chapter (Galore Creek), our exploration program in 1963 at Galore Creek, had been so encouraging that we proposed a major program in 1964 which was calculated to require support by two Hiller 12E helicopters. The pilots selected by their company, Okanagan Helicopters Limited, were Hughie Hughes and Greg Walker – a dissimilar but very compatible duo, as evidenced by the decision of Okanagan Helicopters to remain attached to our project in 1964. Okanagan Helicopters had adopted a general policy of providing annual rotation of their pilots to various projects in order to increase their overall expertise.

I have met few individuals like Hughie, who created such an initial impression of absolute commitment to their working environment. In retrospect, my impressions dating back over 30 years remain unchanged. He was one of those individuals who know themselves – made little of their aspirations, or past accomplishments. Although not uncommunicative, he tended to be quiet, but could always get his message across, as required. One day I was in our office tent when he walked in and I immediately knew that something vital was afoot, as his rare visits were never casual.

"How are things going?" I asked non-committally.

"Not so bad", he replied, "however, the flak on drill site No. 4 was a little heavy today!"

Knowing full well his background as a highly heralded survivor of the Battle of Britain as a pilot, I requested details.

"Well," he said, "I was approaching the drill site with a load when I saw the drill crew waving their arms – obviously trying to steer me off. I veered away just before I heard a loud explosion and felt several fragments hit the fuselage." Fortunately, the helicopter had not been damaged to the stage where it was inoperable. "I really think that we must consider some form of radio-contact between the dispatcher and the drill site." I completely agreed and radio communication was subsequently implemented.

I recall his impressive dedication to wildlife, which I believe I have mentioned elsewhere. Between our campsite at Galore Creek and the staging area at the mouth of Anuk River, the broad flood-plain of the Stikine River was covered by a huge and persistent growth of cottonwood trees. In his frequent flights to the mouth of Anuk River, Hughie had noted several wolf packs including their young in the flood-plain. During the fall of 1963, he had air-dropped various morsels which he considered appropriate. Considering the total lack of handouts in this hostile environment, I'm sure that they were most welcome. A lover of strong cheddar cheese, Hughie was to surprise the rest of our camp by burying a round or two at the camp site in the late fall of 1963. He must have had aspirations of returning in 1964 to dig up his well-aged cheese. Being party to this disinterred treasure, I can only say that it was the most memorable aged cheese that I have ever encountered.

Herman Peterson

During the late 1950s and through the 1960s, I had many flights throughout the northwestern part of British Columbia with Herman Peterson, as reliable a person and pilot as I have ever met. In the late 1980s when I decided to record some of my early experiences in exploration, I thought a chapter on the many fixed-wing and rotary-wing pilots that I had flown with would be particularly fitting as we relied to such an extent on these individuals for much of our access into camps and everyday work.

I initially phoned Herman in October 1989 but was unable to contact him until the summer of 1990. I indicated my interest in having an opportunity to meet him again when next in the Whitehorse area and complete a brief review of his experiences as a pilot, to which he agreed.

In August, 1991 I examined diamond drilling progress at a gold prospect under option to CSA Management Limited and Goldcorp Inc. in the Hope Bay area, near the Arctic coast southwest of Cambridge Bay on Victoria Island. I then flew to Whitehorse via Yellowknife to examine exploration progress at the Jason lead-zinc-silver property near MacMillian Pass on the Yukon-Northwest Territories border northeast of Whitehorse, in which CSA and Goldcorp held a 62% interest. I had arranged to meet Rene in Whitehorse on my return from the Jason property for a visit to the Atlin area.

We finally met each other at the Westmark Klondike Inn, as arranged, after some delay because Rene had decided to go back out to the airport and meet the aircraft, which I had chartered from Air North for my flight to MacMillan Pass. We had little time in Whitehorse, but we did manage a visit to MacBride Museum with its excellent displays of early life in the Yukon including a superb collection of birds and mammals, well described by one of the staff.

On August 24 we rented a U-drive and drove easterly along the Alaskan Highway to Jake's Corner where we turned south along Highway 7 to Atlin. The drive brought back memories of several previous journeys on the same route dating back to January, 1959 when I had travelled with Charlie Ney and John Anderson to meet Herman for our flight into Chutine Lake which I have described elsewhere. It was Rene's first visit to this scenic area which, if anything, has been improved by the emphasis placed on preserving heritage homes and buildings in Atlin.

After checking into Kirkwood Cottages where we stayed for three nights, I walked over to Herman's nearby home on the lakefront. He appeared little changed from the curly haired, energetic and enthusiastic person that I had last seen almost 30 years earlier when he had flown survivors Max Portz and Bob Gilroy from their seven day ordeal astride the floats of an overturned aircraft in the middle of Iskut River to our staging area on the Stikine River at the mouth of Anuk River. I have described that saga elsewhere.

The next day I spent several hours with Herman, recording some of his recollections, which I have transcribed with minor editing. On the following day Rene and I drove out with him to the Atlin airport where he showed us the aircraft he built several years ago and which has also been flown by several of his pilot friends. I took several photos of Herman and the aircraft before we bid him farewell. It was to take me almost a year to forward a copy of the transcript and photos to him because of other commitments. Before leaving Atlin, Herman very kindly provided me with a tape he had made the previous evening of Joe Loutchan playing on one of the early fiddles made by Herman in his basement.

Interview with Herman Peterson, August 25, 1991

As Herman Peterson recalls:

One Lucky Canuck

"The first time I got interested in flying was when I was a little fellow - five years old. I was walking along with Dad - this was just after World War I, when two planes flew overhead. I said to Dad, 'What holds them up there?' He said, 'Well, a skyhook'. I said 'Well, are there people in there?' and he said, 'Oh, yes!' "

Herman was born on December 29, 1913, at La Tuque, Quebec, a date which coincides with the birthday of his wife Doris ("Susie"), and coincidentally with our anniversary! His father was production superintendent at a pulp mill where he had been employed for about 35 years.

While attending school, he had sketches and pictures of aircraft in all his texts. Following graduation from La Tuque High School, he worked in a pulp mill, which did not particularly appeal to him. He was interested in mechanical work, including welding, and this influenced his relocation to Toronto to attend an arc-welding course at General Welding Works.

"I went to welding school in 1940-1 and actually got a job before I finished the course with W.D. Beath & Son."

On weekends he took a pilot training course at Barker Field, piling up the hours and obtaining his private license in 1937. He soloed in a DeHaviland Puss Moth in Cap de la Madeleine near Three Rivers and obtained additional dual certification in a J3 Taylor Cut and a Sears Moth.

At Barker Field he flew with Fred Gillies and his daughter Marion and also with a friend Margaret Little. One day he met a friend from La Tuque who was working with United Steel Corp. Herman invited him to share a flight, which he did, and the experience was so memorable that he enrolled in a pilot's training course and within a short period he was ferrying Beechcraft all over the world, from Wichita, Kansas.

During the early part of World War II, Herman was employed on war-production assignments, but he managed to complete a considerable amount of flying, including passenger flights.

In 1942, he applied for a job as a bush pilot with George Simmons, owner of Northern Airways Ltd. in Carcross, Yukon. He soon received an offer, which he accepted. He and Doris travelled by train to Vancouver and Herman continued north to Whitehorse on a DC-3. Herman was a pilot for Northern Airways until 1950 when Simmons sold out his business. (Interestingly, Herman is referenced on three separate occasions in photos in Pat Callison's autobiographical "Pack Dogs to Helicopters." Pat Callison was an excellent pilot and he and Herman flew together for five years, principally on the Canol Pipeline project and the Alaska Highway construction.)

Herman bought an Aeronca Sedan in the fall of 1950, in Toronto, and flew the aircraft back to Atlin, to form Peterson's Air Service. At that time the road from Jake's Corner to Atlin had just been constructed and was barely passable.

Herman recalls that in 1950, when he moved to Atlin, there were about 300 residents. His flying business during the first three years was "pretty slack." In order to develop his business Herman applied for and was granted all the big game hunting areas in the Atlin Lake watershed on a government license in 1952. He provided several of these areas to applicants, knowing that it would lead to an increase in his flying business.

Herman's first job was a mail contract to Telegraph Creek, being contingent on his obtaining a license from the Department of Transport. As Herman recalls:

"The first trip I made was from Atlin to Telegraph Creek in 1952 in my new Aeronca Sedan. The contract was extended to Iskut a couple of years later."

In 1953 he bought a Fairchild 71. In the same week that he obtained approval to operate the heavier machine in 1953, he sent a pilot to Toronto to pick up the Fairchild, while he remained in Atlin to continue operations with his Aeronca. This was in the fall and during that week Herman was flying the Aeronca near the Telegraph Creek summit, about 40 miles south of Telegraph Creek, in marginal weather.

"The visibility deteriorated near the summit, but I had visibility of the summit and the area beyond, when I encountered a down draft.

"I could just see over the summit, and when I decided to turn back, I realized I wasn't going to make it, and I went straight in, landing in a creek-bed which was full of rocks. The impact tore the bottom off both floats. The aircraft went straight up on its nose and just teetered there and then fell on its back. I got out of it without a scratch."

"I spent the night in the aircraft and the next morning, I started off for Telegraph Creek with a 65-lb. pack, walking through about four inches of snow at the crash site, but there was no snow when I dropped further down.

"Although it wasn't far for someone in good shape, it took me five days with the pack, but I could have made it in a couple of days without it."

He followed the old Telegraph Trail line for most of the distance, which at many locations had been partly obliterated by snowslides.

Herman was to fly the Fairchild 71 for four years. It was first actively employed in the search for Herman and his Aeronca, piloted by Gordon Grady.

"At one point it flew right over me - I could see the rivets in the floats."

The insurance company planned to fly the damaged Aeronca out of the crash site using a helicopter. At the time no appropriate machine was available and the insurance company wrote the Aeronca off and donated it to an individual in Vancouver. The Aeronca was later removed by helicopter to Kinaskan Lake where it was cut into two parts with the intention of flying it down to Vancouver to rebuild it. That was the last that Herman heard about his first aircraft.

Herman sold the Fairchild to the Northland Fish Company in Winnipeg in 1956. The company sent their pilot to Atlin to take delivery late in the fall of 1956. Enroute to Winnipeg from Atlin in the Fairchild, the pilot stopped at Watson Lake for a day or two on account of poor weather. Following take-off he flew into more bad weather near the summit. As Herman recalls:

"There was about 300 pounds of freight in the aircraft and it couldn't gain altitude fast enough. It crashed in brush without significant injury to the pilot. The plane sat in Fort Nelson for a year and it finally wound up on the Fort Nelson garbage dump, after portions of it disappeared."

Following the sale of the Fairchild, Herman bought his first Beaver ("ITU") in which I was to fly on several trips in 1959 and 1960. In 1957 he bought a similar Beaver ("JPM").

Herman's earliest work with mining companies occurred in 1956 when he supported a program for Asbestos Corporation directed by Dr. Riordan, which was based out of Como Lake near Atlin. Herman normally flew his float-equipped Beavers off Atlin Lake for most of the year. However, in order to take advantage of operations involving smaller ice-free lakes in the spring he would haul the aircraft up to the Atlin airstrip in the late fall before freeze-up using a dolly which he had constructed. This was a framed structure supported by four large rubber-tired wheels.

During the winter the Beavers were flown off the airstrip on skis, but when spring conditions arrived, Herman converted to floats and flew the aircraft off the airstrip from the dolly, landing it at Como Lake. I witnessed the take-off of one of his Beavers from the airstrip on June 1, 1959 when we operated for several weeks at the start of our exploration program from a base at Palmer Lake, about 15 miles south of Atlin. Like Como Lake, the earlier break-up on Palmer Lake allowed us to start flying operations from the Atlin area about three weeks prior to break-up on Atlin Lake.

The flight from the dolly was an awesome sight. I couldn't believe that the aircraft could take off from the dolly without a terrible accident. After the aircraft had revved up, the dolly with its burden gained speed as it drove down the airstrip and within a relatively short distance the Beaver smoothly disengaged itself; the dolly careening madly down the airstrip and coming to rest on the side of the strip. Herman later developed a braking mechanism which was activated by the take-off of the aircraft from the dolly - just one of the many innovations which he put to good use not only in his aircraft operations, but much later when he became interested in making fiddles! In 1991, when I visited Herman at Atlin Lake, I learned that he had sold the dolly to Trans North Turboair in 1967 and that it was still being used at the Whitehorse airport after being lengthened and widened to accommodate Otters.

On returning to Whitehorse , I visited Trans North's operation and was kindly shown the modified dolly by Bob Cameron, the manager, who mentioned that it was still in use.

As will be evident to most readers, pioneer bush-pilot operations were not carried out without considerable risk, although my experience as a passenger with Herman was one of complete confidence in his judgement and abilities. Herman always carried out his operations with considerable forethought. I can recollect only one or two times when be berated us for unacceptable performance in support of his flying operation. One of them involved unloading 45-gallon fuel drums from a Beaver at the shoreline of a lake. The drums were tipped over in the aircraft and lined up to roll down to the lake along two poles attached by a clamp at one end to the floor beside the open doorway above the floats. One of the drums prematurely slid off the poles and narrowly missed hitting the float. It never happened again! Herman always gave me the confidence that he would look after his end of the job and he expected us to do the same.

In 1956 Herman purchased a Cessna 180 aircraft ("HST"), which was to have a short life amid tragic circumstances. In the spring of 1957 the Geological Survey of Canada were carrying out survey work in the Mount Edziza area south of Telegraph Creek. One morning, Dr. Angold, a geologist with the GSC, walked down to the creek at their campsite where he was unexpectedly attacked by a cow moose, apparently concerned for the safety of a newborn calf. The moose knocked him down, trampling him and breaking his shoulder. Herman's Cessna 180, piloted by Bob Rae, was in the area and took off shortly after the accident with Dr. Angold and passenger Syd Ward for Atlin enroute to Whitehorse to obtain medical aid. Shortly after take-off, near the summit of Telegraph Creek, the aircraft encountered an unusual downdraft from a north wind and it crashed, killing all three occupants.

By coincidence, a similar Cessna 180 flown by a pilot with Ellis Airlines of Wrangell, Alaska was in the Telegraph Creek area at the time of the accident. The aircraft was involved in a fish count operation and had just taken off from Sawmill Lake when the pilot spotted smoke from the crashed aircraft. Flying about 500 feet above the wreckage it encountered the same downdraft and crashed in trees within 100 yards of the other aircraft. There were no significant injuries to either the pilot or his passengers.

I recall flying with Herman in the same region in 1960 and noticing that as we approached the summit area on a sultry day we encountered a downdraft that dropped us almost 2,000 feet as we cleared the pass. Herman was aware of the potential problem and, not taking any chances, had allowed more than the conceivably required margin to clear the pass.

In January 1959 Herman flew Charlie Ney, John Anderson, and myself to the Chutine Lake area from Atlin on a claim staking jaunt - which I have described elsewhere. In February he also flew John Anderson and I into Little Tahltan Lake in the Barrington River area where we were to stake another molybdenite occurrence reported by the GSC. In 1991 I mentioned these forays to Herman and he remembered that he had supported an operation for John DeLeen, then working as a geologist for Phelps Dodge Corp. in the nearby Barrington River area. Herman had placed two men and their pack-dogs out on a small lake near the head of Barrington River, which was surrounded by trees, in marginal operating conditions for a Beaver.

Herman had mentioned the operation to some pilots in Juneau who had declared that a Grumman Goose would best support the program. John DeLeen had decided to check out this alternative; however, on landing the pilot had strong reservations about his ability to take off. He said to John:

"Maybe you had better stay here and if everything works out right I'll come back and pick you up."

On further reflection, he shook his head and added; "If I get off, I won't come back."

So John went with him and they just managed to clear the trees at the end of the lake. The operation was completed with the Beaver.

Herman bought an Otter aircraft ("SUB") in the spring of 1965, which he was to utilize on a fulltime basis until he sold out his flying business in 1967. As he said:

"I had so much work with the Otter, I was never home. As a matter of fact, I changed an engine in that short period at 800 hours and flew the other aircraft at the same time."

During this period he provided considerable flying support for Oz Hachey, an acquaintance of mine, much of it in the Yukon where Oz was directing exploration work for Noranda Explorations Ltd. In addition he provided flying services to many other mining companies in northern British Columbia, including Cominco in the Iskut River area.

One of Herman's pilots in the early 1960s whom I flew with while directing exploration work in northwestern British Columbia was Stan Gormi, who had been a Polish refugee in World War II, attached to the Royal Air Force as a fighter pilot. Herman remembers him as an outstanding pilot, recalling that he had flown in many missions including operations in late model Spitfires. On one of these engagements his arm was badly shattered by gunfire from a Focke-Wulfe over the English Coast. He managed to crash-land his aircraft in a field, passing out as the aircraft struck a hedge. He flew with Herman for a year and also flew for several years for George Dalziel's Watson Lake Air Services.

Herman used his Otter partly in fire-fighting assignments. The aircraft had a centre-mounted tank which was eccentrically loaded so that in operation it could turn over and dump its load. When the tank was upside-down it was supposed to latch in that position. However, the tank often malfunctioned, failing to latch and dumping its load prematurely. The tank could not be inverted, as designed, prior to re-loading, necessitating a landing for re-loading. In order to rectify the problem, Herman and Clarence Tingley designed a smaller tank capable of holding about 20 pounds of water which was inserted inside the large tank and which provided sufficient weight to lock the tank when inverted. The water then gradually drained from the smaller tank prior to the next circuit.

One of Herman's closest calls occurred near the mouth of Hobo Creek on Williston Bay.

"There was a helluva wind blowing that day and I had two drill rods tied to the float of the plane which didn't affect its performance that much. The water was really rough, with the wind blowing across the lake at an angle. I came over the trees and had about ten degrees of flap on. The only way to get down under those conditions was to slip it in. As soon as I did that, the tail wanted to get ahead of me and I couldn't straighten it out. I had the presence of mind to open the throttle up wide and the plane started to straighten out when I suddenly landed on one float. That's the only thing that saved my life or I would have gone in too."

After this experience, Herman considered the problem. He sent a wire down to the Department of Transport in Vancouver concerning the incident. They responded with several letters requesting some detailed information and later sent two representatives to Atlin to discuss his concerns. Herman recommended that "a ventral fin be employed on all Beaver seaplanes." Eventually this was to become a compulsory requirement on these aircraft.

Another frequent cause of aircraft accidents for pilots flying on visual flight rules are the occasions when the pilot suddenly gets caught in a white-out. This is well described by Herman in an incident at Dixie Lake near the south end of Atlin Lake.

"I was hauling the mail to Telegraph Creek and I had a full load on. It was a Beaver on skis. It was one of those days when you have flash snowstorms, which can be quite blinding. I could see the storm building up right there - so I did a quick left turn at the end of the lake and came in to land. Before I could descend to the shoreline it was zero visibility. I'm not an instrument pilot, but I had sense enough to keep on the direction of the approach and by the time I got to the area of the shoreline it had whitened right out and felt just like being inside a milk bottle. I held it on an even keel with the gyros that I had learned to use on my own. I held it straight until the ship touched down. I sat there and about thirty minutes later it cleared up and here I was right in the middle of the lake. That was the worst situation I ever got into as far as visibility was concerned."

In the latter stages of his active flying career Herman cautiously branched out into the helicopter business. He hired Al Pelletier as his pilot in 1965 and had him pick up a Bell 47-G4 from the factory in the U.S. The helicopter business turned out to be compatible with the established fixed-wing service. Bill Cruise was retained as a pilot in 1966 and flew for him until the business was sold to Trans North Turboair in 1967. In the two-year period the helicopter business logged 1,200 flying hours.

Apart from flying, Herman had loved violin music from an early age. His brother-in-law also had a fondness for violins. He had spent some time in Germany as a radar instructor and while there he obtained some references on violin construction. He built several violins himself and got Herman interested in the craft. Herman also obtained other references but did not actually construct his first violin until 1977. Unable to obtain actual details on the construction of Stradivarius violins from the factory, he built his own machinery, largely on a trial and error basis. He created stencils as a guide for cutting out the woodwork. The actual wood used in construction came from a supplier in the U.S. who imports the wood from Europe. Herman has even constructed a dehumidifier to ensure that the wood is adequately dried in the northern latitudes, prior to finishing and varnishing.

By 1991, when I visited him, Herman had built a total of 13 violins, all of which he had given away to friends and acquaintances, some for the cost of materials.

I worked with many pilots and heard many of their stories, but none struck me as forcibly as Herman Peterson's combination of courage, common sense and reliability, crowned by his uniques interest in violins. A great example of the best of a remarkable breed.

Resource Land Conflicts

Following graduation from University of Toronto in the late spring 1950, I accepted a temporary job with Kennco Explorations (Canada) Limited of Toronto as party chief of a small crew working out of Fort William, Ontario. My direct supervisor was a well-known field scout for Kennco in the Sudbury area named Charlie Baycroft.

Jobs for graduating mining geologists in 1950 were few and far between. Although a heavy focus was placed on the mining industry in the latter two years of our university curriculum, we were also provided with a broad background in petroleum geology. As described under **Jug Hustling in Alberta**, I spent the third summer in my course working with a seismic crew in the Red Deer area. During the latter part of our fourth year, several representatives of major companies in the mining and energy sector carried out interviews at the university. To my knowledge, only Barney Peach received a firm employment offer with Shell Oil. Based on our class marks for the first three years, Barney would be the only one of our group who could graduate with first-class honours – which he subsequently achieved. The Shell Oil representative told me that an employment offer would be contingent on my final year marks --which wouldn't be available until some time in June.

When I accepted the temporary employment offer with Kennco, I made it clear to Charlie Baycroft that there was a possibility of a permanent job offer materializing with Shell Oil during the summer. I explained that if this occurred, it would be in the form of a letter from Shell Oil directed to his address for forwarding. He agreed to the arrangement. As June passed into July and I had not heard from Shell Oil, I assumed that my marks were not good enough, even though I had obtained honours in my final year.

Following the summer program, I moved to Sudbury to write a report for Kennco, which was not completed until October. Unable to secure a permanent job as a geologist, I was lucky enough to be accepted as a miner at Inco's Garson Mine. The mine staff was aware that I had no previous underground experience and was a recent graduate in mining geology. They found a satisfactory haven for me as a mucker assisting an experienced miner in one of the stopes. This rather solitary individual was a former medical doctor in the north, who for some reason not divulged to me had turned to mining during the Depression and never returned to his medical practice.

It was a long, cold winter with few memorable diversions. I made some friends and went out occasionally with former classmates apprenticing with International Nickel as mining engineers. I recall the coldest day, stamping my feet as I waited for a bus to the mine. The temperature dropped to minus 54° Fahrenheit and ground fog was to be a contributing factor to a major tragedy about an hour later. A bus loaded with miners was hit by a train at a railway crossing with a considerable loss of life – 10 to 12 miners.

Sometime during the winter I learned of an unbelievable incident from Ray Freberg, one of my classmates who had been working for Kennco during the past summer on another project. Apparently Charlie Baycroft had visited the crew and mentioned that he had received a letter from Shell Oil directed to me but had decided not to send it on. He gave Ray the impression that it was quite a joke! I immediately phoned Shell Oil, explaining the situation only to learn that indeed an offer had been mailed to me but that the opening no longer existed.

I have often reflected on the circumstances which influence our lives, apart from our genes. In my case there were three major events which were to affect my later life. One was World War II, which saw me evacuated from England to Canada. The second was Charlie Baycroft not forwarding that offer from Shell Oil – which I undoubtedly would have accepted and which would probably have led to a career in the petroleum business. Finally, I would never have met my lifelong love and companion. In retrospect, with all the ups and downs, I wouldn't have wanted it any other way.

Although I appreciated the opportunity to work underground, I realized that it was only going to be a temporary job as I was not cut out to be a miner. Quite unexpectedly, I received a most welcome surprise in early April 1951 from Kennco. Would I be interested in a position as Junior Geologist with Northwestern Explorations Limited, its relatively new western Canada exploration arm, based in Vancouver? Indeed, I would.

My trip west in May on the Canadian Pacific Railway from Sudbury was one filled with a spirit of approaching adventure and "great expectations." I was never to be disappointed in those early aspirations as the highs greatly exceeded the lows. My arrival in British Columbia was to coincide with the start of a period of great optimism within the mining fraternity, significantly influenced by the introduction of helicopter-borne exploration following the late stages of traditional support by pack-horse augmented by float-equipped aircraft where possible.

Within a very short period, I was to meet many individuals who were to influence my career in the mining industry, either by association, friendship or both. Although many of them will be mentioned in my recollections, my earliest associates included Paul Hammond, then exploration manager for Northwestern Ex. living in Penticton with his wife Sybil. Kennco's head of exploration at the time was Bill Dean based in Toronto who made occasional visits to the west. Our Vancouver office included only four of us at the start. John Anderson, who joined Northwestern Ex. shortly after myself, was to become a lifelong friend and my 'best man' in 1953. Gerry Noel, who was to be an usher at our wedding, is best remembered for his optimistic outlook and his geological acuity. He later examined a recently discovered copper showing near Port Hardy on Vancouver Island, staked by prospector Gordon Milbourne. His recommendation led to the closing of a deal which brought about the delineation of the major Island Copper ore body, which produced from 1971 – 1995. Our secretary was Irene Wright (later Nolet) who, apart from running a most efficient office, taught John and me some basic dance steps!

My mineral-industry-related interest eventually lead me to author or co-author about twenty technical papers concerned with geology, geochemistry, mineral exploration, mineral production and safety procedures in mineral exploration between 1965 – 1997. I also made 15 – 20 presentations at various technical conferences on these topics, mostly with the Canadian Institute of Mining and Metallurgy (CIM) and the Vancouver Mineral Exploration Group (MEG). However, I had little other involvement with mining-related associations until after my return from Toronto to Vancouver in 1971, following my promotion to Vice-President Exploration for Kennco Explorations (Canada) Limited.

The government of B.C. under the Social Credit Party (Socreds) was highly supportive of the resource sector throughout the 1950s and 1960s and had even sponsored roads to resource programs. Included were the northern extension of the Omineca Road, north of Germansen Landing into the Toodoggone River area and the ill-fated and costly B.C. Railway extension, proposed from Vanderhoof to Dease Lake and eventually to the Yukon. This boom period for mining was to be severely reversed within two decades with the election of the New Democratic Party (NDP), initially from September 1972 until December 1975 and more recently since its re-election in November 1991 until 2000.

By 1972, I had become a member of the British Columbia & Yukon Chamber of Mines (Chamber), founded on April 23, 1912 by a group of concerned mining men as the Vancouver Mining Club. The Chamber has represented the exploration and development arms of the mining industry, whereas the Mining Association of British Columbia has represented the mines since receiving its charter in 1901.

The Chamber's policy had been non-partisan from its initiation, endeavouring to speak out against any poor legislation proposed for mining, whatever party formed the government. The election of the NDP government in 1972 was eventually to modify this non-partisan approach. The biased policy of the NDP government against the resource sector was initially revealed in a white paper guiding its policy at that time. It resulted in a wholesale revision of the government's mining legislation, including a moratorium on the staking of placer claims. In addition, many new bills affecting mining were introduced without any prior discussion with the mining industry. Dissatisfaction with NDP policy was not confined to the resource sector and it was defeated on December 11, 1975 in a general election which brought back a Socred government. Prior to the election, Ron Stokes, a Chamber member and strong proponent of placer mining, was instrumental in encouraging a group of concerned members to promote the mining industry by knocking on doors in support of the Socreds. It was my first direct involvement in lobbying for the mining industry – although not a significant commitment as I only visited a few West Vancouver neighbourhoods. It was the first and fortunately the last time I recall having a door slammed in my face.

Anti-mining sentiments were not restricted to the NDP government – the late 1970s were to see much evidence of rapidly growing, well-organized and vocal groups of preservationists strongly supported by, or affiliated with, U.S. organizations.

One of my earliest letters written to the government on behalf of the mining industry was dated May, 1974 and directed to the Honourable Leo Nimsick, then Minister of Mines in the NDP government. His mining background was limited to his position as a warehouseman employed by Cominco Limited at its Trail smelter – a position he held just prior to his appointment. The purpose of the letter was to provide him with fundamental information on the importance of the mining industry to B.C.'s economy, its influence on early settlement and its contribution to the province's heritage. It was a waste of time.

My Chamber-related activities were to require a significant portion of my time over the years, commencing in the late 1970s and continuing to the present in 2002. It has been a rewarding experience in many ways and frustrating in others because of strong anti-industry sentiment in government, and of course among environmental extremists. The totally adversarial position of the extremists against the mining sector, in particular, has been incomprehensible to me, irrational in its deceitful and hypocritical nature. These generally well-educated individuals could not maintain their way of life without mineral products and are well aware of the extremely restricted disturbance of mining on the land-base, currently on the order of 0.025% of temporary activity during the development and life of a mine. Facts such as these are meaningless to such individuals, – only the self-gratification of claiming protection of an increasing portion of the planet – with no end in sight as to the amount, and its cost to society, is their objective.

On two occasions I was asked if I would stand for appointment as an executive of the Chamber which calls for a four-year term prior to a two-year term as president. Initially, this occurred in the late 1970s while I was employed as Vice President Exploration by DuPont. At the time I checked with Bob Cathro who was then president of the Chamber and learned that he allowed 25% of his time for the work. On being apprised of this degree of commitment, Gord Wittman, my boss at DuPont following the unexpected early death of Art Baker, was quite blunt: he said "No way!" Years later I was invited once again, and once again I turned the offer down because of other commitments – many of them Chamber-related.

My early period of Chamber involvement overlapped activities with both the Canadian Institute of Mining and Metallurgy (CIM) and the Geological Association of Canada (GAC). In 1974, when Art Soregaroli was scheduled to be President of the Cordilleran Section of the GAC, he accepted a position with the Mineral Deposits Section of the Geological Survey of Canada (GSC) in Ottawa. Quite suddenly, I was asked if I would agree to take his place, which I accepted. The position was to bring me into close association with many of the GSC's gifted members in Vancouver; UBC's geological staff; and eventually the parent body of the Cordilleran Section of the GAC.

I had an enjoyable term as Finance Chairman of the GAC (1977 – 1981), which required my attendance at various GAC Committee and Annual meetings across Canada. I was able to take Rene to a couple of these including one at Banff and another in Halifax, thanks to the support of DuPont. I also acted as Finance Chairman for the Annual General Meeting of the GAC held in Victoria in 1983.

Apart from four papers published in CIM bulletins, my involvements with the CIM were restricted to a short term as Councillor with the Geological Division (1976 – 1977) and as an author and member of the Editorial Committee of Special Volume 15 in 1976 on Porphyry Deposits of the Canadian Cordillera dedicated to Charles Ney, a long-time friend and associate in Kennco. The editor was Atholl Sutherland Brown, then a professor at UBC. The publication ran to 510 pages and included 77 authors – all male. I mention this to illustrate how few women were employed as geologists, particularly in exploration, by the mid-1970s, a situation which has changed dramatically within a decade as evidenced by our summer crews with DuPont which are now fairly well balanced between men and women.

I thought that CIM Volume 15 was quite an accomplishment as did most everyone else involved in the exploration for porphyry-related deposits in the Pacific Northwest. However, it was to be surpassed by CIM Special Volume 46 Porphyry Deposits of the Northwestern Cordillera of North America under the dedicated and capable editorship of Tom Schroeter, published in 1995. I was an associate editor and reviewer of this mammoth volume which required over 30 meetings of the editorial committee, ran to 908 pages and included 145 authors, of whom nine were women.

My association with the Chamber was a symbiotic one. Like all organizations of its kind it relied heavily on a core of volunteers, directed by an executive and advisory committee with only a few salaried employees. My 25-year involvement on the many committees has been an education. Although volunteers came and went, there was always that relatively small core of familiar faces from year to year that generally headed at least one committee and were frequently active on others. However, with time, i.e. in the 1990s, it was to become difficult to recruit volunteers as an increasing percentage of B.C.-based exploration personnel became active out-of-province or on foreign assignments.

The broad range of my committee involvement is reflected in the following assignments:

Executive Committee (1974 – 1999); Advisory Committee (1999 – present)

Safety Committee Chairman (1980 – 2002); member (2002 – present)

Annual Meeting Committee Tech. Program Chairman (1979);

Vice Chairman (1980); Chairman (1981)

Canadian Geoscience Council, member External review Committee for The Future Mandate, Organizational Structure and Recommended Resource Distribution for the British Columbia Geological Survey Branch, Ministry of Energy, Mines and Petroleum Resources; report published March 1990.

Parks '90 Open House & Public Meeting Contacts Sub-Committee Chairman (1991)

Representative, B.C. Ministry of Energy Mines & Petroleum Resources, Mineral Titles Branch, 'Mineral Tenure Group' (1992 to 1999)

Mining Sector Alternate Representative, Commission on Resources and Environment (CORE) – Vancouver Island Table, (1992 to 1993)

Member Windy Craggy Committee 1993

Member, Land Access, Mineral Tenure Sub-Committee to the Premier's Forum (1993)

Representative 'Land Alienation Committee' (1994 to 1995)

Representative, 'Lower Stikine Management Advisory Committee' (1994 to 1999 - subsequently discontinued)

Speaker 'MABC Land Use, Access & Tenure Session', 'Cordilleran Roundup & Teacher Industry Sessions', January 28, 1994

Member 'Land Use Committee' (1995 to present)

Member and Chair Aboriginal Committee, 1997 to 2000; member 2001 to present

Mining Sector Representative, Northern Regional Advisory Committee to Aboriginal Land Claims – B.C. (1995 to present)

Mining Sector Representative, Cassiar-Iskut-Stikine Land Resource Management Plan (1996 to 1999)

Registration Chairman "Cordilleran Roundup and Pathways to Discovery Exploration Methods" '98 Conference, Vancouver, 1998.

Speaker: Paper on "British Columbia Needs Mining", at MABC Teachers' Workshop, 'Cordilleran Roundup', Vancouver, B.C., January 23, 2001.

My assignments with several of these committees were major undertakings, requiring considerable time. Only one was to provide me with a real sense of fulfillment (Safety Committee), though several others produced beneficial results for both the mining industry and others affected by mining sector activities. One was particularly frustrating; the time required to repeatedly state the mining sector's modest requirements in terms of temporary occupancy and disturbance of the overall British Columbia land base (0.025%). This was as the Chamber representative at the Cassiar-Iskut-Stikine LRMP amid a group dominated by preservationists, mining being the only major industry represented after a brief stint by forestry. An inter-related task which preceded the latter was my representation on the Lower Stikine Management Advisory Committee – also a time-consuming activity which nevertheless was successful in promoting an emphasis at the time on economic development on the Lower Stikine, south of Telegraph Creek.

In the following sections I will try to describe my principal recollections of only the more noteworthy committee activities.

Lower Stikine Management Advisory Committee

The protectionist movement in B.C. gained momentum under the Socred government, not the NDP, commencing with the release of *The Wilderness Mosaic*, published in March 1986 as The Report of the Wilderness Advisory Committee. It proposed over a dozen study areas, including the Tatshenshini and Alsek Rivers, and the Stikine River, to be designated as Recreation Corridors. However, at this stage, the report recognized the need to retain access for future mining purposes while ensuring that "wilderness qualities are retained."

In 1988 the Valhalla Society, a B.C. environmental group, proposed a number of additional protected areas in the province in its report *BC Endangered Wilderness – A*

Comprehensive Proposal for Protection. The report included protection of the Tatshenshini–Alsek and Stikine River areas by promoting exclusive use of the Lower Stikine and Tatshenshini rivers by recreational boaters.

Because of my background in the Stikine River region, I was appalled that the government was giving serious consideration to these Recreation Corridor proposals covering not only the Lower Stikine but also 21 candidate waterways in the province submitted by the Ministries of Parks and Forests as lead agencies. I wrote an 8-page letter with attachments dated October 28, 1988 to the Hon. William Vander Zalm, then Premier of B.C., expressing my concerns. It focused on the Stikine River district and its significant mineral and hydro-electric power potential. It also emphasized the critical need to retain access for future mining developments.

Mr. Vander Zalm's response, dated December 9, 1988, was not reassuring as he stressed that "in a corridor where an integrated resource management process is to take place, these recreation resource management objectives are given a priority, when developing management plans." Since the Ministry of Forests was co-ordinating the plan, co-sponsored by the Ministry of Parks, there was little comfort provided concerning unrestricted access along river corridors for industrial development purposes, including mining.

My letter failed to make any reference to a long-standing international treaty concerning access on major transboundary rivers in Canada and the U.S. Mr. Vander Zalm's response prompted me to write the following letter to him on January 27, 1989.

>Dear Mr. Premier:
>
>Re: Lower Stikine River Recreation Corridor
>
>Further to my letter of October 28, 1988 on the above proposal, I have recently obtained a photocopy of Article XXVI declaring the Stikine and other rivers international waterways under the Treaty of Washington dated 1871. The Article states in its entirety:
>
>>The navigation of the river St. Lawrence ascending and descending, from the forty-fifth parallel of north latitude, where it ceases to form the boundary between the two countries, from, to, and into the sea, shall forever remain free and open for the purposes of commerce to the citizens of the United States, subject to any laws and regulations of Great Britain, or of the Dominion of Canada, not inconsistent with such privilege of free navigation.
>>
>>The navigation of the rivers Yukon, Porcupine, and Stikine, ascending and descending from, to and into the sea, shall forever remain free and open for the purposes of commerce to the subjects of Her Britannic Majesty and to the citizens of the United States, subject to any laws and regulations of either country within its own territory, not inconsistent with such privilege of free navigation.
>
>Article XXVI would appear to directly conflict with the objectives of the December 1988 draft of the Lower Stikine Recreation Corridor Management Plan which proposes that the Lower Stikine River would be used principally for recreational purposes and the related corridor would be restricted as to access.
>
>Weekly riverboat trips on Stikine River between Wrangell, Alaska and Telegraph Creek were active until recently and other forms of commercial navigation, partly for recreational purposes, continue. In the early 1960s, I was involved in a major exploration program at Galore Creek employing up to 100 persons/year. We relied on barging up to 200-ton loads from Wrangell to the mouths of Anuk and Scud Rivers for part of our transportation requirement. Barging remains as a transportation option for any future developments in the area.

Recreationists and many other workers have enjoyed the scenic values of the Lower Stikine River Valley for decades while plying the river. I believe that recreationists will continue to enjoy the many attractions of the area, as has been the case with other waterways in the province, many of which also support industrial development and provide the required access for all resource users. These resources must be shared in an environmentally responsible manner and in a spirit of mutual co-operation. The current Recreation Corridor proposal is designed to satisfy only the wishes of a privileged few recreationists over a short summer season, not the majority of British Columbians who will benefit by multiple-resource use and the access provided to the area.

The letter makes reference to the 1988 draft of the Lower Stikine Recreation Corridor Management plan. I was briefly involved in the early discussions which were to lead to the formation of the Lower Stikine Management Advisory Committee (LSMAC). Participants on the Committee included Bob Stevenson, my old friend and successor as Vice President of Kennco following my departure in 1974 to form DuPont Canada's exploration arm. Bob became the mining industry representative on LSMAC which was intended to include "members from government agencies, industries and the public." The Management Plan was reviewed and approved by the provincial government's Environmental and Land Use Committee (ELUC) in May 1989.

As a member of the Executive Committee of the Chamber since 1974, I heard occasional reports provided by Bob at our monthly luncheon meetings on the progress, or, more realistically, lack thereof, of LSMAC on its objectives. Bob was an astute individual and recorded his observations accurately, reflecting his careful attention to details. He was a close friend for over 30 years, so it was a shock to learn of his sudden death at work in his downtown office at the Vancouver Securities Commission on November 26, 1993.

Bob's travel expenditures on behalf of the Chamber as its mining sector representative on LSMAC had been absorbed by Kennco. Because of my previous association with the company, I was not surprised when I was asked whether I would be willing to be Bob's replacement on LSMAC, under the same arrangement. I attended my first meeting of the Committee in Smithers on April 26 – 27, 1994. The Chair of the Committee was Doug Herchmer, Regional Recreation Officer of the Ministry of Forests, the designated co-ordinating Ministry for LSMAC.

From the outset of my involvement with the Committee, which continued through 1999 on an inactive basis, I have been dismayed by its overwhelming protectionist sentiment, reflecting its pro-recreation, anti-industry makeup. Following my first meeting with LSMAC, I wasted little time in recording my concerns, which were expressed in a six-page letter to Doug Herchmer, dated May 25, 1994. I focused my comments on the revised 'Terms of Reference', the available information on the 'Zoning and Guidelines Subcommittee', the 'Hovercraft Operation in use by Cominco's Snip Mine on the Lower Iskut River' and the makeup of the 'Membership' and the motion made at the meeting on 'Canadian Heritage River System designation.'

Considering that the Management Plan had been established five years earlier, a final draft of the Terms of Reference was long overdue and I was able to state that neither Bob nor I had any problems with them. The Zoning and Guidelines Subcommittee, composed of three non-industry members who had been working on the task of designating activity areas for the entire Lower Stikine River corridor for several years, was another matter. At the meeting I pointed out that the zones included large sections of the broad river valley and its confining mountain slopes which were given both a 'primitive' and 'semi-primitive' designation. By definition, motorized access in these areas would be prohibited. Later I was to observe to the Committee that over 95% of the entire river corridor had been designated as 'semi-primitive' or 'environmentally sensitive,' even though such major valleys as the Stikine are commonly major transportation arteries.

Since January 1991, Cominco Limited's Snip Mine had operated in the remote Lower Iskut River region without road access, having considered the construction of a 100 km road from Highway 37 at Bob Quinn Lake as economically prohibitive based on its original ore reserves and capital and operating cost estimates. In retrospect, the road would arguably have been more viable than Cominco's decision to use a combination of fixed-wing aircraft (DC-3 and DC-4) and a single AC 188 – 400 Hovercraft. It may have been safer, considering the number of individuals killed in helicopter and fixed-wing accidents during a nine-year operating period.

Originally, Cominco expected to use the Hovercraft year-round. However, following winter trials, it was decided that the Hovercraft operation was practical only from April through October because of potential problems following freeze-up. The operation continued until 1997, when control of the Snip Mine was acquired by Homestake Canada Inc., the operator of the nearby Eskay Creek gold-silver mine. Homestake claimed that the Hovercraft operation was too expensive, citing significant maintenance requirements and comparing costs to those of transporting of concentrates from the mine by aircraft.

The Hovercraft operation was plagued almost from the outset by criticism concerning its potential or perceived negative impact on the environment – principally on salmon spawning areas within the Iskut River's braided stream channels over-run by the craft. Apart from claims that salmon were directly killed by the Hovercraft's passage, it was claimed that eggs from spawning were covered by the craft's wash. In addition there were claims concerning improper permitting of the craft, and that the noise level affected not only fish but wildlife.

Most of the criticism originated from studies of the operation carried out by Bill Sampson, a local trapper and fisher with a degree in biology who lived with his family on the Stikine River south of Glenora. He also operated a small aircraft from an airstrip on his property which enabled him to fly over former mineral exploration sites, such as Galore Creek, and film any aspects of these operations which he claimed showed unacceptable conditions. Although not a member of LSMAC, he made a couple of appearances at LSMAC meetings between 1993 and 1997 to provide lengthy dissertations concerning the Hovercraft operation.

Criticism of the Hovercraft operation was to lead to numerous studies by consultants retained by both environmental activists and Cominco over its six-year life, none of which conclusively proved any of the adverse claims. Eventual suspension of the operation was heralded as a victory by Sampson, Friends of the Stikine – an environmental group which included members on LSMAC, and the Sierra Club of Western Canada.

A 1994 article in the Canadian Mining Journal featured the Snip Mine operation in its August 1994 issue with a front-page cover photo of the mine in its rugged mountain setting, entitled *British Columbia's Best – Snip – the West Coast's largest, lowest-cost gold mine*. Remarkably, in 1993 the mine produced 149,475 oz of gold at a cash cost of U.S. $152 per oz, when gold averaged U.S. $360 per oz. Snip's performance was to place the mine in the upper tier of low-cost gold producers – a ranking it retained through 1998, in spite of its remote location.

In my letter to Doug Herchmer I pointed out that the 'overwhelming lack of industry representation' was evident in 'the representation of votes cast at the meeting on issues directly affecting industry concerns'. Two motions were involved: (1) a recommendation by LSMAC to the Ministry of Forests for immediate termination of Cominco's Hovercraft operation pending completion of proposed investigations and (2) a recommendation to support the Stikine River's candidacy for Canadian Heritage River System (CHRS) designation. The results were: 6 in favour, 1 opposed and 1 abstention. Needless to say I was the opposer in both cases. Both of these issues were subsequently to involve a considerable amount of my time.

My objection to the CHRS designation was based on not having had enough time to complete a thorough review on possible negative impacts on resource development and access in the Stikine Corridor. I said as much in my letter to Doug. The issue was to be delayed by the announcement on May 2, 1995 of the creation of a British Columbia Heritage Rivers System (BCHRS). The program is supported by an Advisory Board which co-ordinates identification of candidate rivers to be nominated. Guidelines for nomination include preparation of a 'Vision' statement and 'Management Objectives' for the river. Although the 'free flowing' nature of the river inhibits inclusion of dammed rivers, industrial uses or bank modifications are not restricted. Nevertheless, the emphasis is placed on other than industrial use, the key values sought in favoured nominations, being natural heritage, cultural heritage or recreation.

The Chair initially appointed to the B.C. Heritage Rivers Board was Mark Angelo, an instructor on environmental-related disciplines at the British Columbia Institute of Technology, who also acted as chair of the Outdoor Recreational Council of B.C. (ORC), a zealous supporter and promoter of protected rivers and an excellent speaker. I was to cross paths with him on several occasions on totally unsupportable claims he or the ORC made concerning envisaged threats to the environment, particularly the Stikine, by the mining sector. An example is the following letter as published in the June 2, 1995 issue of the Vancouver Sun. The letter as published failed to include several vital facts, through editing – among them that the Snip Mine operation commenced in July 1990.

Stikine/Iskut River concerns

Need shoring up with facts

    Your May 1 story "Fraser tops list of rivers in danger," based on information provided by the Outdoor Recreational Council of B.C., listed the Stikine/Iskut River second.

    Among "a myriad of threats" to the Stikine/Iskut the ORC cited these: (1) a proposal to build five dams on the river system; (2) mine development, a concern because of potential acid rock generation; (3) a proposed six-fold increase in the region's annual allowable cut; (4) "The Iskut, source of at least 40 per cent of the salmon that spawn in the Stikine system, is being drastically altered by the operation of an air cushion vehicle being used to transport ore and service mining property."

    The facts are quite different.

    (1) There are currently no plans by B.C. Hydro or any other known independent agency to develop any of the Iskut/Stikine.

    (2) Currently Cominco's Snip gold mine on the south side of the Iskut River, which relies totally on air and air-cushion-vehicle access, is the only operating metalliferous mine in the entire region and it is not acid-rock generating. None of the region's several major copper-gold deposits is considered to be.

    (3) The proposed increase in the AAC applies to the entire Cassiar Forest District (135,000 sq.km), not the Stikine/Iskut watershed (49,000 sq.km). Currently the Cassiar Forest District has the lowest AAC of any district in B.C. The proposed increase would provide logging in three per cent of the entire district.

    (4)   DFO records for 1979-'90 indicate an average sockeye salmon run for the Stikine system of 92,583 fish. The 1991-'94 average is 236,427 fish or more than double the 1979-90 average.

The 1993-'94 average is 279,715 fish the highest recorded for both those years.
DAVID BARR, P.Eng.
West Vancouver

The proposed motion by LSMAC concerning its recommendation to terminate the Hovercraft operation was made immediately following an unscheduled slide presentation by Bill Sampson, who insisted on its inclusion at the start of the two-day meeting. I recall objecting to its being shown without a Cominco representative being on hand. Cominco had learned of Sampson's presentation which was scheduled to be made a day later – on April 27, and had arranged for Michael Hardin, its senior counsel, to be present. I phoned Hardin on learning Sampson was to be allowed to make his presentation earlier and I arranged to take detailed notes of the slides and the presentation which I later passed on to Hardin. Thoroughly fed up with LSMAC and Sampson, on June 24, 1994 Cominco filed a Writ of Summons and a Statement of Claim against Doug Herchmer, Elizabeth Zweck, as member (she was also acting co-chair at the time and was an employee of B.C. Ministry of Environment, Lands & Parks) and the members of LSMAC – which of course included myself!

Coincidentally, I received a letter from Doug Herchmer dated June 24, 1993 in response to my earlier letter. He reviewed the early history of LSMAC and disagreed with my comments on the "imbalance toward the environmental groups on the committee", stating, " . . . when most, if not all, of the issues and concerns that have occurred are a result of industry ignoring or side-stepping the LSMAC." In Doug's absence, I responded in a letter to Elizabeth Zweck dated July 7, 1994, providing copies of our previous correspondence and specific comments on unacceptable language in a proposed 'Zoning and Guidelines Report' prepared by LSMAC.

On September 12 – 15, I accompanied several members of LSMAC on a field trip down the Stikine River from Telegraph Creek to the U.S. border in a powerboat chartered from Dan Pikula, an LSMAC member at Telegraph Creek, who also owned a former Hudson Bay Post building renamed "The River Song." It provided some restaurant and overnight accommodation including general supplies.

It rained much of the time we were on the river, which rose about three feet during the period. We spent two nights at the Anuk Cabin, a small log building near the mouth of Anuk River on the west bank of Stikine River. We used the building for cooking purposes, most of us in the group being accommodated in tents. Although I had spent much of the 1961-1965 period at nearby Galore Creek, being supplied from a staging area at the mouth of Anuk River, it was so overgrown, the original staging area was unrecognizable.

The overlying threat of the Cominco lawsuit obviously disturbed several members. While at the Anuk cabin, I was approached one evening by Peter Rowlands, a Canadian Airlines pilot, and Rosemary Fox, a Sierra Club representative from Smithers, with their concerns. They asked if I would be willing to intercede with Cominco on behalf of the members. Although I was not personally concerned about the outcome of the lawsuit, I agreed. On returning to Vancouver, I arranged for a meeting with David Johnston, then Vice-President Mine Operations, of Cominco Ltd. to discuss the situation. After explaining my background with LSMAC to him, I suggested that Cominco had made their point with the members and that diffusion of the lawsuit against an advisory group could be in the best interests of both parties. He was silent for some time and then smiling he said, "Yes, we have been considering the same reaction." Within a very short period of time, Cominco dropped the lawsuit. Several months later I was dumbfounded to learn that the Sierra Club had brought suit against Cominco claiming some $25 million in alleged damages to the fish population of the Stikine River. Although the suit was eventually dropped without settlement, I should not have been so surprised, as I had already spent a 2 – 3 year period both as a Chamber member and as a private individual in the mining sector exposing the greed and deceit practised by protection activists. This was specifically in the Windy Craggy Mine affair, which led to its expropriation following the designation of the Tatshenshini-Alsek Wilderness Park on June 22, 1993.

In retrospect, in its ten-year period of activity, with 22 meetings, in such locations as Smithers, Dease Lake and Telegraph Creek, and inspection trips on the Stikine River, including one reconnaissance flight by the Committee, it accomplished nothing substantial. Not surprisingly, almost all of the Committee members were actively involved in the Cassiar-Iskut-Stikine LRMP initiated February 27, 1997 at its first meeting in Dease Lake.

My letters and other reports to the media and government on behalf of the mining sector had one most unexpected result. In May 1998, I was a co-recipient of the Mining Association of British Columbia's "Mining Industry Person of the Year Award 1998". The other recipient was Gavin Dirom, a long-time friend and mining acquaintance. I was nominated for the award by David Johnston of Cominco – our paths only having substantially crossed over the Snip Mine Hovercraft affair.

Land Alienation in B.C.
Commission on Resources and Environment (CORE) and
Land Resource Management Plans (LRMPs)

Impetus for the B.C. land alienation process initiated in the mid-1980s was driven by a humanitarian decision of the General Assembly of the United Nations in 1983. Few people are now aware of how the ensuing study and its recommendations have been manipulated by the protectionist sector to its current narrow focus on restricted development of global natural resources.

The Report of the World Commission on Environment and Development, *Our Common Future*, chaired by Gro Harlem Brundtland, Prime Minister of Norway, was published in 1987. It became known as the 'Brundtland Commission'. The Commission's task was to re-examine the critical environment and development problems on the planet and to formulate realistic proposals to solve them, while ensuring that human progress would be sustained through development without bankrupting the resources of future generations.

In her Foreword, Brundtland takes care to note that protection of the environment must be reasonably balanced with sustainable development. She states that

"When the terms of reference of our Commission were originally being discussed in 1982, there were those who wanted its considerations to be limited to 'environmental issues' only. This would have been a grave mistake. The environment does not exist as a sphere separate from human actions, ambitions, and needs, and attempts to defend it in isolation from human concerns have given the very word 'environment' a connotation of naivety in some political circles."

Early Land Issue Developments in B.C.

Over a decade since this unheeded warning was provided, we have the carnage of a devastated resource industry in B.C. as a grim reminder of the chaos that can result from political irresponsibility by 'sudden (untested) changes' caused by unsound, experimental policies. Almost forgotten is that the protectionist movement in B.C. gained momentum under the Socred government, not the New Democratic Party (NDP), commencing with the release of "The Wilderness Mosaic" in March 1986 and followed by "Parks Plan 90", announced in June 1990. Proposed were 23 river corridors as 'study areas', including the Tatshenshini, Alsek and Stikine Rivers in north-western B.C. as Recreation Corridors. The report of the Wilderness Advisory Committee recognized the need to retain access for future mining purposes while ensuring that "wilderness qualities are retained". I have highlighted the three named rivers because of the time that I was to devote in trying to prevent them from becoming 'protected', because of their importance as major potential travel ways. I could have added the Taku River, although it was not proposed as a study area.

Having spent a gratifying part of my career in mineral exploration projects in northwestern BC, beginning in the early 1950s, I am all too aware of the remote and isolated nature of the region and its spectacular scenery and wildlife. The same can be said for the extension of this Coast Mountains region south of the Skeena River and nearby Portland Canal for an equivalent distance to the mouth of the Fraser River at Vancouver. Few people appear to be aware that the northern half of the coastal portion of our province abuts the Panhandle region of Alaska. This vast region is currently penetrated by only two short roads – both from the Yukon accessing the northern end of Lynn Canal at Haines and Skagway in Alaska in the extreme northwest part of B.C. The most northerly road to tidewater in B.C. is Highway 37A – a short extension of the Cassiar-Stewart Highway built with funds for mine development in the 1960s and 70s from Upper Liard in the Yukon on the Alaska Highway to Meziadin Junction and Stewart. It was later extended south from Meziadin Junction to Kitwanga to link up with Highway 16 from Prince George to Prince Rupert – the only major east-west highway through the northern part of the province. Most importantly, the region and its population of less than 1% of the province's people remains little changed from my first visit to the region in the 1950s – nor is there any expectation of significant change based on the potential for future resource development in the region, considering the transitory nature of such developments.

The Socred government never went so far as to recommend a specific target for increased park protection in B.C., which stood at about 5.5% of the land base in mid-1991. All this was to rapidly change following the election of the New Democratic Party (NDP) government on November 5, 1991. Its parks and protected areas initiative with its target of 12% of the land base by the year 2000 was officially launched in May 1992 under the "Parks & Wilderness for the 90s" plan entitled *Towards a Protected Areas Strategy for B.C.* (PAS). The plan was a joint proposal headed by B.C. Parks and the Forest Service of B.C. For initial guidance it included four categories of large and small study areas, the first was to be designated "as soon as possible", followed by the second "by 1993"; the third "by 1995" and the fourth "by 2000". These time targets were referred to as "Goal 1, Goal 2, etc."

In order to introduce the PAS program, the government arranged for presentations of the plan at centres throughout the province. The Chamber formed a "Parks & Wilderness for the 90s Committee", which I chaired. Our objective, only partly achieved because of the lack of representatives in some districts, was to provide a mining industry individual at each presentation.

The rationale behind the selection of the 12% PAS target was never seriously questioned. It was generally accepted that this was the global target recommended by the Brundtland Commission. In fact, the 'Brundtland Report' simply notes that in the mid-1980s "the worldwide network of protected areas totals more than 4 million square kilometres; roughly equivalent to the size of most of the countries of Western Europe combined." Protected areas at the time ranged from a low of 2.5% in the USSR to a high of 8.1% in North America. The report stated only that, "a consensus of professional opinion suggests that the total expanse of protected areas needs to be at least tripled, if it is to constitute a representative sample of earth's ecosystems." As the earth's land base totals about 125.7 million kilometres, the 12% target exceeds this vaguely justified minimum protected area by 25%.

Although the Parks & Wilderness for the 90s plan publicized in May 1992 included the 'Tatshenshini/Haines Highway' as a lower priority category 3 Large Study Area, provincial and international protectionists convinced the government to accelerate consideration of the study area. On July 14, 1992, the Commission on Resources and Environment Act was passed. Four weeks later, on July 20, the B.C. government made the unexpected and unprecedented decision of referring the land-use planning process for the Tatshenshini/Alsek area to the Commission on Resources and Environment (CORE), headed by Stephen Owen, Commissioner. This action was to have dire consequences for the B.C. mining industry. Although over 600 mines had been developed in the province prior to the government's decision, since 1974 under the Mine Development Assessment Process (MDAP) initiated in 1947, the record of success in the permitting process for new mines had been outstanding. About 180 applications had been received, of which about 55% were permitted while 45% were not pursued or became dormant. Only one application was rejected.

In retrospect, it is clear that the only reason for the government bypassing MDAP in the Tatshenshini/Alsek area was to lay the grounds for preventing the development of the world-class Windy Craggy copper-gold-cobalt deposit owned by Geddes Resources Ltd. More about that earth-shattering precedent later.

Protected Areas Strategy (PAS)

The PAS process aimed to achieve two goals, as specified in the government's 1993 report[1], which was accompanied by a supportive letter dated June 10, 1993 signed by the Hon. John Cashore, Minister of Environment, Lands and Parks; Hon. Dan Miller, Minister of Forests, and Hon. Anne Edwards, Minister of Energy, Mines and Petroleum Resources. The goals specified were:

Goal 1) To protect viable, representative examples of the natural diversity of the province, which are representative of the major terrestrial, marine and freshwater ecosystems, the characteristic habitats, hydrology and landforms, and the characteristic backcountry recreational and cultural heritage values of each ecosection.

Goal 2) To protect the special natural, cultural heritage and recreational features of the province, including rare and endangered species and critical habitats; outstanding or unique botanical, zoological, geological and paleontological features; outstanding or fragile cultural heritage features; and outstanding outdoor recreational features such as trails."

A much later version of the PAS process dated April, 1996[2] added the following:

"More specifically, the PAS is designed to:

reduce jurisdictional complexity and inefficiencies by integrating and coordinating the many protected area programs operating at all levels of government in B.C. (e.g. provincial and national parks, provincial recreation areas, ecological reserves, forest wilderness areas, wildlife management areas);

---

[1] *A Protected Areas Strategy for British Columbia, – The protected areas component of B.C.'s Land Use Strategy.*

[2] *A Protected Areas Strategy for British Columbia, Provincial Overview and Status Report, April, 1996, – The protected areas component of B.C.'s Land Use Strategy.*

establish common goals, principles, criteria and classification systems to allow for a more systematic and ecological approach to the identification and selection of protected areas;

renew legislation to strengthen the definition of protected areas by prohibiting all industrial activities, such as logging, mining, oil and gas development and hydro-electric development; and

ensure all environmental, social and economic effects are considered within participatory land use planning processes before designation decisions are made."

As originally articulated in the 1993 report, the objectives were clear and the mining sector reluctantly had agreed to participate at tables, to the extent that representatives were available and to rely on an 'interest statement' in addition to existing and proposed legislation to represent its interest, if representatives were unavailable.

From the outset of the LRMP process, the mining sector found itself placed in the position of having to try to educate the general public on the needs of the mining sector in order to remain a major contributor to the economy of the province. Few recognized the limited impact that mineral exploration and mining had on land disturbance, though the mining sector has been ranked as the second most important resource industry in the province for over a century behind the forest sector. Few were impressed by the fact that by 1996 only 24,000 hectares or 0.025% of the land base of B.C. remained disturbed from mining activities compared to forestry at 2.2 million hectares (2.3% of the land base). At every opportunity mining sector representatives kept emphasizing its interest statement which included the following major points:

"Mining is unlike other resource-based activities such as forestry, agriculture, hydroelectricity, hunting and fishing, and much of tourism and recreation, in three major ways.

Firstly, minerals are hidden resources, the rarest and most elusive of our natural resources. The odds of an individual prospect being developed into a mine are about 1 in 1000. When we do find them, they are where they are and cannot be moved, except through mining.

Secondly, wherever a viable mineral deposit occurs, its utilization will invariably provide the highest economic return of any land use per unit of land disturbed. For example, B.C. mining activities generate annual direct revenues averaging more than 20 times that of forestry per hectare, nearly 100 times that of agriculture and over 2,500 times that of parks.

Thirdly, mining throughout B.C.'s history has disturbed less than 0.03% of the land base and most of that has been reclaimed. This small proportion of the land impacted should ensure that there is little chance of any real land use conflicts. With modern technology and stringent permitting rules, current mines have negligible offsite or downstream impacts.

The PAS process, as initiated and as it evolved in practice, made no provision for future definition of subsurface resources. It relied totally on a consideration of biodiversity objectives which were largely defined by drainage systems and which had no relationship whatsoever to subsurface resources.

The PAS process was badly flawed from the start and as it evolved, it continued to impact negatively on the forest and mining sectors – the two major components of the province's resource base. Its bias toward protection was completely at odds with the balanced approach of the Brundtland Commission toward environmental protection and resource development. It was obvious that the only significant losers in terms of land use would be the resource sector. However, it was never obvious that the additional 6% to be protected would almost totally ignore mineral potentials. Furthermore the PAS process would virtually eliminate the certainty required for tenure and access to at least an additional 20% of the land base – with some estimates ranging as high as 35%, through designation of Special Management Zones.

In preparation for implementation of the PAS process, a variety of workshops were implemented by the government. Included as strategy workshops were two meetings at Dunsmuir Lodge on Vancouver Island in 1988 (Dunsmuir I) and 1991 (Dunsmuir II), which were reasonably well-attended by both government and industry representatives.

In February 1993, the B.C. Ministry of Energy, Mines and Petroleum Resources contacted various geologists in the province, including myself, as "Mineral Experts", with a request to provide mineral potential estimates for Vancouver Island. The Ministry initiated a Mineral Potential Project for B.C. in April 1992 in response to the demand for accurate, up-to-date resource inventories for B.C. – obviously as part of the developing land use strategy planned by the government. The mineral assessment of Vancouver Island began at that time and was followed by others for the Kootenay and Cariboo/Chilcotin regions. Significantly a high emphasis was also placed on field work by the Ministry in the entire Tatshenshini region in a two-year program commencing in 1992, as there was little geological and geochemical data available for the region, in spite of the high profile provided by the Windy Craggy deposit discovered in 1957.

The mineral assessment process was based on that used exclusively by the U.S. Geological Survey, requiring experts to estimate the number of a specific type of deposit which they felt would be found in each mineral assessment tract. Obviously this could be highly subjective, the degree of certainty being influenced by known resources and discovered mineral deposits, plus experience. Although I originally agreed to be a participant in the Mineral Expert exercise for Vancouver Island, like others involved, I was highly prejudiced against it, as experience had indicated the unpredictability of mineral resource distribution. However, having said that, I also realized that all exploration geologists, including myself, have spent much of their careers in exploring for mineral deposits on that basis, but not with the potential stigma of having contributed to alienation of land from further exploration by such a decision.

In any event, my involvement was abruptly terminated after attending one workshop, following a further request, just after the government's Tatshenshini/Alsek Park decision on June 22, 1993. I faxed the following note on July 3, 1993 to Ward Kilby, the Ministry's Manager, Mineral Potential Project.

"Under normal (?) circumstances, I would be more than pleased to participate to the extent of my qualifications in your proposed workshop(s).

However, when a government, to whom you respond with your recommendations, has already recently denied a reasonable recommendation by its own appointed commission, that the owners of the province's most coveted potential mineral producer (Windy Craggy), lying within possibly the province's greatest potential future source formations for copper, gold and cobalt in a single, well-contained district, should be prevented from continuing their 4 – 5 year old approvals application by a unilateral decision, then any recommendations I may offer on less tangible grounds will surely have no independent value.

You may feel that this is an uncalled for response. However, you should realize that many of us who have spent our entire careers in mineral exploration and providing to the best of our abilities our judgement on the pros and cons of particular mineral property evaluations, have never been so devastated as we have by the government's Windy Craggy decision. I have expressed my outrage to our Premier in the accompanying letter, dated July 5, 1993. It is the latest of at least ten letters which I have written to the government and the media in the last two years on topics related to the unfair treatment of the mineral industry, the products of which most of us take for granted."

There will be more about Windy Craggy in the next chapter.

Clayoquot Sound

On April 13, 1993, the B.C. government announced its land-use decision for Clayoquot Sound, an area of 63,337 hectares on the west coast of Vancouver Island. This was the first of the NDP government's land decisions completed by CORE, but initiated under a steering committee appointed by the former Socred government in late 1990. The need for a land-use decision within the heavily forested area was prompted by preservationist lobbying. Although the Chamber and the Mining Association of B.C. were represented on the Committee with Jack Patterson and Ken Sumanik, the subsequent land withdrawals directly impacted the forest sector more than mining. However, the development of Special Management Zones (SMZs) as buffer zones around the protected areas were a precursor to the additional withdrawals of lands available to exploration and development because of the inherent uncertainty as to tenure and access rights. Before the decision, 81% of the area was available to harvesting and this was reduced to 45%. An additional 17% of the land base was allotted to SMZs, where protection of wildlife, recreation and scenic landscapes take priority over any logging (or mining) allowed. Almost without exception, at Clayoquot the SMZs were quasi parks, inasmuch as much as they were buffer zones around the protected areas. The Clayoquot Sound issue was not only controversial, it demonstrated the "all or nothing" approach of the preservationist sector with the withdrawal of their representatives on the steering committee more than a year prior to the decision. Following the decision, having lost a political battle to preserve all of the trees in Clayoquot Sound, anti-logging activists disrupted logging operations for a lengthy period and demonstrated against Clayoquot logging outside Canadian embassies and high commissions in England, Australia, Germany, Austria, the U.S. and Japan. On July 30, environmentalist lawyer Robert Kennedy Jr joined the fray by touring Clayoquot and criticizing the logging activities. The summer of their discontent culminated on October 14 when 43 environmental activists received 45-day jail terms and fines ranging from $1000 to $2500.

The Windy Craggy Saga

The Windy Craggy saga absorbed so much of my time and that of many others, with such little satisfaction, that recording the events from my obviously biased perspective is not a pleasant chore. The creation of the Tatshenshini/Alsek Wilderness Park in the extreme northwestern part of B.C. by the NDP government on June 22, 1993 was undoubtedly the most ill-conceived and disastrous land-use decision ever made by a B.C. government. In effect, the park remains a monument to the greed and deceit practised by its proponents and the equally green government. The government's decision, which destroyed its credibility as a responsible steward of the province's resources, sent a negative message to potential investors in the B.C. mining sector, but was to take several years to receive general recognition – and then only by a limited sector of the public. Between 1993 and 1999, I wrote over a dozen letters to government and articles, some published in the Vancouver Sun, related to the "Tats" decision. Included was a 28-page chronology forwarded to the government and initially dated January 15, 1996 entitled *B.C.'s Mining Heritage, The Development of Windy Craggy & Tatshenshini / Alsek Wilderness Park – A Chronology*.

The history of the Windy Craggy deposit goes way back to its formation about 220 million years ago following deposition of sulfides of copper, cobalt with associated gold and other minor elements in an ocean floor environment. The enclosing formations were then subjected over eons of time to episodes of faulting, intrusion, metamorphism, subsequent deposition, uplift and erosion to recent glaciation and exposure some time during the past 10,000 years.

Windy Craggy lies within the southeast extension of the St. Elias Mountains which contain the world's largest ice fields outside the polar ice caps. The rugged mountain chain contains many peaks exceeding 5200 metres (17,000 ft) including Mts. St. Elias, Logan (second only in height in North America to Mt. McKinley in Alaska, and the highest peak in Canada), King and Lucania. The St. Elias Mountains extend for 400 km northwesterly from Glacier Bay National Monument, Alaska, through the Tatshenshini – Alsek region into neighbouring Yukon Territory and beyond into Alaska.

In northwestern B.C. the region is transected by the Alsek River and its principal tributary, the Tatshenshini River – dubbed the 'Tats' in recent times. The Alsek River was easily navigable by canoes from the coast in Alaska to its junction upstream with the Tats, but only for a short distance up the Alsek, which accounts for the location of the early aboriginal settlements along the Tats, of which no obvious trace remains. In the latter part of the nineteenth century, aboriginal populations of the Tats/Alsek river valleys were greatly reduced, partly as a result of a smallpox epidemic through contact with European traders on the coast. By 1900, aboriginal settlements had been abandoned and the population emigrated north into the Yukon. Their descendants today are the Champagne-Aishihik First Nation, who live principally in the Yukon.

The earliest mineral claims in the Tats area were staked in the late 1890s at Rainy Hollow on a tributary of the Chilkat River in the south-eastern part of the region. In the early 1900s a 90 km long wagon road was constructed from Haines on the coast in Alaska to access the Rainy Hollow copper-silver deposits. Further to the northwest near the Yukon border, placer gold was discovered at Squaw Creek by a native in 1927. In

1937 Ed Peterson and Barney Turbitt discovered a 46 ounce nugget, believed to have been recovered at Squaw Creek. This was surpassed in 1987 by discovery at Squaw Creek of a 74 ounce nugget, the second-largest ever discovered in B.C.

Although several shipments of high grade ore were made in the early 1900s from the Maid of Erin copper-silver property in the Rainy Hollow area, it was not mined until 1952-1954 by St. Eugene Mining Corporation, an affiliate of Falconbridge Nickel Mines Ltd. Other mineral prospects were explored throughout the eastern part of the Tats/Alsek region which was to be referred to as the Haines Triangle because of its boundaries composed of the B.C. – Alaska border on the southwest, Yukon Territory to the north and land in B.C. to the east of the Haines Highway, constructed in 1944 by the U.S. Army for military purposes in World War II. The road linked Haines on tidewater with Haines Junction on the Alaska Highway. Although the Haines Triangle covers about 1.1 million hectares, the portion lying west of the highway containing the mineral district mentioned above – which was to become the Tats/Alsek park – covers about 954,000 hectares. Six roads aggregating about 100 kilometres in length were also constructed to the west of the highway in the 1900s to access mineral properties. These extended to within 10 kilometres of the Tatshenshini River where a landing strip was also constructed on the O'Connor River for access purposes.

Until the 1950s, little was known about the mineral potential of the main portion of the Haines Triangle lying further to the west, because of its remoteness. There were no geological maps available and over 50% of the region was blanketed by ice fields, glaciers and scree. This was all to change over a 35-year period commencing in July 1958 with the discovery of massive sulfide float near the toe of Tats Glacier following an aerial reconnaissance program initiated by St. Eugene Mining Corporation in 1957, which was to lead to the world-class copper-gold-cobalt discovery at Windy Craggy mountain in 1958. With its general knowledge of the geology of the region, St. Eugene's exploration group were aware of the possible potential for massive sulfide deposits within Triassic age volcanic-sedimentary environments to the northwest. Similar assemblages hosted the rich copper deposits mined at Kennecott, within the rugged St. Elias mountain range of Alaska, over 400 kilometres to the northwest. St. Eugene's exploration targets lay within this unexplored region to the northwest in B.C.

The Kennecott Mines, Kennecott, Alaska

Having worked for Kennecott Copper's Canadian exploration arm for over 20 years, I became fascinated by the early history of the Kennecott Mines discoveries while preparing my paper on B.C.'s Mining Heritage, *The Development of Windy Craggy & Tatshenshini/Alsek Park – A Chronology in 1995.*

The copper deposits at Kennecott were the richest ever mined world-wide, and are a testimonial to the resourcefulness, confidence and perseverance of early mining pioneers. My link to Kennecott was to provide me with a copy of a paper prepared in October 1964 by William C. Douglass on the History of the Kennecott Mines, which I was to reference at length in my own paper.

Prior to the B.C. government's decision on the designation of the Tatshenshini/Alsek Park in June 1993, to my knowledge, no connection was ever made to the nearby Kennecott Mines in a similar geologic, climatic, physiographic and seismic environment. The environmental activists who so avidly criticized the proposed mine at Windy Craggy seemed unaware that virtually every scare-mongering statement they made about the proposal described an issue that had been overcome at Kennecott, as well as at many other mining locations throughout the world, though this was repeatedly pointed out to them and the B.C. government by the mining sector.

Implements of copper were reported to have been in use by the Copper River Indians by early Russian explorers and traders who visited the mouth of the Copper River long before the mineral resources of Alaska came under consideration. It was not until 1884 that the U.S. government directed the army to evaluate the copper deposits.

The Copper River is large, temperamental and glacier-fed from numerous tributaries for over 150 kilometres north from the Gulf of Alaska to its principal tributary – the Chitina River and beyond, the whole being totally unsuitable for navigation. It was to take two arduous seasons for army personnel, headed by Lieutenants W.R. Abercrombie in 1884 and Henry T. Allen in 1885 to eventually reach the copper-bearing region about 100 km easterly from the mouth of the Chitina River. The copper used by Chief Nikolai of the Copper Indians and his people for utensils and bullets was found in the form of nuggets in nearby streams, the source being several copper deposits, one of which was pointed out to the army in April 1885.

Concurrent with the Klondike gold rush in the Yukon, there was a rush of prospectors in 1898 and 1899 to Valdez on the Gulf of Alaska, 125 kilometres northwest of the mouth of the Copper River. They were to gain access to the copper-bearing region via a military road from Valdez established in 1898 by the army, headed by Captain Abercrombie, with geological guidance by U.S. Geological Survey geologists. They noted that the limestone-greenstone contact was the dominant structure along which the Kennecott copper deposits occurred.

Mineral claims staked to cover the copper deposits at Kennicott, Alaska were subsequently acquired by the Kennecott Mines Company in 1908. The company was the predecessor of Kennecott Copper Corporation, which was to become one of the world's major copper producers. Access to the deposits was provided by a railroad constructed from tidewater at Cordova, a length of 350 kilometres over sloughs, streams, deltas, between glaciers and across rivers via pile bridges, which were reconstructed annually, following spring floods. The railroad terminated at an elevation of 700 metres on the margin of Kennecott glacier where aerial tramways were used for transportation to the underground mines at elevations of up to 1700 metres. The railway was completed in 1911 at a cost of $23 million.

Located within a similar glacial environment and within a zone of similar seismic risk to Windy Craggy, the mines operated from 1911 to 1938 without any serious mishaps, except for disruptions caused by heavy snow slides. They yielded 591,535 tons of copper (12.8%) and 9 million oz of silver (2 oz/ton) from 4.6 million tons mined. Based on a weighted average of each year's production applied to average annual copper and silver prices, the total value at that time was about $200 million and the net profit $100 million. An average of 500 men were employed at the mining operation in 1916 – about equivalent to the payroll planned for Windy Craggy.

This excerpt from the 1990 AAA Western Canada and Alaska tourbook partly describes Wrangell – St. Elias National Park, the largest national park in the U.S., and emphasizes the important role of mining in the park.

Ice not only shaped the parks' terrain but also became the namesake of the Ahtna or "ice people" that lived here. These and other tribes forged tools of locally mined copper, which soon caught the attention of outsiders. The first to verify the source of the copper trading was Lt. Henry Allen, who in 1885 explored much of Alaska's interior. Fifteen years later two miners discovered the malachite cliffs above the Kennicott Glacier, which became one of the world's richest copper mines. The subsequent Kennecott Mine became one of the most significant events in Alaska's history. The great wealth and development it spawned affected not only Alaska but also the entire nation. Today, the ruins of the mine are all that remain of this immense enterprise.

Would that our provincial government had given similar prominence to the Tatshenshini-Alsek area's mining heritage. In my chronology I was to conclude the Kennecott Mines saga with the following:

Not generally recognized is that Kennecott Copper still owns the claims covering the old producer which lies in the heart of Wrangell St. Elias National Park adjoining Tatshenshini-Alsek Wilderness Park. Many of the old workings and parts of the old townsite remain. Perhaps it may someday be declared a National Monument!

My early outlook was obviously clouded. In 1998 the claims were transferred by Kennecott Copper to the U.S. Parks Service. Although tourists continue to visit the old mining town of Kennicott and nearby McCarthy, nature is continually taking its toll, and the buildings and old mining infrastructure at Kennicott are steadily disintegrating.

Windy Craggy: Discovery & Early Development

The 1957 reconnaissance program by St. Eugene in the St. Elias mountain range between Kennecott and the Maid of Erin Mine was directed by Jim McDougall, a long-time acquaintance of mine. At the time Jim was a geologist with Ventures Limited and also acted on behalf of subsidiary companies Falconbridge, St. Eugene and Frobisher.

Sufficient encouragement resulted from the 1957 program and in 1958 a follow-up program was planned using a float-equipped Piper Supercub aircraft piloted by Stan Bridcut. In July Jim and Stan landed at Tats Lake and during prospecting located mineralized boulders on moraine which led them to a rust-stained creek draining part of Windy Craggy mountain, about 15 kilometres to the north, later called Red Creek. Prospectors Bill Wilkinson and Meade Hepler were sent out to investigate the Red Creek area which led to the discovery of copper-bearing sulfides. On returning to the Tats Lake camp, a major earthquake of magnitude 7.5 – 7.8 (Richter scale) centred on Lituya Bay, 140 kilometres to the south, shook the area, but caused no visible damage.

Any mineral discovery, large or small, is exciting. World-class mineral discoveries are a rarity and may take years of further exploration to be recognized as such. Although Windy Craggy was eventually deemed to be one of these, it was also to be one of the rarest exceptions – one where politics, greed, deceit and ignorance were all to culminate in the deposit and the area within which it lay being mothballed 35 years later, after an expenditure of over $50 million in exploration, development and permitting costs at Windy Craggy alone.

I refer to it as being 'mothballed', as an eternal optimist who has however more than once been disappointed by political decisions, which are occasionally reversed. At least Jim McDougall was to receive recognition for his good fortune in recognizing the potential value of Windy Craggy. In 1984 he received the Prospectors and Developers Association 'Prospector of the Year' award at the annual conference in Toronto, and I was delighted to have been approached by John Anderson another long-time friend of Jim's to prepare one of the nominating letters required.

The setting of Windy Craggy within the remote and uninhabited Haines Triangle, virtually unexplored to the west of the Tatshenshini River by the 1950s, appeared at first blush conducive to resource development, particularly mining. Unfortunately, its setting inland from Alaska's panhandle to the west and within the world's largest non-polar ice cap, dominated by the St. Elias ice fields in the U.S., was to prove a significant encumbrance to mining development from international environmental activists.

Mineral exploration by the Ventures – Falconbridge group in the Windy Craggy area was spasmodic. However, between 1957 and 1965, prospecting by the group led to the discovery of over 100 in-place and float occurrences of mineralization, apart from Windy Craggy. The corporate priority of the group was nickel and the possibility of Windy Craggy being a volcanogenic massive sulfide deposit was not recognized.

In 1980 Mr. Geddes Webster, a mining engineer previously involved with Falconbridge, was searching for a mining property which would qualify for flow-through share funding. He approached Falconbridge through Stan Charteris, Canadian exploration manager, who suggested the Windy Craggy property. A joint-venture agreement resulted between Falconbridge and Webster's newly formed company – Geddes Resources Limited. Jim McDougall directed an exploration program of diamond drilling and an airborne survey in 1981, which was followed by further diamond drilling, electromagnetic surveys, geological mapping and sampling programs in 1982 – 1983, supervised by Terry Chandler on behalf of Falconbridge. In the 1990s, Terry as president of Redfern Resources Limited was to head a major mine development program to place the former Tulsequah Chief Mine in the Taku River area back into production, with newly discovered reserves, in a prolonged battle with both national and U.S. environmentalists. (By 2001, production remained uncertain after a 5-year permitting period, approval by B.C.'s environmental review process with costs of $7 million, and a subsequent judicial review.)

No further significant exploration programs were completed at Windy Craggy until 1987 when major underground and surface exploration programs were initiated, requiring construction of a 40-man camp, transportation of mining equipment and supplies to an airstrip completed near Tats Lake in 1985 and construction of a road over Tats Glacier to a portal site on Windy Craggy. No difficulties were encountered during the 1987 and 1988 winter programs, other than occasional blizzards which affected travel on the 13 kilometre road from the camp and airstrip to the portal site.

By 1988, Geddes and predecessor companies had spent about $32 million on exploration and development work. With funding becoming difficult, Northgate Explorations expressed interest in the venture and agreed to acquire a 15% equity interest in Geddes for $5 million. Most importantly, the exploration programs had indicated the presence of a world-class deposit at Windy Craggy.

Apart from extensive local and regional geological mapping and sampling, the work done included 7800 metres of surface diamond drilling, 39,700 metres of underground diamond drilling in 121 holes, 4150 metres of cross-cutting and drifting, construction of an airstrip, a 13-km road to access underground workings, extraction of a 200-ton bulk sample, and metallurgical studies. Ore reserves in December 1989 were estimated at 165.4 million tonnes proven, probable and possible grading 1.9% copper, 0.08% cobalt, 0.2 grams gold per tonne and 3.9 grams silver per tonne in the north and south zones at a 1.0% copper cut-off.

1990 was to be a significant turning point in public awareness of the Windy Craggy mine proposal because of the perceived impact of the project on the rafting community using the Tatshenshini/Alsek rivers. My eventual research into the background of this development and later involvement was to reveal how a shared use of wilderness by recreationists and the resource sector, in this case mining, can rapidly be changed through deceitful propaganda by well-organized and well-funded environmental activist groups. It was to lead to a long and bitter personal experience in attempting to show the overwhelming benefits of mining to society, which are mostly taken for granted, compared to its miniscule impact on the environment. In B.C., the NDP government's policies throughout the 1990s were to demonstrate how sudden changes inevitably lead to chaos – particularly when they have not been tested on a small scale.

As part of its filing requirement for review by the government's Mine Development Assessment Process ("MDAP"), Geddes released detailed information in January 1990 in its Stage I Environmental and socio-economic Impact Assessment Report on the Windy Craggy Project. This voluminous study, in five volumes, included Project Description, Environmental Baseline Studies, Access Road Report, Access Road Plans and Profiles and Baseline Study Appendices.

Hundreds of these volumes were prepared as part of the submission to government agencies, public forums and other interested parties. Environmental groups obtained copies for reference and used them as a basis for subsequent criticism of the proposal. They continued to attack the proposed mine development over the next three years.

While Geddes was required to provide quantitative and very detailed technical data on the mining proposal, environmental activists criticized the project in qualitative or subjective terms. In addition, even though their initial objection to Geddes' mine proposal was that rafters on the Tatshenshini River would be subjected to occasional glimpses of an access road and a bridge across the river, they seized an opportunity to promote the creation of a park that would encompass the entire Haines Triangle, a 940,000 hectare area, from the Haines Highway on the east to the junction of the Alaska and Yukon boundaries in the west. This area included both portions of the Alsek and Tatshenshini watersheds in British Columbia. The fact that the mining operation as envisaged at the time would only require about 1100 hectares or 0.1% of the area for infrastructure and access was immaterial to the activists. It was unacceptable to them for largely fabricated reasons, which were subsequently challenged by mining industry representatives using factual data. The mining industry's criticism had no tangible effect on the government.

To apprise both the public and the government that industry and wilderness do co-exist, mining advocates noted the similarity between the Windy Craggy and Tatshenshini/Alsek Wilderness area and that of Westmin Resources Limited's zinc-copper-lead-gold-silver mine at Buttle Lake on Vancouver Island located in the core of Strathcona Park. Base metal mineralization was discovered at Buttle Lake prior to 1918 and the creation of Strathcona Park. In the 28 years of the Westmin operation, it had produced metals valued at over $2 billion at 1995 prices, compared to the $8.5 billion gross metal value projected for the Windy Craggy deposit. In 2001, it was still in operation.

Another area which I decided to research for my Chronology was the history of rafting and kayaking on the Tatshenshini and Alsek rivers. Archaeological data from Yukon and Alaska studies have suggested that Aboriginal people may have inhabited the Tatshenshini/Alsek river region as far back as 8000 years ago. In the nineteenth century, the area was home to both the coastal Tlingit and the inland Tutchone peoples, who settled along the Tatshenshini River at its junctions with the Alsek and O'Connor rivers.

The Tatshenshini River was a key trading route between the Tlingit and Tutchone who lived along the upper reaches of the river. In the eastern part of the region, the Chilkat Tlingit traded with the Tutchone; one of their routes later became known as the Chilkat trail and in the 1940s became the route of the Haines Highway, the present eastern boundary of the Tatshenshini/Alsek Wilderness Park.

For most of the nineteenth century, coastal Tlingits prohibited non-Aboriginal people from using their trading routes in the region, to protect their trade monopoly with the Tutchone people. By the 1880s, however, non-Aboriginals began using the Chilkat trail, and in 1891 English explorer Edward Glave and American packer and guide Jack Dalton, accompanied by two Aboriginals, a guide named Shouk and his companion, Koona Alk Sal, became the first non-Aboriginals to paddle down the Tatshenshini River, using a 6-metre dugout canoe.

In the late 1940s, Arnie Israelson, an American, made frequent hunting trips up the Alsek and Tatshenshini rivers from Dry Bay, using a riverboat powered by an outboard motor.

Focus on the world-class challenge of kayaking the Alsek River in a turbulent section at the mouth of the Tweedsmuir Glacier known as Turnback Canyon also led to early recognition of the rafting/kayaking potential of the Tatshenshini River.

The 'discovery period' for rafting and kayaking on the Alsek River is well documented in a guidebook on the whitewater rivers of Alaska, prepared by Andrew Emdick, M.D. of Valdez, Alaska and dated January 1987. I was fortunate to learn about this excellent guidebook in 1995 and had a most interesting telephone conversation with Dr. Emdick after obtaining a copy of a draft of a chapter on the Alsek and Tatshenshini rivers.

In summary, the draft included the following chronology: between 1961 and 1986, there were a total of 24 separate groups who attempted or completed kayak, raft and one canoe trip on the Alsek River. About 95 individuals, mostly from the U.S., the balance from Canada, Germany, Great Britain, Switzerland and Austria were the main river recreationists. Two fatalities occurred, one during kayaking in 1983 and the other in 1971 when a float plane carrying a father and his son capsized on landing on the Alsek near the confluence of the Tatshenshini. The son survived and was rescued 38 days later, 50 pounds lighter, when his SOS was spotted by a passing aircraft.

The most heralded voyage was that by Walt Blackader, an American, who in 1971 made the first run in a kayak of the Alsek to include Turnback Canyon. His description of the hazardous trip included his famous quote, "I'm not coming back. Not for $50,000, not for all the tea in China".

Most of the trips were serviced by fixed-wing and helicopter aircraft.

Emdick classifies the Turnback Canyon section of the Alsek as Class VI, the maximum in the International Class I – VI system, with Class VI described as 'maximum difficulty, nearly impossible and extremely dangerous'. Other than Turnback Canyon, Emdick classifies the Alsek as "Class III with two or three Class IV rapids, which change little in difficulty with changes in flow".

The rafting potential of the much gentler Tatshenshini ("Class III, never IV" – pers. comm. from Emdick) was not recognized until the early 1970s, which is surprising considering the activities of hard-core kayakers commencing in 1961 on the Alsek. These were the 'professionals' seeking stardom, predominantly from the U.S. but including at least one Canadian.

Emdick stresses the difference between the Alsek and the Tatshenshini rivers, commenting on the different environment, that of the Alsek being "wild and remote, pristine and untrammelled" and further being open terrain "instead of being heavily wooded like the Tatshenshini". He ranks the Alsek with the Devil's Canyon of the Susitna and the Grand Canyon of the Stikine in North America and specifically as "at least equal of the Colorado River through the Grand Canyon". He states further "The 'Tat' is well known as a spectacular trip, but its fame is properly due to its Alsek portion".

Emdick's chapter on the Alsek in draft covered 25 pages. Not only did it provide detailed information on the character of the river and its history, it also contained interesting information on the Tlingit who travelled from the coast up the Alsek to its confluence with the Tatshenshini and up the Tatshenshini for trading purposes, leaving their canoes upstream at times of low water. He also relates that major floods, some originating from Neoglacial period ice surges, began 2800 years ago. In more recent times beginning around 1600 but repeating in 1730, 1803, 1853 and 1910 A.D. floods caused by outbursts from ice-impounded lakes on the Alsek affected the topography and in certain cases the Tlingit people. Several of these were catastrophic with fatalities occurring to natives as far distant as Dry Bay at tidewater, where villages were destroyed around 1853 and 1910.

The 1970s heralded re-discovery of the Tatshenshini by a different breed of river recreationists than the early kayakers on the Alsek. These were commercially-oriented operators and included Richard Bangs, founder of Sobek, the world's largest adventure company, who in 1976 with others was the first recorded to raft the Tatshenshini to the Alsek and on to Dry Bay.

In 1977 John Mikes and his son Johnny, founders of Canadian River Expeditions, Canada's oldest rafting company with extensive experience in handling rafting parties on the Thompson, Fraser, Chilko and other rivers in B.C. made their first trips on the Tatshenshini. Johnny was a good friend of my son Rob at that time who spent the summer of 1975 as an employee of Canadian River Expeditions on rafting trip between Chilko Lake and Lillooet. During our hiking and mountain climbing days, we shared a memorable ascent of Mt. Baker in Washington with Johnny Mikes, and partly with Susan on July 23, 1972. Rob's friendship with Johnny Mikes was also to provide us with a 19-year friend and boarder named Charley, picked up as a kitten by Rob from a garbage dump at Lillooet in 1975, and passed over to us as a small handful on Rob's return to Vancouver.

Through the late 1970s and throughout the 1980s, rafting expeditions increased rapidly as the recreational potential on the lower Tatshenshini and Alsek rivers was widely publicized.

During the latter part of the 1980s a perceived threat to the commercial rafters by a potential mining operation at Windy Craggy, which had been initiated by its discovery in 1958, attracted a sufficient number of environmental activists, well experienced in organizing effective campaigns against resource users. The principal groups involved included Tatshenshini Wild, Western Canada Wilderness Committee and the well-endowed international Sierra Club and its B.C. subsidiary.

Elevation of the recreational potential of the Tatshenshini river to "world class" status was to be one of the most prominent achievements accomplished by the environmental activists. In the span of about three years, a campaign of exaggerated and misleading claims concerning the wilderness values of the Tatshenshini and the threats to it by a mining development at Windy Craggy was to gain the biased support of a provincial government committed to promoting wilderness values, yet claiming to be an advocate of responsible shared use of the province's resources.

Encouraging results from mining and exploration activities in the late 1980s in the Tatshenshini/Alsek River area were not only restricted to Windy Craggy. As previously mentioned, in 1987, at Squaw Creek in the north-eastern part of the area, a 74 ounce gold nugget, the second-largest ever discovered in British Columbia, was recovered by placer miners. It was significantly larger than the 46-ounce gold nugget discovered at the same site 50 years earlier. In addition, Windy Craggy's growing reputation had attracted more than 20 other mineral exploration tenure holders to the area.

Satisfied with its exploration and development progress, Geddes filed a project prospectus with the British Columbia Mine Development Steering Committee ("MDSC") a government agency, in May 1988. This filing was the initial requirement in applying for a mine production permit. In July, it filed a Preliminary Road Corridor Assessment report with MDSC. The report reviewed the need for an all-weather road from the Haines Highway to the mine site. It examined all possible road corridors, rating them on geotechnical, environmental and engineering considerations.

By March 1990, Northgate's interest in Geddes had risen to 36.6% and Cominco had acquired a 19.5% interest in Geddes. Expenditures in 1990 were $12.0 million and included comprehensive geologic reserve estimates, glaciological studies, preliminary feasibility studies and an additional 9300 metres of underground diamond drilling in 26 holes.

Public consultation meetings were held in nine B.C., Alaska and Yukon communities in March.

In May 1990, the Champagne-Aishihik First Nation (CAFN) filed a land claim covering most of the Haines Triangle, even though their predecessors had vacated the land, having moved north into Yukon in the early 1800s. Aware of the growing potential value of minerals in the area, the CAFN insisted that the land use issues be settled prior to having the Tatshenshini area designated as a park. Land claim negotiations between the CAFN and the federal and provincial governments were not initiated until July 1995 over two years after the park designation in June 1993.

Additional results from further studies at Windy Craggy led to the submission of a Stage I Revised Mini Plan to the Mine Development Steering Committee. Changes included a significant reduction in the quantity of potentially acidic waste rock which was to be placed in a submerged impoundment. The plan incorporated the most advanced and effective technologies for preventing acid rock drainage – a critical environmental consideration.

Technical seminars were held in Whitehorse, Yukon; Haines, Alaska and Smithers, B.C. Public consultation meetings were also held in Whitehorse and Haines Junction, Yukon and Haines, Alaska in order to apprise the public of impacts which the mining operation would have within the transboundary region.

By the end of 1990, Tatshenshini Wild and American Rivers enlisted support for the creation of a wilderness park in the Haines Triangle, from over 50 organizations in the U.S. and Canada, claiming a membership of 10 million. Lobbying activists included Al Gore, first as a U.S. senator and later as Vice-President.

The perceived threat of this mine development to the surrounding area was so grossly exaggerated and devoid of factual data, it is remarkable that it was not challenged by anyone other than the mining groups. In Tatshenshini-River Wild, a glossy publication released in 1993, Ric Careless was to write the following:

> "If Tatshenshini is preserved, the entire international complex of the St. Elias protected areas will remain safe and self-sustaining. But if development were allowed to penetrate into the centre of the wilderness, it would forever endanger not just the Tatshenshini, but surrounding national parks and the wildlife, fisheries and superb landscapes of the entire area."

This is typical of the many exaggerations propagated by environmental activists in this and other proposed mining developments. To suggest that the minesite and its supporting infrastructure, estimated by Geddes at less than 1100 hectares, or 0.1 per cent of the 1.1 million hectares of the Haines Triangle would "endanger" the adjoining National Parks totalling 11 million hectares, is preposterous. However, such misinformation by unaccountable individuals can have an overwhelming impact on a receptive, but uneducated public.

In another statement Careless added:

> "Such a threat is already dangerously real. Currently, as this book is produced, a Canadian company is proposing to develop a gigantic open-pit copper mine in the very heart of the Tatshenshini wilderness. The project would entail ripping the top off 6,000-foot Windy Craggy Mountain adjacent to the Alsek River; the construction of two huge

tailings dams, the highest 350 feet; the laying of 150 miles of pipelines, one to transport slurry copper concentrate, the other fuel oil; and building a road to run alongside and bridge the river. This project would spell the end of wilderness in this now pristine place."

Typically, the rhetorical nature of such statements belies reality. Mining activity, although restricted mostly to the eastern part of the Tatshenshini/Alsek region, had taken place since the turn of the century. In addition, there were about 100 kilometres of old mining roads, at least two air strips, plus all of the Windy Craggy infrastructure in "this now pristine place."

In its initial mining proposal, Geddes planned to mine a total of 300 million metric tons of ore and about 600 million tons of waste, at an average daily rate of about 30,000 metric tons of ore and 60,000 metric tons of waste over a 20-year period, mainly by open pit and partly by underground methods of extraction. There are dozens, if not hundreds of mine operations of comparable size world-wide; many of them far greater in both size and rate of extraction than Windy Craggy would have been. One of the largest in North America is the Morenci open-pit copper mine in southeastern Arizona, owned and operated by Phelps Dodge Mining Company which was mining a total of 700,000 tons of ore and waste each day in the 1990s.

Kennecott Copper Corporation, whose predecessor company mined the aforementioned copper deposits at Kennicott, Alaska, in the central part of adjacent Wrangell St. Elias National Park, has operated the world-famous Bingham Canyon open pit and underground copper mine near Salt Lake City, Utah, since 1906. It is registered as a National Historic Landmark and is visited by 150,000 people annually who pay a $10 entrance fee. It mines about 300,000 tons per day and has mined about 5 billion tons to date. In 1995 there were 2,400 men and women employed directly at the operation. The pit was about 800 metres deep and covered 800 hectares.

Attempting to correct the misinformation provided by Careless and other environmental activists that I was to encounter was a frustrating experience. These generally well-educated and erudite representatives of various environmental organizations were paid to attack any new or existing mining industry developments and they were generally very effective. After all, it was their source of employment and factual data were immaterial to their rhetoric. We rarely met on a level playing field as our job was trying to explore, develop and place mines into production, while abiding by all the legislated requirements. We were invariably in a defensive position, trying to correct misinformation and fabrications. It was to lead me to compose a Fact/Myth Sheet concerning Windy Craggy, which was widely distributed through my association with the B.C. & Yukon Chamber of Mines. We never received any complaints or requests for corrections. As I write these recollections in 2000, I am still doing my best to correct similar misinformation by the preservationists.

Throughout 1991, Geddes continued to submit applications as required for numerous permits. Also completed were updated geostatistical reserve estimates calculated by Montgomery Consultants at 297.4 million tonnes grading 1.4% copper, 0.07% cobalt and 0.2 grams per tonne gold and 3.8 grams per tonne silver in the North, South and Ridge zones at a 0.5% copper cut-off. The deposit remained open at depths and along strike.

Following the election in November 1991, little time was lost by the new NDP government in focusing on the Windy Craggy issue.

Early in 1992, the government directed its agency, the Geological Survey Branch of the Ministry of Energy, Mines & Petroleum Resources to conduct a geological study of the area to provide an independent evaluation of the mineral potential of the area. The study was to be completed in an 18-month period and to include results from the 1992 and 1993 field seasons.

The results of the 1992 field program are discussed in MRD-GSB Geological Fieldwork 1992 issued in January 1993 (p. 185-229). The objective of the program is summarized partly as follows:

> The primary objectives of the Tatshenshini project are to inventory all known mineral occurrences within the Tatshenshini/Alsek area, and to compile geological and mineral occurrence databases from which the mineral potential of the area can be evaluated.

The results of the 1992 program as summarized in the following quoted paragraphs are particularly significant, in terms of a potential future contribution to the economy of the Province, if found to be viable. (Above reference p. 217 – 8 and Figure 2)

> "Five new copper occurrences were discovered while mapping the Alsek-Tatshenshini area during the 1992 field season. Four of these are located in a 20-square kilometre area centred approximately 15 kilometres north-east of Tats Lake. Here, as at Windy Craggy, the host stratigraphy includes interbedded pillowed flows, calcareous siltstones and sills of the middle Tats member of the Tats group. Assays of up to 20 per cent copper have been obtained from the most extensively mineralized of these new occurrences, the Rainy Monday deposit. Such new discoveries demonstrate the very high mineral endowment of the Upper Triassic Tats group and the high potential for the discovery of new reserves."

The 1992 program cost over $300,000 and included detailed stream sediment sampling and analyses of the streams and rivers in the Haines Triangle, a recognized prospecting technique for delineating areas of potential mineral significance, which had led to so many discoveries in B.C. since the 1960s (I was later to inquire from Ron Smyth, Chief Geologist of the Geological Survey Branch, as to the results, but was informed at the time that he had been instructed that they should not be released. He was to use his influence and they were finally published in 1996, indicating many target areas warranting follow-up which of course was not possible following the park designation in 1993). In fact, the government's request for an assessment of the mineral potential of the area, partly completed in 1992, which indicated such a high potential based on new mineral discoveries by its own agency, was curtailed in February 1993 and confirmed by the Geological Survey Branch budget on March 30, 1993. The government claimed that it already had sufficient information on which to base the decision it was to make on June 22, 1993.

One would wish to believe that with the encouraging results achieved in the 1992 program that the future for mining in the Haines Triangle would have been well assured. What was to follow was a series of events designed by the government, which based on the eventual outcome, were obviously intended to ensure otherwise.

On July 14, 1992, the Commission on Resources and Environment Act was passed. Four weeks later, on July 20, the B.C. government made the unexpected and unprecedented decision of referring the land-use planning process for the Tatshenshini/Alsek area to the Commission on Resources and Environment (CORE). This action effectively questioned the mandate of the Mine Assessment Review Program (MARP) and the more recently formed Mine Development Assessment Process (MDAP).

CORE completed its report, subsequently released in January 1993 in two volumes entitled Interim Report on Tatshenshini/Alsek Land Use, British Columbia, including appendices.

The overview of the CORE report included the following statement:

The land use question has been brought into focus by the potential for conflict between a major mine development proposal at Windy Craggy Mountain and the high wilderness values in the region. The Commission has been asked to review and report publicly to Cabinet on the major options for land use in the area and on fair, open and balanced processes related to each.

The CORE report considered three major land use options:

1. <u>Wilderness</u>, which would prevent further mineral development, necessitate a compensation consideration for expropriated mineral claims and "the barring of effective access to significant tourism or recreational use of the area due to its remoteness."

2. <u>Mining</u>, which would provide for the creation of integrated resource management and sensitive management areas. The CORE report clearly indicated its bias against this option.

3. <u>Delay</u>, which would impose a moratorium on mineral exploration and development, but not rafting, for a specified number of years until various outstanding issues could be clarified.

In its concluding statement, the CORE report recommended the delay option.

On receipt of the CORE report, the B.C. & Yukon Chamber of Mines (Chamber), formed the Tatshenshini/Alsek Land use Committee to review it. The Committee included 17 members; 13 were experienced professional engineers and geologists. I was a member of the Committee and agreed to review and report on the fish and wildlife values in the CORE report, which appeared to be grossly exaggerated.

A five-page preliminary summary of the Committee's review of the CORE report dated May 11, 1993 was forwarded to the government. More detailed reports were submitted in June, 1993. The summary severely criticized the CORE report for the errors and deliberate omissions respecting mineral values, wilderness values and economic analyses. It recommended rejection of the CORE report and insisted that the review by MDAP of the Windy Craggy project be permitted to continue. The CORE report included a comprehensive mineral policy endorsed by Cabinet, to provide a framework within which decisions regarding projects such as Windy Craggy be dealt with:

In the spirit of the Brundtland Report, Windy Craggy provides a unique opportunity for British Columbia to present to the world that an environmentally responsible mineral development can in fact enhance a vast, surrounding wilderness by providing limited or managed access to a small portion of that area with, at the same time, minimal ecological impact.

This recommendation in the report was ignored, as were those forwarded to the government by prominent individuals and organizations urging reconsideration of the impending announcement of the creation of the Tatshenshini/Alsek Wilderness Park which would deny further mineral developments. Included were comments by Governor Hickel of Alaska and Government Leader John Ostashek of Yukon.

Many articles and letters on the Windy Craggy/Tatshenshini issue were published by the media during this period. One of these by Stephen Hume in the May 7, 1993 issue of The Vancouver Sun condemned the proposed mine development, based on grossly exaggerated or misleading statements of potential damage to the environment. The article was entitled Shangri-La with an underlying quote by Hume "The Tatshenshini offers the NDP a chance to exercise world leadership and rise above the money-grabbing politics of commerce and greed."

Hume's article prompted me to volunteer to write a rebuttal, on behalf of the mining industry. Several Chamber representatives accompanied me to a brief meeting with The

Vancouver Sun's editorial board, where it was agreed that an article of equivalent space – about two-third page, would be accepted. A rebuttal entitled "An Industry Perspective on the TAT" was published in May 14, 1993 edition of The Vancouver Sun. It was generally well received and cited factual data focusing on the most inaccurate and damaging statements in Hume's article. Some time later, I received an unexpected cheque for $150.00 from The Vancouver Sun – it was to be my first and only claim to professional authorship!

My article also provided me with two other opportunities to respond publicly to the most misleading, inaccurate and exaggerated claims made by environmental activists, several of which were repeated verbatim in the CORE Report, as perceived threats from the proposed mine development.

The first of these was a request by Vancouver radio station CKNW for me to appear for a one-on-one debate on the Windy Craggy mine – Tatshenshini/Alsek park issue. My opponent was scheduled to be Johnny Mikes – my son Rob's old friend, who was then operating Canadian River Expedition's rafting operation on the Tatshenshini/Alsek rivers. Familiar with Johnny and his background, I was not unduly concerned about the debate. However, this changed quite dramatically when I arrived at CKNW to find that Ric Careless, with considerable experience in public speaking on behalf of the environmental sector, had taken Johnny's place.

My article also led me to prepare a one-page, double-sided summary of the principal fallacious claims by the environmental activists. It was entitled Tatshenshini – The Facts and was widely circulated under Chamber letterhead. I was later to organize protests at two public events complete with placards and our 'Fact Sheets'. The first of these, on September 17, 1993, was at the Orpheum Theatre in downtown Vancouver. A concert had been arranged featuring Leon Bibb, Ann Mortifee and John Denver to raise funds 'to cover the cost of the successful four-year campaign to save the world-class Tatshenshini-Alsek Wilderness'. The concert was sponsored by Raincoast Books, the Book Company, Canadian River Expedition, Mountain Equipment Co-op and Nahanni River Expeditions. Invitations to attend were provided by as unlikely an entity as MCL Motor Cars, I thought at first, until I realized that they were a dealer for Jaguar, Porsche and Range Rover.

Among those who had alerted me to the upcoming benefit concert at the Orpheum Theatre was Rene, who had noted the announcement in The Vancouver Sun while I was attending a three-day CORE meeting in Parksville. Another was Bruce Downing, a senior geologist with Teck Corporation, who had directed several exploration programs at Windy Craggy and was highly experienced in geochemical exploration techniques. We agreed that he would suggest Chamber involvement at the Orpheum Theatre at its monthly meeting on September 13th. However, the motion was not supported.

Coincidentally on the following day, I received a fax from Bill Dumont, then Chief Forester of Western Forest Products Limited, whom I had previously met at one of the Vancouver Island CORE meetings. The gist of his message was that we should not miss the opportunity provided by the concert to get our message out on behalf of the mining sector. During a phone call to obtain additional information, I learned that Bill had directed a very successful demonstration on behalf of the forest sector a week earlier in the Gastown area of Vancouver on the Clayoquot Sound issue.

Somewhat humbled by an opportunity for second thoughts on 'getting our message out', on short notice, I called a meeting of a selected group of individuals at the Chamber for the following day. In attendance were Bill Dumont, Ted Mahoney (Chief Geologist Royal Oak Mines Inc.), Bill Clark (President of Freeport Resources Inc.), Bruce Downing and myself. All were supportive about our mounting a small demonstration with placards and a brief handout, which I felt should be patterned after one made available by Bill Dumont at the Clayoquot Sound Rally. As described, it coincided with a 'myth-fact' theme which I had earlier had published in The Vancouver Sun in rebutting the article by Stephen Hume.

I prepared a draft of the proposed handout which was expanded with additional comments received from several individuals until sufficient text had been provided to fill a double-sided sheet. It was entitled Tatshenshini – The Facts – a basic theme which was to be featured on our placards.

I also drafted a news release which was faxed to CBC TV, BCTV, CKVU (UTV), Vancouver Sun, Vancouver Province, Financial Post, Canadian Press and B.C. Report. Follow-up telephone calls were most discouraging and no news coverage was to result, which was very disappointing in view of the enthusiasm of the rally participants and the obvious interest expressed by the vast majority of concert patrons.

I spent several hours drafting the placards, with about 25 eventually being made available to participants. Others also helped in drafting the posters and resurrected several from the June 22 announcement of the park designation which had been carried by mining supporters.

About two dozen mining supporters assembled at the Chamber office on Friday evening. Included were: Robert Longe, Dan Pegg, Don Bragg, Cliff Rennie, Bruce Downing, Ross Burns (Royal Oak Mines), Ann Burns, Robert Burns, Ted Mahoney, Gerry Delane, Ben Ainsworth, Gerry Carlson, Doug Perkins, Jack Dunlop, Wendy Dunlop (Last three from Neville Crosby Industries), Dave Jenkins and his wife, Diane Lister, Stan Hoffman, Tor Bruland, and an unknown young lady who joined the group and myself.

In an announcement circulated to various companies to attract support to the rally, I had encouraged people to wear appropriate dress. Several of us appeared with hard hats complete with lamps, which however did not prevent one hysterical woman from confronting our placard-holding volunteers several times, accusing them of 'cutting down all the trees.'

Many people were surprised at the amount of misinformation that had been generated by the environmental activists as reflected in our handouts. We handed out over 500 of these and as line-ups developed, commencing at about 7:00 pm and lasting until 7:45 pm, many of the crowd read them and asked questions about several concerns. I have little doubt that by far the majority of the people attending the concert were there because of the opportunity to hear the featured artists, not to support the Tatshenshini Park designation. Having attended performances by Ann Mortifee and Leon Bibb in the past, I would sooner have been inside the theatre than outside, particularly at the very reasonable cost of $20 per ticket.

In summary, in spite of the disappointing lack of media coverage, we felt that the demonstration was a success. Certainly those associated with known environmental groups and rafting companies who were in attendance, were surprised at the rally. At the early stage of the evening when only a few patrons were in front of the Theatre waiting for companions to arrive and our contingent outnumbered the others, I overheard one man say to another "I don't believe this. I feel as though I'm the enemy!"

On Friday afternoon, October 8, just before the Thanksgiving weekend, Bruce Downing phoned me asking whether I had seen notice in The Vancouver Sun of a proposed rally by the environmental sector against the government's decision on Clayoquot Sound. The rally was scheduled for noon on Sunday, October 10. I had not seen the item. Bruce said that if we planned any attendance on behalf of the mining sector, he would not be able to attend.

Somewhat reluctantly I decided that the exposure could be of value as coverage would certainly be planned by both radio and TV. In addition, I envisaged our attendance as being supportive of a much larger representation by the forest sector in favour of the government's decision to allow logging of about two-thirds of the well-endowed old growth region.

Late in the business day of a long weekend, I phoned several people to see if they would be interested in attending the rally. Most were otherwise committed. However, Jack Patterson, the Managing Director of the B.C. & Yukon Chamber of Mines agreed and thought that he could bring 'two of the kids'. We agreed to meet at the Chamber's office at about 11:00 am to allow enough time to print up copies of our Fact Sheet.

On arrival at the Chamber's office, I was introduced to Jack's daughter, Vicky, and her husband, Bob. While Jack attended to photocopying about 300 handout sheets, I produced an additional, more specific placard for the occasion entitled Clayoquot Sound–Tats Park <u>Unsound</u>. As we walked up to the Art Gallery armed with six placards and our handouts, Jack casually asked, "Do the loggers expect many environmentalists at the rally?" I responded, "No Jack, it's the other way around, the rally has been organized by the environmental sector to coincide with their plan to have a stump from a clear-cut area on Clayoquot Sound transported across the country in support of their outrage against the government's decision on Clayoquot." If Jack was affected by my revelation, he didn't show it.

I was somewhat shaken by the crowd at the Art Gallery – Vancouver's former Court House main entrance facing Georgia Street. This was definitely not going to be like to our experience at the Orpheum Theatre. It was quite obvious that the rally was being attended by individuals almost totally in support of the environmentalists and there was no sign of any adversarial opponents other than ourselves, except perhaps for a small contingent holding a huge banner entitled "Canadian Communist Party". They sat mutely at the back of the gathering, moving only after about an hour, when a march was announced which effectively closed off the ceremonies at the Art Gallery.

Following our arrival, I elected to wander over to the west side of the open area below the steps to the old Court House buildings where several media people had cameras focused on the facing assemblage. Within a very short period I was suddenly confronted by a middle-aged woman, supported by a CKNW radio-station reporter who very angrily said, "What are you doing here, don't you know that we have paid for use of this space?" I replied, "Isn't this rally open to the general public?" to which she admitted it was and I stated, "I am one of the general public". Undaunted, she questioned our group's attendance at the rally and I informed her that the unilateral Cabinet decision on Tatshenshini Park was nothing more than an appeasement, recognizing the reaction that had been received from the environmental sector on the government's decision on Clayoquot Sound.

At this stage Jack Patterson who was nearby, interjected "We would have loved to have received the same sort of decision from the government at Tatshenshini as was provided at Clayoquot."

Shortly after this incident I was confronted by a younger, even more abusive, male who informed me that he had been brought up in the Squamish area in a logging community and had been completely put off by the profiteering of the logging companies. Further comments showed that he had the same attitude to all free-enterprise corporations including mining. As our discussion wore on, I realized that he was an extremist, not interested in any other opinion contrary to that of his own. It took a long time; I made it clear that I was not about to move from my station because of the threatening comments he had made.

Later, another young man hove to and mentioned that he had witnessed my confrontation with the woman questioning my attendance at the rally. Incidentally, I recognized that I had seen her before in the same environmentalist-supported gatherings. He buoyed up my spirits by saying that he was pleased that I had decided to attend the rally. I learned that he was a student at the University of British Columbia and his ultimate objective was to be a history professor. I mentioned that with his background and objective he should have no problem and he concurred that he was witnessing history in the making.

Another contact which I made was an individual of about my own age, garbed in runner's shorts and a 'green' T-shirt, accompanied by a similarly garbed man about half his age. We had a long conversation. The older individual was a 1950 grad from McGill University in Electrical Engineering with a subsequent degree in Civil Engineering. He had worked for many years with a public utility company before circumstances resulted in a breach of contract. Apparently this led to his rejection of a capitalist-based society, and one has to wonder at the events leading up to his present focus considering his background.

Shortly after this exchange I wandered back over to the east side of the Art Gallery concourse area. En route I noted that of the attendees most were in their early to late twenties. All rejected any offers of Fact Sheets. Considering the 100 to 1 ratio of environmental supporters to our mining sector dissidents, I would have expected more confrontations than the few that occurred. In retrospect, the experience provides compelling evidence that, as in most similar situations, there are a few dedicated and vocal advocates and a multitude of identity-seeking followers.

Although there had been several radio and TV representatives on site, and we had volunteered to remain for an expected camera crew from CBC-TV, which failed to materialize, there was no subsequent media coverage of the event.

Our small group was somewhat dejected as we walked back to the Chamber after the rally. En route, a passerby stared at me with my placard and said "Aren't you a little old to be doing this kind of thing?"" I laughed and had to admit that he was probably right.

Alas, memory can be short. In January 1996, I received an announcement in the mail from Friends of the Stikine Society, signed by Maggie Paquet, Managing Director. I had first met her through my involvement with the Lower Stikine Management Advisory Committee (LSMAC), where both she and Peter Rowlands, president of the Society, were representatives. Both were later to be active on the Cassiar Iskut-Stikine Land Resource Management Plan (LRMP), commencing in February 1997.

The announcement invited interested Environmental Non-Government Officials (ENGOs) to a meeting proposed on January 24 at the offices of Canadian Parks and Wilderness Society (CPAWS) on Hastings Street in Vancouver. The meeting was intended to provide information and for planning purposes concerning the proposed Bronson Slope mining project of International Skyline Gold Corporation, on the Iskut River area, a tributary of the Stikine. The project was in the review stage for an Approval Certificate. The brief information provided in the invitation was both inaccurate and misleading and obviously intended to arouse a reaction against the project. I wrote a one-page summary of the Bronson Slope proposal and provided correct information on the four most flawed statements which concerned power provision, the localized rather than regional benefit to be derived by a road extension, the minimal proposed impact on annual allowable cuts by the road access and the outdated nature of transboundary access proposals.

I had recently completed a mineral resource and infrastructure coloured map of the Stikine River region on behalf of the Chamber with the assistance of Teck Corporation, which was used for general reference purposes, with copies later provided to the Cassiar Iskut-Stikine LRMP table. Armed with copies of both the map and the one-page statement, I went to CPAWS' office address and stood in the hallway opposite the office to await arrivals. Few people can resist a free map, and all those who arrived for the meeting accepted my handout. Eventually, John S. Brock, President, and Wayne Roberts, Chief Geologist for American Bullion Exploration Ltd., then owners of the proposed Red Chris mine development near the village of Iskut, arrived on the scene. Like myself, they had unintentionally been placed on the faxing list for ENGOs. They were not in the office for very long, as Maggie Paquet indicated that they were not welcome. She exited shortly after and was naturally quite surprised to see both myself and my handouts, copies of which she nevertheless accepted.

As I write this recollection in late 2000, the Bronson Slope copper-gold-silver-molybdenum deposit with an inferred resource of 90.2 million tonnes with a gross value calculated at $1.2 billion, located one kilometre south of Cominco Ltd.'s highly successful Snip gold-silver mine, is no longer active within the permitting process. Snip closed down after operating between 1990 and 1999 and has completed its reclamation. Red-Chris is still modifying its original proposals. The nearby Isk Wollastonite project, also in the permitting stage since 1997 is dormant.

Resource Land Conflicts Continued

Although I had been involved in several preliminary land-use issues in B.C., as previously described, my direct participation in the government's 'Protected Area Strategy for British Columbia' (PAS), commenced in January 1993, when I agreed to act as an alternate on behalf of the mining sector to Bruce McKnight, then with Westmin Resources Ltd., in Vancouver. At the time, Westmin had a major interest in the outcome of land-use planning processes initiated by the Commission on Resources and Environment (CORE) on Vancouver Island in November 1992, through its ownership of the Myra Falls base metal mining operation in Strathcona 'B' Park.

Following the earlier described environmental-protection initiatives, including the Brundtland Commission, the Wilderness Mosaic, Parks Plan '90 and the Dunsmuir I & II round tables on the environment, the NDP government passed the CORE Act on July 13, 1992 and appointed Stephen Owen, the former Ombudsman, as the first Commissioner. He was to be a very busy individual until the CORE Act became obsolete in April 1995.

In addition to the Vancouver Island CORE table, the major Cariboo-Chilcotin and Kootenay-Boundary region CORE tables were also initiated in late 1992. The Vancouver Island CORE table overlapped the previously discussed controversial Clayoquot Sound land-use decision announced by the B.C. government on April 13, 1993, which protected a 63,337 hectare region on the west coast of Vancouver Island. Finally, CORE was to be entrusted with the Tatshenshini-Alsek area land use decision, which so devastated the B.C. mining industry with the proclamation of Tatshenshini-Alsek Park in northwestern B.C. on July 22, 1993.

Although the PAS process was being employed at the Vancouver Island CORE table from its outset, the PAS report outlining the process and strategy was not officially released until June 10, 1993. Its deplorable ignorance concerning evaluation of subsurface resources was exposed in its 'Summary' under 'Transitional Issues', abstracted as follows:

> Implementation of the Protected Areas Strategy will require the consideration of a number of transitional issues. Recreation areas, forest wilderness areas and wildlife management areas allow various levels of industrial activity such as mineral exploration and development and therefore do not meet the new Protected Areas Strategy definition for protected areas. Yet there is general agreement that, with very few exceptions, they meet the criteria for protected areas. **The resource potentials of these areas have largely been assessed, with the exception of the subsurface mineral and energy resources which must be evaluated by the Ministry of Energy, Mines and Petroleum Resources** (emphasis mine). Existing recreation areas and forest wilderness areas will be recommended for upgrade to full protected areas status wherever the mineral or energy potential is determined to be low. Where the potential appears to be significant, regional or sub-regional land use planning processes or area-specific management plans will be used to evaluate the areas and consider recommendations for final protection designations. Wildlife management areas will be studied under the same level of detail as other study areas and the relevant land use planning processes will be used to recommend whether all or part of these areas can be confirmed as protected areas.

Existing land and resource tenures in candidate protected areas will be assessed to determine whether they are compatible with protected areas status. If these tenures are not compatible, options will then be presented to resolve any outstanding issues.

Finally, the Protected Areas Strategy presents an opportunity for the public to review, debate and comment on appropriate protected areas legislation and management to achieve the vision, objectives and goals of the Strategy.

A fundamental flaw in the PAS process concerning subsurface mineral resource evaluation, never corrected, is revealed by the assumption that the Ministry of Energy, Mines and Petroleum Resources (MEMPR) would be capable of evaluating these resources within the seven years provided for completion of CORE and LRMP table deliberations and recommendations. Not only are vast areas of B.C. covered with post-mineral drift, glacial debris, scree and ice, experience has repeatedly shown that significant mineral deposits are discovered in previously mined and written-off areas, in addition to mineralized areas not exposed or capable of surficial detection or evaluation because of cover by later rock formations. Discovery of mineral deposits through the ages with little exception has been made by prospectors, small and large mining companies – the 'juniors' and 'seniors' in modern times, albeit guided in many cases by important geological mapping provided by government agencies, but not exclusively. Technological advances in mineral detection capability, particularly diamond drilling and geochemical and geophysical ground and airborne techniques since World War II have had an enormous impact on the detection of new mineral deposits and mines, unrecognizable by surface exploration.

Increasingly, with most significant outcropping mineral occurrences having been detected over 150 years of exploration in B.C., future discoveries, mostly lying within the true subsurface, will be made by geophysical methods, and confirmed by diamond drilling and even subsurface development. Improved technological advances in geophysical techniques since World War II have led to the discovery of the tops of viable mineral deposits at depths of over 500 metres below surface. More recently depths of up to 1000 metres to such targets are forecast.

In a letter dated June 29, 1992 to the Hon. Anne Edwards, Minister of Energy, Mines and Petroleum Resources, criticizing the obvious flaws of the proposed Protected Areas Strategy by the newly elected NDP government, I provided tabulated data which showed an average of about 17 years between 'discovery' and the start of production by 42 of the most significant metallic mineral deposits in B.C. Since then, the three most recently permitted mines, all starting production in 1997 (Kemess, Huckleberry and Mount Polley), followed discoveries made at least 30 years earlier. My letter was copied to then Premier Mike Harcourt, John Cashore, Minister of Environment, Lands & Parks and Colin Gableman, Attorney General in the NDP government. Anne Edwards' constructive response supported my concerns by recommending "appropriate mineral sector evaluations prior to park designation." Her letter was also copied to Dan Miller, Minister of Forests and Stephen Owen, Commissioner on Resources and Environment.

In summary, the government was well aware of the mineral sector's historical timeframes for discovery to production of significant mineral deposits. They completely ignored them. Moreover, they irrationally and with no supporting evidence assumed that their Geological Survey Branch was capable of providing such assessments within a few years. Not only did the government ignore historical precedents established by private risk takers, when faced with the results of their own Geological Survey Branch's assessment of the $300,000 program within the Tatshenshini/Alsek area, as previously discussed, which showed " a very high mineral endowment of the Upper Triassic Tats group and the high potential for discovery of new reserves", they simply shut down the planned balance of the 1992-3 program to assess the overall mineral potential of the area, and created the park in June, 1993.

As previously discussed (Early Land Issue Developments in B.C.), from the outset of the mining sector's involvement in CORE and LRMP tables, the mining sector took every opportunity to emphasize its interest statement concerning the need for unrestricted access and security of mineral tenures in all but fully protected areas. It was consistently ignored.

The Vancouver Island CORE table held most of its meetings in Courtenay and ended its negotiations quite abruptly in November 1993, when union representatives of the 17,000 member B.C. logging industry left the table. The table met for 47 days of meetings during the year and included 14 sectors, steering committees of up to 10 people and numerous CORE and government staff which brought attendance numbers to over 100 people. It was the first and perhaps only table to include a Youth Sector composed of half-a-dozen ardent teenagers. As was to be the norm, except for the Cassiar Iskut-Stikine LRMP table, there was no official aboriginal representation, although a short presentation was made at the outset by a First Nation representative giving the reason for non-participation, which was that participation at the CORE table could prejudice future land-claim negotiations.

The CORE process followed the PAS format which was to be used at both CORE and LRMP tables, with little variation. The initial preparation emphasized several negotiation-based concepts to be followed, namely 'shared decision-making' and 'interest-based' opposed to 'position-based'. Guidance for procedures to be followed included: terms of reference, ground rules, work plans, and statements of interest. The entire process was directed by CORE staff, including Technical Working Groups, a mediator and facilitator.

From the outset, progress at the table was significantly delayed by inadequate policy direction from government and control by the mediator. This was partly due to the table being a 'first' and to its large size. Inexplicably, the mediator permitted six representatives at the table for the conservation sector, each of whom had the authority to speak on issues, whereas other sectors were invariably represented by one voice, with back-up by an alternate(s), in adjacent chairs. I recall a spokesperson for the conservation sector insisting that "we don't all agree on the various issues and need to have all our views expressed," when questioned on the need for six spokespersons.

Considerable time was also spent on negotiating regional boundaries, which were generally governed by drainage basins and of course had no relationship whatsoever to mineral resource deposition. Apart from providing and debating the contents of individual sectoral statements of interest, considerable time was also required to prepare a vision statement for the entire table which would indicate how we wished Vancouver Island to appear 20 t0 30 years in the future. As an individual who had been so impressed at the age of 65 at the natural, rather than closely controlled, evolution of the province, with all its biological, cultural and resource-based diversity over a period of almost 150 years, I was and remain an unbeliever in the entire process. Following 18 days of table meetings, the vision statement remained incomplete.

The 'scenario-building' stage of the process, which followed, was to lead to a polarization of views and interests which culminated in the split of the table into a conservation/recreation group and a multi-sector group. Eventually, the latter was composed of forest managers/manufacturers, forest industry independents, forest employment unions, mining, local government, committees and small business, agriculture and aquaculture. Guidance for scenario-building was based on the current Vancouver Island land use in terms of social, economic and environmental characteristics. Sectors produced 'indicators' of their interests and worked through committees on a land designation system. The committees were composed of alternates and met while the main table was in progress. Following the definition of the land use plan, the technical working group (TWG) claimed to be able to calculate the impacts of changes in terms of indicators, such as direct, indirect and induced job losses and reduced government revenue caused principally by reduced timber harvesting. Obviously, similar predictions for the mining sector were far too speculative for use as were environmental indicators, although such qualitative indicators were developed by the TWG.

The PAS process followed by CORE and later LRMP tables initiated negotiations which were to be based on scenario-building results which ranged from conservation-based scenarios providing 5 – 12% protection (averaging about 5 – 8%) and development-based scenarios ranging from 1 – 3% protection. The objective being to reach an overall 12% protected target for the province.

Early in the process, it was recognized by some participants, but not by the government, that it would be difficult, if not impossible, for such a diverse group of stakeholders to achieve consensus in mediated negotiations designed primarily for two-party negotiations, where each one had the desire and power to make a decision and abide by its outcome.

The Vancouver Island CORE table was so large that there was rarely any opportunity for casual banter to relieve some of the tensions that developed. One I recall was occasioned by a conservationist who criticized clear-cuts that were seen by a kayaker while paddling off the west coast of Vancouver Island. A union member from the forest sector mused, "Ah, but there's nothing like the smell of a clear-cut, first thing in the morning!"

Following a year of meetings, the Vancouver Island CORE table had made no significant progress in negotiations leading to consensus before the process came to an abrupt halt in November 1993, when the forest sector labour group left the table. What then evolved was one proposal by the TWG and three others from the table. These included one from the Forest Independents, and one from each of the conservation and Multi-Sector Coalition group, the latter including mining.

The proposed protected areas in these plans ranged from 12 – 35%. Economic impact studies on these scenarios indicated that the Multi-Sector Coalition Plan would have resulted in a loss of about 1000 jobs (mostly forestry), and close to $10 million in annual government revenues. By contrast, the Conservation Plan would have caused job losses of over 14,000 and more than $100 million in annual government revenues.

The proposal subsequently submitted by Mr. Owen to the government and later approved by Cabinet provided for 13% fully protected and 8.1% in 'regionally significant lands' – the equivalent of Special Management Zones which were to become so abundant in LRMPs and so unacceptable to the mining industry because of the uncertainty they created for would-be investors.

Land Resource Management Plans (LRMPs)

The Commission on Resources and Environment Act became redundant and was officially withdrawn on May 7, 1996 following the development of LRMPs. There were no significant differences in methodology in the PAS process under LRMP tables. Insofar as the mining industry was concerned the entire process was flawed from the outset, because priorities established on land-use were based on surficial features and the subsurface was all but ignored. The insidious nature of the PAS process was not recognized by the general public, nor publicized by the media. The original intent of the process as stated was to withdraw selected lands from multi-resource users to ensure that a total of 12% of the provincial land base would be set aside as protected areas (parks) to include outstanding features, ecological sites and representative ecosystems. This plan, as developed in late 1992 and officially proclaimed in early 1993, designated control of the process to the Land Use Coordinating Office (LUCO) with the year 2000 as the target date for the additional 6.5% of protected area required.

With the proclamation of the Tatshenshini-Alsek Park under the CORE process on June 22, 1993, the objective became unattainable from the outset as 1% of the remaining 6.5% protected-area target was immediately delegated to an already over-represented alpine to sub-alpine ecosystem within the province. Protectionist-oriented LUCO had no problem with this and the well-funded national and international protectionist sector highly supported the subsequent evolution of quasi-protected regions, given special status, as an outgrowth of the earlier buffer-zone ideology initiated in 1972 with the creation of Mount Edziza Park in the Stikine River area.

The history of the development of these buffer-zones is interesting and in the case of Mt. Edziza Park reveals the deviousness practised by the Ministry of Parks in elevating the status of a recreation area to full park protection. Virtually every recreation area created since 1972 was eventually to receive upgrading to park status – the lone exception through 1999 being Mount Edziza Recreation Area (4000 hectares) and Stikine Recreation Area covering 226,265 hectares of the Upper Stikine River above Telegraph Creek. I was to write detailed accounts of mining sector concerns within both of these regions for the Cassiar Iskut-Stikine LRMP table in 1997 and 1998 as the mining sector representative for the B.C. & Yukon Chamber of Mines. However, only Mount Edziza Recreation Area was to survive the LRMP decision in 2000.

The earliest detailed geological mapping of the Mt. Edziza region was completed in the early 1970s by a Geological Survey of Canada party led by Dr. Jack Souther, who was directly responsible for mapping the Edziza Volcanic Complex. The youngest of these volcanic rocks are less than 2000 years old and include 30 cinder cones lying south of Mt. Edziza, a 2,787 metre high shield volcano. Volcanic flows emanating from Mt. Edziza and its volcanic domes and small caldrons formed on a pre-existing forested upland plateau of mineralized older rocks and covered an area of about 130,000 hectares. While mapping, Dr. Souther examined a mineral tenure map of the district and was astonished to see that several of the cinder cones had been claimed by mineral tenures. Upon investigation, he learned that the staking had been completed for the British Columbia Railway, then under construction to Dease Lake. The staking was designed to provide a ready source of ballast for the railway bed. The GSC agreed to support a series of Canada-wide lectures by Dr. Souther proposing a park to cover the recent volcanic formations on Mt. Edziza. Coincidentally, Dr. Souther had an opportunity to examine the Red Dog (Spectrum) property gold veins and he completed several polished section studies of specimens. It was not his intention to include in the park any of the mineralization within near-surface older rocks (pers. comm., 1997). However, the B.C. Ministry of Parks created the Mount Edziza Recreation Area covering 100,770 hectares on the same date (July 27, 1972) as the park proclamation, providing a 1 – 10 km wide buffer zone around the park area. On March 21, 1989, all but 4000 hectares of the recreation area, covering the Spectrum gold property on its margin, was surreptitiously annexed to Mount Edziza Park, nearly doubling its size to 228,700 hectares.

The annexation was reportedly made without any consultation or agreement by the Ministry of Energy, Mines & Petroleum Resources (MEMPR), the forerunner of the Energy and Minerals Division of the current Ministry of Employment and Investment. A report by Paul Wodjak, P.Geo., Regional Geologist, MEMPR, dated September 1993, submitted to the Prince Rupert Interagency Management Committee, entitled Evaluation of Mineral Potential for Mount Edziza Recreation Area states that the Spectrum property "was excluded because Ministry of Parks was reluctant to engage in legal claims for compensation of mineral claims. The Spectrum property comprised the only mineral claims in the Recreation Area at that time." By 1997, when I wrote the Cassiar Iskut-Stikine LRMP table, the Spectrum property had been examined by 12 different organizations since its discovery in 1957, and exploration work valued at $8 million (1997 dollars) had indicated a geological reserve of 242,000 ounces of gold and an excellent potential for expanding the reserve to the 1.0 million ounce level.

Another prominent group of buffer zones around protected areas which I have previously described were those in the Clayoquot Sound area of Vancouver Island, established in April 1993 as Special Management Zones.

The Protected Area Strategy (PAS), announced by the B.C. government on June 10, 1993, a mere 12 days before the Tatshenshini/Alsek Park designation, provided a summary definition of Recreation Areas, Forest Wilderness Areas and Wildlife Management Areas, all covered by existing legislation and discussed under a particularly onerous section, insofar as the future of mining was concerned, on 'Transitional Issues'.

The 'Transitional Issues' section consisting of six paragraphs, included the following, which revealed the bias placed by the environmental-protectionist drafters against mining – historically B.C.'s second most important resource industry for almost 150 years. As discussed, it revealed the abysmal ignorance of the government concerning B.C.'s mineral potential, historic exploration and development history, and factors influencing definition of mineral resources within the subsurface, that are completely alien to the PAS process' surficially defined land-use zones. It reflected the principal flaw in the process which was to lead to the subsequent decline of a vibrant mining industry within a decade, with annual mineral exploration spending in the province reduced to 10% of its pre-NDP government level, a level incapable of sustaining the industry.

"Of the existing protected area designations, only national and provincial parks and ecological reserves fit the new Protected Areas Strategy definition for protected areas. Existing legislation for recreation areas, forest wilderness areas, and wildlife management areas allows various levels of industrial activity such as mineral exploration and development.

Most recreation areas and forest wilderness areas have been designated as a result of extensive public review. There is general agreement that, with very few exceptions, they meet the criteria for protected areas. The resource potentials for these areas have largely been assessed, with the exception of the subsurface mineral and energy resources. The Ministry of Energy, Mines and Petroleum Resources will conduct mineral and energy potential evaluations based on existing resource inventory data, augmented, where appropriate, by Ministry field studies. These evaluations will be completed within the timeframes for the relevant regional or sub-regional land use planning processes.

Each area has a unique history of evaluations and administrative conditions. These special conditions will influence how each area is considered for final designation. The existing recreation areas and forest wilderness areas will be recommended for upgrade to full protected areas status wherever the mineral or energy potential is determined to be low. Where the potential appears to be significant, regional or sub-regional land use planning processes (Commission on Resources and Environment or Land and Resource Management Planning), or area-specific management plans will evaluate the areas and propose recommendations for final protection designations. The

Ministry of Energy, Mines and Petroleum Resources will undertake field assessments in these areas, where necessary, to provide further assistance in resolving areas of conflict."

The mining sector had reluctantly agreed to an increase of over 100% in the protected area target from 5.5% to 12.0% by the year 2000. However, at the time it was unaware that within 7 years an additional 20% of the land base would lie within Special Management Zones, including recreation, forest wilderness, wildlife and many others, all considered to be unacceptable for would-be investors in mining because of uncertainty as to future tenures and access, in addition to the normally accepted high risk of mining ventures.

The first two sentences of the above-quoted second paragraph are misleading. There was no public review for the Tatshenshini-Alsek Park and the Mount Edziza Recreation areas, and many others of less size. The statement concerning studies to be provided by the Ministry of Energy, Mines and Petroleum Resources (MEMPR), is an insult, considering the $300,000 spent by the Geological Survey Branch (GSB) of MEMPR in the 1992 season to show that the Windy Craggy mine was not only a world-class deposit, but lay within a world-class mineral belt. As previously discussed, the entire Tatshenshini-Alsek region was declared a park 12 days after release of the PAS report.

More pertinent, however, is the flawed assumption that any government geological agency would be capable of evaluating the mineral or energy potential of any district to the extent of providing a qualified statement that no significant mineral occurrences could be present. Both the GSB and the Geological Survey of Canada have provided excellent geological base maps and studies of existing mineral camps. However, their task except in Quebec does not include finding and developing mines, which is the occupation of prospectors, geologists, geophysicists and geochemists plus experts in mine evaluation, and their organizations. Only a handful of mineral deposits of the 648 past and current producing mines in B.C. from 1852 to 1998, having a minimum of 10,000 metric tons per year of annual production could be credited to discoveries by government agencies.

The rating of mineral potential provided as 'low' or 'very high', in certain government maps made available at LRMP tables, was generally discredited, based on experience by mining sector personnel. We were all too aware of the major diamond discoveries to date in the NWT after 20 years of unsuccessful reconnaissance work by DeBeers geologists; the uranium discoveries in the Carswell River area of Saskatchewan underlying barren cover rocks; the numerous new discoveries within 'old' mining districts, etc., etc.

Cassiar Iskut-Stikine LRMP

As I have previously written, my representation on the Lower Stikine Management Advisory Committee (LSMAC) had already shown the biased nature of other individual and sectoral representatives on the committee toward protectionism. Accordingly, there was no reason to expect any major change from a larger table dominated by local residents. However, I realized that LSMAC would probably become a dormant entity while the LRMP table was in progress. Based on experience at preceding tables, of which about 20 had been completed or were in progress by early 1997, when the Cassiar Iskut-Stikine table was initiated, a period of 2 – 3 years was anticipated for completion. In addition, Kennecott Canada Exploration Inc., which had so generously borne the cost of my representation on LSMAC, indicated a willingness to continue direct financial support for my representation on the LRMP, by bearing all related travel expenses for the table meetings.

At the time, I strongly believed that a continuing presence by the mining sector at these tables was essential – envisaging a continuing educational role emphasizing mining's importance to the B.C. society and its economy. This has been demonstrated historically, particularly within the sparsely populated northern regions of the province with their dependence on natural resources. I also realized that my background as an explorationist had included by far the greater amount of my time in the Stikine River district, with my 1959 – 1960 direction of reconnaissance-based surveys over northwestern B.C. from north of Atlin to the mouth of the Nass River in the south. In addition, my 1960 – 1965 management of the Galore Creek project, my major exploration and development project, which I have described elsewhere, which remained a significant potential asset for the region and the province requiring continuing certainty of access, which I believed I could encourage.

The Cassiar Iskut-Stikine LRMP covered about 51,000 square kilometres of northwestern B.C. and was defined by the Stikine River drainage system and all its tributaries, its western boundary abutting the international boundary with Alaska. Also added to the LRMP was the headwaters of the Unuk River immediately to the south of Iskut River, a highly mineralized southeasterly extension of the mineralized belt lying on the east margin of the Coast Crystalline Belt throughout western B.C.

The region covers about 5% of the B.C. land base and in the late 1990s contained about 1000 residents – unchanged from the early 1950s, or 0.025% of the B.C. population, mostly in three separate communities: Dease Lake, Telegraph Creek and Iskut, dominated by Tahltan aboriginals. At the time, the Tahltans had not entered into treaty negotiations with the federal and provincial governments; however, the LRMP region included the major portion of their traditional territories. The Tahltans were also the first aboriginal group in B.C. to elect to be represented at the LRMP table, on the understanding that such attendance would not prejudice any future land claim negotiations.

When the table negotiations commenced in February 1997, the Cassiar Iskut-Stikine LRMP already contained two existing mines (Snip and Eskay Creek gold-silver properties), three developed prospects (Bronson Slope copper, Red Chris copper-gold, and Isk wollastonite), all in the Environmental Assessment Process (EAP) review with the object of proceeding to the mining stage. In addition, there were 32 other developed prospects and 437 known mineral showings.

In 1999 I wrote an article which was published in the September 20-26 issue of The Northern Miner[1] emphasizing the historic dominance of B.C. in respect to mineral potential and world-class mineral deposits, in the North American Cordillera, compared with its neighbours: Alaska, Yukon and Washington State. The total value in 1998 dollars of B.C.'s historic mineral production was $156 billion. In order to emphasize the extraordinary value of the known mineral potential in the Cassiar Iskut-Stikine LRMP, I calculated the gross mineral value of the top ten known developed prospects at almost $70 billion, again in 1998 dollars.

Prior to initiation of the LRMP in 1997, the region's land base was already 18% protected through Mount Edziza and Spatzizi Plateau Wilderness Provincial Parks, plus a 4.4% Recreational Area covering Stikine River north of Telegraph Creek. Historically, 13 previous provincial Recreation Areas had all become parks. As such the Cassiar Iskut-Stikine LRMP at its outset was already one of the highest protected of the 30 planned

---

[1] NDP blamed for poor B.C. image, The Northern Miner, Sept. 20-26, 1999

LRMPs in the province – this in spite of probably having the greatest known, but not necessarily recognized, mineral potential per unit area of any of the LRMPs.

Almost all of the table meetings were held at either Dease Lake, Iskut or Telegraph Creek. The 2 – 3 day meetings were all held at the end of the week, usually on Friday and Saturday, in order to balance the need to take some time away from work with the need for some time off. Some meetings were also scheduled on Thursdays and with earlier meetings continuing until 9:00 pm Thursday evenings. In order to arrive at meeting destinations on time from Vancouver, I arranged to take an early morning flight to Smithers, arriving there about 1-½ hours later at 9:00 – 10:00 am. In the early stages of the LRMP, I rented a car at the airport and drove the 600 – 750 kilometres to Iskut or most distant Telegraph Creek, with Dease Lake in between. The return trip generally brought me back to Smithers, late on Saturday for an overnight stay generally at the Hudson Bay Lodge, followed by a flight back to Vancouver around noon Sunday.

The first Cassiar Iskut-Stikine LRMP meeting was held on February 27 – March 1, 1997 at the Northern Lights Community College at Dease Lake. In attendance were 58 people from the public and 13 government members, including Tom Soehl, LRMP Coordinator of the Inter-Agency Management Committee and Stuart Gale, LRMP Facilitator. As the first meeting, it attracted a record number of participants, mostly local residents. Within a year, it was to drop to about 35 – 40 in total, and within two years to 20 – 21 public and about 8 – 9 government participants, a level which it was to maintain for a final 18 months. Of the non-resident public participants at the first meeting, only five were to remain after two years: Vern Betts of Homestake Canada Inc. and myself representing the mining sector; Rosemary Fox, a former Sierra Club vice-president from Smithers; Glenda Ferris, a self-declared environmental activist from Houston, B.C., who was to act as principal spokesperson for the Tahltan First Nation, and Geoff Phillips of Terrace, B.C. All but two of the government people were non-residents, the exceptions being Gerrit Apperloo, Dease Lake manager of the Ministry of Transportation and Highways, and Kathy Bisset with the Ministry of Forests in Dease Lake.

Comments made publicly at the first meeting by several participants indicated an unwelcome attitude to the presence of non-residents. Principal among these were Bill Sampson, a fisher/trapper and UBC biology graduate who lived with his family on the east bank of the Stikine River, south of Glenora. I have mentioned him previously in regard to my experience on the Lower Stikine Management Advisory Committee. Bill attempted to have non-residents banned from membership at the LRMP, specifically myself as a so-called "Professional lobbyist for mining" and other industry representatives. When this was not supported, he suggested at least participation by non-residents should be based on a six-month trial period. I flatly rejected the proposal, which in any event should never have occurred, as the Coordinator should have quashed the topic when it first arose, based on LRMP guidelines which provided for participation by any tenure holders. The Tahltans also indicated their opposition to any restrictions on participants.

Participation at this and other LRMP meetings in Telegraph Creek and Iskut – both Tahltan villages, were subjected to a 'welcoming' statement for over a year (8 meetings) from a Tahltan Band Council representative. The first of these – the Dease Lake welcoming address, was provided by Chief Yvonne Tashoots on behalf of the Tahltan Joint Councils. The statement reviewed the Tahltan's history within 'Tahltan Territory' or the "Traditional Territories of the Tahltan Nation" and provided no doubt that all other participants at the LRMP table were considered their guests in that "over the generations of time, Tahltans have fought, given their blood, negotiated agreements, provided safe access and traded with outsiders within our traditional territory". I felt most uncomfortable, as presumably other 'guests' must have, in being addressed in this fashion. An identical statement could be made by most Canadians about their country of which the Tahltans are but one of the many First Nations. Canada has 630 native bands.

From the outset of the LRMP table, I recognized the tremendous imbalance in its make-up, in terms of 'conservation' and 'resource industry' participation. As time would tell, the entire land-use process was biased toward conservation and protection, mostly at the cost of the two principal resource users – mining and forestry. Within four months there were no forestry industry participants remaining at the table, following moratoriums imposed on certain key logging areas. By contrast, mining industry representation was eventually to grow to an unprecedented four private sector individuals – myself representing initially 140 mining tenure holders. Included were my old DuPont friend John Kowalchuck, then manager of Kenrich Mining Corporation and his alternate Raul Versosa; Bart Jaworski, senior geologist with Whitegold Resource Corp., and Vern Betts and his alternate Marlin Murphy representing Homestake Canada, which owned the Eskay Creek and Snip mining operations in the LRMP. Both Kenrich Mining and Whitegold Resource had important developed prospects in the Iskut and Unuk River watersheds which they wished to represent personally. Unfortunately their participation was to terminate by mid-1998 through lack of corporate funding as the B.C. mining industry continued to suffer from its inability to attract exploration and mining development interest because of the lack of investor confidence in B.C. under the NDP regime.

I was a member of the Steering Committee from its outset until the B.C. & Yukon Chamber of Mines officially terminated its representation at LRMP tables in January 1999. I also sat on a sub-committee on community-based initiatives. I could only comment that there was little that could be done to foster interests in mineral exploration and mine development that was not already reflected by mineral tenure holders within the area. However, this community-based initiative reflected a major concern of many residents who have had to rely largely on seasonal, or even non-resident employment for income, a common situation throughout the sparsely populated northern regions of the province.

As I have previously discussed, the LRMP process as developed at the table was virtually identical to its forerunner – the CORE process. Included were presentations and discussions on the principles of consensus-based planning; boundary definitions of the LRMP and its multitude of sub-divisions for reaching decisions on two major scenario developments (conservation and development-oriented) beginning after the first year of the process and considered the essential requirements for reaching consensus. All of these developments were driven by input from the Technical Working Team (TWT) consisting of government employees. Definition of Ground Rules, Terms of Reference, and Scheduling of meetings received more input from the table members. Many presentations by the TWT and others covered past practices, required negotiating skills and development of socio-economic indicators by consultants for the LRMP; and field trips within the LRMP to provide a personal exposure to selected features. This lengthy preliminary phase of the LRMP process was considered essential to adequately understand the complexity of many potentially conflicting issues and most importantly, for so-called 'bonding' purposes by table participants.

My previous experience with the Vancouver Island CORE process; the government's deviously implemented Tatshenshini-Alsek Park decision and my participation on the Lower Stikine Management Advisory Committee had left me with a most jaundiced outlook concerning my involvement as a mining sector representative with the Cassiar Iskut-Stikine LRMP. The best I could hope for from the outset was to provide the table with reliable information on my experience as an exploration geologist in wilderness areas. This included the importance of mining to society and the miniscule level of negative impacts on the environment per hectare disturbed, versus the benefits – the highest of all resource users.

Although I travelled alone to a few of the LRMP meetings, I was most fortunate in the generous support provided by Homestake Canada in driving from Smithers to the LRMP meeting destinations and return. My principal companion and driver was Vern Betts, with only a couple of trips in the company of Marlin Murphy. Vern and I would generally take the same flight to Smithers, where he would spend a little time at Homestake's office at the Smithers airport. Our 6 – 8 hour drives would be broken up by coffee and lunch breaks with a minimum of time involved in acquiring the necessary commodities and getting back to driving. Nature calls were the only other regular diversions. Traffic was never heavy and we eventually knew every curve on the road. Mercifully, we generally found plenty to discuss, besides the LRMP process. I remember on one of my final trips musing about the combined miles flown and driven from Vancouver to our meeting destinations and return. "Vern", I remarked, "Do you realize that by now we have travelled over 40,000 kilometres, the equivalent of around the Earth to these meetings?" He was as surprised as I was.

Considerable publicity has been given to the beneficial aspects of consensus-based negotiations. However, as I and others have previously described, the LRMP process as applied to subsurface resources is so obviously flawed, that it is difficult to believe that the promoters of the process, albeit protectionist-oriented, are either incapable or unwilling of conceding its inapplicability.

The August 2000 issue of Cloudburst, a newsletter of the Federation of Mountain Clubs of B.C., contains a copy of a letter written by Brian Wood, Co-chair of the Recreation and Conservation Committee, to Tom Soehl, the Cassiar Iskut-Stikine LRMP Coordinator, complimenting the LRMP table on reaching consensus (in May 2000). Over 150 letters were written by B.C. mining sector individuals condemning it. Mr. Wood's letter conveys the typical misunderstanding of mining's impacts on the environment that is regrettably shared by much of the well-educated public. He describes the "memorable extended trips, usually mountaineering and/or backcountry ski trips" by his members and his extensive "canoeing, rafting, mountaineering and a month-long ski traverse north from the Great Glacier." There is not one word of any negative impact from mineral exploration/mining on his own and his members' enjoyable trips. He then indicates his members' shock and dismay to learn of the "recent objections to this plan by the mining and minerals sector", noting that "clearly, as this was a consensus process, every party to the process had to give up many areas or interests that were important to them, and consequently we feel that there should be no disturbance of this agreement."

Contingent on the government's final reaction to the recommendations provided by the LRMP and the many written objections from the mining sector, Mr. Wood would be hard pressed to name one table representative other than the mining sector before its withdrawal, who gave up anything. Consensus is supposed to be a 'win-win' resolution, not one that misleadingly provides socio-economic gains for only one principal resource user (recreation/conservation) so unjustifiably and hypocritically, to the deprivation of another. Within the Cassiar Iskut-Stikine LRMP area, in which I had been able to demonstrate its highest mineral potential per unit area of any of the LRMPs and CORE tables in the province, there was already 18% of the land base protected in two major parks plus 4.4% in a recreation area, before the LRMP was initiated, one of the highest protected to land-base ratios for any LRMP in the province. At so-called consensus, with the inclusion of uncertainty to the mining sector provided by Special Management Zones

(SMZs), recommended at 22% of the land base, plus 26% in protected areas, only 52% of the land base could be considered of potential interest to mineral investors. Obviously a significant 'win-win' resolution for the recreation and conservation sectors, at the expense of other resource users – notably mining.

The Tahltan presence at the LRMP had a marked, but unpredictable influence on the table and its decisions. From the outset, the Tahltans made it clear that any recommendations from the LRMP would not prejudice any of their future land claims. With a First Nation population comprising about 60% of the total within the LRMP, the representatives emphasized that "any effort by the provincial government and special interest groups to recommend land use designations for resource development and other uses within our traditional territory – **must be conducted with meaningful and substantive, Tahltan participation in the planning recommendations and decision-making process**" – (emphasis mine).

Not generally recognized by some 'outside' table members are characteristics commonly shared in the aboriginal community. These include a deeply-rooted ancestral reverence and respect for their elders; patience in achieving their aims; and a surprising sense of humour, which however may contrast with a sensitivity to criticism, whether intentional or assumed. Several of these characteristics were to be demonstrated at the table, one of which was particularly memorable.

I have previously mentioned the presence of representatives of Friends of the Stikine at both the Lower Stikine Management Advisory Committee and the Cassiar Iskut-Stikine LRMP. Prominent at the LRMP table were chair Peter Rowlands, Maggie Paquet, and directors Bill Sampson and Dan Pakula. The January 22 --23 LRMP table meeting at Dease Lake was unusual in the presence of about 8 Tahltan band members headed by Chief Louis Louie of Iskut, rather than a single spokesperson. It was obvious from the outset that this was no ordinary welcome. The welcoming committee announced that in December, Friends of the Stikine published a newsletter which included an article on the LRMP process by Peter Rowlands. Apparently various table members had expressed concern that the inaccurate and disrespectful nature of the article constituted a significant breach of the LRMP Ground Rules. The Tahltan Joint Councils, in particular, were deeply offended by the article, and at the beginning of the meeting both chiefs and a majority of councillors from both band councils read from a prepared statement. It expressed their sentiments and demanded to the table that Friends of the Stikine representation at the table be terminated.

Although not reflected in the minutes of the meeting, it appeared that the most offensive comment in the article was a reference to the Tahltans as the "Cowboys of the North". Although Peter Rowlands apologized somewhat tearfully to the Tahltan First Nation and the table, it was agreed that membership of the Friends of the Stikine would be terminated, based on a serious breach of the Ground Rules. The termination applied only to Peter Rowlands and Maggie Paquet as official representatives of Friends of the Stikine at the table. The whole process struck me as an over-reaction, as I could not believe that Peter would deliberately write anything intentionally disrespectful of the Tahltans, and in spite of my past differences with this environmental activist, I spoke in his favour at the table.

Several months later at one of our table meetings, Norman Day, a Tahltan LRMP community representative and Band Councillor at Telegraph Creek, sidled up to me during a coffee break. "Dave", he said, "I have just got back from a wedding of one of my nieces. I took a good photo of her beside a bunch of the guys all wearing cowboy hats." He chuckled and then said "I was thinking of sending a copy to Peter Rowlands!" I guess I gasped, but said the obvious, "Norm, I don't think that would be a good idea", fully aware that he wasn't seriously considering the possibility. In November 1998, Friends of the Stikine were reinstated as table members. I wasn't so much surprised as disgusted at this turnaround, which so reflected the memorable Shakespearean phrase "Full of sound and fury – signifying nothing." As will be further described, a similar reaction from the table to the Friends of the Stikine's obviously unintended 'jibe', was eventually to be levelled against the B.C. & Yukon Chamber of Mines.

Mitch Cunningham and his wife Jacquie joined the LRMP table following their arrival from Alberta to run the Red Goat Lodge near Iskut in mid-1997. One of their innovative developments was the raising of llamas which were partly tethered for grazing purposes on the margin of the highway, where they were literally a traffic-stopper for curious tourists and others. I remember Mitch most vividly as an amiable individual who, like other table members, became quite exasperated at what he considered the pedantic pace of the 'preparatory' stages of the LRMP table. More than once he bellowed out, "when are we going to get to the meat and potatoes?"

This critical part of the LRMP process arrived as the "land use scenario development stage" at the January 23 – 24, 1998 LRMP meeting at Dease Lake in the all too familiar biased guidelines affecting subsurface resources established for all LRMPs by Technical Working Teams. This was an introductory exercise designed to familiarize the table with the subsequent analyses, mostly based on uncertain socio-economic parameters, which would guide final land-use agreements and potential consensus between 'interested' parties. I say uncertain, as in the case of the Cassiar Iskut-Stikine LRMP, vast areas remained virtually unexplored and neither the TWT nor local residents had set foot on most of the area to be arbitrarily zoned by watersheds into sub-areas for consideration based mostly on limited surveys and knowledge.

The first working meeting for scenario development, held on February 27 – 28 at Dease Lake, considered the Klappan River watershed and was followed by four additional meetings to consider the four other selected watersheds comprising the LRMP, ending on September 25 – 26 at Iskut – the location of the first meeting following the 1998 summer recess.

As I have previously noted, those of us within the mining sector continued to emphasize the main flaw of the LRMP process in its treating knowledge concerning subsurface resource potential on the same basis as surface resources which can be assessed visually. The problem is exacerbated by a basic requirement of the LRMP process in rating area-specific land-use interests and issues from very-low to very-high for comparative purposes, using a series of maps and overlays. Examples provided at the table for wildlife abundance were generally considered reliable, even though they were based on habitat suitability as opposed to specific counts. An interesting example of the questionable reliability of even these data was provided in the area containing the Red-Chris copper deposit near the community of Iskut. A nearby area was mapped as having a high caribou ranking, even though a knowledgeable resident pointed out that there were none there as all the caribou in the area had been killed by hunters. Ultimately even this factual information had no effect on specific wildlife ranking.

Subsurface resource evaluations at any time are based on the previous result of mineral exploration within areas, which in itself is dependent on numerous factors, including accessibility, exploration activity, extent of post-mineral cover which can range up to 100%, and extent and applicability of exploration using geophysical and/or geochemical surveys to search for hidden resources. Obviously, a favourable and reliable political and land-tenure climate for carrying out mineral exploration activities, with the high costs and risks involved, are of paramount concern. Maps and overlays developed by the B.C. Ministry of Mines followed the prescribed rating requirement, but industry representatives, including myself, pointed out their unreliability based on historic experience. In a subsequent widely circulated reference, I prepared in March 2000 on the Growth and Recent Decline of the B.C. Mineral Industry – Recommended Remedies I noted the following:

Unlike other resources considered in land-use planning, mineral potential must be evaluated increasingly in three dimensions. New mineral discoveries will be at increasing depth and will become more dependent on highly sophisticated and integrated geological, geochemical and geophysical survey and analytical methods, followed by advanced drilling and mining and metallurgical techniques that will lead to the development of future mines in both old mining districts and yet unrecognized ones.

Significant mineral discoveries have been made in the past with modern concepts and technology in areas considered "unfavourable", e.g. the emerging Canadian diamond mining industry currently centered in the Northwest Territories; the uranium discoveries below sandstone formations in the Athabasca Basin of Saskatchewan, and many other mineral discoveries at both local and regional scales.

By the end of May, 1998 the Cassiar Iskut-Stikine LRMP table had completed the five proposed Scenario Development exercises in sufficient time for a summary to be prepared in June. Although the table had been advised that the exercises were not binding, with few exceptions they directly influenced the subsequent decisions made by the table and also served to focus on areas with particular conflicts between resource-users, requiring settlement. The resultant data led to the development of two land-use scenario summaries for analysis:

1) <u>Development-oriented</u> providing for 'unfettered' mineral-exploration/mining-related activities in only 8.2% of the land base; 71.8% in SMZs and 20.0% fully-protected in parks, and 2) <u>Conservation-oriented</u>, with only 1.5% available for 'unfettered' mineral-exploration/mining; 69.1% in SMZs and 30.4% fully-protected in parks.

The results of this exercise, with its implanted ranges for table consideration, were a disaster for the mining sector, and a far cry from the more than 90% of lands open for exploration and mine development within B.C. prior to the NDP's drastic and irresponsible experimentation in land-use planning.

Following receipt of the Scenario Development summary with its map and tabulated data, I forwarded copies on June 29, 1998 to 18 individuals representing interested mining companies with mineral tenures within the LRMP and also to Mary Lou Mallot and Graeme McLaren of BCMEM. These were the same individuals that I had been providing with regular reports following each LRMP meeting. I requested that any specific concerns should be directed to Tom Soehl and myself. Only Dave Yeager of International Skyline Inc. responded; although a meeting including several key mining industry and MEM individuals was held in Vancouver in June to discuss the results and our future strategy.

Several events, both during the summer break for the LRMP and at fall meetings, were to have a major impact on the mining sector's involvement in the LRMP process. Included were the following:

(1) The results of the two Cassiar Iskut-Stikine LRMP scenario summaries were so devastating to the mining sector that it led to a letter dated September 17, 1998 from Lindsay Bottomer, president of the B.C. & Yukon Chamber of Mines, being forwarded to the Honourable Dan Miller, Minister of Energy and Mines and Minister Responsible for Northern Development. It bluntly summarized the many previous criticisms from the mining sector to the government concerning the flawed nature of the land-use process in respect to subsurface resources. It focused on the experience at the Cassiar Iskut-Stikine LRMP, coming within one year of the highly advertised

establishment of a wilderness/protected area larger than Nova Scotia or Switzerland in the Muskwa – Kechika region of northeastern B.C. with its relatively unexplored base-metal, oil and gas resources, and both with a high potential for future development.

Copies of Lindsay Bottomer's letter were made available to the Cassiar Iskut-Stikine table, even though the government later reprimanded the Land-Use Coordinating Office (LUCO) for forwarding the letter to the table, as it was considered a private letter. Considerable time was subsequently taken throughout the fall meetings by the table and its procedural sub-committee, which concluded that the letter breached the Ground Rules and the Code of Conduct of the Cassiar Iskut-Stikine LRMP, and in preparing a response to Lindsay's letter at the September meeting, which however was not officially delivered until mid-October.

(2) Dan Miller responded to the Chamber's letter, by arranging a meeting with Chamber representatives on October 14 which covered a number of mining industry concerns, apart from the LRMP process. As a further result, a message on government policy concerning mining was delivered at the Cassiar Iskut-Stikine table meeting at Telegraph Creek on October 24 – 25. The effect of this message was only to anger the dormant protectionist sector of the table, which regarded it as an unjustified intrusion in the table process.

(3) At the September 25 – 26, 1998 table meeting at Iskut, during discussion of scenarios, I commented that conflicts existed between terminology for the proposed Large River Corridor zones and the enacted Mineral Exploration (MX) Code in respect particularly to access. Several members disagreed, Glenda Ferris stating that the MX Code had been considered.

On my return to Vancouver following the meeting, I enlisted the aid of Gavin Dirom, P.Eng., a member of the MX Code development committee over a two-year period. The Cassiar Iskut-Stikine Large River Corridor Zones drafts were subjected to a critical review, requiring nine pages, which revealed an overwhelming bias throughout the draft against mineral exploration and mine development activities, not reflected in existing acts and codes pertaining to the mining industry and other industrial activities in the province. I forwarded our comments under cover of a memorandum dated November 4, 1998 to the table and they were included in the data available for the November table meeting.

Jim Bourquin, one of the leading protagonists for the establishment of Large River Corridors as Special Management Zones which would permit rafting activities, but restrict road access essential to future industrial activities, attacked our critique at the November meeting which he claimed "had ruined a year's work".

(4) From the start of the CORE process and subsequent LRMPs, the mining sector repeatedly indicated the inherent problem of Special Management Zones which had increasingly been a source of confusion and uncertainty in the eyes of potential mineral investors because of priorities potentially assigned to their management which conflicted with existing legislation. The above provides an excellent example.

In 1998 the government created a Special Management Zone Committee with the object of ensuring that resource development was approved efficiently in these areas, in view of the virtual absence of mineral-related activities in SMZs, which occupied about 20% of the land base in completed LRMPs. The Chamber had reluctantly agreed to participate on the committee, but had become increasingly disenchanted with its progress by late 1998.

(5) In addition, the deteriorating business climate in B.C. throughout the 1990s which had so affected the resource sector because of NDP policies, led to a

reaction covering both large and small businesses in the form of a B.C. Business Summit Meeting held in Vancouver on November 8 – 9, 1998. Although well attended and highly publicized, recommendations made to the government went virtually unheeded. Not one of the principal remedial proposals made on behalf of the mineral sector had been implemented by late 2000. At that time an opinion poll found that 67 per cent of decided voters supported Gordon Campbell's B.C. Liberals, while only 17 per cent supported the NDP.

(6) Finally, although I was aware that Lindsay Bottomer had decided that no response to the table's letter of condemnation to his own letter of September 14 was warranted, I kept being made aware from my table affiliation that a response was expected with an apology – which of course I did not support. He finally agreed to provide a single paragraph letter which indicated the Chamber's right to communicate with the government in a private letter. I carted this with me for delivery at the November 27 – 28th LRMP meeting at Dease Lake. In addition, in order to clarify any rumours about the mining sector's involvement at the B.C. Summit Meeting, I included a copy of the MABC press release of November 8, 1998 on 'Mining Calls for Moratorium on Land-Use Planning'.

I once again drove north from Smithers with Vern Betts, realizing that this was going to be a difficult meeting. On arrival on the morning of November 27, I was virtually met at the door by Stuart Gale, Facilitator for the LRMP who asked, "Have you got a response from the Chamber?" I said, "Yes, but you're not going to like it." He read it, and then said, "Can you soften it?" I replied, "I doubt it, but I'll try."

Soon enough, with copies of the letter for all table members, the reaction set in. Among the earliest, that provided by Eric Havard of Eagle River Guide Outfitting was probably the most memorable. With a grandiose gesture, he crumpled the Chamber's response, throwing it to the floor and said, "Here's what I think of the Chamber's reply."

Other equally aggravated responses followed, several of which were quite humiliating and insulting. Of all the table members present – about 35, only one – Steve Quigley, spoke out in favour of the need for a mining presence at the LRMP. Although I was mortified by the reaction of the table members, after two years of sharing in the process and explaining the beneficial aspects of our mining activities to the area and the province, I was quite prepared to continue to the end of the meeting. It was with some surprise that I noticed that Vern Betts had arisen from across the table and approached me saying, "Let's go Dave- I've had it." I didn't argue – it was an unexpected early departure – which I welcomed.

On our long drive back to Smithers, we discussed the meeting as we cooled off. At the time, I didn't realize that I had attended my last LRMP meeting. However, the results of the meeting, added to our overall frustration with the lack of support for the mineral industry's recommendations calling for remedial measures for its survival as a significant resource in B.C., eventually supported a move to abandon participation at the LRMP process. Even though I initially argued in favour of participation, I became convinced that continued efforts within the process were counter-productive.

My final report dated January 6, 1999 to the Chamber and regularly copied individuals included a copy of the official summary of the table meeting, as in the past. In addition I included a copy of my memorandum of November 30, 1998 to Shari Gardiner, Chair of the Chamber's Land-Use Committee, which was also copied to other Chamber representatives.

Included in my memorandum to Shari were some of the more vilifying comments made by table members following the report tabled by the Procedure Committee on the Chamber's letter of September 14 from Lindsay Bottomer:

We were "hypocritical" by electing to maintain our position at the Table, while publicly proposing a moratorium on the process;

We were accused of having deliberately attempted to gain our 'objectives' by 'holding the Table to ransom' without any example provided to support such a statement; and

We were described as "arrogant" and "whining all the time".

I also noted that "undoubtedly, our 9-page critique of the River Corridors SMZ proposal has caused a pretty severe backlash as its proponent, Jim Bourquin, stated that it had 'destroyed a year of work by the Committee'. Let's hope that they are beginning to recognize their bias and the problems they are creating with such proposals – although I doubt it."

The B.C. mining industry officially withdrew from the LRMP process on January 22, 1999. The release provided to the news media described the principal problems to the mining industry with the flawed process. Included were the failure to live up to the 12% target for protected areas, which had already been exceeded at 18% in Parks and Park Study Areas. In addition, the Chamber noted that a further 20% of the land base had been blanketed with 225 Special Management Zones (SMZs) and that no new exploration money had been spent in any of the SMZs and "that it is impossible to raise money to explore when there is no certainty that subsequent development would be permitted."

Gary Livingstone, President and CEO of the MABC also observed in the news release that although the government had provided assurance that 12% was the limit for park creation, the government now had no intention of stopping at 12% and that "the economic consequences of going further will be significant." He also noted that since 1990, 14 operating mines had closed and only 7 opened. I later observed in articles in the media that none of the new mines had resulted from discoveries since 1990, nor were any obvious economic discoveries made in B.C. since the NDP took office in November 1991. Gary also noted our most frequently stated solution to the mining industry's dilemma:

"There is an easy solution to all of this," says Mr. Livingstone. "First the government should live up to their original commitment to stay at 12% parkland. Second, they should go back to the original version of Bill 12, The Mining Rights Amendment Act, and return ultimate authority to it. This would effectively assure mining access to all areas of the province outside of parks."

As my parting 'contribution' to the Cassiar Iskut-Stikine LRMP Table, I prepared a 14-page summary accompanied by six figures, entitled *Mining Revisits B.C.'s Land Use Plans*. Sufficient copies were made for the table members and were available for the table meeting in February 1999. The paper briefly traced the evolution of mining as an essential contributor to our way of life since the Stone Age, and particularly its influence with forestry on B.C.'s economic development through the 1980s. It traced the negative impact of the NDP government's land-use policies, focused almost totally on protectionism since its election in November 1991. In this nine-year period, growing uncertainty as to access and mineral tenure rights caused a drop of mineral exploration expenditures from $263 million in 1990 to about $20 million in 1999, with no known economic mineral discoveries. The paper then traced the development of the land-use policies, initiated with the disastrous Tatshenshini-Alsek Park decision at the cost of the world-class Windy Craggy mine development. The inequities of the land-use process as reflected by the CORE and LRMP process and the related development of SMZs preceded a final summary of essential requirements for a sustainable mining industry and proposed remedial measures immediately required.

As had been my experience during my two years at the Cassiar Iskut-Stikine LRMP Table, I received very few comments back from table members on the written submissions that I made on a wide range of LRMP mining-related issues. This was not surprising, based on the protectionist-oriented bias of the majority of members, in addition to the emphasis by others mostly concerned with the priority search for community-based initiatives. Oral presentations made by Dani Alldrick and Dave Lefebure – geologists with the Ministry of Mines, generally drew negative verbal responses, similar to those expressed for any kind of government-imposed industry presentations which were generally considered as unwelcome and interfering. Considering our often repeated statement about the benefit provided to the LRMP area and its residents by site-specific mining activities, and unobtrusive exploration programs in the past, this was frustrating and most disappointing.

In order to emphasize the latter, at one of our last meetings held at Iskut, I took an old Kennco Explorations (Western) Limited map with me and taped it to the wall near one of the large air-photo displays of the entire LRMP area which I had obtained through MEM. The map showed the locations of stream sediment samples collected at over 950 sites on creeks and rivers in the heavily glaciated portion of Stikine River drainage lying between Telegraph Creek and Porcupine River, covering an area of 8000 sq. kilometres. I explained briefly to the table that the program led to the discovery of the Galore Creek copper deposit and had been carried out over an area of 50,000 square kilometres in 1959 – 1960 by a crew of 20 men supported by helicopter and occupying over 170 two-man campsites in 1959 alone. In that period, apart from our stay in Atlin and at a base camp in Telegraph Creek, we had contact with only five other individuals over a four-month period – providing an indication of the remoteness of the area. It is doubtful, if summer traffic within the area would be any different today, apart from the fishing and recreational boating activity on Stikine River.

With the Cassiar Iskut-Stikine LRMP Table having provided a Consensus Recommendations Package for public review in May 2000, I joined about 300 others from the mineral industry in providing written comments to Tom Soehl, the Process Coordinator, in a two-page letter dated May 25, 2000.

Dear Tom:

Re: Public Review of Consensus Recommendations for Cassiar Iskut-Stikine LRMP

As you are aware, I was the mining representative for the B.C. & Yukon Chamber of Mines (Chamber) at the Cassiar Iskut-Stikine (C I-S) LRMP, from its first table meeting on February 27 – March 1, 1997 at Dease Lake until the Chamber's decision to withdraw from the LRMP process January 22, 1999.

My relevant career in the Canadian mining industry has included the following:

(1) Mineral exploration in every Canadian province and territory, except PEI, over a 50-year period commencing in 1946, most of which has been in B.C.

(2) My interest in the Stikine River area dates from the late 1950s, when I acted as project manager in a two-year, 20-man exploration program covering 50,000 square kilometres in northwestern B.C. This led to my co-discovery of the Galore Creek (Stikine Copper) copper-gold deposit in the Scud River area in 1960, where I provided on-site exploration management of major programs from 1961-5. Overall project costs to date exceed $50 million and the property is currently ranked as the highest-grade, undeveloped porphyry copper deposit in B.C.

(3)   As Vice-President Exploration for DuPont of Canada Exploration Ltd. (1974-84), I was responsible for promoting the production decision at the Baker gold-silver mine, which produced from 1981-3 in the Toodoggone River region, adjacent to the Cassiar C I-S LRMP area.

(4)   My long-term interest in the B.C. mining industry led to recognition of the outstanding mineral potential of the C I-S LRMP region, described at the LRMP table, but obviously never accepted by the Table, and most recently, in my widely circulated paper, forwarded to the B.C. government in March, 2000[1], a copy of which is being forwarded under separate cover.

I have reviewed and commented for the Chamber, as have others, on the C I-S Consensus Recommendations Package (C I-S document), and take this opportunity to express my continued strong rejection of a flawed process which has never adequately recognized the value of subsurface resources to the people of British Columbia. It is particularly deplorable that a group of about 20 non-government C I-S table members, charged with the responsibility of recommending a fair and balanced plan for sharing our resources, should continue to ignore the potential benefits of mining to society within one of the most highly mineralized regions of B.C. This is particularly frustrating within an area only accessible by land via a major mining road 'the Cassiar Stewart-Highway', on which their very livelihood depends. Most contemptible has been the government's claim to having reached consensus at the Table, after directing its own table representatives, who are generally considered to represent the 'voice of the people', to abstain from voting if they were to break consensus.

The C I-S document reveals the same bias against mining activities that was evident in the recommendations proposed at the table by the "Large River Corridors Committee" and which I noted in a detailed critique prepared for the Table with the assistance of Mr. Gavin Dirom. Throughout the C I-S document, mineral exploration and mining are grudgingly described as "permitted" and "accepted activities", which provides no certainty to potential investors who must raise the huge capital commitments to discover and develop a mine.

The C I-S document is also filled with presumptuous and arrogant statements, including 'preferred' access routes for mining, e.g. a restrictive demand by unknowledgeable drafters of potential future access for the Galore Creek deposit, on which prior studies had considered about a dozen potential routes. A more negative access restriction can be seen in the blatant proposal of a protected area, without access, in the Unuk River valley immediately upstream from the U.S. border.

The uncertainty currently provided by the LRMP process to mineral tenures or mining proposals within Wildlife Management Zones, has been created by years of fallacious information from preservationists on claimed impacts on wildlife habitat by mining. This was particularly evident to me at the C I-S LRMP Table, and you will recall my announced intention of providing the Table with alternative evidence. Copies of my brochure on "Coexistence of Wildlife and Mining-'related Infrastructure", co-sponsored by many mining companies, the Chamber and the Mining Association of British Columbia, were provided to all Table members. Not a single adverse comment or written objection

---

[1] "Growth and Recent Decline of the B.C. Mineral Industry – Recommended Remedies" by David A. Barr, P.Eng., March, 2000 (i.e. 10 top developed mineral projects in the C I-S LRMP have a potential resource value of about $70 billion, compared to the total historic mineral production value of B.C.'s mines, at $156 billion in 1998 dollars).

has been received from the Table members or others, and yet this uncertainty concerning mining is permitted to continue.

I devoted two years in which I attended 14 Table and one Steering Committee meetings at the C I-S Table with the object of trying to inform the Table members of the benefits of mineral exploration and mining to the district and B.C. As in my experience at the Vancouver Island CORE Table, our presence was made redundant by a flawed and biased process which, to date, has failed to recognize the impracticality of adequately assessing mineral potential.

As history has shown, many mineral deposits are discovered, increasingly few can be predicted, and fewer still make mines. Those that are successfully developed have by far the highest value to society per resource area temporarily disturbed than any other. As a pertinent example, the Eskay Creek Mine, which occupies about 22 hectares with its overall infrastructure, has produced annual revenues of about $6 million per hectare, with significant employment and benefits provided to C I-S LRMP resident communities. It took over 50 years by many individuals and companies to discover the mine.

When, in a recent poll, 96% of 1200 widely distributed B.C. residents indicated support of the B.C. mining industry, how can residents who have been so dependent on a non-intrusive resource be so opposed to encouraging this temporary use of less than 0.1% of the land base?

Although it would appear improbable that the C I-S LRMP Table is capable of recognizing the flawed nature of the C I-S document, I share in others' recommendations that the government undertake an immediate halt to any further alienation of lands within the C I-S region and throughout B.C., pending completion of a thorough audit, cost/benefit analysis and environmental review of the process. It should be completed by an independent group of professionals totally unconnected with the process.

Yours truly,

David A. Barr

Enclosure

cc: See Page 3 (25 senior B.C. government including 3 mining and business representatives)

My experience at the Cassiar Iskut-Stikine LRMP made me realize at an early stage the need to provide the Table members with written references concerning key mining-related issues, rather than relying on verbal comments, which all too often were never recorded in minutes of the meetings. I also developed the practice of providing principal mineral tenure holders in the C I-S LRMP area, in addition to others, with information which had developed at each meeting on mining-related issues. During our 21-month participation at Table meetings, at least 14 such memoranda were issued, increasing to 15 tenure holders at our final meeting in November 1998, from 4 tenure holders in February 1997. Incidentally, there were about 150 mineral tenure holders at the start of the C I-S LRMP and this had dropped to about 80 by May 2000.

In an accompanying memorandum dated September 19, 2000 directed to several key individuals for possible future reference, where back-up to our industry involvement at the C I-S LRMP Table might be required, I summarized eleven references, most of which were available directly to Table members, or in correspondence folios at Table meetings.

(1) Mineral Resources/Infrastructures Map

Anticipating the need for maps of a convenient scale for reference by Table members of both the C I-S, Mackenzie and Lillooet LRMPs, also the Northern Regional Advisory Committee (NRAC), depicting major access roads, protected areas and known mineral deposits/mines, I enlisted the generous aid of Teck Exploration Ltd. for this purpose. The C I-S map was available in March 1997, just after the start of the Table meetings and was also distributed to the NRAC Table at that time.

(2)   Mine Visit

At the initial LRMP meeting at Dease Lake, Feb. 27 – Mar. 1, 1997, I proposed that a field trip (and possibly a meeting) be scheduled in the fall for all Table participants at the Eskay Creek minesite to familiarize them with an actual mine. This was supported by Vernon Betts of Homestake Canada and by Stuart Gale, LRMP facilitator. A visit to the mine was eventually made by about thirty Table participants and some of their family members in September 1998.

On October 22 – 23, 1997, I had an opportunity to visit the mine in the company of Vern Betts and was most impressed by this superbly designed and operated mining operation. It is contained within a compact area of 22 hectares (including access road to tailings and waste disposal site), with its 60 km restricted access road from Hwy. 37 occupying about 150 hectares.

(3)   Satellite Photo Coverage of LRMP for Table

The Vancouver Island CORE table included photo satellite coverage with sufficient resolution to define operating mines within a sea of clear-cuts. In order to provide convincing evidence of the relative lack of current disturbance by mining in the C I-S LRMP, I proposed obtaining similar coverage to be funded either by the LRMP or the mining sector. After obtaining some estimates, I contacted Dani Alldrick, a friend and Senior Geologist with the Ministry of Energy & Mines (MEM) to see whether any assistance could be provided by the ministry. Through his efforts, this was obtained at a scale of 1:100,000 in ten large photos which cumulatively occupied about 120 square feet of wall space at our various Table locations. They emphasized the total lack of roads within the area and the clear-cuts from logging, mostly in the upper Iskut River drainage. The three villages were barely visible, and the location of the Snip and Eskay Creek mines had to be pointed out.

(4)   History of Mount Edziza Park and Spectrum (Red Dog) Developed Prospect

On September 26, 1997, a field trip was arranged to the Spectrum (Red Dog) development mineral prospect within the Mount Edziza Recreation Area, on the east side of Mount Edziza Park. It was accessed by two helicopters and attended by 24 individuals over a 7-hour period. The object of the exercise was to provide Table members with a view of part of the park and the relative setting of the Spectrum property within a recreation area, where access was virtually impeded by the park boundaries. Following the trip, I prepared a 3-page letter dated October 15 on the history of the park creation and the development of the Spectrum property. This was forwarded to Tom Soehl and was available to the table at the October 24 – 25 meeting. It emphasized the proposed creation of the park by a GSC geologist; the subsequent 100% buffering by BC Parks; almost continuous exploration of the Spectrum property since 1957 by 12 different companies, only to have it engulfed within a recreation area.

On October 29, I met with Rod Salfinger, CEO Arkaroola Resources Ltd., current owner of the Spectrum property, and reviewed potential access to the property based on a 140 metre wide strip established by survey through the recreation area which would obviate a previously discussed need to access part of Mt. Edziza Park. A memorandum dated October 31 was widely circulated and included Tom Soehl, Graeme McLaren, MEM and Gordon Enermark, Senior Financial Analyst, MEI.

(5)   Information Analysis and Inventory Handbook

Reference maps provided to the Table for filing in the Information Analysis and Inventory Handbook included a highly misleading map designating exaggerated tourism use and tourism viewscape areas. By letter dated October 29, 1997 to Tom Soehl – since referred to Tourism and MEI by Tom, I recommended that a more reliable version be provided to the Table.

(6) Principal developed Mineral Prospects

Insofar as the socio-economic analysis of the C I-S LRMP in respect to mining was concerned, by letter dated October 30, I provided data to Gord Enermark, indicating 11 major mineral deposits in the Developed Mineral Prospects category based on Dave Lefebure's excellent MEI report dated May 28, 1997. Based on these data and 1998 commodity prices, I later produced an article published in the Northern Miner and widely circulated – including the C I-S Table, showing the world-class nature and superiority of B.C.'s historic mineral production within the northern Cordillera. It emphasized that the 1998 gross mineral reserve value of the 10 top developed prospects in the C I-S LRMP was equivalent to about 40% of the value of B.C.'s historic mineral production ($156 billion).

I was to maintain a reasonably close liaison with Gord Enermark concerning evaluation of mineral reserves. By letter dated April 10, 1998, I forwarded to him comments on evaluations he had prepared of major developed prospects for the socio-economic analysis, based on gross reserve values using 'gold equivalent values in ounces'. I also emphasized the value of the Mt. Klappan anthracite coal reserves for which no value had been attributed. See '(7) Importance of Coal Reserves within C I-S LRMP'.

(7) Importance of Coal Reserves Within C I-S LRMP

In a letter dated January 6, 1998 to Tom Soehl and the C I-S Table on "Summary of GAP Analysis Results, Stikine River Corridor & Spatzizi Plateau Extension – Areas of Interest", I emphasized the threat of a proposed southern extension of Spatzizi Park to cover an access corridor which could be critical to the future of the major undeveloped Klappan coal reserves. I also wrote a separate letter to Larry Pituliy, Consultant to Gulf Canada Resources Limited, owners of significant coal tenures at Klappan, dated January 7, urging that a letter be written by Gulf with a copy to Mr. Soehl on this issue.

A letter dated January 19, 1998 with a 16-page report from D.G. Harp, Gulf's Vice-President North America, to Peter Ostergard ADM of MEI, copied to Dan Miller (then Minister, Employment and Investment) and others, pointed out that Gulf's coal reserves alone contained 2.8 billion tonnes of in-place anthracite coal, based on reserve estimates following $60 million in exploration expenditures, and that Gulf was actively pursuing its development interests.

(8) 1998 Provincial LRMP Workshop, Prince George, BC

On March 11 – 12, 1998 I attended the 1998 Provincial LRMP Workshop at Prince George as a representative of the Chamber and the Mining Association of B.C. I participated in a panel chaired by Graeme McLaren on 'Economic Development Considerations in LRMP'. The panel included presentations by Gordon Enermark, Rolf Schmidt (Sr. Land Use Geologist, MEM), Jim Britton (Mineral Planner, MEM) and myself, all of which focused on mining.

Copies of the agenda and my presentation notes, which included 15 overheads, are available. Copies of the complete proceedings were reportedly to be printed for reference. The meeting attracted about 90 participants from the province and afforded an opportunity for Derek Thompson (Asst. Deputy Minister – Land Use

Coordinate Office) in his introductory comments to emphasize the recent commitment to Northern Development in B.C. I had an opportunity to state the mining sector's acceptance of 12% protected area goal, but no additional land alienation, which could deny access to mining and inhibit certainty to tenure and investment.

My latter comments and those of Derek Thompson were not appreciated by other members of the Cassiar Iskut-Stikine LRMP at the meeting, notably Glenda Ferris representing the Tahltans and Eric Harvard of Eagle River Guide Outfitting.

(9)   Brochure on 'Coexistence of Wildlife and Mining-Related Infrastructure'

Widespread misinformation generated by environmental activists concerning implied threats to wildlife and their habitats by mining-related activities, prompted me in September 1997 to consider providing visual evidence of coexistence between wildlife and mining-related activities. The Chamber assisted in mail-outs of a letter I provided requesting photos from existing mining and exploration companies supporting our claims. The request generated an excellent response with 50 photos provided including 16 species at 10 different mining locations. Eventually I attracted nine separate mining companies who agreed to act as sponsors, in addition to The Letter Shop who produced 6500 copies of a 4-page, double-sided colour-brochure also bearing the support logos of the Chamber and MABC. An introductory statement was followed by 31 photos, showing 19 different wildlife species at 12 different mining locations and areas. One thousand of the brochures were donated to MABC's Mining Education Committee for use at B.C. schools and sponsor companies also received about 2,000 copies for their own distribution. Although table member Norm McLean of B.C. Parks indicated his wish to have an opportunity to rebut the brochure, when produced, I never received any negative comment either verbal or written from any table member following its distribution.

(10)   Summary of GAP Analysis Results, Stikine River Corridor &
Spatzizi Plateau Extension – Areas of Interest

I have alluded to the Spatzizi Plateau Extension proposal as it affected the important Klappan coal resource. In my same January 6, 1998 letter to Tom Soehl, which was distributed to the Table with a copy of Gulf Canada Resources' letter on its Klappan coal tenures, I summarized the history of the Stikine River Corridor with comments on the importance of the mineral resources to the LRMP region and the essential need for unrestricted access for mining purposes as provided by existing legislation. I highlighted the Stikine for obvious reasons, not the least of which was the threat to access provided by the Large River Corridors Sub-Committee, developed by the LRMP. Although this sub-committee included Vern Betts as a mining representative, the Committee's guidelines as developed were obviously biased against the resource sector – particularly mining. A draft of the proposals led me to point out to the full Table at its meeting on September 25 –26, 1998 that this bias existed and in addition that the guidelines were frequently inconsistent with provisions of existing legislation – namely the recently enacted Mineral Exploration Code of B.C. Glenda Ferris stated that the Sub-Committee had considered the MX Code and that there was no conflict.

On returning to Vancouver, I discussed the Large River Corridors draft with Gavin Dirom who critically evaluated the draft with only a minor input from myself. Nine pages of critical comments were forwarded to Tom Soehl under cover of a memorandum from myself to the C I-S LRMP dated November 4, 1998. It revealed the bias against the mining industry in such terms that Jim Bourquin, a Table member who originally proposed the Large River Corridors as a Special Management Zone to support his proposed river rafting endeavours, angrily complained at our next Table meeting, "You have ruined a year's work."

(11) Land Use Planning and the B.C. Mining Sector

On May 13, 1999 I provided a presentation on the above topic to the Minerals North Conference at Stewart, B.C. It was attended by many government representatives and at least two members of the Cassiar Iskut-Stikine table: Glenda Ferris and Jim Bourquin. Copies of the 11-page paper with figures were available for reference purposes. I also arranged for copies to be made available to the C I-S Table at their next meeting. I considered this the most important and informative item which I provided to the Table as it covered the history of the development of the Protected Area Strategy process and all the principal concerns of the mining industry, particularly the flawed nature of CORE and LRMPs in respect to subsurface resource detection and development. It included specific reference to the C I-S Table.

After our official departure from the LRMP process in January 1999, I still spent much of my time in activities related to promoting public awareness of the plight of the mining industry in B.C. Most of this was directly related to problems created by the B.C. government's land use process, with the additional burden imposed by aboriginal land settlements in progress with both the federal and provincial governments.

In January 1999, prior to officially withdrawing from the LRMP process, I received an invitation from Paul Wodjak, with the organizing committee to provide a presentation on "Land Use Planning and the B.C. Mining Sector" at the 11th Annual Minerals North Conference being held at Stewart, B.C. on May 12 – 14, 1999. I was pleased to accept, planning to build on my farewell contribution to the C I-S LRMP Table, entitled Mining Revisits B.C.'s Land Use Plans, dated January 31, 1999 and delivered to the Table in late February 1999.

In organizing the talk, I felt that it was imperative to provide visual evidence, rather than relying on verbal comments, to graphically trace the growth of the mineral industry and its recent decline. In addition, the same objective held for the multiple issues leading to the mining industry's decline during the decade of the 1990s. Finally, realistic remedial recommendations to reactivate the industry were required.

I was to be indebted to Dani Alldrick, geologist with the Ministry of Energy & Mines, and other MEM members for assistance in generating data on the historic growth of the B.C. mineral industry over a 150-year period commencing in 1852. A series of colour slides were prepared showing the location of 648 producing mines in the province, in 20-year increments at a cut-off of 500 tons of total production. These were symbolized in four categories: coal, precious metals, base metals and industrial minerals. They dramatically portrayed the importance of access for mineral exploration and development on the subsequent location of operating mines and communities, particularly in the first 100 years through 1950. Through 1900, there was only one mine in operation in the northern half of the province, of 60 in operation. Discounting the Portland Canal area, where ocean access to the southern part of the Alaska Panhandle was provided, only five of 250 operating mines were located in the northern half of the province by 1940. The continuing remoteness of this northern region, with its few isolated communities containing less than five per cent of the provincial population through 1999, continues to reflect its relatively recent mineral exploration history, compared to that in the south. The few mines that have as yet developed belie the very high mineral potential of the region, as I was able to demonstrate in a subsequent paper[1] and in a Northern Miner article.

---

[1] 'NDP blamed for poor B.C. image, the Northern Miner', September 20 – 26, 1999;

Because of restricted access, the comparatively high mineral exploration and mining costs in this northern region were the only significant concerns to the investment community prior to the restrictive land use policies developed by the NDP government in the early 1990s.

Later in 1999, I became interested in researching B.C.'s mineral endowment compared to that of its immediate neighbours in the Cordillera. My initial objective included the production of a graph illustrating the historic value of B.C.'s total solid mineral production in 1999 dollars, for comparison with its neighbours. I approached Jim Lewis, Economist with MEM who within one week provided annual data starting in 1836 broken down into metals, industrial minerals, structural materials and coal. Needless to say, I was most impressed and grateful. With the assistance of Gavin Dirom, I superimposed graphical data on mineral exploration expenditures through 1999 commencing with $20 million Canadian in 1999 dollars in the late 1950s and peaking at over $300 million Canadian in the 1980s. This coincided with total solid mineral values of over $4.5 billion Canadian in the early 1980s. By 1999, the total value of mineral production had dropped to the $3.0 billion level, compared with a drastic reduction in mineral exploration expenditures to $25 million, or 8 per cent of its peak value in the 1980s.

I was able to secure similar data for Alaska and Yukon from the 1880s to 1997 and to a lesser extent Washington State, which indicated an overwhelming superiority in mineral endowment within B.C., based on value of total mineral production and known world-class mineral deposits. B.C.'s total value in historic Canadian dollars was $64.4 billion compared to about $8.6 billion for Yukon, with $2 billion allowed through 1967, and $9.7 billion for Alaska (in U.S. dollars). In order to provide current comparisons, I also obtained values in Canadian dollars for 1997 output using an exchange rate of 1.3856 for Alaska and Washington production. These still showed B.C.'s dominance with B.C. at $3.1 billion; Alaska $1.2 billion; Yukon $0.2 billion and Washington $0.9 billion.

A comparison of known world-class mineral deposits in these jurisdictions, based on the gross value of their past production combined with future reserves showed eight in B.C. ($94.7 billion); two in Alaska ($31.3 billion); two in Yukon ($5.1 billion) and none in Washington.

The above comparisons were included as part of my article submitted to the Northern Miner. I circulated copies to about one dozen individuals with a copy of the original submission, including those that had helped me obtain data. For the umpteenth time I included the Hon. Dan Miller, Premier of B.C. at the time and Minister of Energy and Mines and Minister Responsible for Northern Development. Characteristically, I received no response. A copy also went to Gordon Campbell, Leader of the Opposition, who responded with an appreciative letter, dated October 22, 1999, indicating that he had forwarded a copy to Richard Neufeld, the Liberal Party critic for Energy, Mines & Northern Development. Mr. Campbell noted in his letter that "the decline of the mining industry in British Columbia is truly one of the saddest legacies of this NDP government," and "we're committed to restoring the mining industry in the province through regulatory and taxation reform, as well as addressing the issues that surround security of tenure."

By late 1999, discouraged with the continuing problems affecting the B.C. mining industry, I decided to complete an in-depth study of the industry, with the object of providing copies to all the Ministers of Mines of the provinces and the Northwest Territories, in addition to our senior government and opposition leaders, plus selected mining associations and others. The object was to provide a reference which would discourage other jurisdictions from embarking on flawed land-use policies affecting subsurface resources.

The final paper covered 29 pages plus nine colour plates containing 21 figures; three appendices covering mining industry recommendations as proposed at the B.C. Business Summit held in Vancouver on November 8 – 9, 1998; a summary position paper by the Chamber on the role of minerals in aboriginal land claim negotiations; and a copy of my previously issued colour brochure entitled *Coexistence of Wildlife and Mining-Related Infrastructure*. The paper was my previously mentioned *Growth and Recent Decline of the B.C. Mineral Industry, Recommended Remedies*. It benefited from a covering letter by Bruce McKnight, dated March 20, 2000 with the following lead paragraph:

"I would like to bring to your attention and to recommend for reading and future reference the above-noted paper by David Barr. It is a comprehensive and well-researched document covering the 150-year history of mining in British Columbia, its recent decline and some suggested solutions. Through its systematic documentation of successes and failures in B.C. mineral policy, this paper will remain a valuable reference and should be of interest to anyone concerned about a viable mineral policy – in any jurisdiction."

The paper also benefited from critical reading by Gavin Dirom and Ken Sumanik, plus contributions from Dani Alldrick, Larry Jones, Mike Fournier, Tom Schroeter and Jim Lewis – all with the B.C. Ministry of Energy & Mines. One hundred copies of the paper were produced and by November 2000, only a few remained at the Chamber.

In retrospect, I probably spent far too much time in agonizing about the inability of the NDP government to recognize its inequitable and unfair treatment of the B.C. mining industry, and to join others in trying to redress the continuing damage caused, not only to the industry, but to B.C. society and its economy. One would like to believe that an industry which historically had contributed so strongly to B.C.'s growth, settlements and generally admired heritage, while having such a minimal impact on the land, would continue to be encouraged.

As a related footnote to this lengthy chapter, in July 2001, I received a letter from Colin Smith, P.Eng., President of the Association of Professional Engineers and Geoscientists of British Columbia, indicating that I had been selected by the Association as the 2001 recipient of the C.J. Westerman Memorial Award. I was overwhelmed by this totally unexpected honour, which on later enquiry, I learned grew out of a nomination by the Chamber. The award is "for a professional geoscientist who combines a solid professional career with outstanding service and dedication to advancing public recognition of geoscience." Although I will cherish this award, I realize that there are many others in the profession who, though equally or more qualified than myself, did not happen to be nominated.

The Sherwood Mine Strathcona Park, B.C.

The reader might question the authenticity of this story. A mine within a park? Surely not in B.C.! Yes, even in B.C., as this episode will relate. However, the event that I am recalling somewhat unenthusiastically, does not concern the pros and cons of mining within a park. In this case the mining property had already been expropriated and the issue at hand was the determination of the amount of compensation to be paid to the owners.

Since the value of the expropriated property would be argued by lawyers before an expropriation board unfamiliar with mining property valuations, reliance had to be placed on technical information provided principally by expert witnesses - mostly geologists. This was to be a frightening situation, pitting brother against brother, so to speak.

My involvement commenced in July 1990, when I was contacted by William Pearce, then Senior Counsel for the Legal Services Branch of the Ministry of Attorney General for B.C. He explained his need to obtain an independent consultant's valuation of the Sherwood mine property situated in the Bedwell River area in the southern part of Strathcona Park, which was subject to expropriation proceedings. I had been recommended to him by Graham McLaren, then attached to the Land Management Branch of the Ministry of Energy, Mines and Petroleum Resources. Graham had worked for DuPont of Canada Explorations Ltd. as a summer student while attending the University of British Columbia and while I was general manager.

My initial impulse was to turn down the assignment. I felt that under the circumstances my friends and acquaintances in the mineral industry would regard my preparation of a report on behalf of the government as the action of a turncoat. Furthermore, I did not have access to up-to-date costs for placing the property into production. After further consideration, I felt that as a geological consultant and a professional engineer, my report would be completely unbiased and that any party should have the right to an independent consultant's opinion. I decided to obtain more detailed information on the data available on the property, recognizing that although it was called a mine, it had never reached production. In addition I resolved to limit my involvement to that of an evaluation, rather than a valuation and to recommend that the latter should be completed by a mining consultant based on my determination of an indicated reserve and the possibility for further discoveries on the property.

I met with Bill Pearce in Victoria on July 30 at his office at the Legal Services Branch of the Ministry of Attorney General in Broughton Street. I was impressed by his concise review of the history of the development of Strathcona Park and the Sherwood Mine and it has provided me with a most useful reference for this narrative. He provided me with a letter the same day in which he summarized the history of the park's development and the general guidelines for my report. He noted that under the B.C. Expropriation Act, compensation is based upon the "market value" of the interest in land, which is defined as "if it had been sold at the date of expropriation in the open market by a willing seller to a willing buyer."

Before discussing my subsequent role in preparing my report and appearing as an expert witness before the B.C. Expropriation Board, some review is essential of the past history of Strathcona Park and mining developments in the area. A prominent feature sub-paralleling much of the park's 60 km long eastern boundary is its western boundary which lies within a few kilometres of Nootka Sound, on the west coast of Vancouver Island. Captain James Cook of the Royal Navy landed there in 1778 on his epic voyage in search of the Northwest Passage.

The park was created on March 1, 1911 under the "Strathcona Park Act" and is B.C.'s oldest provincial park. It contains a number of interesting features, including Mount Golden Hinde, 2200 m - the highest point on Vancouver Island; Mount Washington ski area - the largest and most popular on Vancouver Island and Della Falls - Canada's highest waterfall, with a total drop of 440 m in three cascades into Drinkwater Creek across from the Sherwood Mine. Joe Drinkwater, a resident of Alberni and pioneer prospector, named Della Falls and nearby Della Lake after his wife.

The park was named after Donald Alexander Smith, a Canadian pioneer and one of the principals involved in the construction of the Canadian Pacific Railway. As Lord Strathcona, he drove the last iron spike on the CPR at Craigellachie in the Selkirk Mountains in 1885 uniting Canada from east to west.

The park contains one other dominant feature which belies the popular misconception held by preservationists that industry and wilderness cannot co-exist. I have written much to correct this falsehood and have cited the significant Westmin Resources mining operation at Buttle Lake in an adjoining Class B park where mining is permitted - as a shining example of shared resource-use. The mine is important to the B.C. economy and particularly to the Campbell River district as a source of employment. The tens of thousands of visitors who annually access the park, and some the mine, could not do so without the road to the mine site at the south end of Buttle Lake. Mining, as in other parts of the Province, played an important role in the early exploration and development of the Strathcona Park district, along with logging.

The Spanish explorer, Malaspina, sailed into an inlet west of Strathcona Park in 1791 and named it Cevallos after one of his officers. Almost 150 years later, in 1936, a bushwhacker named Alfred Bird discovered a mineralized quartz vein beyond the head of the inlet which was to herald the start of the Zeballos gold boom.

Reports of the Minister of Mines, British Columbia for 1898 and 1899 contain brief references to placer-mining activity on Bear (Bedwell) River in the 1860s and to Chinese placer miners abandoning the district in the late 1880s, but provide no information on the operations or the quantity of gold recovered. The headwaters of the Bedwell River lie just west of the head of Drinkwater Creek, and the river flows westerly from its headwaters for about 15 km before trending southerly for 8 km to its outlet at the head of Bedwell Sound which lies about 25 km northeast of the community of Tofino on the west coast of Vancouver Island.

The discovery of gold in the Zeballos area was to stimulate renewed exploration for gold and gold mining activity over a large portion of Vancouver Island until the early part of World War II. During this brief four-year period, many gold discoveries were to be made in the Zeballos area and 100 km to the south-east in the Bedwell River and Drinkwater Creek district. Included in the latter were the Musketeer and Buccaneer mines on the south side of Bedwell River, which collectively produced 6908 ounces of gold and 2994 ounces of silver from 11,500 tons milled in 1941-42. These two properties were situated about 11 km west of the Sherwood Mine discovery made by Wally Sherwood of Alberni in 1938.

Although I was familiar with the high-grade nature of the gold ores in the Zeballos Camp, I was also aware of the very narrow gold veins, averaging less than one-foot in width, which were common in the district. The Zeballos Camp produced 276,000 ounces of gold at an average grade of 0.38 ounces gold per ton milled from six mines during the period 1933-1953. The Privateer Mine, by far the largest, accounted for about half of the production. Not generally recognized is that the narrow width of the gold-bearing veins necessitated considerable dilution during mining by waste rock in the wall of the veins. Several of the mines upgraded the ore prior to milling by hand-cobbing obvious waste rock - this procedure being affordable at the low price of wages paid during the depression years - but not in the post-World War II period.

In 1979 while employed with DuPont, we became interested in the potential for gold in the Cordilleran region as the price of gold increased dramatically in the late 1970s toward its peak of $900 per ounce in 1980 following freeing of the U.S. $35.00 per ounce fixed price of gold in 1969. 1 prepared a paper entitled 'Gold in the Canadian Cordillera' and presented it at the CIM Annual General Meeting in April 1979 and at four other locations in B.C. and Ontario prior to its publication in the CIM Bulletin in 1980. Although the Zeballos district was mentioned, it was obviously not especially significant, ranking 17th in terms of actual gold produced at the time in the Cordillera.

However, it was not until 1990 that I was to realize just how restricted actual lode gold production from vein-type deposits had been on Vancouver Island. It was actually revealed in the research which I completed for my October 17, 1990 report on the Sherwood Gold Mine property. A table in my report showed that 96 lode gold vein-type deposits mined in B.C. from 1894-1987 had produced 13.4 million ounces of gold. Five of these, including the Zeballos Camp (treated as one deposit) had produced only 288,689 ounces of gold, or 2.2% of the total. It was to be a significant revelation, but one which would be difficult to convey at the subsequent expropriation hearing to a non-technical board which was hearing diametrically opposite 'evidence' from expert witnesses representing claimants and the respondent.

Following its foundation in 1911, the boundaries of Strathcona Park were established between 1911 and 1913. There were no valid mineral claims reportedly in the park at this time even though several properties such as Big Interior Mountain (1899) and Big I (pre-1906) had been staked or recorded earlier. Prospecting within the park was prohibited until April 15, 1918 but was then permitted until April 5, 1957 when the Strathcona Park Act was repealed and Strathcona became a Class "A" Provincial Park in which prospecting and mining activities were once again prohibited.

On May 13, 1965, Strathcona Park was reclassified from Class "A" to Class "B". The principal difference in the status of Class "A" and Class "B" parks in respect to exploration for and exploitation of mineral resources is that no interest in a Class "A" park shall be alienated except as authorized by a valid and subsisting park use permit, which shall not be issued unless in the opinion of the Minister issuance is necessary to the preservation or maintenance of the recreation values of the park involved. This would appear to eliminate exploration and exploitation of minerals in all but the most unusual circumstances. With respect to Class "B" parks, a park use permit, or a resource use permit, may be issued permitting exploration and exploitation of mineral resources provided such activity is not detrimental to the recreational values of the park concerned. This distinction has permitted the continuation of the important Westmin mining operation in the central part of Strathcona Park, as described earlier.

In 1966, Cream Silver Mines Ltd. was incorporated under a B.C. Charter and in the same year it acquired 66 mineral claims by staking in the Buttle Lake area south of the Westmin Resources Limited ("Westmin") property and lease area, and north of the

Sherwood claims, the object being exploration for similar massive sulphide base metal deposits with significant precious metal values, similar to those mined by Westmin.

In March 1, 1973, a moratorium was put into effect which provided that further exploration and development of mineral claims within any provincial park could not occur without the authority of the Lieutenant Governor-in-Council. Westmin, which had also staked a large number of additional claims in the park in the 1960s was permitted to continue.

In 1979, Cream Silver Mines Limited ("Cream Silver"), prevented from developing their claims by the 1973 moratorium, sued the B.C. Government.

In 1985, the Tener case, involving crown-granted mineral claims already in place in Wells Grey Park before the park was formed, was decided in the Supreme Court of Canada. The Tener decision was essentially that the provincial government would have to pay compensation if they continued to prevent Mr. Tener from developing his crown-granted claims.

On April 30, 1986, the B.C. Supreme Court ruled that the ruling in the Tener case could not be extended to the Cream Silver case. Cream Silver filed an appeal, which was indefinitely adjourned. The basis of this appeal was that a Crown-granted claim was an interest in land which was compensable under the Expropriation Act, whereas a mineral claim, being a chattel (personal interest), was not.

On March 14, 1987, the government made a policy decision to re-classify lands within provincial parks where mineral claims were located to "Recreation Areas" with a view to letting mineral claim holders explore for the minerals within their claims under stringent conditions with the expectations that most of the claims would not "prove out" and would be considered worthless. For those claims that did "prove out", it was anticipated that a decision would be made in the future to either allow the mine to proceed or to re-classify the subject area from "Recreation Area" to "Park", depending upon the potential sensitivity of the proposed mine to the recreational values in the Recreation Area and to the environmental impact of a mine.

A public statement issued by the Hon. Bruce Strachan on March 11, 1987 discussed the government's intentions concerning the future direction and management of Strathcona Park. The new policy provided the following principal requirements and guidelines for Strathcona Park.

1) Negotiations toward reasonable quit-claim settlement will be attempted for all existing claims.

2) When any of the existing claims lapse, they will automatically become Class "A" Park.

3) The Recreation Area surrounding the present claims will only allow development deemed necessary to access and service the existing mineral claims. There will be no new claim staking in this area.

4) In managing the mineral exploration process, the government will pursue all legal avenues to ensure that both direct surface impacts at the claim site, and the long-term effects on recreation throughout the park are minimized.

In the above guidelines, no limit was specified as to the length of time provided for mineral claim owners to further explore their properties. The statement also noted that:

While the basic decision on park boundaries and status has been made, the government remains committed to undertaking a further public consultation process to prepare the management plan for the park

To this end:

Public consultation will begin in May 1987 and meetings will be held in several communities on Vancouver Island and in Vancouver.

Initially, the public will be presented with an analysis of the park resources and issues and invited to express opinions on the management options.

Based upon the preliminary public input, a draft plan will be prepared and presented for further public review prior to finalization.

As previously noted, Cream Silver Mines had mineral claims located within Strathcona Park and obtained a resource use permit to do some exploratory drilling. This gave rise to a public protest resulting in the appointment of the Strathcona Advisory Committee in April, 1988. At that point, the Crown notified Sherwood Mines Ltd. that no resource use permits relating to mineral exploration would be issued while the Strathcona Advisory Committee was reviewing the future plans for the management of the Strathcona Recreation Area. In June 1988, the Advisory Committee, in a report to the Crown, recommended that no future mineral exploration or mineral development be permitted in the Park beyond the existing Westmin Resources Ltd. lease zone. An Order-in-Council was passed on May 19, 1988, placing a moratorium on mineral exploration and development within the Strathcona Recreation Area which was later extended in August 1988 to allow the Crown sufficient time to decide what it would do with Strathcona Recreation Area. The government finally resolved the issue by passing an Order-in-Council on November 25, 1988, which prohibited the issuance of resource-use permits for mineral exploration in the Strathcona Recreation Area and in June 1989 the Recreation Area was re-classified to Class "A" Park.

This convoluted history reflects the important influence of government policy concerning land withdrawals, which was to explode on the B.C. scene along with the growth of environmental extremism in the late 1980s and throughout the early 1990s. It ushered in the provincial Protected Area Strategy, the short-lived Commission on Resources and Environment and finally the Land Resource Management Plans which were all directed toward land-use issues. As I will describe later, it partly overlapped a federally-inspired program of Flow Through Share funding in 1986-88 which particularly influenced massively increased mineral exploration expenditures throughout Canada by the junior mining sector.

The British Columbia Department of Mines responded quickly to the need for guidance in geological mapping and mineral development reporting which was to follow gold discoveries in the Bedwell River area in the late 1930s. Hartley Sargent, a late acquaintance of mine and a highly respected geologist with the Department of Mines, compiled data for Bulletin 13 'Supplementary Report on Bedwell River Area' released in 1941. It remains the principal source of reliable information on geology and mineral occurrences in the Sherwood Mine area and was to be extensively quoted during the ensuing Expropriation Board hearing.

Following his discovery, Wally Sherwood soon interested Pioneer Gold Mines of B.C. Limited in his claims. At the time, Pioneer was recognized as a highly successful gold mining company based on its interests in the Pioneer Mine in the Bridge River Camp, one of B.C.'s principal lode gold mines, having produced 1,333,531 ounces of gold from 2.4 million tons of ore mined between 1908-1962.

The early work by Pioneer is described as follows by Hartley Sargent in Bulletin No. 13:

Supplies are brought up Great Central Lake, and to the end of the logging railway on Drink-water Creek, on steamers and rolling stock operated by the logging company, Messrs. Bloedel, Stewart and Welch. From the railway to the property, about 6½ miles by trail, supplies are taken on pack-horses.

An examination of surface exposures and a few cuts on the Sherwood Vein was made in November 1939 for Pioneer Gold Mines of B. C., Limited. It was impossible to begin to do more than preliminary work that year because of heavy snow. In 1940 a start was made as soon as snow conditions permitted. A warehouse was built at the railway, and necessary work was done on the Drinkwater Creek trail as far as the site of a base-camp, at approximately 1800 feet elevation, 4½ miles by trail from the end of the logging-railway. The base-camp, built of logs, is on the north-eastern side about where the main trail crosses Drinkwater Creek, and is less than 100 yards from a log cabin built years ago and used in connection with work on the Della and Big I properties.

A branch-trail was built on the steep north-eastern side of Drinkwater valley just north-west of the creek draining Love Lake. It leaves the main trail about a quarter of a mile south-easterly from the base-camp, and climbs by a series of switchbacks, which in the first mile are built on a ridge between Love Lake Creek and a parallel creek 500 to 700 feet to the north-west. In about 2 miles the trail climbs approximately 2,400 feet to the mine-camp, at 4,200 feet elevation. By late in June trail construction had gone far enough to permit taking supplies on pack-horses from the railway to the mine-camp.

From this camp, consisting of tents, most of the work on the Sherwood Vein and on the No. 1 and No. 2 veins of the P. D. Q. property has been done. Underground work on the Sherwood Vein was carried on vigorously and three levels were driven. During the summer a second temporary camp was built for use in driving No. 7, the lowest level. When work was suspended for the winter in December, it is reported that work on the three levels (No. 1, No. 3 and No. 7) amounted to 1707 lineal feet, all driven by hand.

Sargent's description provides little indication of the difficult access and working conditions involved, as such was the norm in virtually all the early exploration and development work in remote parts of the province. A later and more humorous account of a brief visit to the area in 1945 by Robert Gayer (see below) is more revealing. References to the underground work on the three levels, "all driven by hand" reflects the absence of any powered drilling equipment. The level excavations with crosscut dimensions of about 3 x 6 feet were driven by miners with sledge hammers striking drill-steels held by assistants against the work face, - creating a pattern of holes several feet deep on the face. These were loaded with explosives and connected fuses and detonated to provide each section of advance on the level.

Pioneer continued its work in 1941 and completed a further 420 feet of drifting, cross-cutting and raising on the Sherwood mine between March 1 - November 15 before terminating its option. In 1942, Sherwood employed two men who hand-sorted muck derived from the Pioneer work, segregating out 22 tons of ore which was shipped to the smelter at Tacoma, Washington and reportedly assayed 3.25 oz gold/ton and 5.75 oz silver/ton. At the time the price of gold was U.S. $20/oz and contingent on smelter, transportation and labour costs, the smelter return might have been enough to offset the overall costs.

In 1945 Cangold Mining and Exploration Company Limited optioned the Sherwood claims. An average of ten men were employed between May - October. Although considerable surface work was done, it appeared that a decision had already been made to gear the property for production as construction of a 50-ton mill and a surface plant was proposed as early as possible in the 1946 season. The Vice-President and Managing Director was Robert B. Gayer of Vancouver - an individual I was to re-discover in 1991 during the Expropriation Board hearing. He was to be most helpful in providing me with copies of early reports and newspaper clippings of the Cangold venture.

An early account of Mr. Gayer appears as Chapter Eighteen in 'Spilsbury's Coast', a classic west-coast book by Jim Spilsbury, a colourful entrepreneur, pioneer, painter, aviator, inventor and matchless raconteur, and his collaborator Howard White. It not only describes Gayer, but is an account of a visit to the Sherwood mine site during the Cangold period by Spilsbury to install a radio-telephone. Harbour Publishing has kindly permitted quotation of the the following paragraph, which describes a far more volatile person than the Gayer I met:

The way Gayer explained it, he was the director and general manager of a large mining concern on Vancouver Island, Cangold Mining and Exploration Company Limited. They had an operation on top of a five-thousand-foot mountain up near the head of Great Central Lake, and they simply had to have good communication with the outside world. This of course meant radio-telephone, and only the best would do. "None of these half-assed things you press a button to talk and then listen to the other guy and you can't stop him," says Gayer. "We already have one of those goddamn things, and I wouldn't give it hell room. I want full two-way telephones and to any part of the goddamn world, any time of the night and day! Time is money goddamn it in my business and I won't fart around with anything less. Now, can you do it?"

The copies of the reports and newspaper articles which I obtained from Mr. Gayer were subsequently entered as 'Exhibit 47' at the Expropriation Board hearing. The newspaper clippings had been carefully mounted for copying in chronological order. The collection of 50 articles from eight different newspapers or journals cover the period February 2, 1945 to September 20, 1946. They are highly promotional and optimistic in title and outlook, announcing plans for the Sherwood Mine to be in production by September, 1946. The last reference to its activity that I located was in a Ministry of Mines Report which indicated that operations were suspended on October 10, 1946 because of adverse weather conditions and that a crew of fifty men was employed.

What was to be the final epic in the spasmodic life of the Sherwood Mine commenced somewhat casually in 1982. By that time the property consisted of 19 contiguous Crown-granted mineral claims covering 715.95 acres. Fortuitously, the claims had been Crown-granted in 1946 and 1947 by Mr. Sherwood - a significant decision at that time as the Crown-granting improved title to the level that if expropriated, compensation would be due on an added interest in the land as well as in the subsurface minerals.

In 1982 W.C.R. Construction became interested in the Sherwood property and subsequently acquired a partial interest in Cinta Resource Corporation (Cinta), which was wholly owned by Mrs. Merna Tattersall and which had obtained a 25% net profit interest in the property. In July 1984 Mrs. Tattersall obtained a 69% controlling interest in Sherwood Mines Ltd. through Cinta by the acquisition of 186,000 shares of Sherwood at a cost of $167,400 from Lillian Hart, the widow of Mr. Sherwood.

Casamiro Resources Corporation (Casamiro) was incorporated with a B.C. charter in 1984 with Mrs. Tattersall as president and reportedly having a controlling interest in Sherwood under terms of a letter of intent dated November 19, 1984.

At this time Casamiro was not in a position to carry out any development work on the Sherwood claims. Earlier amendments made the Park Act (1965) subject to the Environment and Land Use Act (1971). In addition, they permitted the provincial cabinet to regulate and control the extraction of natural resources within any recreational area or park. At least from 1978 to 1986 it was the government's position that no new exploration or development work, with rare exception, would be authorized within park boundaries.

However, in 1985, the Supreme Court of Canada unanimously agreed that the Crown's refusal to grant a park use permit for mineral development to Mr. Tener on his crown-granted claims in Wells Grey Provincial Park amounted to expropriation under the Park Act (1965) even though the Crown had not involved the machinery of public taking.

This decision led the government to modify its restrictions on mineral exploration within provincial parks by an Order in Council dated March 14, 1987. Certain lands in Strathcona Park including the Sherwood Mine claims were re-classified from Class "B" Park to Recreation Area. However, resource-use permits were still required.

Through my association with CSA Management Limited as Exploration Manager, commencing in 1985, 1 was involved in the formative stages of a Flow-Through Share fund for the junior mining sector named First Exploration Fund 1986. 1 subsequently acted as Exploration Manager for all three First Exploration Funds which raised $104.5 million in the 1986-88 period.

Each of the First Exploration Funds had an Exploration Review Board chaired by Alastair Sinclair and composed of Marcel Vallee, Ralph Westervelt and myself, all with geological/consulting backgrounds and Terry Holland representing the fund. The Board's objective, in addition to providing a degree of credibility to the ultimate composition of the Fund's stable of companies, was to review applicants for funding and to determine whether their projects warranted support. In practice it was a very subjective exercise. Accordingly the reviews were initially designed to ensure that the proposed budgets were reasonable, with a preference given to projects with tangible targets as opposed to those at grass-roots levels. Each project approved was also subject to an on-site audit by an outside consultant whom I arranged to retain on a contract basis. In all, I eventually obtained agreements with about 50 different consultants across Canada on behalf of the three funds.

By coincidence, one of the first projects reviewed by the First Exploration Fund 1987's board on February 25, 1987 was Casamiro's Sherwood Mine proposal along with another minor project. Casamiro's submission on that day was one of eight companies involving 13 projects. Funding for five of the companies, including Casamiro, involving 8 projects and $4,736,000 was approved by the Board and three other companies' proposals were tabled for various reasons. This was not an unusual board-meeting load. During 1987 a total of $46,771,037 was approved for 60 mining companies involving 105 projects at 18 board meetings. The amount approved to Casamiro for the Sherwood mine project was $600,000 of a total $640,000 proposed in a two-phase program which was ultimately reduced to $525,000.

Our process for board approval of funding amounts to companies for proposed projects could not be remotely compared to the effort placed by a professional engineer in preparing most property evaluations and reports, which generally include property visits. The amount of time spent by the Board in approving funding for Casamiro was about 1½ hours. The following comments quoted from the minutes of the Exploration Review Board meeting on February 25, 1987 are reproduced in their entirety, as a most pertinent example:

The Sherwood gold mine property program represented all of the work proposed under phase 1 and part of the phase two work. David Barr indicated that the costs were extremely reasonable and would be incurred over a 60 day period. He suggested that as a phase one project is $600,000, the Review Board should perhaps consider only approving the first phase with the results of that work determining whether the subsequent phase should be pursued. Marcel Vallee felt the program was not much better than a grass-roots program.

Terry Holland suggested that the company was not highly capitalized and $600,000 may well be the maximum capitalization possible.

Alastair Sinclair confirmed his support to approve only the phase 1 work. He felt that there was a practical problem in that the tonnage which had been reported may well be exaggerated. He did feel that there was some potential but that it was limited. Ralph Westervelt agreed that the data available on the project did not appear to be that exact.

Terry Holland then moved that the Exploration Review Board approve the First Exploration Fund 1987 and Company, Limited Partnership entering into a subscription Agreement with Casamiro Resource Corporation for the reduced amount of $600,000 which represents phase 1 of the Sherwood Gold Mine project. The motion was subject to the receipt of a proper budget and the financial and legal items which were outstanding. The motion was seconded by Ralph Westervelt and passed unanimously.

By contrast I spent 108.5 hours between July 31 - November 17, 1990 in preparing my report for the Ministry of Attorney General on the Sherwood Gold Mine, which included a property visit on October 2. My subsequent services through March 1992 on behalf of the Ministry were to total over 700 hours, many of which felt like days!

I was to spend many sleepless nights during the Expropriation Board hearing cogitating on various studies or tasks required to effectively represent my convictions, as reflected in my report. One of these was the realization that I had supported the funding to Casamiro for work on the Sherwood property only 3½ years earlier as a board member of First Exploration Fund 1987 and without any further effective work having been performed on the property, I was now recommending no further work. I convinced Bill Pearce that this area should be discussed during my direct testimony as a witness.

Although the above-mentioned minutes were entered as evidence and I explained the different nature of the approvals, one of the Expropriation Board members in particular was not convinced and I'm sure he believed that I was a knave and had ulterior motives for changing my mind.

Casamiro planned to complete a rehabilitation, drilling, raising, mapping and sampling program within the area of the old workings. The work envisaged was to be completed in a two-month period commencing in June. The program as proposed was based on a report prepared for Casamiro in December 1986 by R. Terry Heard, P.Eng. In his report Heard estimated "50,000 tons of possible reserves having an average grade in excess of 1.5 ounces gold per ton" based on "published data in Minister of Mines reports."

Although Casamiro had applied for a Resource-use Permit from the Ministry of Environment and Parks in the early spring, 1987, it was not issued until November 6, 1987. Frustrated by Ministry delays, Casamiro spent $325,000 prior to the issuance of the permit in leasing and shipping mining equipment to the property and camp preparation. Only a minimum amount of rehabilitation was completed, as reported by an on-site audit on December 11, 1987 by Nick Carter, P.Eng. I had arranged for a consulting agreement with Nick, a long-time acquaintance, partly because of his location in Victoria on Vancouver Island.

Casamiro was obviously deeply frustrated by its experience in attempting to have an exploration program completed on the Sherwood Mine, as was First Exploration Fund 1987. Casamiro's objectives for its interest in the property were further affected by the government's Order-in-Council on May 19, 1988 which placed a temporary moratorium on the issuance of resource use permits for any mineral exploration and development in the region. The moratorium was extended indefinitely by a further Order-in-Council on November 25, 1988 and by the establishment of the region as part of Strathcona "A" Park on June 12, 1989. Casamiro and Cinta then filed an Application for Determination of

Compensation, which was contested by the Crown and eventually upheld by the courts. This lead directly to the hearing before the Expropriation Board commencing on October 7, 1991 to determine the amount of compensation, with concluding argument heard on May 25, 26, 27 and June 4, 1992.

There were to be in excess of 200 exhibits filed in the proceeding. After including the reports of all experts, over 500 documents in total were submitted in evidence. The written argument of the claimants and the respondent contained 138 and 171 pages respectively.

While walking back from one of my nine days in what felt like solitary confinement, during direct by Bill Pearce and Ted Hanman and cross-examination by Gary MacDonald of Ferguson, Gifford - barristers and solicitors for the claimants, I turned to Bill and Ted and laughingly said, "You know, I could write a book on this case." Bill said, "You should!" Now that I have provided a summary of the background leading to the hearing, I will try to confine myself to the highlights - or lowlights - of the actual hearing and its aftermath.

From letters dated August 9 and 16, 1990 from Bill Pearce, the terms of reference were established for my consulting work on behalf of the Ministry of the Attorney General. Arrangements were made for Wright Engineers Limited of Vancouver , with their well-established and respected background in mining capital and operating costs, to prepare an economic valuation of the Sherwood property.

Their valuation would be based on a "fair market value" but would also include a "break-even analysis." The latter is a calculation which has been little used by the mining industry in the past, as most valuations are completed at the pre-feasibility or feasibility stage to determine the rate of return which a reserve of a particular tonnage and grade will generate with all other costs assumed. The break-even analysis is designed to establish the variations of production levels based on ranges of reserve tons and grades that will produce a zero net present value for a project. This would occur where the annual inflows (revenues) of cash would exactly offset the outflows (costs). Projects lacking sufficient tons and grade would not reach the break-even level.

Essentially, my role was to determine the tons and grade of a reserve based on available information and the probability of discovering additional reserves to reach an acceptable return on the required investment.

During one of my early conversations with Bill Pearce, I had commented on my belief that the owners of expropriated mineral claims with no obvious marketable value should at least be compensated on the basis of their past expenses, plus interest and possibly a penalty for loss of opportunity. I had previously cited this opinion on several occasions in writing. Bill disagreed on the basis that the actual cost of work done has no direct bearing on determining the value of the property.

Casamiro actually completed very little work on the property as it was prevented from carrying out any significant work in the year before expropriation occurred by the delay in being granted a resource-use permit.

The Expropriation Act's basis for compensation was based on "fair market value" determination, and did not affect the conclusions and recommendations in my report dated October 17, 1990 completed for the Ministry of the Attorney General. Nevertheless, I was strongly of the opinion that the provisions of the Act appeared inapplicable to properties such as the Sherwood Mine. Accordingly, I wrote the following letter on December 3, 1990 to the Honourable Russ Fraser, Minister of the Attorney General for British Columbia, with copies to the Hon. Jack Davis, Minister of Energy, Mines & Petroleum Resources; the Hon. Ivan Messmer, Minister of Parks and others.

Dear Sir:

Recently, I completed an assignment concerning an evaluation of mineral claims in a Class "A" park in British Columbia which were judged to have been expropriated by an Order-in-Council prohibiting the issuance of resource permits for mineral exploration within the park. Although not part of my terms of reference, the assignment aroused my interest in settlements provided under the Expropriation Act ("Act"), whereby compensation is based upon the "market value" of interest in the land, which is defined as "if it had been sold at the date of the expropriation in the open market by a willing seller to a willing buyer."

Necessary reliance by those engaged in the high-risk minerals exploration and development business on an Act which appears to have been formulated to provide settlement provisions for real estate with its more stable short-term market, appears an unreasonable and unfair condition. Consider the following major differences.

1. Historically in Canada, development of mineralization discovered on a mineral claim into a viable producing mine occurs at a rate described as ranging from about 1:1,000 - 1:5,000.

2. Successful exploration is generally the result of persistent effort with a time span of about ten to twenty years between the start of exploration and the attainment of production.

3. Within the exploration and development period, the market value of shares traded on a stock exchange by a successful company may be extremely cyclical and highs and lows within a year may commonly have a tenfold range.

4. The fair market value of any individual property based upon the Act's definition can only be compared with transactions concluded on similar properties which can vary considerably depending on the relative negotiating ability of the buyer and seller. Depending on the stage of the property's exploration, the valuation may be highly subjective and even cash flow evaluations based on properties containing proven reserves with applied discounts, operating and capital costs, and recoverable metal values considered to lie within acceptable levels of confidence have led to many unprofitable situations.

5. Over 95% of mineral exploration carried out in British Columbia has been completed in areas where the real estate value of land is not considered assessable for other than mining purposes.

A topical example of the above criteria is provided by the history of the Eskay Creek deposit in the Iskut River area, which has attracted so much recent media attention. Exploration activity dates back to 1932 when a syndicate directed by the late Tom McKay staked the area. The property holdings were optioned to Premier Gold Mines, who undertook a program of trenching and diamond drilling on the numerous outcropping gold-silver zones in the area. Most of the work was concentrated on the southern and central portions of the property. It was during this period that most of the showings were discovered and named, such as the #21, #22 and Mackay zones. Premier Gold Mines relinquished its option in 1938; however, McKay maintained active interest in the property through publicly and privately financed companies and option agreements. Prior to the Calpine Joint Venture (Calpine Resources Inc. 50%: Consolidated Stikine Silver Ltd. 50%), 11 companies explored the property, undertaking various diamond drill campaigns totalling over 13,000 feet in 34 holes, plus underground development on the Mackay and 922 zones. The #21 Zone was trenched and drill-tested by Premier Gold Mines in 1937 and Kerrisdale Resources in 1985.

Calpine announced its discovery in November, 1988, during an initial ($300,000 budget) phase of a $900,000 program, to earn a 50% interest in the property. Five of six holes were planned to test the #21 Zone and its possible extensions. One of them intersected a highly significant interval containing 96.5 feet grading 0.73 oz gold/ton and 1.1 oz silver/ton. Calpine shares rose from $0.25 to $2.50 during November 1988 on the strength of this announcement.

By September, 1989, following further drilling on the extension of the #21 zone, the stock reached $9.75 for a 40-fold increase in less than one year.

Although I have not made extensive comparisons with treatment provided in expropriation of mineral properties in other parts of Canada, enquiries were made to determine settlement practice in Yukon Territory and Alberta, the nearest jurisdictions to British Columbia.

In Yukon Territory, where settlement provisions are under federal jurisdiction, two alternative modes of settlement were authorized by an Order-in-Council for mineral claims extinguished in Kluane Park in 1973 (Appendix A). All 21 property owners elected an option whereby:

"Compensation would be made for the expenses incurred in the acquisition, proving and development of the mineral rights (but not including the operation costs of a producing mine) less any monies received from the Federal Government through cost sharing assistance plans, plus an incentive bonus of up to 20% of the net out of pocket expenses of the claim holder."

In Alberta, similar settlements have reportedly been provided by the Government for expropriation of mineral lands, and the basis for this is contained in 'Mineral Rights Compensation Regulations' under the Mines and Minerals Act, a copy of which is attached as Appendix B. The latter provides, inter alia, for compensation payable in respect of money fairly and reasonably expended in the exploration for or the development of minerals at the site, and an allowance, in an amount determined by the Minister, for simple interest on the amount expended.

As a determination of fair market value for expropriated mineral properties appears to be rarely attainable to the satisfaction of the expropriated party, I believe that the options cited above for Yukon Territory and Alberta provide a basis for a settlement that would be considered fair and reasonable under most situations in British Columbia and would reduce or eliminate the burden imposed on expropriated property owners by litigation proceedings and costs. Exceptions are those rare mineral properties at more advanced stages of development with probable or proven reserves which would appear capable of a viable mining operation.

Yours truly,

The letter received favourable responses. In 1992 I included my letter in a submission to the "Report of the Commission of Inquiry Into Compensation for the Taking of Resource Interests" (the Schwindt Report), which included so many adverse submissions from members of the mining industry that Schwindt's recommendations were not adopted. Although the government indicated a commitment to resolve the compensation on resources issue in the Spring Session 1993, it remains in limbo as I write in December 1997.

I was officially retained to provide a geological assessment and valuation report on the Sherwood Mines Ltd. mineral claims by letter dated August 9, 1990. I spent most of the next two months reviewing the extensive references available on the property; similar narrow gold-silver bearing veins on Vancouver Island, particularly three properties having recorded profitable operations in the Zeballos mining camp and non-economic properties in the Bedwell River area. In addition I completed three examinations of the Sherwood Mine area. The first of these was on October 2, 1990 followed by two others on August 2 and September 2, 1991.

I was not alone during my visits to the property. A number of individuals who were to be involved in the eventual Expropriation Board hearing in Victoria were also present. We arrived on October 2, 1990 at the helicopter base near Campbell River on Vancouver Island that morning. The group included Bill Pearce and Ted Hanman of Cox, Taylor & Bryant - a legal firm in Victoria, who was to assist Bill during the hearing. Others included Denis Moffat and Ron Lampard of the Ministry of Parks; Robert Mouat of Wright Engineers who was also to act as an expert witness and George Alvarez, a Ministry of Mines Inspector.

We were fortunate in having a relatively clear day for our outing. We flew to the property by helicopter and were much impressed by the rugged topography in the Drinkwater Creek area. I took many photos of the surrounding area and some of the underground workings. Several were to be included in my report. During the subsequent hearing, Mr. Coates, Q.C., one of the Expropriation Board panel, on hearing evidence which I provided on the underground examination, suddenly blurted out, "Did you take any samples?" in an imputative tone - or so it seemed to me. The site having been declared a park, sampling was of course not permitted and I assured him that I had not.

We had difficulty accessing the No. 3 level at the property as our progress into the drift was barred by a mound of caved rock. Luckily I found an old shovel nearby and I was able to clear away enough of the pile so that we could proceed. The walls of the drift were coated by a rusty crust deposited by percolating water running through fissures from above and it was difficult to find any fresh surfaces. However, one of the photos I took of Bill in the foreground and George behind clearly showed the width of the narrow quartz vein, which was about 8 - 10 inches wide. The photo was included in my report and may have prompted Mr. Coates' enquiry.

I also took several photos from the top of Della Falls looking north-easterly across the Drinkwater Creek valley toward the Sherwood Mine which showed the prominent rock slide which we called "The Cut" defining the western extremity of the mine workings against a prominent fault traceable southwest through Della Lake. The photos were also included in my report and I was to cite the photographic and government geologic mapping evidence of the faulting as the reason that no extension of the Sherwood vein was ever located to the west of The Cut. Interestingly enough, a 'rusty' coloured patch of vegetation visible as a small clearing to The Cut in one of the photos was to be claimed by Nick Carter as evidence of an extension of mineralization. My photo was taken in the fall and the 'rust' was merely fall coloration as shown by a similar photo taken in the summer without coloration during my second visit in 1991.

A critical component of any mining property evaluation is the proven, probable or possible reserve that has been indicated or inferred from previous exploration - plus the geologic reserve that may be postulated on the basis of favourable target areas defined by geologic, geochemical or geophysical surveys. This determination was of course the object of my report. The gold-bearing Sherwood Vein was discovered in The Cut in 1938 by prospecting and apart from some trenching completed on surface to expose the vein to the east of The Cut, other exposures of the vein were only those in the underground workings.

All the underground work was completed by Pioneer Gold Mines in 1940 and 1941. It included 849 metres (2590 ft.) of drifting[27] on three levels: No. 1 (1443 m. elev.), No. 3 (1375 m. elev.) and No. 7 (1,214 m. elev.) and minor drifting on two sub-levels (No. 5 and No. 6) between the No. 3 and No. 7 levels. In addition 88 m of raising[28] was completed on the vein between the No. 5 and No. 7 levels. I considered the most reliable grade information to be that available from sampling and assaying results summarized in the Pioneer Gold Mines' 1941 annual report. It reported an average gold content of 1.26 ounces per ton across 30 inches for 185 feet of drift on the No. I level and 1.36 ounces per ton across 15 inches for 254 feet of drift on the No. 3 level.

There is a paucity of data available on the extent of any mineralization on the No. 5 level and only minor information on gold contents in the raise between the No. 5 and No. 7 levels. However, Mr. B.W.W. McDougall, a mining engineer, reported on the property in 1944 for Cangold Mining and Exploration Co. Ltd. " . . . no ore of apparent commercial consequences is indicated on this level". This opinion is corroborated by sampling completed by later workers.

Dr. Hartley Sargent, a geologist with the B.C. Dept. of Mines, mapped the Bedwell River region in the late 1930s and early 1040s and examined, mapped and sampled the mineral deposits of the area in considerable detail. In his 1941 report he commented as follows on the Sherwood Mine 'ore shoot' and its exposures on the No. I and No. 3 levels:

In both adits the greatest ore widths are at the portals and these widths diminish inwards. Values are cut off abruptly in both levels in such manner as to strongly suggest an ore shoot raking westerly at an angle of about 70º. It is also plainly indicated that a considerable and perhaps the main portion of this shoot has been removed by the erosion of Drink-water Valley.

I referenced the above in my report as part of Dr. Sargent's description of the Sherwood Mine mineralization as it clearly substantiated that there was no extensions of the shoot in the balance of the workings both to the east, away from The Cut, or at depth, as then known.

In my research, I was able to locate eight previous ore reserve estimates on the Sherwood Mine, prior to my own. All of these were presented and discussed in my report in a chronological order and included the following summary:

Author	Date	Tons	Gold (ozs.)	Silver (ozs.)	Class
B.W.W. McDougall	Oct. 1944	27,830	13,195	–	Probable & Possible
R.T. Heard [1]	Dec. 1986	50,000	75,000		"Guesstimate"

---

[27] Drifting: Excavating a horizontal underground passageway along a vein or related structure.

[28] raising: Excavating a vertical underground passageway along a vein or related structure.

[1] These reserves are one and the same as discussed below. In reviewing the above, it is evident that ore reserve estimates completed by McDougall (1944), Heard & Carter (1989), G. McLaren (1989) and Barr (1990) are in a separate range of contained ounces of gold ranging from 13,915 - 22,064. These contrast sharply with a much higher group of four estimates attributable to Heard, Heard & Carter and Heard et al,

Min. of Mines Map[1]	1988	49,500	73,755		
Heard & N.C. Carter	Feb. 25, 1989	46,630	19,070	33,574	Geological
Heard & Carter [1]	Mar. 10, 1989	49,500	73,755		
Heard & Carter	Mar. 19, 1989	47,620	48,334	88,573	Inferred
Heard et al	Nov. 1989	56,914	78,200	133,805	Inferred
G. McLaren	Nov. 1989	45,966	22,064	49,643	Inferred
D.A. Barr	Sept. 1990	30,510	19,938	34,999	Probable & Possible[1]

Note: (1) Shown separately

Since the value to be placed on the property would presumably be highly influenced by the quantity of gold and silver delineated in reserves and the degree of reliability of these reserves being present, i.e. whether proven, probable (indicated) or possible (inferred), a critical examination was required of significant differences shown by the various estimates. Why then were the two groups of estimates so different?

The lower range of reserve established by the four estimates described above, including the February 25, 1989 estimate by Heard and Carter, all followed reserve estimation parameters accepted by the mining industry. Most importantly they all included averaged assays from samples taken at reasonably regular intervals in the drifts and allowed for dilution of those values at zero grade from the averaged, relatively narrow widths of the veins, to a one metre or 1.5 metre width for mining purposes. This is standard practice in the industry.

The first reserve estimate reported by Heard in December, 1986 of 50,000 tons grading 1.5 ounces gold per ton was not really an estimate and would more properly be called a "guesstimate". It was to lead to a bizarre situation with an unexpected "chicken or egg" dilemma, which I finally resolved after considerable research. Heard's actual reference follows:

Reserves have not been addressed due to a paucity of data points. Using the published data in Minister of Mines reports, which is at this time unsubstantiated, one can generate 50,000 tons of possible reserves having an average grade in excess of 1.5 ounces gold per ton.

Heard's report and the above reference was soon noted by the B.C. Ministry of Mines geological division in one of its preliminary maps, without recognition of source.

In their first joint report dated February 25, 1989, being document No. P346 dated March 1989 in a list of documents provided to the Ministry of the Attorney General by Ferguson Gifford, the legal firm acting on behalf of Casamiro, Carter and Heard conclude that:

The geological reserves calculated for the Sherwood Vein are considered to be a much more realistic estimate than the 45,000 tonnes grading 51 grams/tonne gold reported for the property by the Ministry of energy, Mines and Petroleum Resources on Preliminary Map 65, 1988.

ranging from 48,334 - 78,200 ounces of gold, or up to 3.5 times the range of gold contents of the former group.

The implication of course is that the reserve was <u>derived</u> by the Ministry, whereas it was the metric equivalent of the 50,000 tons grading 1.5 ounces gold per ton guessed at by Heard only two years earlier!

As noted in the preceding tabulated summary of ore reserves <u>all</u> the four significantly higher reserve estimates were attributable to Heard and Carter. What happened to influence the drastic change to 2.5 times the inferred gold content within one month of their initial report and 4.0 times the amount within two months?

The answer is quite simple. They failed to allow any dilution of the actual assays within the vein in extending the values out to a 1.5 metre mineable width. In addition they extended the reserve down to the No. 7 level where, as noted earlier "no ore of apparent commercial consequences is indicated on this level." Industry practice in such instances would only permit an extension of half the distance from one sampled level to the next.

Considerable time was to be devoted by the claimants represented principally by Carter, Heard and Ross Glanville, the latter a mining engineer and financial analyst and by the respondents represented by myself and Robert Mouat plus many other witnesses on the actual and potential reserves of the property. An equal or greater emphasis was to be placed on comparisons of the Sherwood property with other 'comparable' properties in determining the probable market value.

In my initial report I was only to concentrate on 13 other narrow vein-type gold prospects and two former operating mines (Bucaneer and Musketeer) in the Bedwell River area, neither of which was profitable. One of my conclusions was that:

The Sherwood Vein has demonstrated the greatest average width and estimated ounces of contained gold within probable and possible reserves, which include the former short-lived mining operations on the Musketeer and Bucaneer properties in 1941-42. Neither of these two operations were viable at that time, the Bucaneer announcing reasons for closing when even lessees were unable to win operating profit from narrow ore bodies outlined.

My final conclusion, which was to remain unchanged after completion of other detailed studies on 'comparative' deposits in British Columbia and days of testimony, follows:

Economic parameters developed by Wright Engineers for a viable gold-silver bearing narrow vein deposit indicate that none of these types of deposits, even those of the Zeballos camp that were viable in the 1930s and 1940s would be viable in today's economic climate. Consequently, the writer concludes that it is highly unlikely that the Sherwood property contains a gold-bearing vein type deposit that would be viable under present economic conditions, given the results of past exploration and production of these types of deposits in the Bedwell River area. Accordingly, a prudent investor would be expected to seek a more promising area in which to anticipate an acceptable return for the risks involved in participating in an exploration venture based on gold-silver bearing vein deposits.

Following completion of my report in October, 1990 there was a lull for me in activity related to the Sherwood property until Examinations for Discovery of various future witnesses in the hearing were arranged. These took place in Vancouver in January and Victoria in June, 1991 with Gary MacDonald appearing for the claimants and Bill Pearce for the respondent. All the key witnesses, including myself, were examined and, of course, this generated lots of research for both sides to be used in the future hearing.

On August 2, 1991 I visited the Sherwood mine area once again, this time in the company of Bill Norquist, Robert Mouat and Dave Wortman of Wright Engineers; Denis Martin of Piteau Associates and Ron Lampart of the B.C. Ministry of Parks.

My final visit to the property was made on September 2, 1991 with Bill Pearce, Nick Carter, Terry Heard and Ed Skoda. The principal object of this visit was to have an opportunity to visit the nearby PDQ claims, forming part of the Sherwood Mine property and to examine several other veins on the Sherwood claims which had emerged as possible target areas for other vein-type gold mineralization based on early reports of activity on the property in the 1940s. The veins were extremely narrow, being in the order of 2-4 inches in width, with no visible sulfide mineralization. While traversing by myself later in the day, I located a small fragment of rock with a barren quartz-vein and stuck it in my pack as a memento. This mental lapse was to cause me some loss of sleep later - following my response to Mr. Coates' question as to any samples having been taken on my first visit – a year earlier when I was underground - for assay purposes.

The Board hearing commenced in mid-October, 1991. It was held in an old home on Government Road, 2 - 3 blocks south of the Empress Hotel, which had required little to no modification. Each week for the next three months I was to spend Tuesday to Friday at the hearing on behalf of the Crown initially to listen to testimony provided by the claimants and finally in December to spend nine days in 'isolation' as I gave testimony in chief for three days and under cross-examination for six days. It was a lonely and stressful time.

The testimony provided by witnesses led to requests from Bill Pearce for further studies particularly of comparable properties. I benefited considerably from the generous assistance provided by the B.C. Dept. of Mines whose offices were located near the hearing. I was able to very rapidly compile a list of production from 41 quartz vein type deposits on Vancouver Island. It showed that a total of 309,356 ounces of gold were produced from 719,583 tons mined (0.43 oz. gold/ton). Of this amount, 275,686 ounces (89.1%) had been derived from 14 'mines' in the Zeballos Camp, of which only three had been profitable. The word 'mines' is in quotation marks to denote that three of the mines had each produced less than 100 ounces of gold and three others less than six ounces of gold each.

Although not related, a total of 841,823 ounces of gold had been produced as a valuable by-product from the Island Copper mine near Port Hardy in the northern part of Vancouver Island by 1990 from 263.5 million tons mined (0.0031 oz. gold/ton). This was to increase to 1.2 million ounces by mine closure in 1995.

The record of production from vein-type gold deposits on Vancouver Island is such that I would recommend against anyone spending the time and investment required in searching for such deposits on Vancouver Island in preference to other more favourable geologic terrains in B.C. or elsewhere. I have no doubt that the time and funding invested to date in attempting to develop vein-type gold deposits on Vancouver Island far exceeds the financial return.

With respect to the Respondent's 'comparables' on which Mr. Mouat's valuation was based, the board stated:

Accordingly, the board concludes that the limited information given relating to these former gold producing mines, all closed down many years ago, are of no assistance as comparables to be considered in determining the market value of the Sherwood mine. If cash payment is related to the prospect of finding gold bearing ore in commercial quantities, the startling difference in the payment made by Cinta, and the cash payments disclosed by Mr. Mouat, alone confirms the board's opinion that these five mines are not comparable.

Although I was somewhat critical of part of the final action taken by the board in determining "fair market value" as will be described, I was heartened by their conclusions concerning the status of reserves based on the conflicting evidence they received.

Pages 62 and 63 of the Reasons included the following conclusions:

The board concludes that the market value of a mining property bears directly upon the probability or possibility that it may one day become a profitable mining venture. The words, 'probability' and 'possibility' are very significant in law. Probability connotes something higher than a 50 per cent possibility: see Farlinger Developments Ltd. v. East York (1975), 8 L.C.R. 112, 123-4 (Ont. C.A.). On the basis of all the evidence the board cannot conclude that there is a probability that the subject property contains ore reserves of sufficient quantity and grade to justify the expenditure necessary to bring the Sherwood mine into production. The probability that there are sufficient reserves has not been clearly established even by the claimants' own witnesses.

Notwithstanding our finding as to probability, the evidence does establish a possibility that further exploration and development might reveal an ore body. It is this possibility which fuels the interest of the mining industry and it is this interest that creates a market in speculative mining properties. The respondent's witness, Mr. Barr, stated at page 27 of his report:

In retrospect, the property should have been subjected to a more conventional exploration program, having been almost totally dependent on an underground exploration program within one year of its discovery with no known detailed geological mapping, geophysical or geochemical surveys or diamond drilling to provide guidance to the presence of other potential ore-bearing structures or additional shoots within the Sherwood Vein.

After reviewing all the evidence, the board concluded that a reliable indicator of market value was the purchase of Mrs. Lillian Hart's shares in the Sherwood Mine property by Cinta Resources Corp. on July 12, 1984, if properly adjusted to reflect the four years and four months from the purchase date to November 25, 1988, the deemed expropriation date of the claims. With simple interest applied this totalled $289,800. This was then increased by 25% as an adjustment to provide for the entitlement for the claimants to be issued resource use permits in 1987 which were not available in 1984. This raised the inferred market value to $362,000.

Although presumably satisfied with this approach and value, the Board then turned to an alternative source of support and entered into a most questionable exercise - one which paralleled that for which they had criticized Mouat's selection of 'comparable' properties, citing his lack of expertise in various areas. They arbitrarily selected their own from the list of properties appended to Mr. Mouat's report and arriving at "an adjusted value range of $270,000 to $425,000." The board finally settled on a market value for the Sherwood Mine in the range of $350,000 to $375,000 and gave "the benefit of the doubt to the claimants by determining a market value of $375,000 as of November 25, 1988.

Interest on $375,000 compounded annually from November 25, 1988 until paid was also allowed taking into account any monies paid by the respondent on behalf of the claimants and an additional 5% per annum on $325,000 from November 25, 1988 until April 16, 1993 - the date of the board's decision, plus costs. The entire amount approached $800,000.

Interestingly enough, the audited financial statements for Casamiro for the year ending November 30, 1987 listed deferred exploration costs for the year of $436,730, placed a value of $524,938 on exploration and development of mineral properties and identified the 25 per cent interest of Sherwood Mines Ltd. as having a value of $408,745.

Had the B.C. government legislated an expropriation act for early-stage mineral properties as I have previously reviewed for Yukon and Alberta, the settlement could have been reached without the estimated $1.2 million spent by the claimants and respondents in legal costs and those required for the Expropriation Board and expert witnesses.

Several weeks after the end of the board hearing, the witnesses for the respondent were all invited to a "debrief" at a law office in Vancouver. To celebrate the anticipated jocular mood of the occasion I entered the large boardroom with a paper bag over my head. I felt quite foolish when I was immediately recognized, but fortunately only a few of the 20-25 people who eventually participated at the gathering were present. I brought a small tape recorder with me which I placed on a table nearby and recorded the event. With the exception of a few comments made by Bill Pearce, the Master of Ceremonies for the occasion was Ted Hanman. There could not have been a better choice. His quick wit was appreciated by everyone - even those that were specifically targeted for their comments during the hearing.

Although it has been a difficult chore to try to cull his presentation, I have extracted the following as perhaps the most memorable. In his introduction to the gathering he said,

When you look back on all the preparations for appearances at the hearing they all are kind of anti-climatic. You struggle with them . . . frequently for days, weeks and months and then you're on and then eventually you're done and off you walk away and think, Jesus what did I accomplish?

There were just a couple of humorous passages that I took out of the transcripts (I interjected "just a couple?") . . . Well more than a couple these were the ones when you were on your knees Dave . . .

Now, see if you can guess who these are from. We start out with Gary MacDonald at his best, in cross-examination and he's beating up on this witness quite severely and the witness is in full retreat and is about to admit that everything he said was wrong . . . He's talking about the cost of tunnelling into this mountain, and so it starts off:

Question: How far are you going to drive these headings?

Answer: Well, I don't really know... you have to really sit down and think about that, because nowadays I don't think anybody would do it... eh? I can't imagine a mining operation nowadays, climbing up that hill with horses and driving those headings by hand, the way they did... it's impossible!

Question: I agree, this property would be approached in entirely a different way in 1988... the question to you was... (the witness then interjects before the question is completed)

Answer: Yes... you'd walk away faster.

Question: Would you hazard an opinion for the Board (here Ted interjected with 'God, I love those kind of questions...') to the fact that Pioneer may have ceased operations on this property, due to war-time conditions?

Answer: It could be... In Mexico they tell you it's because of the Revolution.

The witness was David Wortman, a long-time mining engineer with Wright Engineers Limited who had visited the Sherwood Mine with our party in 1991.

There's lots of others... unfortunately Dave Barr's succinct knives in the heart were buried in nine days of transcript... so I gave up on him... and Robert Mouat's were buried in statistics, so I gave up on him.

Later in his reminiscences, Ted focused on a couple of presentations that Bill and he wished to make.

Now and then Counsel gets involved in a case like this and you really understand virtually nothing... about the topic you're involved in... you rely so much on the witnesses who come and help you with it... and at the end of the day the lawyers recognize... certainly Bill and I did here... that the witnesses put in huge amounts of time and effort and really did a superb job in trying to educate some pretty stupid people... Bill and I... about very complex topics and in turn tried to educate some even stupider people... the Board... To that end we've prepared a couple of small tokens of appreciation... we couldn't recognize everyone because it was just not possible... but there's a passage in one of the questions and again it's a tribute to Mr. Wortman where he talks about the helicopter pilot being somewhat out of his mind for landing on this landing pad and letting everyone out on that day when there was Wortman and Dave and I and Bill and Robert who visited the site... and as we were climbing into the helicopter... I took from the helicopter landing pad... and the Parks people might want to know this if you're ever back there... a large shackle which was hanging at the top of... trust me... is from the mountain and I stuck it into my pocket and I've carried it with me ever since, knowing that there would be a day like this... and on behalf of Wright Engineers I wonder if Robert would come forward and... Robert you can use it as a paperweight...

Now the guy who holds the record here for suffering longest... is Dave Barr... now by the end of his testimony we were really concerned about his general health... he said, 'yes' to almost everything that was asked of him... the good news is that he left them (the Board) so generally confused that they couldn't figure it out. We have something here for him... and I know that Bill here who spent endless hours working with him and dealing with him... there was a point in the testimony that most of you won't be aware of... where there was grave concern mounted by the other side that we actually knew that there was extensive amounts of gold in this mountain because we had gone in and sampled it... secretly and hadn't told them. Now we maintained until the very end that we hadn't sampled anything... Now, lo and behold, Dave came to us one day shortly before he was going on the stand and confessed up to Bill... (Bill continued)... Dave went in with Wright Engineers in August of last year and walking around he picked up a piece of rock on the ground and during the hearings they made a big issue of the fact that they weren't allowed to take any samples and meanwhile poor Dave confessed to me that... 'Oh by the way I brought this piece of rock back... what do I do with it?... ' So I said, just give it to me and I'll put it in my briefcase... If they mention it on the stand... just say 'Yea I took it'... and I'll produce it. Meanwhile poor Dave lived in fear for two or three weeks that they would ask a question and he'd have to say 'Yes I took it... a sample'. So as luck would have it they never asked the question. Anyway, we would like to present you with a plaque with the missing sample which grades out to about four ounces of gold to the ton! (Laughter).

The plaque and 'sample' grace my office wall and are a constant reminder of a commitment I made after the hearing, "That's it as far as being an expert witness is concerned." A couple of years later Bill Pearce phoned me to enquire whether I would appear on behalf of the Crown concerning similar settlements involving mineral claims in the Alsek Tatshenshini Wilderness Park expropriated in June, 1993. He did not realize how emotionally involved I had been in writing to the government and in the media against the park decision, and I, of course, declined. The subsequent settlement to Royal Oak Mines for the Windy Craggy Mine expropriation was to cost the government (and us taxpayers) about $166 million in cash and subsidies. It was also to be recognized as the principal cause of the rapid decline of investor confidence in the mineral resource sector of British Columbia, which is still being felt as I write on December 31, 1997.

Survival Adventures & Searches

I have been fortunate indeed to have had the opportunity to spend a considerable amount of time in mineral exploration in the Coast Mountains region of British Columbia. Much of this work was done in the 1950s through 1965, with the aid of helicopters which permitted access to remote regions, some of which may not have been visited before. The area is one of rugged grandeur and spectacular scenery in alpine settings. To the explorationist, the area lying below timberline at an average of 4,000 feet above sea level, is generally extremely rugged and ranges from difficult of access to impenetrable.

The Coast Mountains district, apart from being rugged, also has some of the highest precipitation recorded on the West Coast. In the summer months, there can be lengthy periods of relatively clear skies; however, more commonly the reverse is true and adverse weather can seriously inhibit travel by aircraft within the region.

My exploration work in the Coast Mountains was quite diverse although most of it was completed in the region extending from the B.C.-Yukon border southerly to the head of Observatory Inlet, about 80 miles north of Prince Rupert. The area lies immediately east of the Alaska Panhandle and includes one of British Columbia's most well mineralized belts on the eastern margin of the Coast Plutonic Complex. The area is largely drained by Stikine River and its principal tributary, Iskut River. My work included: (1) staking programs - two in mid-winter in the Barrington River and Chutine Lake areas; (2) over 50 property examinations; (3) over 50 follow-up of geochemical anomalies in 1959-60; (4) acting as co-party chief with Dick Woodcock in a major helicopter-supported geochemical reconnaissance program in 1959 which extended from north of the B.C.-Yukon border to the Scud River area, and party chief of the extension of the same program southerly to the head of Observatory Inlet in 1960. The area covered by this program exceeded 10,000 square miles and led to the staking and later discovery of the Stikine copper deposit; (5) supervision between 1961-1965 of a major exploration program at Stikine Copper.

In 1954, I also supervised the first helicopter-supported exploration program undertaken by Kennco Explorations. The work was based at a tent camp on Bute Inlet and included prospecting and geological mapping of an area of about 3,500 square miles extending northerly to Knight Inlet and Klinaklini River, southerly to Toba Inlet and Toba River, easterly to Chilko and Tatlayoko Lakes and centred on the Mt. Waddington Homathko Icefield regions.

My experiences in these programs were to convert me to the use of helicopters in remote areas and to engender a healthy respect for the wilderness. Much of the work was in the company of relatively inexperienced persons like myself, and I was impressed at the rapidity with which assistants, mostly university students, developed self-sufficient attitudes. In our reconnaissance program, they worked from fly-camps which were moved at five- to seven-day periods throughout the season to locations extending as far as 40 miles from base camp.

Survival

There must be few explorationists who have not experienced an unexpected night out with some discomfort caused either by being overextended in their traverses or, more commonly, by a proposed rendezvous with an aircraft that did not materialize. I experienced this on three separate occasions, excluding the frequent times when weather had delayed a flight, when I have been accommodated at a base camp or in the back of a pickup truck. Two of these were in the Coast Mountains portion of the Stikine River district which, incidentally, includes a Night-out Mountain, about 12 miles south of Telegraph Creek.

Most night-out situations result in little more than minor inconvenience, depending on the degree of mental and physical preparation. The Safety Manual for Mineral Exploration in Western Canada, which was developed following my recommendation to the British Columbia & Yukon Chamber of Mines in 1980 says:

> Many fatal accidents can be directly traced to a lack of proper preparation, usually related to a mental attitude called the 'It can't happen to me' syndrome.

The section dealing with 'Survival' indicates adequate procedures to be taken to minimize risk.

Early explorationists and others living in the wilderness learned how to coexist with wilderness conditions the hard way or, more commonly, by following the advice of more experienced companions.

Barrington River

My first night-out experience was in July 1959 when I was dropped off by helicopter on a ridge above Barrington River at an old Amax Exploration Inc. molybdenum property lying about 10 miles due south of our base camp at Little Tahltan Lake. At that time, being that close to base camp, we did not leave a safety kit in the event of an overnight stay. When the helicopter failed to pick me up at about 5:00 p.m., I lit a fire and settled down to wait. When nightfall came, I had already climbed up to an old cache where a weather-beaten tarp covered some cans left by the last occupants. Wary of what I ate because all of the cans had been subjected to freezing and were probably at least a year old, I selected some preserved fruit for both dinner and breakfast. I got up several times and placed more wood on the fire and actually enjoyed several hours of sleep. When the helicopter failed to appear the next morning, I set off knowing that the only real obstacles were two tributaries to Barrington River. One I waded with the assistance of a stabilizing pole and the other I crossed on a large snow bridge created by the remnant of a slide on the mountainside. I arrived back at camp without incident in the late afternoon to learn that our helicopter had encountered mechanical problems and would not be returning to base camp until the following day, at the earliest.

Galore Creek

September 26, 1962 - although almost 30 years ago, I remember these events as though they happened yesterday. My memory is aided by having written my recollections some 15 years later while en route back to Vancouver from a trip to Winnipeg.

We had arrived at Galore Creek for the summer exploration program which would lead to recognition of the emerging major copper deposits at the head of the basin which contained our base camp. By late September, our summer field program had been completed and our on-property complement had been reduced to about 20 personnel completing diamond drilling and geophysical surveys.

Nearing the end of our summer program, we still had several third-order stream sediment geochemical anomalies, resulting from our 1960 program, which had not been followed up. One of these was a sample obtained on a small creek, a tributary to Scud River, and located east of the mouth of Galore Creek, about 10-12 miles north-northeast of our base camp. I decided to follow it up, given an opportunity, so on the morning of September 26, 1962, Gerry Rayner and I were flown in to a centrally located site in a Hiller 12E helicopter. Arrangements were made for a pick-up that evening on a ridge near the mouth of Galore Creek, between 5:00 and 6:00 p.m.

Gerry and I spent an energetic day clambering up several creeks in the anomalous area and completing geological and geochemical reconnaissance work. In the latter part of the afternoon, it started to rain and later, we could see that it was snowing at higher elevations. Thoroughly soaked, we arrived at our pick-up location, a relatively bare ridge at timberline containing several juniper bushes. Getting colder, as we waited for the helicopter, we succeeded in setting fire to one of the bushes and were soon reasonably dry and warm.

At about 7:00 p.m., we realized that the helicopter must have had a problem and as it was rapidly getting dark, we headed down the ridge about half-a-mile into a well timbered area. We both set to work building a lean-to as we had no tent, sleeping bags or emergency supplies. This was an oversight as we had taken such equipment with us when we were plane-table mapping on the South Butte, a precipitous rock buttress forming part of the south wall of Galore Creek basin, and surrounded on all sides by glaciers, but only about two miles south of our base camp.

I left Gerry to complete the lean-to and started collecting branches for a fire. To cut a long story short, the wood was all so wet, even that partly protected at the base of the trees, that our combined efforts were unsuccessful. By this time, we were both thoroughly soaked again and sat down in the lean-to, which started to drip as the evening wore on. We eventually lay down, huddled together, trying to generate some warmth. At about 10:00 p.m., we were so frozen that we got up and thumped each other for several minutes to regain circulation. We repeated this energetic activity at half-hour intervals when shivering became too unbearable. Later, we even decided to start walking slowly up the ridge although it was almost pitch black and raining. For some reason, my night vision seemed a little better than Gerry's, but it was the blind leading the blind. We were soon walking in wet snow and not too much later decided to turn back rather than risk a serious fall. Miraculously, we regained our wet lean-to and spent the rest of that interminable night shivering and thumping each other.

Gerry dozed briefly several times and perhaps I did although that is not my recollection. My eyes ached from straining periodically to catch an indication of approaching dawn. It finally came at about 5:30 a.m. Neither of us lost any time in starting up the ridge, as we had decided that neither of us could face another night under similar circumstances. We had little to carry and only one orange left between us from our lunch the previous day. We soon reached fresh snow and we later learned that one inch of rainfall had been recorded at our base camp rain-gauge during the latter part of the previous day and throughout the night. The rain and snowfall tapered off shortly after we started our trek. We had decided to traverse above timberline in a southerly direction, parallel to Galore Creek, and eventually cross the east fork of Galore Creek via the glacier, whose terminal lake at that time was a little over one mile from the confluence of the east and west forks of Galore Creek. We had never flown over the area nor had we ever traversed any portion of the route.

After leaving our lean-to, we climbed steadily and gained an altitude of about 5,000 feet, which we maintained for several hours as we crossed the heads of five or six gullies trending westerly into Galore Creek. After about one and a half hours, we stopped briefly to eat our orange. We had to lose about 1,000 feet of elevation in order to gain access to the toe of a glacier at 4,000 feet, which trended south-easterly for about two miles from an ice field about five miles southerly from our lean-to location. As we climbed the glacier, we were soon enshrouded in fog, with visibility restricted to about 100 feet. There were few crevasses to contend with in the lower part of the glacier, but as we climbed they increased in frequency until finally we came to one crevasse which extended from one wall of the glacier to the other. The walls were precipitous and did not provide any safe means for crossing the crevasse. In its central portion, the crevasse narrowed to six feet and contained several narrow snow bridges, the thickest of which was about two feet. Because of the upward slope of the glacier and the lack of traction, we could not risk trying to jump across. The thickest snow bridge looked very marginal.

We stood and discussed our options. If we turned back, we would have to drop back down to the toe of the glacier then climb up and try to make our way south along the summit ridges and icefalls. Alternately, we could try heading down below timberline and bushwhack our way through the thick undergrowth and across the numerous deeply incised creeks. We never seriously considered either alternative.

We finally decided on the thickest snow bridge. Even now, after all these years, I still feel the fear as I lay down and started inching across, pulling myself by digging in my fingers with Gerry hanging on to my ankles as long as he could. The crossing did not take as long as I am taking to write about it. When I gained the other side, it was Gerry's turn. I reached for his outstretched hands as soon as I could as he pulled himself across.

We continued upward and although we encountered several other crevasses, we bypassed them without problem. About noon we reached the summit area at an altitude of about 6,000 feet, and we eventually started downward, still in fog but fortunately we broke out of it quickly, only to realize that we had wandered over 90 degrees off-course and were heading down an ice field which was the source of a northeast trending glacier heading back to the area we had traversed. We turned back, begrudging our lost altitude but grateful that we had recognized our change of course within a drop in elevation of about 500 feet. Back in fog, we relied on compass bearings in the extensive featureless ice field area which we followed for about one and one-half miles before starting to lose elevation again as we approached the north side of the valley containing the east fork of Galore Creek.

We dropped into a steep gully which descended to an elevation of about 2,500 feet immediately above the north margin of the glacier occupying the east fork of Galore Creek. We were both extremely tired at this point and anticipating no problem in gaining access to the glacier. As the gully narrowed, we reached a point on the north wall of the glacier from which the dry creek bed dropped vertically for 60 feet. Unwittingly, we had descended a hanging valley whose abrupt termination marked a much earlier position of the glacier. Gerry gallantly agreed to scout a route leading out of the valley at the lowest possible elevation. I was very grateful and, as I recall, we only had to climb several hundred feet before finding an accessible route on to the glacier.

The next two hours were relatively anti-climatic. We crossed the glacier, ascended the ridge on the south side of the valley and traversed through timber until we reached our base camp on the west fork of Galore Creek at about 4:30 p.m.

Those in camp were speechless when they saw us. They explained that on the previous day one of the drillers had severe stomach cramps and in case it was appendicitis, he had been flown by our helicopter to Wrangell and the helicopter had not returned. We had unreliable radio communication and they had no information on the expected time for return. Concerned for our safety, Jim Peaver, a contract geophysicist, and a tall young diamond driller with Midwest Diamond Drilling Company who had lost one lung, had decided to walk out to Stikine River via the Anuk River immediately west of our camp to seek help. Although the distance from our camp to the mouth of the Anuk river along the actual path of the river is about ten miles, their traverse required them to cross the west fork of Galore Creek and gain access to Anuk River via a pass at an altitude of 4,500 feet, a climb of over 2,000 feet vertically, followed by a descent down Anuk River valley, portions of which are almost impenetrable because of thick stands of willow and alder.

We learned that the two had left about two hours before our arrival and might still be within the basin of Galore Creek. We fired off a rifle; later we learned that they did not hear it.

Following a meal, Gerry and I were asleep by 6:30 p.m. and we slept through until noon of the following day. In the afternoon, the helicopter returned and I immediately flew down the Anuk River with the pilot. At about 4:00 p.m., we spotted the pair in the river. They were about four miles from its mouth, totally exhausted and soaked. They had gained the pass and descended to timberline the previous evening where they slept without incident. In a period of about nine hours after continuing down Anuk River valley that morning, they had traversed about four miles, much of it in the river itself. Gerry and I had a far easier time, apart from the obstacles encountered, traversing about 13 miles in 11 hours.

Johansson Lake

The third night-out occurred in 1971 when I was visiting some of Kennco's mineral prospects in the Toodoggone River area with Charlie Ney, our Chief Geologist. Following a visit to the Chappelle property, which was to be developed to production in 1981-3 as the Baker Mine, we left by helicopter to return to Smithers via Johansson Lake and another nearby Kennco mineral discovery – also placed into production as the Kemess Mine in 1997. The pilot landed at Johansson Lake near an unoccupied outfitters cabin to fuel up.

With a full load of fuel, we took off across the lake in gusty wind conditions. It soon became obvious that the pilot was having difficulty gaining altitude. He turned in a different direction without any improvement and suddenly headed around back to shore. He almost reached it, but not quite. We made what is termed 'a hard landing,' partly on the shore and partly in the water. The impact was severe enough that the rotor blade struck and partly severed the tail boom, and in its buckled condition it thrashed for several minutes, repeatedly hitting the water. I was in the centre between the pilot on my left near shore and Charlie on my right in the water. Charlie made a move to open the door and get into the water, fearing a fire as the fuel line had been cut. I restrained him and a little later the rotor blade stopped.

We emerged from the helicopter, shaken rather than injured and I took several photos of the pilot and Charlie looking dejectedly at the damage. We tried unsuccessfully to call out on the radio. We surveyed the contents of the nearby cabin with an eye to an overnight stay. Although there was no food, tents had been hung to dry and we were to use them that night as covers. The pilot had some emergency provisions which we later sampled.

After completing a check of our surroundings, I recalled that three smoke fires about 100 feet apart at the apices of a triangle are an international distress signal. I checked with the pilot to see if the distances were correct, and still shaken, he agreed somewhat disinterestedly. There was a large clearing near the cabin and I spent about an hour gathering branches and topping the piles off with moss plus some aviation fuel.

We had little to do but wait, knowing that Alpine Helicopters would be trying to locate us when we failed to return to Smithers. Late in the afternoon, we were all in the cabin when we heard the distant drone of an aircraft. I rushed outside and started lighting the fires, which took longer than expected. There was little smoke from the first fire and the third one was being lit when the aircraft was overhead. Looking up, I gazed with considerable surprise at a brilliant red flare emitting a trail of smoke which suddenly appeared several hundred feet above and off to one side of the aircraft. I had not known that our pilot had a pencil flare in his pocket. The aircraft made a sweeping turn and landed at the lake. A small float-equipped aircraft, it was fully loaded with three passengers and the pilot. We explained our predicament and the pilot was able to make radio contact, verifying our location and predicament with Alpine Helicopters' base in Smithers. Unfortunately, they were unable to send out an aircraft to pick us up until the next day.

While talking to the passengers and pilot of the aircraft, I asked, "Did you see the fires?" They looked at me curiously and the pilot said, "What fires?" Quite deflated, I pointed to the still smouldering fires nearby. The pencil-shaped "flare-guns" with their spring loaded mechanism, which when released, can discharge flares up to a height of 200 feet, have been extremely useful as a distress signal. Cartridges for the guns also include 'bangers' to attract attention and possibly scare off bears.

Max Portz and Bob Gilroy, Iskut River

In July, 1964, while working in the Galore Creek area, I met Max Portz, 48, and Bob Gilroy, 39, at our Anuk River camp following their rescue from the middle of Iskut River where they had spent eight days on the overturned floats of their Piper Pacer float-plane which crashed after hitting either a log or a sand-bar while attempting a landing on the river.

Bob was a bush pilot, mining engineer and prospector whereas Max had spent ten years in the RCAF as a fighter pilot and had experience as a diamond driller, miner, logger, psychiatric nurse, game warden, and staff instructor in a jail. Max and Bob had decided to spend part of the summer prospecting the Iskut River area although it was Max's first prospecting experience. The pair were friends of Doug Chappell, an experienced pilot and principal in Trans Provincial Air Services based in Terrace. I had met Doug several years before and during the early 1960's, we relied on Doug and his company for our fixed-wing requirements in servicing our exploration program at the Galore Creek copper property. The eventual rescue of the two was, to a large extent, attributable to Doug's concern for the pair after having helped them get established at their prospecting camp on Iskut River near Bronson Creek.

On the afternoon of July 14, they left their camp somewhat hurriedly to make a brief trip in the floatplane to a prospecting site about one mile down the river. On landing there was a violent jolt and the aircraft somersaulted in the middle of the river. Although the cabin of the aircraft was totally submerged, they managed to force a door open and each of them grabbed hold of a float as they rose to the surface of the river which they estimated to be flowing in excess of 15 miles per hour. The aircraft eventually came to rest on a sand bar about 100 yards from either bank and about the same distance from an island downstream.

Resting on the upturned floats, which were six feet long and about 15 inches wide with an inverted V-shape cross section and ridges running the length of the float for stabilizing purposes, they stripped, dried out and took stock of their situation. Initially, they were not particularly alarmed as they were confident that, with considerable daylight left, there was a reasonably good possibility of their being spotted and picked up before nightfall by another aircraft.

Apart from their light clothing, they were poorly equipped, having only personal items on them including a compass and a small nail file. The flooded cabin contained axes, flares, rope, emergency rations, firearms, and jackets; however, the floor of the cabin was about six inches to one foot underwater and several attempts made later by Max to access the cabin underwater were unsuccessful because of the near freezing water temperature. However, with the aid of the nail file, he was able to cut off several fragments of fabric, the largest of which was two feet by 15 inches in size. These were used to try and signal aircraft and to protect them from the cold.

Throughout their first night there was a light, damp wind which chilled them to the bone. They sat back to back, changing positions frequently to take the brunt of the wind in turn. They also linked arms to keep from falling off the floats. After the second night, they changed their method of sleeping, determined to get more rest.

> One of us would lie down flat on his back with his head in the other's lap. The one sitting up would double himself over the person lying down. In this way the person underneath could sleep without running the risk of falling off and remain a bit warmer. When the man on top couldn't stand the cold any more we would change positions. We cut part of our shoe laces off and tied the fabric flags around our necks and backs to protect us from the wind and the rain.

> We cut our socks off at the ankles (using the nail file again), split them, and using the string I had, laced them together, making woollen undershirts; they felt wonderful and warm. One night, about 8 p.m., it started to rain, and drizzled all night and if it had not been for our fabric shawls and our woollen sock undershirts, I don't think we could have stood it.

I have quoted above from a lengthy article written by Max which appeared in the February 13, 1965 issue of the Star Weekly, a Toronto paper, some seven months after their remarkable experience. Max makes frequent reference in the article to the strength that he gained during his ordeal from his religious beliefs. He mentions that he "prayed almost continually" during lulls in their conversation. He appears to have been fortunate in having a stoic, uncomplaining companion in Bob Gilroy, who suffered from asthma. The two spent the first few days recounting their past experiences. When this topic was fully exhausted, they created subjects to discuss.

> We high-lead-logged the slopes of the Iskut, tractor-logged the river bottom, built a road up the south shore for the ore trucks that were hauling from a couple of mines we had staked, had a flat bottom barge to take grizzly bear hunters up the Iskut from the mouth of the Stikine, where we had a beautiful log cabin inn complete with service station for planes, trucks and boats. Business was terrific. We even made up a song to the tune of 'North to Alaska.'

Although they had considered attempting to reach shore, several factors mitigated against any such attempt. The water was swift and very cold and Bob, who was the best swimmer of the two, doubted his ability to swim that distance under the conditions. Initially, they had planned having Bob make an attempt to reach shore while attached to Max on the float by the 250 feet of nylon rope in the cabin of the aircraft. Serious consideration of this proposal appears to have dissipated when they were unable to access the cabin.

They saw several bears in the river, one a very large grizzly which came within ten feet of the floats early one morning. They also saw a wolverine.

As the days passed, they occasionally heard and saw aircraft flying by, one of which flew directly at them one evening, only 20 or 30 feet above the water. Although they waved their fabric flags violently, they were not spotted, nor were they seen when the aircraft returned some time later. They spoke very little the rest of that Sunday evening, the seventh night of their ordeal.

By Monday morning, with their morale at its lowest ebb, they recognized that they were rapidly losing their strength and for the first time, expressed their fears to each other.

During the day, they noticed an increase in air traffic and hoped that the best possibility for the initiation of a search lay with Doug Chappell who was probably the only one knowing their precise camp location. Their faith was to be vindicated for Doug had visited their camp and was alarmed to note that Max's sleeping bag, which had been hung up outside to air, was soaked and that dishes were only partly cleaned and cooking utensils left out to rust. On checking the inside of the tent, he spotted a partly written letter from Max to his wife and saw that Max had a daily entry up to July 13 and nothing following. He flew to Stewart and alerted the RCMP, Department of Transport, and RCAF Search and Rescue in addition to all the bush pilots operating in the Stikine and Iskut rivers.

After another night of rain causing intense shivering, the weather improved. Although they had not eaten for eight days, they were both drinking more water. At about 1:30 p.m. on Tuesday afternoon, a Beaver aircraft piloted by Herman Peterson of Atlin flew up Iskut River about one-half mile away but failed to spot them. It was the fourteenth aircraft that they had seen in eight days. They recognized it as one which had flown a Cominco crew into nearby Bronson Creek above their camp. Being too weak to stand for any length of time, they correctly estimated when the aircraft would return down river and were ready with their flags which they waved violently. The aircraft continued on then suddenly turned and, losing altitude, flew low beside them and waggled its wings several times. They collapsed, overjoyed, on the floats knowing that at long last their ordeal was about over. On board the aircraft with Herman was Bruce Mawer, a Cominco geologist, and his partner, with their last load of supplies being moved out following completion of their program in the Bronson Creek area. Not mentioned in Max's article is that Bruce, who had been looking down at the river, spotted not the flags but a brief flash of light, probably reflected from the aluminum floats.

The plane returned sometime later after unloading the camp equipment and Mawer's partner at a convenient location on Stikine River in order to facilitate rescuing the pair. Mawer got out of the Beaver and stood on one of the floats to help Max and Bob climb into the aircraft. After picking up Mawer's partner and their equipment, they flew north to our staging area at the mouth of Anuk River where they were fed and rested until they were picked up by Doug Chappell. They then flew back to their campsite on the Iskut River, packed everything up and flew to Stewart. Max reached Vancouver the next day via Terrace and was welcomed by his wife who had flown to Vancouver from their home at Rossland to meet him, the trip arranged by Max's considerate employer. Max spent a week recuperating from exposure at a hospital in Rossland.

In November 1989, over 25 years after this remarkable survival epic, I was writing the above summary. I phoned directory assistance in Rossland, B.C. and obtained the telephone number for M. Portz, one of five Portz' listed. His wife answered and I was saddened to learn that Max has passed away in 1988 following a five-year bout with cancer. I explained my interest and learned that Max had suffered after-effects from his ordeal which required surgery to strip nerves in his legs and feet damaged by severe chilling and exposure. He fully recovered after this treatment and enjoyed an interesting and eventful life.

Iskut River Area and Bowser Creek Areas, 1955-56

A prospecting program in the rugged Iskut River area of north-western British Columbia was the scene of a survival episode shared by three students in the summer of 1956. I learned of this in 1980 while casually discussing our company's interest in the area with Al Lonergan at a Mining Exploration Group luncheon meeting in Vancouver. Al was one of the students involved.

"Today whenever I read or hear about people who have survived three to four days without food, I wonder why everyone gets so excited," Al quietly observed, while recalling the event.

In 1956, Al was a student at the Montana School of Mines. He had previously worked as a miner in the Salmo area of British Columbia and was known to Roland E. (Roly) Legge, a prominent Vancouver mining engineer who subsequently hired him for a summer exploration program based out of Stewart in the north-western part of the province. Later in the season, Mr. Legge decided to establish a small prospecting party at Julian Lake, a small lake near the terminus of Lehua Glacier on one of the east tributaries of Snippaker Creek, which drains into Iskut River about 12 miles to the northwest. A one-month program was funded by the McMann brothers of Calgary who had earlier worked as diamond drillers in the Slocan area.

I had completed geochemical surveys in the same area as part of our 1959-60 program in north-western B.C. while working for Kennco Explorations (Western) Limited and the Snippaker Creek area was also the site of the tragic helicopter accident on July 3, 1980 which claimed the lives of four exploration personnel and a pilot in DuPont's exploration program. (See 'DuPont of Canada Exploration Limited – Darkest Days')

Al, John Legge (one of Roly Legge's two sons) and Brian Ledingham were flown into Julian Lake on July 10 by Howard Fowler, a pilot and associate of Mr. Legge's. I recalled that he had also been an associate of Wendell Dawson, one of the discoverers of the Granduc copper mine, whom I had met while examining one of Wendell's properties in the nearby Unuk River area in 1954.

"Howard Fowler dropped us off at this small lake in two loads. We had enough provisions to last us a month. We didn't have a radio, but we knew where we were."

The party had planned to establish their base camp at the lake and backpack to several fly camp locations up to eight miles away, returning to the base camp for supplies on a weekly basis.

"Roly Legge asked us to prospect this area, which as far as we knew had not been prospected by others in the past. There were grizzly bears around and some kermodes (a white sub-species of black bear found in the northern coastal area, rarely seen at present - author's comment), but no other animals except mountain goats. The three of us worked together and I was considered the leader because I was the oldest and working my way through school." Al was about 29 years old at that time. "We looked for gossans, working the sides of the glaciers."

Because of the presence of bears, and with no secure food cache, they wrapped their remaining provisions at base camp in a piece of canvas and tied it up in a tree with rope. "Unfortunately, we didn't realize that grizzly cubs can climb trees whereas the older grizzlies can't. We were away on our third trip for about six days, and when we came back there was a grizzly and two cubs in the area. When we went to get our food, we found that the canvas had been torn down and everything was pretty well eaten with remains scattered around. There was some canned food but the cans had been punctured. Only some tea and coffee remained."

At first they did not worry too much. They had been prospecting for three weeks and Howard Fowler was scheduled to return with additional supplies on August 7th, about a week later.

"We sat and waited, but there was no traffic in the area. We had about ten shells for our rifle. After a day or two, we were getting kind of hungry and we went up to look for some goats. We saw some but after a couple of shots they were gone."

They went out hunting several times in the next few days, but were unsuccessful and they were getting tired. When they only had three shells left, they decided to retain these for protection from the grizzlies. By this time, the weather had deteriorated and they were to be subjected to almost continuous rain and fog for a period of about two weeks.

They stayed in their tent. Al recalled having said, "We'll wait here. We won't go hunting anymore. We'll just sit here and save our strength." A little later, Brian and John informed Al that they considered that the "situation was kind of desperate" and they wanted to walk down to the Iskut River.

Al retorted quite bluntly, "Well, I can't argue with you - but you're going to die. Look, since I'm head of this expedition, I feel responsible and I'm not going to go."

Al then went and got a piece of paper from his notebook. He said to them, "Look, before you go, I'm going to write down that I'm not responsible because I've advised you not to go and this is your decision. I'm going to get you to sign this so that I'll be absolved of any responsibility."

Al told them that he knew that they would never make it out and then started to write the note. "They both looked at me, then looked at each other and then one of them said, 'he and I are going to talk about this some more,' and then they went for a little walk beyond the camp." They later came back and said to Al, "No, we'll stay."

Al expressed his conviction to them that as soon as the weather cleared, Roly Legge and Howard Fowler would be in to look for them. He told them that from his experience, the best thing was to conserve their strength and take it easy. And they did, finding no difficulty retiring relatively early and sleeping through until after noon each day.

They tried eating grass and boiling some berries, but could not stomach either.

A few days later, on about August 10th, just after lunch they heard the drone of a helicopter but the fog was so dense they could not see it and after a little while it left. They later learned that an alarm had been broadcast in that part of the Alaska panhandle about their predicament and a pilot from Wrangell had tried to locate them.

On August 18th, they heard some voices and saw a party of four men backpacking by the lake. The legendary Tom McQuillan and three others were en route back to Tom McKay Lake, about 12 miles to the northeast where they were scheduled to be picked up by an aircraft. Although low on provisions, they shared their meagre supplies with the starving party before they continued on their way.

The weather did not clear for another three to four days, but an aircraft from Wrangell arrived with food supplies that same day and took them back to Wrangell in two flights.

The members of the party suffered no significant after-effects from their experience. John Legge ate too much before boarding the aircraft for Wrangell and was sick for about a day. Al estimated he lost 20 to 25 pounds or about 15% of his original body weight.

When Al again recited this event for me in 1989, some of the details of his experience were difficult to recall. I went to the Geological Survey of Canada office in Vancouver and bought a couple of 1:50,000 scale maps of the general area to determine the actual lake where they had established their base camp. It may not have been named at the time. Al agreed to try to obtain some additional information from John Legge. Much to my surprise, the information provided consisted of photocopies of two newspaper clippings, one from the Vancouver Province and the other from the Northern Daily News in Prince Rupert, both dated July 12, 1955, describing a remarkably similar incident but of much shorter duration involving Roly Legge and Howard Fowler.

Legge, 52, and Fowler, 40, had left Vancouver in late May on a prospecting trip in the Bowser Lake area, about 35 air miles south of Stewart. On completion of their program in early July, they were based at a campsite at the west end of the lake. Coincidentally, a period of foggy weather set in, and the many attempts that they made to fly out in Fowler's Super Piper Cub, trying to find a hole in the fog, used up their fuel supply. "We were also out of grub," Fowler eventually told his rescuers, "and we knew either one of us would have to walk out for help pretty soon if some plane didn't happen to come along and spot us." They spent the next four days conserving their energy, sleeping alternately, and waiting for the fog to lift. Then Legge decided that it was time to set out for help, carrying only one can of meat to eat during his subsequent three-day trek. His route covered an estimated 56 miles, of which 30 miles were over hard-packed snow on the Salmon Glacier and a final nine miles along a dirt road from the former Premier Mine into Stewart.

Doctors who attended on Legge at Stewart marvelled at his feat. "Battered and nearly starved, he was in a state of shock and exhaustion when he reached Stewart, but otherwise in good physical condition." The Province article notes that, "Fowler was in good condition and was not hospitalized. He slept 20 hours a day while Legge hiked out."

Fowler's rescuers were Captain Jack Anderson, 33, pilot with Queen Charlotte Airlines, who later formed his own company and partly supported our Baker Mine exploration program in the early years, and Emerson Wallace, 34, Chief Pilot for Granduc Mines. They flew from Prince Rupert on the evening of July 12 in a QCA Cessna 180 and located Fowler without incident by his pup tent on Bowser Lake. He was flown back to Stewart by Wallace in Fowler's Super Piper Cub after it was gassed up. Anderson accompanied them in the Cessna.

Gary Carstensen, Tofino to Zeballos - late 1940s

Gary Carstensen was an engineer for Okanagan Helicopters Limited, who supported our helicopter-based exploration program at Bute Inlet, B.C. in the summer of 1954, which I directed for Kennco Exploration (Western) Limited. (See 'Babies, Helicopters, Perennial Snow & Ice, B.C.')

When recalling survival episodes, I have often related one of Gary's, which must rank among the highest for the relatively short period involved. In the late 1940s, he was employed as an engineer with Queen Charlotte Airlines on Vancouver Island. One of his flights involved a schedule from Tofino to Zeballos in a Stranrear bi-wing amphibious aircraft with a 6-8 passenger capacity. His functions included fuelling up the wing tanks after checking in the passengers, immediately prior to departure of the flight. Normal procedure was for Gary to exit from the hull-door onto the wing of the aircraft for fuelling. When completed, the hose was released and Gary re-entered the aircraft through the hull door. His slamming of the door was the signal for the pilot that fuelling was complete and that the passengers and Gary were secure and ready for take-off.

On this eventful but very gusty occasion, the fuelling was almost complete, and Gary was still on the wing of the aircraft when the hull door suddenly slammed shut. The pilot powered the aircraft, without further ado, whereupon Gary dropped the fuel line and grabbed a strut, afraid of being swept off the wing.

The aircraft accelerated and became airborne with Gary having wrapped his arm securely around the strut and hanging on for dear life. None of the passengers were aware of the prearranged ritual for take-off, or Gary's predicament. Fortunately the event occurred in late June and the relatively warm air-temperature permitted Gary to retain his precarious hold as the aircraft gained its cruising altitude of about 3000 feet above sea level. Gary had no particular recollections of his flight, other than his desperate need to hang on for 30 minutes.

In our relatively brief association during our six-week program at Bute Inlet, nothing that Gary ever did or said could detract from my overwhelming admiration for his ability to survive such a remarkable experience. In 1990, while trying to verify my recollections, I phoned Gary's home, only to learn from his wife that he had died in 1980 from a cerebral hemorrhage while still employed with Okanagan Helicopters Limited.

Searches

I have described several survival episodes, some of which I have a personal connection with and others in which I was directly involved. Exploration, for whatever purpose, must anticipate the possibility of communication or other supports being cut off, temporarily or otherwise. It is essential to provide for the survival of those involved until they can fend for themselves or alternately until others can recognize and respond to their emergency.

With significant technological advances in the last two to three decades, many of us have tended to assume that 'others' will look after us in the event of an emergency, rather than ourselves, as was the credo of our forebears. Nevertheless, most survival manuals still emphasize the basics of self-preservation: be aware of potential problem areas; know how to anticipate them and deal with them in an emergency.

Search on Mt. Seymour, 1950s

My first involvement in a search occurred in the 1950s while working with Kennco Explorations (Western) Ltd. in Vancouver. Charlie Ney, who lived in North Vancouver and had hiked and skied in the North Shore mountains for many years, asked me if I would like to join in a search for a young woman who had been reported missing while hiking with a companion on Mt. Seymour one weekend in late September. There was considerable concern for her welfare as she was known to have been hiking clad only in a light dress and running shoes and she had not been found by nightfall on a chilly evening.

Charlie and I arrived at the first-aid hut on Mt. Seymour the following morning and we were both assigned a search area by the search master who, as I recall, was a member of a local Search and Rescue group. Several hours into the search with another individual, we were traversing on the eastern side of Suicide Bluffs about two kilometres north of the skiing area parking lot when we learned by our portable radio that the woman had been located below the bluffs in the valley floor. She had taken shelter overnight amidst the large boulders strewn over the area beneath the cliffs. We were asked to provide assistance in preparing for her evacuation by helicopter.

It took us about half-an-hour to reach the location where four or five other search and rescue party members were already on site. They had already provided the woman with warm clothing, food and water. She was suffering from hypothermia, which was not surprising considering her apparel and the long cold night she had survived huddled among the boulders.

Within 15 minutes we heard the helicopter approaching; however, I was surprised to see that it was a twin-rotored Piasecki which hovered about 150 feet above us, in an area which contained several large fir trees and no possible landing sites. For some reason I had expected a smaller helicopter. Soon a stretcher was lowered on a cable and we quickly placed the woman in the stretcher as the cable was winched back up to lower a paramedic. He checked the harness over, and within a minute she was hauled up to the helicopter, followed by the paramedic. While the operation was in progress I frequently looked up at the helicopter and was much impressed by the pilot's skill in maintaining a stationary position, which could easily be gauged by the lack of its movement relative to the tall trees nearby.

Within 15 minutes the helicopter had delivered the woman to the Vancouver General Hospital from which she was released following a short period of observation.

Search for Don McKinnon and Companions, Mt. Baker Area, 1962

Sometime later, while I was still working with Kennco in Vancouver in 1962, Don McKinnon, one of our employees, failed to report for work following a weekend. We phoned his residence sometime during the day but received no response. By Tuesday morning, following several telephone calls, we learned that he had planned a skiing weekend at Mt. Baker in Washington. We telephoned the Park Ranger at Mt. Baker that morning and learned that there had been a heavy snowfall on Monday and that there were still a couple of cars in the parking lot well covered with snow. Both cars were dug out and it was confirmed that one was Don's and the other had a Washington licence plate.

By Tuesday evening four or five Kennco employees, including myself, John Anderson and Charlie Ney, had travelled down to Mt. Baker preparing to initiate a search. By the following morning we were joined by a party of searchers sent out by a local contingent of the U.S. Air Force, as the other car had been identified as owned by a USAF officer.

Within an hour of skiing in the active ski area looking for signs of ski tracks heading out of the patrolled section, we heard a shout from a hill slope above us. We responded and waved at a lone figure who skied down towards us. It was Don, obviously highly excited and haggard, but otherwise in good physical condition. Before returning with him to the ski chalet he briefly described what had happened and showed search personnel the route they should follow to rescue the USAF officer and his companion.

While skiing on Sunday afternoon in overcast conditions with poor visibility, Don met the two air force men at the top of a ski lift and decided to make a run with them. Apparently all three were unfamiliar with the run but assumed that it would lead back to one of the lifts.

After considerable time they realized that they must have missed a turn-off back to the active ski area. Unwilling to start the long uphill climb, retracing their route, they decided to continue down hill. Much later they had another discussion, realizing that they had probably made a poor decision as they were so far downhill by this time that their only possibility of getting back to the ski area that day lay in locating a logging road which they could follow back out to the main highway. They were in fact completely lost, as subsequent information was to reveal that there were no logging roads in the portion of the Mt. Baker area that they were descending. As dusk approached, they stopped in increasingly rough terrain. At Don's insistence and applying his bush experience, they built a snow shelter and passed Sunday night in relative comfort considering the conditions.

By the following morning they decided that they had to retrace their steps. This was obviously a most demoralizing decision as they had dropped thousands of vertical feet from the ski area. Following a long day without nourishment, in which one of the officers had to make increasingly repeated stops in his exhausted condition, they prepared for another long night. They did not have the energy to build a shelter; however, they were able to build and maintain a fire. Don recollected the length of that interminable night, without sleep, in which the most exhausted of the three expressed repeated concerns about his condition and the possibility of not surviving. By that time he was suffering from frostbite in both hands and feet. As Don prepared to leave at daylight, the affected companion indicated that he was unable to continue. Don and the other air force man left without him, continuing upward, but stopping far more frequently because of the deteriorating condition of Don's companion. By midday he told Don that he didn't have the energy to continue. Don plodded on, nearing the active ski area before dusk prevented further travel.

Once again, he lit a fire and managed to survive the night without undue concern, except for the condition of the most exhausted officer. He set off at daylight, when we were initiating the search above him. After we encountered him, both of his companions were rapidly recovered by helicopter, the most seriously injured ultimately losing all the toes on both feet.

More than 30 years later, in 1993, I tried unsuccessfully to locate Don, to obtain his firsthand recollections. Although it is probable that the three would have been spotted by an airborne search team as they were in an open area, there is no guarantee that the weather conditions would have permitted flying. Don's superb condition and will to live led to the rapid recovery of his two companions. Don worked with me at Galore Creek in 1963 and spent most of his subsequent career in surveying.

Search for Steven Eby, Mount Seymour, November 1993

On Thursday, November 18, 1993, I learned that a lone hiker was missing on Mt. Seymour following a hike on the previous Sunday. The circumstances reported were not uncommon, and reflected particular similarities to one of the two other searches I had been involved in 30 years earlier.

Steven Eby, 28, a Montana resident, had only recently moved to Vancouver to attend film school at the University of B.C. and was unfamiliar with the Mount Seymour area and had few friends in the city. Described as a skilled outdoorsman in excellent physical condition, Eby was living alone at the time. He had expressed interest in hiking at Mt. Seymour to a friend and asked him if he would like to accompany him on the Sunday hike, but his friend had declined.

Eby was not reported missing until late Tuesday evening, November 16, when his friend became concerned and telephoned the police. An immediate search of the parking area led to identification of his car, which contained hiking equipment. This led to speculation that he was probably not equipped for an overnight emergency. Within two hours after the report was received by the police, a search had been initiated by local search and rescue team members.

Wednesday was unfortunately a miserable day, and little was accomplished. By Thursday morning, Eby's mother, Julie Medland and her son Brian had arrived from Toronto and were to maintain a continuous vigil at the search and rescue command post established at the ski-patrol building at the Mount Seymour parking lot. They were later joined by Eby's girlfriend from Montana.

On Thursday, two helicopters completed searches from the air while two boats were used to search the west side of Indian Arm lying immediately east from Mt. Seymour. The helicopters carried heat-sensing devices, which successfully detected searchers but no-one else. About 30 volunteers from the North Shore, Lions Bay and Coquitlam rescue teams combed the area assisted by police dogs, but without success.

I telephoned the North Vancouver detachment of the RCMP on Thursday evening volunteering to assist in the search if required. I was subjected to a perceptive 'grilling' by an officer interested in my outdoor experience, physical condition, experience in searches in which he emphasized, "You could be out there for ten hours, and maybe overnight, and we don't want you pooping out!" This latter may have had something to do with my advanced age! I was 65 at the time. The conversation suddenly broke off as he grunted, "OK, be at the Seymour Mountain ski patrol hut at 7:30 tomorrow morning."

I immediately started to accumulate appropriate clothing and equipment, which necessitated a quick drive to Save-On to obtain some trail mix. I also located a 1:50,000 scale topographic map showing the Mount Seymour Provincial Park area, which I thought could be useful.

I left for Mt. Seymour at about 6:50 a.m. in darkness, knowing my way fairly well because of my participation in the North Shore Triathlon in early June. It was based out of the Andrews Recreation Centre on the Seymour Parkway and included three cycle loops which extended to the Mt. Seymour turn-off.

As I started to drive uphill, I noticed the taillights of two cars ahead of me, which faded and brightened in fog patches. Several kilometres up the road the lead car suddenly swerved across the road, blocking the route uphill and blue and red flashing lights appeared. I waited while a young RCMP officer questioned the driver of the van behind him, eventually letting him drive on. Next it was my turn. "What business do you have on Mt. Seymour?" he asked.

"I'm on my way to assist in the search for that missing hiker."

"Who asked you to come up?"

"I spoke to the North Vancouver RCMP last night and they asked me to be at the Ski Patrol hut by 7:30," 1 replied.

"You may proceed, Sir!" he said in a most authoritative tone. I chuckled to myself as I drove on, wondering if volunteers were really welcome.

The temperature at the parking lot as I parked and walked toward the Ski Patrol hut was at the freezing level with a light drizzle and relatively heavy fog, the visibility being about 150 feet. Several individuals were donning gear and heading to the Ski Patrol hut. An area was cordoned off for several large vans belonging to Search and Rescue groups, and the RCMP, which also included a mobile diner which provided hot and cold drinks, sandwiches, soup and even pasta for the tired searchers on a 24-hour basis.

The North Shore Rescue van, over 24 feet in length, is an impressive rig which functioned as the command centre for the operation. Apart from a diverse array of first aid and rescue equipment, it contains radios, a map table and a laptop computer which provides rapid information to assist the search managers.

After checking in as a volunteer I found a seat on a bench in a room crowded with other searchers waiting for assignments. About 30 individuals were sitting and standing, mostly in small groups. One of the senior members of the North Shore Rescue group gave a lengthy presentation on the status of the search at the time and the priority search areas being considered. He was soon joined by Ian Todd, the search manager and an 18-year veteran of the group. Todd first determined who were familiar with the Mt. Seymour area and had them occupy one side of the room. Next he asked for experienced search and rescue personnel, who were directed to another section, leaving a balance of less experienced volunteers. Patiently, he formed the search parties into groups of three and assigned them specific search areas. I felt like the unwanted kid at a hockey rink when pick-up teams are being selected, but eventually was assigned to search the Mystery Lake area off the top of Mt. Seymour's main tow over to a steep gully rimmed partly by cliffs which extended to Goldie Lake and then a lengthy trail back to the parking area - a crudely triangular traverse of about six kilometres in length. My companions were both experienced members of rescue groups, Al Billy with the North Shore group, and Al Leonard, team leader, from the Maple Ridge Search and Rescue group. Al Billy acted as our leader and was the only one equipped with a radio. All of us carried ice axes and, in addition to personal equipment, I was issued a bag containing a rope weighing about 15 pounds.

We were fortunate to obtain a lift by skidoo part way up a trail paralleling the Unicorn Run which lies east of the Mystery Peak chairlift. After some further climbing in visibility down to about 100 feet and walking in about six inches of snow, we fanned out and headed easterly to locate Mystery Lake. The area had been assigned a priority on the basis of a hiker reporting having seen a man walking alone along the Mystery Lake trail at about 3 p.m. on the day Eby was lost. Although the area had been searched previously it was still considered to warrant further search.

We looked for any evidence of a single track and while fanned-out, 100 feet or more apart, mostly within hailing distance of each other, we searched along the base of the many cliffs in the area and at the base of any large trees which might provide shelter.

After about an hour we eventually located the headwaters of the gully which heads south-easterly from east of Mystery Lake to Goldie Lake. There was no obvious trail; however, there were occasional red and yellow plastic markers on trees defining the route which gradually develops into a well-defined gully. In this area we located a single set of closely-spaced footprints, heading northerly, not more than one day old. Al reported the find to the command centre and an exchange of information followed. Alas, the footprints were those left by a Parks searcher, who unlike the community search and rescue personnel, are permitted to search alone.

Finally we reached an area where the slope levelled off and a clearing appeared ahead. It was Goldie Lake. After a further check with the command centre we followed a distinct trail, with markers, leading back to the parking lot, about 2-1/2 kilometres distant. Out of the fog, the bottom of a rope-tow appeared and I realized that this was the Goldie Tow where over 30 years earlier I had helped our children learn the rudiments of skiing!

Feeling somewhat fatigued after about 3½ hours of steady bushwhacking in a continuous wet snowfall which had added several unwelcome pounds to my load, we reached the ski patrol hut looking forward to a rest, some nourishment and, for myself, the hoped-for announcement that my services would no longer be required. It was about 1:30 p.m. and we had traversed about eight kilometres. Ian Todd looked at us as though we were new troops and said: "We need you to accompany this couple right away up to Dog Mountain. Rod and his companion were on the Dog Mountain trail on November 14 in the late afternoon and spotted footprints heading back toward the parking lot, after the snowfall started. Rod noted that the footprints disappeared somewhere between Dog Mountain and First Lake and he feels he can recall the general area, if he can re-traverse the route. He may have to go right to Dog Mountain in order to better recall his impressions of where he last saw the prints on the route back toward the parking lot."

Fortunately, Al Billy spoke up and said that we needed at least half an hour to eat and rest before continuing on. At this stage I seriously considered bailing out, as my presence didn't seem that vital. However, I felt obligated as a volunteer to continue.

We left at about 2:00 p.m. I cheerfully accepted re-packing of the heavy rope, while inwardly agonizing as everything that I donned seemed to have gained so much weight. I had also acquired a welcome battery-powered lamp, thinking of the approaching darkness. We trod up-slope to the Dog Mountain turn-off and were soon stretched out on the trail. We reached First Lake after about 20 minutes and continued on to the west past several of the bridges over mushy areas where Rod was trying to recall the last set of footprints which he had seen.

En route, the party stopped at Rod's request, as he recalled a section where the footprints could have strayed off the trail. Two of us tramped through some scrub in a partly cleared area containing a small creek. Within several feet there was a poorly defined sudden drop of the creek over a bluff to a point about 15 feet below. We accessed the area, realizing that in the darkness Steven Eby could easily have fallen below. Not seeing any signs of a body, we re-accessed the trail and continued on.

Eventually we came to a split in the Dog Mountain trail and we took the southern branch which leads to Dog Mountain within about 1,000 feet. Dog Mountain could more properly be named Dog Bluff as it barely protrudes above the surrounding terrain. We took brief shelter in a cluster of fir near the 'summit' before starting to retrace our steps, allowing Rod and his companion to lead the way at their own pace in their effort to recall their earlier impressions. By this time it was about 4:00 p.m. with nightfall approaching. Within half an hour we had retraced our steps back to the first of the series of bridges Rod recollected. Al Billy and I used blue and orange tape to mark off the area of the first possible bridge where Rod thought he had last seen the absence of boot prints during his earlier hike. We did the same at the last bridge, confining the probable area to within about one kilometre. By now we were using lamps, both hand-held and on helmets.

Where I had felt that I could not possibly continue on to Dog Mountain earlier on, by now I had my second wind and gradually became oblivious to my sore knees. I was like a horse heading back to the barn. However, the time it took for us to reach the parking lot from First Lake was nearer 30 than the 20 minutes on our hike out. We reached the First Aid hut at about 5:30 p.m. following our 10 kilometre trek, to be rapidly and, for us, most emotionally greeted by Steven's mother and son and later Steven's girlfriend, all expressing their obviously heart-felt gratitude for the role of the searchers. I swallowed hard when Steven's girlfriend said, "I'm going to give him a real tongue-lashing when he's found, considering all the problems he has caused." We all made appropriate positive comments, but my gut reaction at this stage was that Steven would not be found alive.

After some time, probably 15 or 20 minutes, with my heart still pounding at about 140 beats per minute from the day's efforts, I took a small bowl of soup and an equally small portion of pasta from the mobile kitchen unit and departed.

Driving back through the fog to North Vancouver, I realized how exhausted I was. I found this difficult to accept considering that within the past week I had completed both the 12-k Masters Cross Country Race (First in 65-9 age group) and the equally demanding 15-k Khatshahlano Run with nowhere near the feeling of such fatigue. I finally rationalized my situation by assuming that the emotional impact of hoping to save someone in terribly dire straits could easily make the difference, in addition to the unaccustomed burden of a backpack, small as it was.

Since participating in the search for Steven Eby, I have learned of two different individuals, both of about my own age, who have recently suffered fatal heart attacks while participating in fall hunts. On October 9, 1993, Jim Gaunt, a practising geologist for more than 30 years, died from a heart attack while on a moose hunting trip in the Gowganda area of Northern Ontario. He was 58. In November, 1993, Gordon Read, age 57, a vigorous hiker and outdoorsman and a close friend of Ken Sumanik, an acquaintance of mine, died from a heart attack in the East Kootenays on an elk hunt, in the process of loading a killed cow elk into a pick-up truck.

Most fatal accidents related to hunting involve the victim being mistaken for prey. Much has been written about the 'easiest' way to dispose of a victim, i.e. take him out on a fall hunt, as it is so hard to prove intent. However, the high incidence of fatal heart attacks on fall hunts can only reasonably be attributed to the lack of the conditioning of hunters to the degree of prolonged physical effort demanded by long and difficult traverses in the bush. Obviously the extra effort required to dress and transport the remains of the kill to a road is another factor.

I well recall an article I read about 20 years ago about a fall hunt for elk in northern Wyoming. Apparently an access road to a park ran along its northern boundary. As the hunting season approached in the fall, elk travelled southward to safe haven in the park. Near midnight of the deadline for the hunting season, total bedlam prevailed on the access road as impaired individuals and vehicles, either mechanically immobilized or out of gas, stalled. They were bodily ejected from the access road by the following horde seeking advantageous spots on the 'firing line' facing north. After midnight, straggling elk had little or no chance to reach safe haven, considering the formidable firepower of their well-armed predators. The article describing the hunt featured a photo of a scene near the firing line showing the body of one of the hunters sprawled on the ground, a victim of a fatal heart attack, while his companions butcher an elk nearby.

The November 27, 1993 issue of the Vancouver Sun contained a well written chronological account of the search for Steven Eby from November 16-21, noting the enormous efforts made to find him and the reluctant decision to call off the search on November 22 concluding that all that could have been done had been done. The search for Steven was the largest ever mounted in the Lower Mainland. It was also one of the most frustrating and disappointing considering the numbers of searchers involved. On Saturday, November 20, a sweep search was performed with searchers spread out 100 paces apart traversing across the mountain. On Sunday, the final day of the search, selective searches were again instituted but nothing developed.

During the five-day period, search teams had come from Kamloops, Hope, Chilliwack, Comox, Lions Bay, Coquitlam, Maple Ridge, Whistler, Surrey, Sunshine Coast, Squamish, Salmon Arm, Vernon and Bellingham. There were about 140 searchers plus dozens of other volunteers - cooks, computer technicians, radio operators, helicopter pilots and boat operators.

In the three-year period since 1990, a total of 20 individuals died in the North Shore mountains and canyons, all but two of their bodies having been recovered. The popularity of hiking and skiing in the North Shore mountains is not without a cost in time, effort and funds. In the first 11 months of 1993, there were 32 search and rescue operations. Statistics provided for 1992 by the provincial emergency program, printed in the November 27 issue of the Vancouver Sun, indicated 104 rescue teams, 3,500 volunteers, 481 searches, 789 victims of which 703 were found alive and 36 dead (including suicides), 50 not found (including hoax calls), and 89 percent found alive. The same issue highlighted the following safety tips developed by North Shore Rescue after years of experience:

Always tell someone where you are going, when you expect to be back, and who to call if you do not return on time;

Always hike in a group and carry a map of the area, a flashlight, food, extra clothes, matches, a whistle, and a large garbage bag for shelter;

When skiing, obey all ski hill rules and never ski out of bounds;

If you do get lost, stay where you are and huddle under a tree for warmth. The rescue team will soon be out looking for you, even in the dark.

I also reflected on the dedication of the searchers, their professionalism, and that of their organization, which would inspire confidence in anyone. I recalled a conversation between Rod and Al Billy near First Lake on our trek back to the parking lot. Rod, like myself, was obviously impressed by the search and rescue organization. He appeared seriously interested in getting involved and asked Al Billy what the requirements were to qualify as a member of the search and rescue teams.

"Well", said Al, "the procedures are generally similar, but I can only really describe those for the North Shore group." He took his time and lucidly enumerated the steps taken to induct would-be team members. I remember being very impressed by his account, so much so, that in recording this memorable day, I later phoned him to ensure my recollections of his description were correct. I'm glad I did, because it was a good opportunity to express my admiration for the enthusiasm and competence of those with whom I had come in contact during our day together – as well as to get my facts straight.

The first step is the completion of a well-thought-out application form in which the applicant provides a summary of outdoor experience and qualifications; first-hand knowledge of local trails - which could be from non-existent to excellent; first aid background which as a minimum should include basic St. John's Ambulance and for many includes industrial first aid; team activities, i.e. school, rescue, military organization; references - particularly those from existing team members; related equipment owned and recreational activities. Interviews are then arranged with successful applicants during which detailed questions are asked to obtain an indication of how the applicant might react under various search-rescue related situations. Included is a subjective assessment, e.g. would the applicant voice a concern faced with a dangerous situation? Accepted applicants from interviews are designated as 'members-in-training'. The training period lasts about one year and requires attendance at weekly lectures. These are routinely attended by all members and commonly include a speaker providing general up-dated information, a slide show, or specific data on new techniques. In addition, the trainee obtains a Radio Operator's ticket; maintains or improves First Aid qualification and has an opportunity to participate in search and rescue programs in which performance is noted.

Some members-in-training gracefully withdraw if they find that the activity or commitment is unacceptable. Al indicated that he could not recollect any trainee being turned down during his five years with the group. Turnover averages about two or three annually in the 40-member team.

In early March 1994, after writing the above, I forwarded a copy to Allen Billy requesting he review the draft as I had relied heavily on his statements. He responded promptly and I was interested to learn that for several years he has been preparing a book on the oral tradition of the North Shore Rescue Team. According to Allen, "the book is developing into a collection of stories, perceptions, rescued subjects, people involved in other agencies, etc." Many of the stories are based on interviews designed to help develop certain "themes", e.g. first search, the spouses' perspective, medical rescues, etc. He requested my authorization to excerpt part of my draft, which or course I willingly provided.

Allen also provided me with an important sequel to the search for Steven Eby. Although not publicized in the media, "Steven's body was found early in the New Year by a woman walking her dog near the mouth of Holmden Creek at Indian Arm". The location is almost four kilometres due east of Mystery Lake and almost 1,150 metres (3,500 feet) in elevation below, as Indian Arm is an extension of Burrard Inlet, at sea-level.

After receiving Allen's response to my draft, I purchased a copy of a 1:50,000 scale 'Port Coquitlam' topographic map, 1989 vintage, which indicates that the intervening area between Mystery Lake and the mouth of Holmden Creek at Indian Ann contains sections of very steep topography, which were obviously no insurmountable problem to Steven. Considering the lateness of his departure from the Mt. Seymour parking lot, it appears highly unlikely that he could have reached Indian Arm during daylight on that evening. Consequently it appears reasonable to speculate that he survived a night out and reached it probably the following day, but in an exhausted condition.

Whatever the events leading to his death, they are particularly tragic as he was obviously so close to habitation as shown by the location of his body and nearby buildings on the west shore of Indian Arm as shown on the Port Coquitlam topographic map.

Travel, Trekking and Climbing

It wasn't all rocks and minerals and frustrating encounters in the political realm. Life has blessed us with many opportunities to marvel at the world's wonders. These have included wonders uncovered by travel, trekking and climbing.

Prior to our marriage, Rene's furthest travels from Ontario had been to Wine Harbour in Nova Scotia, the early home of her mother's family. Rene had many recollections of her trip there as a child and we were to revisit the general area while living in Toronto in the early 1970s. Following her graduation from the University of Toronto's School of Nursing in 1952, Rene's next longest excursion was from Ontario to Vancouver, influenced by her brother Bob's resident internship at the Vancouver General Hospital following his graduation from the University of McGill.

Following our marriage, I looked forward to traveling with Rene to different parts of our world – and less frequently, with our family. Among so many places to go, Rene was most attracted to Great Britain, which we visited together initially in 1969 and an additional eight times by 2001. Our next most frequently visited locales were the soothing Hawaiian Islands for short winter vacations, made more enjoyable by the presence of our family on two occasions and close friends on others.

By 2001, we have visited over fifty countries, including six of the seven continents, (the exception is Antarctica) and sailed five of the seven seas. Our trips have ranged from somewhat demanding adventures to the luxury provided by cruise ships with their convenient home-away-from-home accommodation and the exposure to new sights and experiences via the ease of frequent shore excursions. They have all been memorable, as reflected in our numerous photo albums and diaries which have provided the basis for several written accounts.

A Khumbu Trek – 1986

My chosen career as a geologist was inspired more by a love of the outdoors than by any early attachment to rocks and minerals. Much of my early prospecting, exploration and property examinations took place in Ontario and Quebec between 1947-51. Although the work was interesting and challenging the surroundings generally consisted of bush and lakes with rare glimpses of distant hills. Mosquitoes, black flies and sweat were constant companions in most of the summer months. All that changed in the Spring of 1951 when I had that golden opportunity to go west... far, far west.

The difference between working in the Precambrian Shield and the Cordillera was that between night and day as far as awe-inspiring vistas were concerned. I loved working in the mountains and counted myself among the most fortunate to have been one of the few intruders among them virtually every summer from 1952-65 and on a more spasmodic basis following our return to B.C. from eastern Canada in 1971.

Our second stint in B.C. was to include an unexpected and more intense and enjoyable relationship with mountains. This was directly due to our son Rob's immediate interest in hiking and climbing on arriving in B.C. We made many local hikes together in 1971 and as he gathered friends I was fortunate to be included on several significant outings. Rob was a climber where I was a scrambler and only a few of my subsequent 'climbs' involved rope work, except for safety purposes on glaciers.

One Lucky Canuck

Our first notable climb was with Rob's friend John Mikes in June 1972 on Mt. Baker via Coleman Glacier from an overnight hut which we shared with Susan who also joined us during the hike up to Coleman Glacier. It was a beautiful sunny day and we thoroughly enjoyed the outing and the slide back down the glacier on our nylon jackets. I remember that we were mildly berated by some more serious and well-equipped climbers for having only two ice axes between us.

In June 1973 we joined the North Shore Hikers on a climb of Mt. Weart in the Garibaldi Park area as a prelude to an attempt on Mt. Rainier. Dave McAuslan was with us and the three of us shared a tent at Camp Muir and a rope with Jack Whitlock, our leader, on our memorable climb of Mt. Rainier in sub-zero weather at the higher elevations in mid-June.

In September, 1974 Rob, John Mikes, Dave McAuslan and I climbed Mt. Garibaldi after an earlier attempt, unsuccessful due to adverse weather. In July, 1978 Rob and I joined his friend Chris Blann and Chris' father, Brian, on another memorable climb of Mt. Hood, again completed under superb weather conditions.

Our last successful climb together was in September 1983 when I joined Rob and his friend Brent Hawkins on a climb of Mt. Whitney, California, the highest point in the U.S. except Alaska at 14,496 feet. Our final attempted climb was a year later on the Black Tusk, which Rob had climbed a couple of times before. It was to be my second and final attempt as dense fog prevented us from advancing much more than half a mile from our vehicle toward our objective.

Although Rob has had many other hikes and climbs with friends, I have also been most fortunate to have shared many memorable hikes with my life-long love and best friend. Apart from many enjoyable hikes in the North Shore mountains with our family and, occasionally, on our own, Rene and I have counted as highlights many treks during vacations. These have included numerous peaceful jaunts through the British countryside to more ambitious hikes, including Mt. Teidi at 10,600 feet in May, 1977, the highest point in the Canary Islands; Mt. Snowdon, Wales, with hail, fog and snow in May 1983; Ben Nevis, Scotland – equally frigid at the top and the Chamonix area, Switzerland, both in September 1987.

This preamble is intended to provide a setting for our most ambitious and memorable trek in the Khumbu (Everest) region of the Himalayas of Nepal in October 1986. I had yearned to see these great mountains for years but never expected to fulfill the dream.

Ten years later I'm not sure exactly what motivated the trip, which was to be one of our lengthiest vacations, lasting from October 1 to November 4 and including almost two weeks in India prior to the trek and a few days in England before returning to Canada.

We returned to Delhi on October 13 following several days aboard the "Royal Athena", a luxurious houseboat on Lake Dal at Srinagar in the Kashmir to spend part of the night at the Taj Mahal Hotel prior to leaving for Kathmandu, the capital of Nepal.

The day before leaving Srinagar we had completed a lengthy hike, bushwacking through shrubs and thorn bushes from the edge of Lake Dal, to the top of a surrounding ridge about 1100 feet above the lake to visit Sankaracharya Temple. It provided us with dramatic views of Lake Dal and its flotilla of several hundred houseboats, anchored cheek by jowl. Returning from our hike we had an Indian lunch and then left for a two-hour shikara ride around the lake. It was a cool reprieve from hiking and afforded fascinating glimpses of life around the lake which is supported by many islands of lush earth and grasses, bordered by floating masses of lily pads and lotus. The islands are mostly communal with several small factories and diverse stores including tailors, paper maché and woodwork outlets.

Late in the afternoon Rene suddenly became very ill with diarrhea and vomiting. Unfortunately she didn't feel much better the next morning. However Gulam, our travel agent contact came to our aid, providing Rene with several 'Imosec' (loperamide hypochloride with antibiotic) pills, which gave her immediate relief.

As part of our itinerary, arrangements had been made for us to visit the Indo Kashmir Carpet factory, one of the largest of many such plants in the district, which are important to the local economy. Rene insisted that she felt well enough for a visit to the factory, so we flagged down a 'scooter' (pedicab), which dropped us off at the factory in Srinagar.

The Indo Kashmir Carpet factory employed 5,000 people in 1986, including about 250 families, which form the foundation of the carpet industry. Carpet designs are obtained from museums where they are photographed and then reproduced on paper with respective colours denoted by a letter or symbol. The required pattern for weaving is then abbreviated by thread line for the width of the carpet. A narrow carpet, two feet or so wide would have one person working on it, whereas a wide carpet – up to 12 feet or more in width would have five or more – invariably all one family!

In most of the carpets there are 400 stitches to a square inch and a 4 x 6 foot carpet takes an average of 1.5 years to complete. The coded pattern of instructions that we were shown, were reproduced on sheets of brown paper which were 3-4 inches wide and as long as required depending on the pattern detail. A 4 x 6 foot carpet could have 50 sheets of instructions in a small bundle stored for reference in the factory library which would be provided to the weavers, as required, noted as a loan in a ledger. As most carpets have patterns that are mirror images running down the length of the carpet, the sheets used would run consecutively from 1-50 for one-half of the 4 x 6 foot carpet and from 50-1 for the other half.

As many as 20-30 spindles of thread (wool or silk) are used and knotting threads are suspended between two enormous rollers 8-10 inches in diameter. One of the weavers chants the instructions as he and his companions, many of which are children, weave so rapidly tying knots that their fingers are a blur. Weavers acknowledged given instructions with a responding "a-a-ah!". Each knot consists of a coloured thread rolled twice around two base threads (warp), pulled down to the last completed thread line of the carpet between the rollers, and cut with a small scythe. When each line is completed it is only 1/20 inch wide and it is then securely tamped down.

The factory we visited had no artificial light, only a few windows and was quite dark considering that it was a sunny day. As would be expected, weavers commonly develop eye problems. In addition, there are a preponderance of children in the trade, as very nimble small fingers are required in knotting. Hours of work in 1986 in Srinagar were 8-9 per day, six days per week.

When the carpet has been woven it is taken to a room where it is evenly trimmed to the required thickness by hand with scissors after washing and drying. The common decorative fringe is then added to the carpet by women.

The store attached to the Indo Carpet Factory has hundreds of carpets on display. Because of our approaching departure for our flight to Delhi in the afternoon, we only had about 15 minutes to select one of the carpets for shipment. It was woollen, 3 x 5 feet for floor use as opposed to the silk/wool carpets intended for use as wall hangings. It cost about $575 Canadian (7280 rupees), including shipping by air to Vancouver, which was to take about six weeks. I suspect that we probably paid far too much at the time after seeing other more finely woven carpets. I'm sure that I was strongly influenced at the time by the knowledge that it had probably required one person (child) a year to complete!

We flew from Srinagar to Delhi via Air India Flight 428. En route we had glimpses of the Himalayas through gaps in the thick cloud cover. At Delhi we were met by Mr. Zutski, head of Zutski Travel and an assistant and were driven to the Taj Mahal Hotel. Following a very modest dinner served in our room we decided to take a stroll around the hotel grounds before getting some sleep prior to a 3:45 am departure from the hotel for the airport.

Our visit to the hotel coincided with a wedding and reception planned for that evening. The swimming pool and the broad surrounding terrace had been decorated for the event. All the pine trees in the terrace area were strung with blue and white bulbs and tables and chairs had been provided outside for the wedding guests. These arrived in the nearby foyer carrying presents and dressed to the nines – the women all wearing brightly coloured saris. As we watched the procession two heavyset men wearing shirts without jackets rose from where they were sitting and offered us their seats. Suddenly one of several security guards and another individual in plain clothes armed with a sten gun waved two others up from their chairs and the party formed a large protective ring near the entrance. I then noticed that the plainclothes men had obvious outlines of revolvers under their shirts. A small procession of elegantly dressed men and a woman appeared and walked through the protective cordon into a room at the foot of a broad stairway in the main lobby.

We decided not to wait any longer for the bridal party and walked up the stairway to the next floor. There we encountered a group which included three beautifully clad young women – the bride dressed in a long red and gold gown with an ornate jewelled headress and heavily jewelled earrings and gold bracelets. She appeared to be on the verge of collapse and was literally supported by her companions. We then saw an event board which announced that Mr. Arun Singh, Minister of State for Defence and his wife, plus retinue were hosting a dinner. Following this entertainment we departed for bed.

Our flight on October 14 to Kathmandu was uneventful until the aircraft was diverted to Patna in northern India, about 150 miles to the south, because of poor weather. En route we did have several excellent views of the distant Himalayas, including Annapurna, but were prohibited from taking any photos.

We finally arrived at Kathmandu, dropping through cloud cover at about 10:00 am and were rapidly passed through immigration and customs. Outside the terminal with our bags we were approached by a young Nepalese cab driver. Our destination was the Yellow Pagoda Hotel in the central part of the city and about 6 miles from the airport. I asked him the fare and he said $3 U.S. I then asked him the price in rupees and he said "60". The conversion rate was 21 Nepal rupees to $1 U.S. at the time, so I enquired whether that was Indian or Nepalese to which he retorted "Indian." That settled, we engaged in some sight-seeing as we drove into Kathmandu.

The Yellow Pagoda was not a pagoda at all, but a very ordinary three-storey building with tiny rooms and a dingy adjoining bathroom. In truth we were spoiled coming from two weeks of accommodation in some of the most luxurious hotels in the world. "Oh well", we said, "its only for one night." Were we ever to be surprised!

Immediately after checking-in we walked over to the 'Himalayan Journeys' office, which was only two blocks on the main road from the hotel to confirm our flight to Lukla the following morning. Although I had defined the itinerary for our tour in India and Nepal it was arranged through Westcan Treks Adventure Travels in Vancouver. The Indian portion was superbly prepared by Zutski's Travel and the Nepal portion through Himalayan Journeys. Apart from our accommodation in Kathmandu the only itinerary arranged for us in Nepal was the Everest trek scheduled from October 15 - 24 with departure from Kathmandu for Bombay confirmed for October 27.

While completing research on our Everest trek, I was particularly concerned by a reference to adverse weather in the Lukla area frequently affecting flights from Kathmandu. On occasion this had caused large backlogs of trekkers at Lukla with no policy on a first-come first-served basis. Consequently many trekkers had missed connecting flights back to their points of origin. Because of this I obtained a confirmed reservation for both of us on the earliest possible flight to Lukla from Kathmandu following the season opening for trekking in 1986. The confirmed date was October 15. In addition I allowed five extra days following our scheduled return to Lukla for possible weather or other delays in order to be back in Kathmandu in plenty of time for our return flights to Delhi, Bombay and London.

On entering the Himalayan Journeys office, which was crowded with clients, we were in a state of complete shock when we were quite casually informed at the desk that all flights on October 15 to Lukla had been cancelled, including our own, as the official opening of the trekking season had been delayed to the following day. No effort had apparently been made to reserve space for us on a later flight. A senior representative interjected that perhaps we could fly out of Lukla in about a week's time. I quickly exploded, saying that "we have travelled half way around the world to meet a confirmed flight and no effort has been made to provide a later connection". After considerable hassling on specific flights we were finally 'confirmed' for a flight on October 17 and told that we might even fly out on the 16th. The representative wanted to know when our return flight was due to depart from Kathmandu. I was still in a state of shock and highly suspicious of our status and said in quite a loud voice, "I'm not sure that I want to tell you – all you'll do is add more days!" The nearby travellers laughed sympathetically.

We were both very dejected by this development and not at all convinced that some other delay would not unfold, jeopardizing our trip. I actually spent several hours that afternoon considering alternative treks, such as Annapurna which would not require a flight to starting points as they are accessible by road.

That night I woke up several times to the barking of dogs outside the hotel and again early in the morning to the sound of roosters. Rene was still not feeling well and had eaten very little since her first bout of nausea three days earlier. We located a drug store and were surprised to find Imosec pills identical to those provided by Gulam at Srinagar. The cost 15 rupees for four pills – about 20 cents each. At the same store we bought a plain chocolate bar for 20 rupees! The pills were supposed to be a prescription drug. We took a walk through nearby Durbar Square with its fascinating historic temples, statues and monuments, taking many photos of the square area and the extensive marketplace on its margin. In the Kathmandu Valley there are three Durbar Squares, each denoting the origin of the three surrounding settlements – Kathmandu by far the largest, adjoining Patan and Bhaktapur, about 9 miles to the east.

As part of its introduction on 'Exploring the Valley', Insight Guides' excellent reference on Nepal, our Bible during our visit, states:

The typical visitor arrives by air, heads directly to his Kathmandu city hotel, and explores the Western-influenced urban center from its Durbar Square. He will perhaps make excursions to Patan or Bhaktapur, marvel at the prayer wheels of Swayambhunath or Bodhnath, then depart for his next destination.

This is precisely what we did, although the succeeding reference instructed us to do otherwise if we wished "to feel its pulse and get its spiritual roots" which required a visit to the surrounding hills. We simply didn't have time.

However, we did see more of Kathmandu and the surroundings than many visitors who are reported to drive from the airport via tour bus to the posh Oberoi Hotel for a day or two stay and then board the bus back to the airport!

At the time, Kathmandu was the capital of the world's only Hindu monarchy and its medieval core remained little changed from centuries of habitation. In the mid-1980s its population was about 300,000, the metropolitan area about 800,000 and that of Nepal about 17 million – 50 percent under the age of 21! The language was 60% Nepalese, 20% Indian dialects and religions practised were 90% Hindu and 8% Buddhist. Nepal had mercifully not been exposed to significant non-Asian visitors prior to the mid part of the 20th century.

In the afternoon we took a taxi out to Swayambhunath Stupa, a Buddhist temple about two miles by road west of Kathmandu near the western edge of the Valley. Stupas are large bell-like Buddhist temple mounds which may contain relics within their base. The Swayambhunath Stupa lies on top of a hill which we accessed by about 300 steps. This sacred site is reportedly about 2,500 years old, well before the advent of Buddhism, but contains more recent additions.

The top of the mound is surmounted by a square stone structure with the eyes of the Buddha painted on each of its four sides. The question-mark 'nose' beneath the eyes is the Nepalese number 'one' – a symbol of unity. The stone structure is itself capped by a highly ornate gilded beehive mound. The circular base of the stupa contains about 100 recessed prayer wheels which are spun by the devout in a clockwise direction.

There were several monkeys around the Stupa, and on the approaching steps, who reportedly pilfer anything available. They are very agile and have been seen sliding down railings beside the steps. Numerous beggars occupied prominent locations on the steps and filth was rampant. At the base of the mound there were also numerous inscribed stones and tablets considered inviolable. We were to see many of these during our trek.

On our return to Kathmandu we decided to walk part way back to the hotel, at least as far as the bridge over Bisnumati River, on the west side of the city. The area is highly rural with fields of grain and vegetables and widely spaced small square homes, typically two storeys high with only a few small windows and tiled roofs. At one spot we saw two young women winnowing grain. When I stopped some distance from them I prepared to take a photo. One of them immediately ran into the house, while the other laughed and continued to pour grain from a metal basin. We also walked past a crowd of about 30 people laughing together and watching children taking turns on a huge swing suspended from the branch of a tree. Near Kathmandu we saw several other children flying kites high in the sky, unbelievably among a couple of golden eagles.

Before returning to the hotel we again checked with Himalayan Journeys concerning our flight to Lukla and were told that we were confirmed for an 8:15 am flight on October 17. That evening we had dinner at the coffee shop of the Annapurna Hotel. Rene had poached eggs and French fries while I had spaghetti Neapolitan, which I thought delicious, and a small bottle of Star beer.

The next morning I woke up with "Kathmandu Quick Step" which was to persist throughout the day. We still managed another walk through Durbar Square. We also bought about a pound of cashews and sultana raisins and another chocolate bar for a trail mix on our trek. We spent about an hour that afternoon meticulously washing and drying the nuts and raisins and removing the stems from the raisins.

We got up at 5:30 am on October 17, arriving at the domestic airport at about 6:30 am to find that we had been "re-confirmed" on a flight leaving at 10:00 am. We flew in a Twin Otter with relatively clean windows and had magnificent views of the surrounding countryside as we flew east toward Lukla. The eastern margin of Kathmandu appeared as a knife-edge contact between new closely-spaced buildings and the fields on which the city encroached. As we left the Valley we flew over many small villages and terraced fields in the foothills of the Himalayas. We even caught a glimpse of Mount Everest with its characteristic billowing snow tapering off from the peak to the east. Our flight to Lukla took only 40 minutes compared to about 6 days of strenuous up-and-down hiking required from the nearest road at Jiri from Kathmandu. The Lukla airstrip is a wonder to behold. It is perched high above the gorge of the boiling Dudh Kosi river at an elevation of about 9,400 feet and runs uphill at an inclination of about 15 degrees from the edge of the u-shaped glacial valley for a length of about 800 feet to the centre of the village of Lukla. The aircraft touched down near the foot of the airstrip and with power turned off, rolled to the top of the airstrip coming to rest with a neat turnaround. Some time after our arrival I took a photo of the aircraft taking off. With the strong prevailing updraft and without any trekkers it was airborne within 500 feet from start-up.

We met Lynn Mitchell, about 25 years old, from Australia who, with Rene and I, were the only remaining trekkers from the original group scheduled to start the trek on October 15. Four others also joined our party to travel with us as far as Dingboche, (14,300 feet), at the base of the great south wall of Lhotse (27,923 feet), Everest's nearby neighbour about 8 miles to the north-east on the Nepal/China border.

Included were Angus Robson who runs a restaurant in Seattle and his wife Setsuko. The other pair were Jim Linardos who works for the City Bank in Taipei and his wife Jane. Jim is an ardent climber having reached the summit of McKinley in Alaska in 1984 and is a friend of Jim Wickmeyer who was to make an attempt to climb K2 with a partner in 1987.

The seven of us sat on the ground near the airstrip for a picnic lunch before leaving for our first campsite at Phakding (8,700 feet) an easy downward 3-hour hike. Before leaving, we met Zumba the sirdar (leader) of our Sherpa party and Pemba his son-in-law and assistant. Others on the party were a cook, assistant cook, yak and nak packer and four porters. The porters were all young women, the smallest weighing about 100 pounds who, like the others, carried a 60-pound pack.

Yaks may be wild or domesticated members of the ox family. Most of the ones we used or saw were black in colour, very shaggy, their bodies and legs being somewhat similar to muskox. We had never heard of naks. However, we were informed that they were a cross between cows and yaks. There certainly is a distinct difference, naks having a less shaggy appearance, with shorter rough wool and occasional black and white coloration. Insight Guides insists that naks are merely female yaks which give milk. I should have paid far more attention to their underpinnings, but they were rarely in view.

Before starting our trek I took a photo of Rene standing wistfully alongside the trail, which could be seen running off to the north on the east side of the Dudh Kosi River. The early portion had been roughly cobbled with rocks because of the marshy nature of the ground and we had to watch our footing. Fortunately this gave way shortly to firmer ground and the opportunity to gaze around at the scenery as we walked.

Our tent had been set up before we arrived and we had each been issued a sleeping bag (for an extra charge) and one candle. We had opted to rent sleeping bags rather than suffer the inconvenience of having to lug our own around the world. This is common practice with most trekkers; however, our rentals – which would have been donated by climbing parties to the sherpas, both had inoperable zippers. They would both zip up from the bottom, but would not remain in one spot when pulled all the way up. Fortunately we had safety pins to rectify the problem.

Past experience had shown that at altitudes above 10,000 feet I suffered from increasingly severe headaches; Rene was fortunately unaffected. This had certainly marred my enjoyment of our climb on Rainer and Hood. By good fortune, just prior to our planned climb of Whitney in 1983, our daughter Susan had read about diamox, a diuretic recommended in a medical journal. Dr. Follows had enthusiastically prescribed it and it had been so successful that I had given all my back-up supply of 222s to other climbers on the mountain who were suffering headaches.

After a very modest dinner in our tent at Phakding we read for a while and, unused to the lumpy ground which was all too evident beneath our very thin rubber mattresses, we both succumbed to a sleeping pill. This was in addition to malaria and diamox pills, both of us taking diamox because of Rene's untested response to altitude above 12,000 feet at the time.

We awakened to a chilly morning of 5 degrees centigrade to quickly don four layers of clothing on top – a T-shirt, woollen shirt and sweater followed by a shell. Hot water was delivered to our tent in a basin for a wash and shave, a procedure to be followed each morning of our trek. Following a breakfast of porridge, an omelet, toast and tea we were ready for the trail by 6:50 am.

The floor area of our tent measured 7 x 5 feet and, being a very plain nylon wedge-type, it even lacked a wall. With the side less than 4 feet high we either knelt, squatted or laid down. From the outside the restricted nature of the inside space was deceptive as the tent was covered by a large fly which belled out behind the tent. We never got wet, except by accident as it never rained during the entire trek. The accident was caused by a leaky teacup. We pointed this out to one of the porters. However, it returned the next day – the victim being Rene who spilt water on her sleeping bag and the floor before hurling it some distance from the tent with an appropriate comment. It was not to return.

We followed the Dudh Kosi river on its right bank for a short distance before crossing the river on a make-shift bridge and re-crossing about one mile later. We encountered one particularly precipitous section where the trail was partly washed out necessitating a minor detour. Further along the original trail was impassable and we were told that the section from Jorsale to the juncture of the Dudh Kosi and Bote Kosi rivers, just below Namche Bazaar had been completely re-routed due to flooding. The new route was on the east side rather than the west side of the river and was very rough and steep with no appreciable gain in elevation but requiring about 1200 feet or more of total elevation change in a four-mile section.

Apparently the trail has been subject to frequent washouts. Patrick Morrow (1986) mentions that in September 1977 an avalanche fell from the upper slopes of Ama Dablam (22,350 feet), which is about 10 miles up-river from Jorsale, into a pastoral lake "triggering a 30-foot wave that roared down the valley, washing away seven bridges and part of the trail".

At Monjo we stopped for a snack in brilliant sunshine in a small field near the river with spectacular views of Thamserku (21,700 feet) to the east and Khumbila (18,900 feet) to the north, immediately above Namche Bazaar. One of the most lasting impressions I have of our trek is the constant natural beauty of the surroundings; the magnificent scenery and truly majestic mountains. One of our constant companions throughout our trek from Namche Bazaar to Dingboche was Ama Dablam with its symmetrical peak and lower shoulder. At one point the trail was in the valley at an elevation of about 13,500 feet and at a horizontal distance of 13,200 feet from the peak.

Just beyond Monjo we met two terribly sunburned young Frenchmen. We talked to them very briefly and they said that they had spent two months in reaching the 28,000 foot elevation of Everest before having to abandon their climb to the top. Sometime later we caught up to Pat, a strong, but quite slender woman, 38 years old, whom we had met briefly at the domestic airport in Kathmandu. She was from the Seattle area where she guided women climbers in the Pacific Northwest after having taught school, most recently to emotionally disturbed children. She was on her way to attempt to climb Island Peak (20,300 feet) which lies just south of Lhotse. Another woman whom she knew had come down to Phakding from Namche Bazaar and wanted to join her on her attempt on Island Peak. Pat believed that she was too inexperienced and would only rope up with a partner of her equivalent ability. Carrying a 50-pound pack, she was somewhat discouraged by the ups and downs of the new trail when we left her. Later when we were resting on the trail just below Namche, she came up the trail, took off her pack and said to Rene, "I need a hug." When we left her for the final time we each gave her a hug at Namche as she walked off to find a hotel room.

Our camp was pitched on a flat area near the top of Namche in front of one of the many tiers of buildings on the terraced mountain slope between the Dudh Kosi and Bhote Kosi rivers. Two of the party's yaks were resting beside our tents and we joined them, having climbed 4000 vertical feet during our 6.5 hours of actual trekking.

After a short rest we decided to see a little of the town before dinner. Outside our tent we watched several children careening down a long mud slide on the side of a nearby slope. I took a photo of them. They sat on pads with their feet on either side of a pole stuck through the centre of a can. They attained considerable speed, frequently falling off and rolling down the slope and laughing hysterically as they picked themselves up for another slide.

At a small store we bought a toque knitted from Yak wool for Rob. The toques take about one day to knit, are made in Namche from wool spun at Namche from sheared Tibetan yaks and sell for 60 rupees ($3 U.S.).

Our dinner at Namche was served on the second floor of an unheated building at Namche. Dinner included noodle soup, tasty local roasted potatoes, boiled cabbage, buffalo meat in pastry, meat and lentils in sauce, canned pineapples, choice of tea, coffee or hot chocolate.

At 8:00 pm as we were starting to settle down for the night in our tent, a wailing noise started which sounded like model aeroplanes. It came from a couple of long Buddhist horns and was followed by Indian pipes. We were not sure of its significance but it was certainly not a concert. After about half an hour as we were chatting, drums started to beat and they too stopped after a short interval. I jokingly said, "Will bells be next?" They were – about 15 minutes later.

It froze during the night and we had both put on extra clothing before starting our morning chores. We went back into the unheated building for breakfast, to find a woman in bed in the same room saying her prayers – rotating a small prayer wheel in her right hand and counting beads with her left. She was oblivious to both us and the cold. We could see our breath as we ate breakfast.

Leaving Namche at about 7:30 am we managed to get off on the wrong trail. En route we gained about 400 feet in elevation and passed by a magnificent tent camp on a flat area above Namche. Fourteen bright red conical tents with walls were aligned with military precision and spacing and included a large dining tent and a toilet tent. We gasped with astonishment and envy. It was serving an obviously large group of photographers. From our elevated perch, just before starting our gradual descent to cross over to Phunki Thanghka on the east side of the Dudh Kosi river, 1000 feet in elevation below, we had a view of distant Everest, Lhotse and Ama Dublam with many lesser peaks on either side of the river valley.

We had not been above the tree line until reaching this elevation and would descend back through brush and scrub timber before reaching our objective for the night at the famed Buddhist monastery at Thyangboche perched on a ridge at an elevation of 12,800 feet opposite the confluence of the Dudh Kosi and Imja Khola rivers.

We had learned before starting our trek that because of the early use of firewood by travellers in the Khumbu and other parts of the Nepalese Himalaya for cooking and heating purposes, regulations require that all trekkers and their porters, cooks and guides must be self-sufficient in the national parks. As we entered the south boundary of Mt. Sagarmatha National Park Headquarters at Jorsale, most of our trek took place in this sensitive forest area, which supports a growth of blue pine, birch and rhododendron at higher elevations. As required, our party carried fuel for cooking purposes and heat was provided by clothing. Throughout the area yak dung was collected and dried for fuel purposes and the smoke from these fires in villages had a memorable pungent odour.

Although Mt. Everest is named after Sir George Everest, Surveyor General of India 1923 - 1943, to the people of Khumbu Humai the mountain remains Chomolungma, Tibetan for Mother Goddess of the Wind. Sagarmatha, the Nepalese name for Everest for centuries, means 'churning stick of the ocean of existence'.

At Phunki Thanghka we passed several children playing near their home and I took a quick photo of them – a boy of about six and a sister of about five carrying a two-year old sister on her back. We had seen a similar group at Durbar Square – two girls between seven and eight with younger kin supported on their backs by securely wrapped cloths. Both were begging, the older attractively adorned with lipstick! Although we passed many Nepalese children on our trek, only a few asked for 'bom-bom' (candy) or pens. The candy had been carried earlier by trekkers but Nepalese parents discouraged any donations of candy because of potential tooth problems and a distaste for begging.

After crossing the narrow suspension bridge over the Dudh Kosi river at Phunki Thanghka I took a photo of two laden yak running across the bridge. We stopped there for lunch where we were entertained by yaks passing by and water-driven prayer wheels turning. We noted that all our dishes and pots were washed by hand with sand and gravel in a little creek which was also a water supply. We were to camp near the same spot on our way back.

After lunch we started off on the long climb with a gain of over 2,000 feet to Thyangboche monastery. Near the top we were passed by a column of yaks and their porters with yak bells tinkling, shouts from the porters and guides who periodically pelted the yaks with small stones to keep them moving. We didn't attempt to, but we could not have kept up with them. At the top of the ridge we turned around to look back at the long trail beneath us and the distant panorama to the south. On the trail we spotted an unbelievable sight which brought a smile – three British women in a line marching in step up the path with shirts billowing in the wind and folded-up umbrellas acting as canes.

Thyangboche monastery lies within a scenic setting in marked contrast to its stunning views of the surrounding snow-clad peaks. The monastery buildings lie on one edge of a large cleared grassy field, partly fenced and surrounded by pines, azaleas and rhododendron which must be extremely colourful when in bloom.

We walked up to the monastery on arrival but were intercepted by a relatively young monk who invited us to accompany him. He asked us to sit down in a small office and then brought a tray with tea which he poured for us into tiny silver cups. He then showed us all sorts of items including rice paper drawings, copper tea kettles, a copper pot, beads, etc. and said, "For sale." We thanked him without buying any items, gave him 5 rupees and left. The custom that we had read about purchasing and providing the head lama with a silk scarf appears dated as nobody to our knowledge participated in this practice.

After a brief look around the severely decorated monastery we visited a small two-storey lodge where we had another small cup of black tea costing one rupee per cup. The place was packed with trekkers – all in the 20-35 age range and none as ancient as us. Many of them and others we met had walked in to the area from various points accessible by road, including Lumi, taking 8-12 days. Most of the mountaineers walk in to the Everest region to acclimatize to the altitude.

That evening after a frigid meal of unmemorable items, low cloud which had developed in the evening on the mountains cleared and the surrounding peaks were sharply illuminated by a full moon and myriads of stars. Sometime later, I talked to Zambu our sirdar concerning our return routing. He had planned a return to Thyangboche from Dingboche - our destination for the following day and then Namche Bazaar with a long third day back to Lukla. I suggested Phunki Thanghka and Monje, both of which were almost 2,000 feet lower in elevation in the valleys plus the provision of a more even trekking time on all three days. He was very amenable, although I recognized later his preference of Thyangboche to Phunki Thanghka because of the very limited space for pitching tents at Phunki, necessitating his porter's early arrival there to claim the space.

We slept fitfully that night as the temperature dropped to minus 20° centigrade. Following our custom, we were served tea at 6:30 am with a basin of warm water. We quickly washed and by the time I had finished shaving, ice was beginning to form on the surface of the water. We left at 7:30 am and walked through a light forest of pine and rhododendron, balsam, birch and cedar with frequent stops down to the suspension bridge crossing the Imja Khola river at an elevation of 12,330 feet. The bridge inspired the taking of at least four photos, with Rene, yaks and the magnificent scenery. Many of the trees we passed were covered with a growth resembling Spanish moss. Occasionally we saw walls built of prayer stones along our path. Quite suddenly as we started the long ascent toward Pangboche (13,000 feet) we were welcomed by the warmth of the sun and we started to peel off our layers of clothing as we gained elevation.

As we passed Pangboche with its cluster of stone houses and rock corrals for yaks, about five feet high our path was partly blocked by five small children and one small baby. They asked for "bon-bon" and we gave each of them a lifesaver and took their photo. Shortly after we heard a pattering sound and finally spotted an ultralite plane flying high overhead up the valley. I took several photos of this event and days later on returning to Lukla, we learned that the unfortunate pilot had crashed at Pheriche and had been flown out to a hospital in Kathmandu where he was recovering from fractured legs, an arm, ribs and punctured internal organs.

After lunch we trudged totally alone up a trail on the west side of the Imja Khola river taking many photos and finally reaching a fork of the river where waters draining the Khumbu Glacier – one of the great obstacles for Everest climbers, mix with those draining the south slopes of Nuptse, Lhotse and the north face of Ama Dablam. We stopped to admire another water-driven prayer wheel and a memorial to climbers who have lost their lives in the Everest region.

Beyond, the trail rises quite steeply to 14,000 feet along an east facing slope with some exposed sections, which we passed through with care, a strong cold wind chilling our faces. Quite suddenly I felt very fatigued although Rene was fine. We took shelter behind some rocks for a spell and Rene insisted on taking my pack for the remaining half-mile trudge into Dingboche and our camp at 14,300 feet where we arrived at 3:15 pm. We were immediately given some tea and urged to lie down for a while on mattresses in Char Che View lodge, a stone building with several tiers of bushes, a centrally located fireplace where rhododendron wood is burned, cardboard matting-covered walls and an earthen floor. The bunks rent out at two rupees per night.

That evening we had our first dinner in a warm building as we sat on a bunk near the fireplace where the meal was cooked. Two of the women porters ate a quantity of boiled potatoes for their dinner, taking them from the pot in their skins and dipping them into a bowl of sauce. The Nepalese have the British to thank for potatoes; they were introduced into Nepal in the 1850s and have remained a staple part of their diet.

We again slept in our tent next to Lynn's isolated at the edge of a corral in solitary splendour. Despite much higher elevation than at Thyangboche it didn't seem any colder. After breakfast we started on our way back and detoured on a trail overlooking nearby Pheriche from a ridge at an elevation of 14,400 feet. Pheriche is a small settlement containing yak corrals and about five buildings. It maintains a small first-aid station for trekkers in trouble at high altitude and lies about 6 miles and a recommended two-day hike from a tiny sporadically occupied hamlet at Gorakshep (18,000 feet) near the site of Everest base camp on the west side of Khumbu glacier.

Near Pangboche we once again met Pat. She had found a friend to accompany her on her attempt on Island Peak and had located a yak to pack her bag. We arrived at Phunki Thanghka at about 3:30 pm to find our sherpas still erecting tents on a small open bench, almost bare of grass but containing yak dung everywhere, with pads even set out to dry on a number of large boulders beside our tents. I gingerly picked up a thorny stem to remove several of these threatening objects from the entrance way to both of Lynn's and our tent, only to receive several thorn pricks for my efforts. Rene picked up two stones to remove a large yak pad drying on a boulder forming part of a low wall which we could easily have stepped on, and had a difficult time getting it unstuck, much to the delight of two watching sherpas.

The next day was October 22 and once again we awakened to cold temperatures and clear skies, but slept far better than on the previous nights. After breakfast we walked steadily without any appreciable rest stops from 7:20 am until noon when we reached Namche Bazaar for lunch. Considering the number of photos I had taken on the way up the valley on this section I was surprised to find that I still managed to take 6 or 7. Included was one of a 3-4 year old girl carrying a 1-2 year old boy with a bare bottom on her back. En route we also passed by several stalls on the side of the trail where various wares, mostly of Tibetan origin, were exhibited for sale. We had stopped there briefly on the way to Dingboche. A vendor had indicated a price of 600 rupees for a prayer-wheel that I admired and I located a similar one for 250 rupees at another stall. On returning to this lower priced stall I once again asked the price and was again told 250 rupees. The vendor laughed when I suggested 100 rupees and we finally settled at 180 or about $9 U.S. The carved yak bone drum hollowed-out and containing a small roll of rice paper with an indecipherable message is the most interesting part, with undoubtedly debased silver on the handle, cap and part of the drum. I had thought that the small recessed semi-precious stones were polished turquoise and possibly jasper. The turquoise may be, but the jasper is definitely enamel paint! What do you expect for $9?

After lunch at Namche we walked back along our original route intending to spend the night at a Japanese lodge just south of Monje where we had previously had lunch. Alas another Himalayan Journeys party heading north had already set up five sleeping tents plus a large cook tent, taking up all the available space. We continued on to the next possible site, which was also claimed and finally settled on a field on a bench of the Dudh

Kosi river just past the bridge crossing at Chomoa. We arrived at 4:20 pm having hiked for 8:40 hours. We were both really tired having hiked 10 - 11 miles, climbing 2400 feet and descending 4200 feet.

Pemba, the assistant sirdar, was very helpful to us on that day, walking with us for much of the re-routed very steep section of the trail below Namche, where it was easy to slip as we found out. Rene took a fall through lack of traction on her older, more comfortable running shoes. She changed these to her newer ones and just to be sure that she didn't slip again I took her hand while leading down several steep sections while Pemba held her other hand while walking behind. Looking back several times I laughed as it must have looked quite comical to other trekkers, somewhat akin to "tip-toeing gently through the tulips."

That night we had our dinner in front of our tent again and by 9:00 pm we were asleep, neither of us waking before 4:00 am when I got up to find it quite warm outside with the moon showing in occasional patches of clear sky. By 6:00 am it was again perfectly clear. We left Chomoa at 7:30 am and stopped at Namaste Lodge south of Phakding at 9:30 am for a very early lunch. The stop was prompted for another reason. It is owned by Zambu's daughter and son-in-law and afforded Zambu an opportunity to chat with them and his 3 year old granddaughter. While talking with Zambu, I learned that in the Everest area a porter earned 60 rupees ($3 U.S.) per day and a low-class guide earns 100 rupees per day. At Namaste Lodge the price for a dormitory bunk was 2 rupees per night and a hot shower 10 rupees. By contrast the most costly item on the menu was a large bottle of beer at 50 rupees and the lowest, chapati with jam at 6 rupees.

A short time after passing through the tiny settlement of Ghat we stopped after crossing a bridge over a creek on the east side of Dudh Kosi river. Soon we saw a relatively slight Sherpa carrying a young American woman of about equivalent size seated on a board on his back, which was supported by a tumpline to his forehead. I took a photo of them as they crossed the bridge. The Sherpa stopped near us for a rest. He was pallid and sweating profusely. I talked to the woman, who was about 25 years of age, and her husband as we all rested. She was dressed in a ski-suit for warmth and was suffering from mountain sickness which had started just above Namche Bazaar. Unable to afford a helicopter, the husband had hired the sherpa at Namche to carry his wife to Lukla for 150 rupees ($7.50 U.S.) – a two-day hike. I commented to the husband that the porter was certainly very strong and he laconically replied, "Not as strong as the sherpas!" As we walked on toward Lukla the 125 pound porter with his burden kept up with us, climbing 1050 feet vertically during the two hours it took to reach Lukla.

During our seven-day trek we met several other individuals affected by mountain sickness between Namche Bazaar and Dingboche. One young woman was very ill just above Phunki Thanghka – supposedly psychologically affected in addition. Both Setsuko Robson and Jane Linardos succumbed to mountain sickness near Thyangboche and had to return to Lukla rather than hiking on to Kala Patar, their original objective.

Near Lukla we were hiking along the trail with a moderate grade on a relatively steep slope above the Dudh Kosi river when a team of yak approached us with their guide. We stepped to the outside edge to let them by and one of the yaks just missed hitting me with its horn but I was struck in the chest by its load and almost knocked off the trail. I was quite sore for several days and had trouble breathing. However, if it had to happen it was in the right location and time – a stone's throw from Lukla.

We arrived at Lukla at about 2:00 pm and had lemon tea at New Panorama Inn, which is owned by Zambu. We learned that instead of being accommodated in a lodge at 2 - 3 rupees each per night we would once again spend the night (or nights) in our tent in the same field where we had lunch on arriving from Kathmandu. Lynn's tent and ours were the only ones visible in town!

To celebrate our successful trek we dropped in at the Lukla Coffee Shop after changing our boots for running shoes and cleaning up. We each had a large glass of Raksi, a rice wine which had a similar taste to Saki, as one would expect. We talked for over an hour with two women who had just completed a 21 day trek from Jiri to Kala Pattar with a retinue of porters, cook and guide. One of them had trekked previously in Nepal and the other had climbed Whitney without any ill effects from altitude. One was from Salt Lake City and her companion with previous experience in Nepal was a Canadian from North Bay. She had insisted, after a similar experience to our own, on the party being provided with a cook tent and a toilet.

That evening we went back to the Himalayan Journey supply building for dinner, which we shared with a group of young Nepalese men and women. The dinner was similar to our others which had little variety. After dinner we walked back to our nearby tents to find Pemba standing guard, which was apparently necessary at Lukla. Rene asked if the toilet was open. On checking we found it locked. Pemba waved vaguely to the dark wide-open space of the field and said, "You can go up there somewhere", which we did, having become accustomed to an absence of toilet facilities during our trek.

Following breakfast the next morning we were rushed up to the check-in shed for boarding passes for one of two flights due in at about the same time that morning. As we waited outside in the cold we were relieved to see that the young woman who had been carried to Lukla and her husband had both been issued boarding passes. The woman still looked very wan. As we waited a siren suddenly sounded from the nearby airport conning tower and we were informed that it indicated that the aircraft had left Kathmandu for Lukla and would arrive in about 35 minutes. There followed a flurry of activity and all the waiting passengers and porters rushed to the airstrip where baggage and freight was separated into two piles. At the prescribed time another siren sounded and the first aircraft arrived shortly after. Rene and I had two seats together in the Twin Otter next to a fairly dirty window, which was unfortunate as the sky was free of clouds and we would have had superb views of the Himalayan mountain chain on the flight. The take-off from Lukla was quite exciting with the aircraft suddenly releasing its brakes at full throttle and the aircraft careening down the slope at ever-increasing speed. Just at lift-off Rene's nails dug into my thigh, emphasizing her joy at being airborne!

Although we were to spend several most interesting days visiting surrounding sites at Kathmandu before leaving for Delhi, Bombay and England, that was the end of a most memorable and adventurous trek. In the seven days of our trek we climbed just over 15,000 feet, descending the same amount from Lukla to Dingboche and return. For every two feet of elevation gained between the two sites there was a foot of elevation lost, with all the ups and downs.

The most striking aspect of our brief exposure to the Khumbu region and its people during our trek was the likeness of one day to the next. Although there were different sights and experiences the trek was a relatively solitary experience, shared by only a few others. Each twenty-four hours involved cold nights, a day's trekking with magnificent scenery under ideal weather conditions, simple meals and fairly Spartan camping conditions. The ideal trekking conditions must be emphasized for it was never too hot, nor did it ever rain.

None of our other vacations really approached our trek in Nepal. They have included a great array, ranging from periods of real rest mostly in Hawaii, and more recently on a couple of cruises with limited exposure to different lands, customs and people under luxury conditions of accommodation to another extreme – our four-week trip to China in 1990 by air, vehicle, train, ship and foot which was both a cultural revelation and an emotional shock, but never a rest. Our trek in Nepal took place under ideal climatic conditions for hiking, within sight of the highest mountains in the world, and under relatively Spartan living conditions – our retinue of Sherpas and others were doing all the really hard work.

Mount Rainier, Washington – 1973

The weekend of June 9 - 11, 1973 saw Rob, David McAuslan and myself in the Rainer Park area for our climb of Mt. Rainier. We arrived at the park entrance at 10:30 pm after a 5-hour drive from Vancouver. We stayed the night at a motel with another climber, Jack Whitlock, who had been kind enough to suggest that we join him in an attempted climb together with 12 other members of the North Shore Hikers Club.

We were up in time to arrive at the park ranger's office along with a multitude of other climbers who, as it turned out, had travelled to Mt. Rainier on the same day to practice climbing techniques at lower elevations. It was an extremely foggy day, quite cold, and we took a little time getting ready to start off from the parking area at Paradise at an elevation of 5400 feet. We had first to satisfy the ranger's office of our ability to make the climb and meet equipment requirements. With this done we weighed our packs which were from 30-60 pounds. Immediately upon leaving the parking lot we were tramping in snow and we wound our way up through the trees on the lower slopes of Mt. Rainier in quite thick fog, which persisted to an elevation of about 9600 feet. We arrived after about 5 hours at Camp Muir, a camping area, at an elevation of 10,000 feet. There we pitched tents on the snow, made dinner and were in our sleeping bags by about 9:00 pm as we had planned getting up at 2:30 am for a 4:30 start up the mountain. None of us got much sleep as the wind buffeted the tent throughout the night. By 4:30 all 16 of us were starting to move up the mountain on four separate ropes, aided by crampons and ice-axes. The next seven hours saw us climbing through spectacular scenery, mostly glaciers and a modest amount of rock. Five different people were unable to make it to the top for various reasons, one from fatigue and four others on one rope who slid down a glacier for about 200 feet suffering minor injuries. We considered ourselves most fortunate to eventually reach the summit with its volcanic crater at an elevation of 14,400 feet. We did not remain long as one of our party had frozen feet, the temperature being at the freezing level with a 35 mile per hour wind. Our descent to Camp Muir took 3 hours and following a rest there we continued down in 1.5 hours to Paradise, reaching the parking lot at about 6:30 pm. We had dinner in Everett and eventually arrived back home, extremely tired, at 1:30 am.

Golden Ears, BC – 1973

I had hoped that my daughter Susan and son Rob and I would collectively make the trek up Golden Ears, at 5598 feet the highest of the prominent rock knobs which form the Blanshard Group in Golden Ears Park near Vancouver. Rob had already climbed the mountain in the spring with several companions and had made other plans for Sunday. Susan had developed a form of bursitis on the back of her left heel - which although extremely painful, had not prevented her from hiking up with me to the base of the West Lion via Hyland Creek, the weekend before. However, the minimum 10 miles round trip and about 5000 feet in elevation change was prohibitive in her condition.

I left home in West Vancouver shortly after 7:00 a.m. on September 11, 1973, travelling via Hastings Street, past Port Moody and Port Coquitlam, turning north on a narrow road pointing to Golden Ears, about 3 miles east of Haney. The lack of signs caused the usual apprehension, but eventually I passed by the park gates along a relatively new, wide, well-graded and hard-topped road.

I turned off at the entrance to the Alouette Lake campsite and noted several cars parked near the gate, one of which was occupied by a bearded individual - possibly another hiker. Rob had mentioned being able to drive three miles beyond a gate crossing the road just beyond Alouette Lake. Examination quickly proved that vehicle access around the gate was now out of the question, so I parked the Blazer on the roadside near the gate but not in the turnaround area. A prominent sign warned against this.

The thought of an extra six-mile hike and the rather forebidding day with cloud cover down to about 2000 feet gave me second thoughts about starting off; however, I rationalized that it would be good to hike part of the way in any event. I started off at 8:30 a.m. with a light day-pack from an elevation of about 500 feet.

About 100 feet beyond the gate I heard a holler behind me. A park truck was in the turn-around area next to a station wagon - the only vehicle in the vicinity besides my own. "Is that your station wagon?" the figure called, "Nope," I responded and continued onward.

About one half mile down the extremely wide and well-graded hard-packed gravel-surfaced road from the gate, still mentally cursing the park bureaucracy that made it necessary to add 6 miles to the hike, a park station wagon came down the road toward me, driven by an unsmiling, unwavering and obviously preoccupied official.

I walked on past the bridge over Gold Creek and on up the road to a fork in the road, the most well-travelled portion of which bore to the right. I walked on for about one quarter mile conscious of the roar of Gold Creek, still very evident off to my right. Deciding that this was the wrong route, I retraced my steps past the fork and took the left turn. An aluminum trail marker just beyond the fork was reassuring.

About two miles further on, I came to a second gate across the road, this one, rusty, obviously of a much earlier vintage than the first. Beyond the gate the road became far too rough for two-wheel drive access. It dropped down into the bed of Gold Creek within about half a mile. A short trail through the bush led to a cable bridge (Burma Bridge) across the river. The bridge was composed of a thick steel cable for footing and two narrower-diameter cables as hand holds. The three cables were roped together by single strands at intervals of about three feet. The bridge was quite safe, although it tended to sway.

Beyond Burma Bridge a short trail led to an old corrugated railway bed which I followed for about one-quarter mile through the bush, across an old log bridge over a tributary of Gold Creek and an old logging road heading up into a logged-out area. The road climbed quite steadily from an elevation of about 1000 feet through a logged-over area which supported fairly dense second growth. At an elevation of about 2500 feet the road switch-backed abruptly and steeply and switched again near an old hollowed-out tree which appeared to have been used in the past as an explosive magazine for the logging operation. It had probably provided temporary shelter for many hikers in the intervening years. At an elevation of about 2800 feet, the road ended and a couple of aluminum markers on old stumps showed the point where a trail took off from the road directly up the slope beside a narrow gully.

The trail winds steeply and fairly directly up the slope through slash and then into virgin timber, rising 1000 feet to a ridge crest at an elevation of 3800 feet, within about one-half mile. The trail turns left from this point rising up the ridge into a rocky section overlooking the main crest of the Blanshard Group. From an elevation of about 1000 feet, I had been in a fairly thick fog with visibility ranging from 50-200 feet, so it was with some delight that I noted an improvement in the visibility. The low clouds filling the valleys were being elevated by the southwest side of the ridge crest and were spilling over into the Pitt Lake watershed to the west.

I walked along the trail to the main ridge, which meets it at an elevation of about 4300 feet near a couple of small moist meadows. I could see that the ridge beyond the rock knoll was in cloud and I was undecided about continuing. However, even though it started to rain intermittently, I was curious about the degree of markings along the route and reasoned that it would be difficult to get off the main ridge crest because of the increased steepness of the slopes.

In general the trail is well defined and easy to follow, even in fog. Apart from occasional pieces of flagging on stunted firs, there are many cairns along the route of all shapes and sizes. Several of them brought a smile as they revealed the humour of the builder. Someone had built a tiny cairn carefully balanced from a few well-shaped small rocks next to a large one, some two or three feet high. I also saw an extremely delicately constructed cairn composed of about seven rocks of decreasing size, balanced on top of each other. The architect must have realized that his monument would be short-lived.

About a quarter mile past the end of the rock knoll, I lost the trail in a snowfield. I decided to continue up the snowfield on the same general bearing which I had been following and after climbing some 400 feet in thick fog in the wind-cupped snow I came up against a rock ledge with a small crevasse which was a little too wide to leap over. About 50 feet to the side I was surprised to pick up a fairly well-defined trail climbing up the side of the rocky spur. Once again I debated about continuing, but on checking my altimeter I was surprised to see that the elevation was 5200 feet; only some 400 feet below the summit. I reasoned that, as I would have no trouble retracing my steps from this point, I could venture on a little further.

I climbed up the rocky ridge and eventually came to its summit. I spotted another higher ridge looming, a darkened grey mass through the fog, and encouraged by the presence of another small cairn, I made my way up through talus and on to yet another rocky knoll. At the top I noted a couple of scraggly stunted trunks of denuded firs some 8 feet high which looked like a good marker and once again, I spotted a higher ridge ahead. I thought that this must surely be the summit, but once again after climbing about 100 vertical feet, I noted a higher mass protruding from the fog ahead. Another 80 vertical feet climb brought me to a bare summit with a 2-foot high cairn and the unmistakable signs of man's edification of high places. Two flag poles, one with a green ribbon and one with a red ribbon, protruded at low angles from the cairn. A green wine bottle contained several notes. It was 1:30 p.m.

I took several quick pictures and after checking and double-checking my approach route, I started back down after a quick cup of tea and a handful of nuts and raisins.

I hadn't dropped more than 80 feet in elevation when I realized that I was already off-course. I retraced my steps upward and tried again, only to wind up further down above a steep cliff which I reasoned I must have climbed. My conviction got the better of good judgement and I forced myself to climb down. The lowest portion of the cliff involved about a 40-foot descent averaging about 70° to a partly bush-covered ledge below. I came close to panic but forced myself again to face the ridge and descend in the safest possible way. Although my fingers were numb, the handholds were excellent and the rock sound, so after what seemed like an interminable period, I shakily reached the ledge. I walked along it both ways and it faded out to precipitous slopes. I realized then that there was no way that I could possibly have climbed up this section and that the actual ridge route had to be more than 90 degrees from the bearing I had taken. I looked back up the route that I had descended and decided that there had to be an easier way up. At one end of the ledge I found a longer but less precipitous slope and had no trouble climbing back up. I then veered around the ridge slope and was tremendously relieved to spot one of my ascending footprints in a muddy portion of the ridge. I followed the barely perceptible trail down, to lose my way once again a few hundred feet lower down. Guessing that I was incorrectly biasing my course, as in the previous case, I deliberately swung off course again and this time spotted the small cairn which I had seen on the first knoll above the edge of the snow field. Half way down the snowfield the cloud suddenly dispersed and the rest of the ridge below was fully exposed including a portion of Pitt Lake to the northeast.

The rest of my hike was somewhat anticlimactic, although increasingly tiring. On the other side of the Burma Bridge it started to rain, increasing to a torrential down pour which continued unabated for about half an hour. It quit as I approached the bridge over Gold Creek and it was here that I was startled by two fairly large deer that climbed up the slope onto the road and went bounding up the road when they saw me. I got back to the truck at about 5:40 p.m. after having covered at least 16 miles and about 5800 vertical feet of climbing.

Haleakala Crater, Maui – 1991

A memorable day! After considerable deliberation and a modest amount of planning, Rene and I joined Eric and Lyla Henry in a day trip across the Western Crater floor area of Haleakala. Our trek was preceded by several days of concerned discussion on our requirements and scheduling. The trip was prompted by a brief telephone conversation some days earlier with Rob in Vancouver who suggested we should make the trek, as he had done on his own at least ten years previously.

Lyla and Eric were both enthusiastic about the prospects and within a day they had purchased a couple of small daypacks at Longs Drugs in Lahaina, and we were sufficiently impressed to buy a similar pack.

We drove in our own car rentals to Haleakala, leaving Kihei at about 8:10 am and negotiating the picturesque winding drive, with a brief stop at park headquarters to drop off our own car at the 8,000-foot level parking lot and Lyla and Eric's at the Visitor Center at the 9,800-foot level at about 10:00 am. We parked near a truck pulling a trailer which had hauled up a pack mule and horse owned by a young, muscular outfitter who kindly took a photo of our foursome at the trailhead. We learned that he took parties of up to six persons on horseback for three-day outings to Kapalaoa Cabin in the crater for $150 (US) each with accommodation, meals, and transportation all provided. He was to later pass us during our lunch break on the Sliding Sands Trail with his two charges heavily laden with provisions being delivered to the cabin prior to the arrival of a client group. His love for his animals was obvious during our brief encounter when he apologized to his lead mule for the heavy load. The affection was evidently mutual as demonstrated by caresses between himself and his lead mule when he stopped to talk to us on the trail. He informed us of the mule's much higher intelligence (Grade 8) than that of the packhorse (kindergarten).

We were blessed with clear skies above 6,000 feet, with haze and some clouds beneath, which had been somewhat typical during our ten-day stay. We departed for our trek with several outer layers of clothing because of the cool temperatures. Within five to ten minutes of our entrance into the crater, we were marvelling at the scenery, the solitude, and our good fortune to be where we were.

The Sliding Sands Trail descends easterly into the crater via a series of irregular switchbacks immediately to the south of several prominent cinder cones. The public appears to have been highly respectful of the requests, by frequent signs, to "Keep on the Trail" which has contributed significantly to the remarkably unblemished nature of the overall appearance of the crater slopes and floor. The principal trails, constructed in 1937, bear evidence of the thousands of trekkers visiting the area by trails entrenched up to several feet in depth through the cinders and by the smoothed surface of the volcanic rock. Although the trail is rough in sections where lava flows occur, I was impressed by the professionalism of the original surveyors who selected and maintained such an even and acceptable trail gradient, providing switchbacks as required. This is most evident on the Halemauu Trail near Koolau Gap with its elevation change of about 1,300 feet from the floor of the crater to its rim where one can capture four to five switchbacks in a normal photo frame.

In referencing our hike after our trek, I read the following excerpt from *Haleakala: A Guide to the Mountain* by Cameron & Kepler (1988), which would deter all but 'hardy' trekkers as it recommends allowing eight to ten hours of hiking time compared to the six hours or two miles/hour recommended by the Park Ranger whom I contacted beforehand and our actual time of 7.5 hours.

Sliding Sands Trail (9,780 feet) to Halemauu Trailhead (7,990 feet): 11.5 miles, 8-10 hours so allow an entire day. This outstanding hike for the hardy combines the features of hikes 1 and 3, but adds the central puu (pp. 46, 47, 51), rock sculptures (p. 54), a cinder desert and the Silversword Loop (p. 54). Start early and carry food and water; a canteen refill is available beyond Holua Cabin. The shortest route traverses the central craters on the Ka Moa O Pele trail (3.9 miles from the start). You end up six miles by road and 2,000 feet elevation below your starting point, so arrange for car pools, drop-offs at the summit, or pick-ups at Halemauu Trailhead. Many people camp the night before at Hosmer Grove (p. 32) to assure an early morning start. Though extremely satisfying, this hike needs careful timing, so check sunrise and sunset times, remembering that at this latitude darkness generally falls before 7 p.m.

Sliding Sands Trail is the preferred route into Haleakala Crater because of its remarkable scenery and ease of access from the Visitor Center parking lot at 9,800 feet. It is not the best way out however, for its gradual 3,000 foot drop creates a tedious climb back, a difficulty magnified by thin air and loose cinders.

Following our lunch break, we soon arrived at the turn-off from the Sliding Sands Trail leading to the Halemauu Trail and Holua Cabin. We walked through a hostile black lava field on the crater floor with interspersed sections of cinder to make the 200-300 foot ascent of the western slope of Halalii cinder cone where we had a brief water stop. Beyond this divide the trail switchbacks to the east where it bifurcates at a major juncture - the right, or easterly trail leading 0.4 miles to the Bottomless Pit on its route toward Paliku Cabin and the left, or north-westerly trail providing access to Holua Cabin via the Silversword Loop Trail. Just past the juncture en route to Holua Cabin, Lyla and I diverged off the trail to examine the remnant of a breached volcanic 'bubble' about 30 feet in diameter, a portion of which provided a sparse shelter from rain or snow, and which had been the source of a small lava flow.

As we trekked toward Holua Cabin, descending very slightly, we passed by the Silversword Loop - taking the more direct trail as we had already admired many of these hardy and spectacular plants which were near extinction less than a century ago but which now number over 50,000 (Cameron & Kepler, 1988). As we neared Holua Cabin, we noted more frequent evidence of the "brilliant yellow Evening Primrose (*Oenothera Striccta*), a hardy alien from South America which thrives along road edges and trails . . . Hawaii has no native wildflowers in the conventional sense; any plant that looks like a wildflower was most likely introduced." Near this locale Eric commented, "Who in God's mind would want to introduce a dandelion?" of which we saw several during our trek!

Near Silversword Loop we approached a tall, sweating, young man, heavily laden with a bulging backpack en route to Kapalaoa Cabin, about four miles distant. He had descended Halemauu Trail from the parking lot - the route we were to follow. We had noted the clouds swirling on the crater rim in that area and he remarked that he had appreciated their cooling effect.

Near Holua Cabin, we observed an abrupt emergence of delicate ferns at the 6,900-foot elevation and as we approached the cabin, we stopped to watch quail-like birds on the path. We also admired and photographed a Hawaiian goose (Nene) at the cabin. There were a few hikers at the locked cabin who were resting in the grassy meadow near the crater rim. Two of them later departed to explore, via flashlight, the prominent volcanic tube nearby. We were later to exchange greetings with them at the parking lot, not a little impressed by their apparent relief at seeing that we had safely arrived at our destination!

The ascent from the crater floor up the west rim to the top of the crater must be experienced to be believed. In spite of having walked about ten miles at high altitude and feeling relatively fresh after a rest at Holua Cabin, the climb in the sun was draining; we relished the shady areas. I had carried an altimeter with me on the trek, and could not believe that we had only ascended 500 feet when I heard Eric's expectant cry of being near the summit - realizing that we had not reached half way. The last thing anyone wants to hear in such circumstances is that, "you're not even half way," or words to that effect. So I said nothing. There was no problem and Eric even ran the last stretch to the parking lot where we were unexpectedly greeted by Denny and Beth Parkinson who had also visited the crater that day and knew our schedule. However, they had not expected us to be almost two hours behind our estimated time of arrival at about 4:00 pm; we arrived at 5:45 pm.

One of the unexpected highlights of our trek was being witnesses to the relatively rare phenomenon of the 'Spectre of the Brocken' near the top of Halemauu Trail. The 'Spectre' is a full circle-rainbow around your own shadow projected on to the swirling clouds beneath. We stopped to admire this strange apparition before continuing on the final stretch of the trail which led to the parking lot at a relatively gentle gradient.

En route up to retrieve his vehicle, Eric suggested that we should wait for sunset at the summit - which we did - anticipating the widely publicized beauty of the sunset view. We were most impressed by the experience, with the abrupt softening of light and the subdued violet to orange hues which accompanied sunset at this altitude. I had stowed four cans of Bud Light in a cooler in Eric's car trunk at the Visitors Center parking lot and these certainly helped us relax at the end of our eventful day.

Astounding China – 1990

Our first trip to Asia occurred in November, 1979 when Rene and I learned of a reasonable three-week group tour which included highlights of Tokyo, Kyoto, Hong Kong, Bangkok, Kuala Lumpur, Singapore and Bali. The closest point to China that we visited was Lokmachau in the New Territories at Hong Kong, a border point in the Shamchun River Valley. It was a fascinating sampler; however, we were not to return for our brief but intense exposure to China until 1990.

In June 1989, when the People's Liberation Army tanks rolled into Beijing's historic Tiananmen Square, one of the major victims was China's emerging tourist industry. The square is Mao Zedong's creation and he is honoured at the nearby Mao Zedong Mausoleum. Zhou Enlai, another revered Chinese leader, was the focus of the Tiananmen incident in 1976 when thousands gathered at the square to mourn his death and protest the tyranny of the Gang of Four. By the end of 1989 major tourist hotels in China's principal cities had suffered significant reductions in occupancy. Guilin, with a population of 300,000 and highly dependent on tourism, reported 80,000 visitors by mid-June 1990 compared with 450,000 for all of 1998.

The subsequent hard-nosed efforts of the Chinese government to restore confidence in travel to China, reached us in mid-June 1990 in an unexpected invitation from Chinapac International Tours and Travel Services (CITS) of Vancouver to join a 21-day tour to China, starting on August 13, 1990. A private tour proposed for the Killam and Anderson families of Vancouver had been expanded to include a total party of twelve of which the Killams and Andersons comprised six. We considered the overall price of $2995 per person particularly attractive and immediately planned to participate.

We met most of our travel group following check-in at Vancouver International Airport on August 13. Included were Cheryl Killam and her daughters Laura and Sisley - both students; Joan Anderson and her son John - also a student, who were to be joined in Shanghai by John's sister Samantha who was taking a Mandarin course in China. Others in the party were Joe Jaworski and Barbara White, teachers from Sioux Lookout, Ontario and Stan Weese, retired and his friend Malcolm Crane, an analyst in Vancouver.

Our flight to Shanghai on Air China CA 992, which took 11 hours, followed the west coast of B.C. and Alaska and featured excellent service provided by six flight attendants. Travelling with us was another Chinapac Tours group, which included a recently retired dairy farmer from New Brunswick who was taking his first trip of more than one week, unaccompanied by his wife. He was sitting next to a woman who had telephoned to see if she could share accommodation with him to avoid the single supplement. The items one quite casually learns during travel, short of a life history, are often remarkable!

My recollections of our visit to China, being recorded in 2001, are still as vivid in many respects as they were a decade ago. However, my current narrative is highly influenced by 120 pages of diary notes by myself plus 11 pages by Rene; an album containing 500 photos, culled from over 1000 taken during our trip, a 12-page 'Asia Pacific Report' supplement in the June 1, 1990 issue of The Vancouver Sun; and, finally, our Lonely Planet 'China, a travel survival kit' guidebook, 1988 Edition. It accompanied us on our trip and was directly responsible for our visit to the remarkable stone sculptures at Baoding, near Dazu, requested and agreed to by Chinapac Tours, prior to the tour.

In retrospect, it is not surprising that our relatively brief visit to China ranks in my memory as the most impressive of all our travels. It is natural to make comparisons between one's very limited heritage and environment and that of others. In essence, the only similarities I was aware of between China and Canada were that both countries occupied a similar part of the globe – both lying in northern latitudes and both with almost the same land areas of 37-38 million square miles. There the similarity ended. The most striking difference was Canada's population of 27 million in 1990 compared with China's 1.1 billion and growing at more than 15 million a year with its objective of keeping its population below 1.2 billion until the year 2000, and its well-advertised one-child-per-couple policy. By 2000, China's population was to reach 1.25 billion (337 per square mile) compared to Canada's 31 million (8 per square mile). Not surprisingly, one of the most powerful impressions during our travels was the number of people visible wherever we went. Although the three major cities in China (Shanghai, Beijing and Tianjin) currently contain over 34 million people – exceeding Canada's total population, the majority of the population, totalling over 900 million, is rural.

Shanghai

We arrived in Shanghai where reconstruction was in progress at the airport using large bamboo scaffolds as a cribbing around the buildings. The sultry heat of 37º centigrade was a contrast after the air-conditioned aircraft. This temperature was to remain relatively constant during our travels.

We were driven to the nearby Cypress Hotel for dinner, even though we had finished a sizeable meal on the aircraft only two hours earlier. We were not prepared for the volume of food served to our party of eleven – all sitting around one circular table with the food piled on a centrally located 'Lazy Susan'. This was a common practice throughout our trip for most meals served in hotels or restaurants. Our common liquid at lunch and dinner, much appreciated by most because of the heat, was beer.

Driving through the city to our destination, the Peace Hotel, we were overwhelmed by the crowds, traffic, new construction, squalor and apparent poverty throughout much of the outlying areas though the core area appeared to be cleaner and more prosperous. Particularly impressive was the number of bicycles, reportedly over 6 million in a city of 12 million. The bicycles commonly carried more than one person and I later photographed one family, with father cycling, mother on the crossbar and two young children, one on the handlebars and the other on a seat over the rear wheel. The bicycle was also a major vehicle for commercial transportation – carting unbelievable loads on two-wheel carts attached behind. Cars were few, with taxis, over-crowded buses and vans vying with bicycles. Also impressive was the relatively cheerful demeanor of most of the crowds and the proud attention given by adults to small children.

Throughout the city, particularly in the outlying areas, we noted many individuals lying or sitting on deck-chairs which we were told by our guide they carried from their nearby homes to watch life passing by on the streets.

Our room at the Peace Hotel measured 6 x 6 metres with a 5 metre ceiling appearing severe because of lack of furniture, but mercifully well air-conditioned.

Our first full day in China was to be the forerunner of a three-week period of almost ceaseless activity and awesome impressions. It included visits to the Yu Yuan (Happy) Garden and Jade Buddha Temple in the morning, followed by visits to a workers' housing unit, carpet factory, jade factory, silk factory and a trading centre in the afternoon before being taken to the Shanghai Hotel for dinner featuring at least 12 different dishes on a very large Lazy Susan. We eventually became fairly adept at not knocking glasses or other items off the table as the Lazy Susan was swung around.

The Yu Yuan Garden was built between 1559 and 1577 by the Pan family, who were rich Ming Dynasty officials. It is the only classical Chinese garden in Shanghai and contains ponds, rockeries, bridges, sculptures and a 400-year-old gingko tree amidst its ornate buildings and pathways. The Jade Buddha Temple was built between 1911 - 1918 and housed about 700 resident monks in 1990, having been largely inactive from 1949 to 1980 during the cultural revolution. Among several gold-plated Buddhas and other ferocious-looking deities within its extensive hallways, the center piece is a 2-metre high seated white jade Buddha – possibly made of alabaster, which is encrusted with jewels and weighs about one tonne. It was reportedly brought to China in 1882 by a monk from Burma. The carpet factory was reminiscent of those we visited in India in 1986, except for far better lighting and not being dominated by families and their children. The carpet weavers that we saw in Shanghai used art work displayed on nearby walls to guide them. Our visit to the jade factory was short as the workers had left for home because of the excessive temperature (37° celsius). The sales area, which we visited briefly, contained fabulous carvings in all shades of jade, sodalite and ivory, priced at up to 100,000 yuan ($25,000 Canadian). The silk factory had a vast floor space occupied by looms with female workers. The lighting was poor and the area hot and humid with a temperature of about 35° celsius. The silk-making procedure was fascinating. The source of the silk is silkworms which are fed on bamboo trays by peasants. When cocoons are spun, they are shipped to factories where larva cases, which we saw, are boiled and the filament is unwound by hand into long strands. Each cocoon yields about 200 metres of barely visible filaments, which are wound into initial coils representing the output of 400 cocoons totalling about 90 kilometres of silk threads and weighing about 60 grams (2 oz.) The showroom displayed several elegant silk screens embroidered with colourful pictures of all sizes with different coloured threads on the obverse sides and no evident duplication of colours from the reverse. We purchased a circular 12 centimetre diameter screen within a wooden frame on a wooden stand which could be rotated to show a white kitten viewing a green locust on one side, with brown and white colouration on the other, costing 150 yuan. The workers' housing unit we briefly visited was said to be representative of rental housing provided for workers and retirees. The typical rental provides accommodation in apartment-style buildings for 4-6 people in one suite of about 52 square metres (500 square feet) of floor space. Basic rental is about $3 plus $7 per month for gas, electricity and water. The average wage in Shanghai is 200 yuan ($45) per month.

The Peace Hotel lies at the intersection of the Bund and Nanging Liu Avenue, the tourist crossroad in 1990. The Bund is an Anglo-Indian term for the embankment of a muddy waterfront and represents a commonly used walkway along the side of the Huangpu River. We joined Malcolm and Stan for an after-dinner walk along the Bund before collapsing into bed after our first day in China.

Suzhou and the Grand Canal

The city of Suzhou , which lies about 145 kilometres (90 miles) west of Shanghai in the Yangtze Basin has a history dating back over 2500 years. With the completion of the Grand Canal in the Sui Dynasty (589-618 AD), Suzhou flourished as a centre of shipping and grain storage. By the 14th century Suzhou was the nations' leading silk producer and had attracted the Chinese aristocracy who constructed elaborate gardens and villas, many of which exist to this day and remain, with the Grand Canal, principal tourist attractions together. The city has a population of about 600,000.

On arrival at Suzhou after a 3-hour drive from the Peace Hotel, including 45 minutes to reach the outskirts of Shanghai, we picked up an intense local guide, who informed us that our impression of Suzhou should be that, "it is small, ancient and beautiful," presumably in contrast to Shanghai. Our first local visit was to Wangshi Liu ("Master of the Nets") Garden lying within the heart of Suzhou's residential area. It is the smallest and arguably the most famous of all Suzhou's gardens which totalled more than 100 in the 16th century. It was laid out in the 12th century, abandoned and then restored in the 18th century. It includes well-furnished Ching dynasty (1663-1796) buildings, rock grottos, small ponds, the smallest of 300-odd bridges in Suzhou and an 800-year-old cypress tree. The tranquility provided by the beautiful and serene garden setting belied the conditions which must have existed in Suzhou just 40-50 years earlier when it was the site of a desperate battle in the Sino-Japanese War (1937-45) followed by the Communist Revolution (1945-49) when more than 500,000 men were involved in bloody fighting lasting over a month in the most decisive battle between the communists and nationalists. As in other parts of China that we visited, little visual evidence remained of these conflicts.

Our principal attraction was the Grand Canal, which is the third great waterway of China after the Yangtze and Yellow rivers, and the longest canal in the world. Begun 2400 years ago, it created an inland waterway stretching 1794 kilometres (1113 miles) from Hangzhow in South China to Beijing in the north. It was, and remains, a remarkable achievement requiring the labour of millions of people, at an enormous cost in lives. Opened officially in 610 AD it required the crossing of four major rivers (Yellow, Yangtze, Huai and Quiantang) and was designed to be 40 paces wide. In its heyday some 15,000 junks, sampans, and barges carried grain, timber, salt, fish, cloth, pottery and other goods along the canal from the south. Today parts of the Grand Canal are impassable or not used, but it is being restored in sections both for transportation and irrigation purposes. The widths of the parts of the canal that we saw on our trip ranged from as little as 20 metres at a bridge crossing to over 150 metres. Canal depths are reportedly up to 3 metres. Our trip provided a relaxed view of life along the margins of the canal, among the scows, barges and small hand-powered boats, and included youngsters swimming at various locations in water that would be considered highly polluted by western standards. One of more than 50 photos I took during our trip shows a young woman eating with chopsticks from a bowl and standing on the front of a barge loaded with crushed rock, which passed our riverboat. Several others show lines of up to ten barges tethered together in order to conserve fuel. The volume of cargo being transported and its diversity was noteworthy.

On leaving Suzhou in our van to return to Shanghai, our guide Jennie informed us that as we had another three-hour drive back to our hotel and were scheduled to be at an acrobatic show at 7:15 pm, there was little time to spare after our arrival at the Peace Hotel and certainly not enough for dinner. Furthermore, as the show was so good we also needed to rest on our drive back to Shanghai in order to enjoy it. Cheryl immediately spoke up, "We can't rest if the driver keeps on honking his horn!" This brought immediate guffaws from us – but of course, no reaction from the diver. The area between Suzhou and the outskirts of Shanghai, like much of China, contains abundant rice fields broken by clusters of buildings – mostly worker's homes lying along the margins of canals and irrigation waterways. The farmers generally work alone, though clusters of workers are occasionally seen. The main road we travelled had a poor asphalt surface, composed of poorly compacted earth mixed with crushed rock and lime, as evidenced by workers that we saw widening sections of the roadway. Stripped to the waist, they frequently smiled and waved as we passed by.

The acrobatic show that evening featured over one dozen acts in a two-hour period. Rene commented, "It's more like a circus." It included a magician, acts performed by dogs, lemurs, elephants, tigers and a panda in addition to superb balancing and tumbling acts. Most of the pandas we were later to see were quite inactive and usually dozing. The performing panda sat in a broad chair at a table containing various food items and proceeded to eat these with a large knife and an unpronged fork. The most dramatic of the balancing acts included running jumps by four successive acrobats from a very flexible board on to the shoulders of the individual directly beneath to form a single tower of five balancing acrobats.

Guilin

Of all the cities we have visited, Guilin is arguably located within the most dramatic setting because of the spectacular karst topography in the surrounding area. Conditions promoting karst topography are well-jointed, dense limestone near surface in areas of moderate to heavy rainfall and good groundwater circulation. Removal of limestones, dissolved along vertical and horizontal cracks over eons of time produces the natural bridges, underground caves and spectacular remnant pinnacles heralded in various parts of the world. Visualize relatively closely-spaced stubby pinnacles of rock rising for hundreds to thousands of feet above the land surface in an area of several hundred square kilometres. This is the condition which currently exists in the Li River valley in the Guilin area, Kwangsi Province of Southern China.

Guilin, founded in the Qin Dynasty (201-227 BC) has always been famous for its scenery and has been extensively eulogized. Originally developed as a transportation centre, its population peaked at over one million during World War II as people sought refuge there. In 1990 its population stood at over 300,000 and tourism continued to rank as its major industry, despite the downturn caused by the Tiananmen Square incident in 1989.

We flew from Shanghai to Guilin in a Shanghai Airlines Boeing 757, taking just over two hours to fly 1700 kilometres to the southwest over the typical rural setting previously described between Suzhou and Shanghai. The view of the landscape around Guilin from the air was awesome, with prominent limestone pinnacles we were later to visit protruding within the city itself, several crowned by isolated buildings. We were provided with a pleasant room in the Osmanthus Hotel with a magnificent view of a tributary of the Li River, a busy bridge across it, and the nearby mountains.

In the afternoon we visited the Tunnel Hill Commune on the outskirts of Guilin where we were given a presentation on the commune by its manager, with our assigned tour guide, Liu, acting as interpreter. The commune covered an area of 75 hectares and contained a population of 4800 of which 1800 were workers living in 740 households, some of which were over 300 years old. We visited two of them (one new, one old) and both were at the very bottom of living standards measured by Western values. We also visited a small medical centre staffed by two women, one of whom provided a couple of our group with massages. We had an opportunity to cross a nearby river on a picturesque stone footbridge where we photographed some of the local scenery including workers in fields near 'Pagoda Hill', a limestone spire rising several hundred feet above the field with a seven-tier, 50-foot high pagoda perched on its summit.

The next morning at 6:30 am from our window we could see people cycling and exercising in the distance. I left for a run and reached the bottom of Solitary Beauty, a 450 foot high limestone pinnacle accessible by a series of steps with an observation area at the top which provided a magnificent view of the surrounding area.

Following breakfast we left by van on a one hour drive down to Zhujing to catch a boat on the Li River for a 4.5 hour trip to Jangshow - about 60 kilometres further to the south. The scenery was spectacular, being dominated by limestone spires which reach over 1500 feet in height with shear to overhung faces at some locations into the river. Undercutting by the river is 3 metres or more. Caves are abundant with rare stalactites exposed on some partially eroded caves at the edge of the river. Villages appeared sporadically along the route, but activity is abundant with water buffalo browsing, people fishing - frequently with cormorants, others gathering a bright green weed by diving with or without masks, and storing the weed on bamboo rafts. The weed is commonly used as feed for water buffalo, which are domesticated and used for work on farms. Bare or sparsely clad youths can be seen swimming and occasionally helping on rafts. Near the southern end of our journey the region is more subdued with bamboo flourishing at the river's edge and the ground commonly terraced for several hundred feet back from the river.

Although our attention was almost totally consumed by the superb scenery and the life along the river, we were far from being an isolated tour boat. We learned that on average there were about 100 boat trips per day carrying about 7000 tourists from Guilin or Ahujing to Jungshow. One of the photos I took shows six other tour boats on a stretch of the river ahead of us.

On arrival at Jungshow I paid 2 yuan to photograph an old man holding two cormorants at either end of a 2 metre long pole, in order to provide a close-up view of others I had photographed on small fishing boats. We wandered along a narrow street lined with numerous small stalls where the hawkers attract attention of the tourists yelling, "Hello... look!"

I finally yielded at one of these, which contained a quantity of dusty, partially termite-chewed plaques obtained from old temples. One in particular attracted my attention, even though details of eight finely-carved figures on a latticed background were indistinct because of the layer of grime with which they were covered. I was informed that it was early Qing (Manchu) Dynasty, and it certainly looked as if it could be 300 or more years old! The man I dealt with wanted 350 yuan for it, and when I said, "Oh no!" he replied in English, "How much?" Rene suggested I offer 100 yuan, and he unconcernedly countered with 160 yuan and then accepted 150 yuan ($40 Canadian). Our discussion had attracted quite a gathering, as we found common in our modest dealings in small stores in China. As we left with our purchase, we were followed by three other hawkers, each bearing other plaques and I had a difficult time assuring them that my extravagant moment had passed! After returning to Vancouver I carefully cleaned off the plaque to expose eight exquisitely carved figures in deep red-coloured mandarin cloaks, with black-eared hats, all of the figures partly covered with a veneer of gold.

On the following day we visited Fubo Hill, Reed Flute Cave, Seven Star Park, and a Pottery Factory, all in the Guilin area. Fubo Hill lies on the west bank of the Li River. Like Solitary Beauty, its summit provides a superb view of the surrounding area. Its principal attraction is the Thousand Buddhas Cave at the base of Fubo Hill. The cave name is an obvious misnomer since there is only room for several dozen statues – all life size – in the cave. These reportedly date from the Tan and Song Dynasties (400-900 AD) being mostly Buddhist and minor Tao in derivation, so a maximum 1300 years old. Nearby buildings on Fubo Hill housed a 6-foot diameter, 300-year-old pot, used for cooking rice for over 1000 people, and a 300-year-old bell. Nearby, we watched two couples practicing their modern dancing steps in the park at the base of Fubo Hill.

Reed Flute Cave is named after clumps of reeds, which once covered the entrance to the cave, used by locals to make musical instruments. The cave is certainly world-class in stature, with its innumerable grottos, the largest of which held more than 1000 people during World War II when used as an air-raid shelter. That particular grotto, "Crystal Palace of the Dragon King," measures about 100 x 60 x 20 metres high. Most of the grottos are covered by a variety of stalagmites and stalactites, many of which bear local names, of objects they resemble. We found that the overall effect of the presentation provided in the grottos was somewhat diminished by garish coloured neon lights illuminating selected portions. Our route through the caves covered an excellent walkway, about 800 metres in length within on oval-shaped circuit.

Seven Star Park lies on the east side of the Li River near Liberation Bridge. The park contains Camel Rock, an obvious resemblance, and afforded an opportunity for Rene to be photographed holding not a camel, but a large eagle, on her arm. Our final visit to a poultry factory was inconsequential, considering our future ramblings. That evening we were driven with our luggage to the Guilin airport for a delayed flight in a totally packed China Southwest Airlines Boeing 737, which finally left at 9:15 pm, 1½ hours late, to arrive at Chengdu two hours later.

At Chengdu we learned that our schedule for the next day had been changed some time earlier, without our knowledge, to provide for a trip to Emei and the nearby Grand Buddha at Leshan, requiring a 6:00 am breakfast for a 7:30 am train ride the following morning. Even at this relatively early stage of our travel in China, the pace of our expected excursions was obviously beginning to affect our group. On arrival at our van at the Chengdu Airport, after a lengthy walk, we were introduced to Sheila, our local guide. She announced that we would be overnighting at Emei and spend two of the three days that had originally been planned for the district in that area. I pointed out that our itinerary had us returning from Emei to Chengdu on the same day and that our itinerary for Chengdu would need to be changed, as two full days in Chengdu had been provided to visit other nearby attractions. Sheila blamed Liu for misinforming us, and even though I had checked with him a few hours earlier to confirm our itinerary, he blamed our group, unreasonably citing the amount of baggage we carried. I pointed out that we were complying with our baggage allowance, to which he had no response. The altercation was in sharp contrast to the bus ride to the airport at Guilin, where we sang songs to him after he had crooned some Asiatic songs to us.

Sichuan Province

Sichuan, centrally located in the south-central part of China is its largest province, being about the size of France and it is also the most heavily populated with over 100 million people. The largest city is Chongqing and Chengdu is the capital of the province. The region is rich in natural resources and following the Cultural Revolution it was selected as a focal point for debunking the commune system by decentralizing agriculture and promoting a capitalist system with greater autonomy in decision-making and establishment of free markets. The mountainous regions are the natural habitat of the giant panda. Attractions which we were to visit in Sichuan province required access not only to Chengdu, but in addition, Emei, Leshan, Dazu, Chonqing and points along Yangtze River between Chonqing and Wushan.

Chengdu to Emei

At the South Railway Terminal in Chengdu we followed the sign marked, "Cushioned Passengers First" and boarded the last car in the train containing double-facing seats, a small table in between, fans, seat covers and overhead racks. Our route to Emei crossed through a very rural area with moderate pollution, which is common throughout heavily populated areas in China both rural and urban because of coal-burning fires, with smoke sifting upward from kitchen areas, normally placed adjacent to houses in rural areas. The region is criscrossed with irrigation ditches supporting rice and a variety of vegetable growth. Fishermen are occasionally evident along the larger watercourses as evidenced by nets with bamboo supports. Cattle, water-buffalo and goats are common, generally isolated from each other with an ever-present individual standing nearby.

Our trip from Chengdu to Emei, about 120 kilometres to the southwest, took about 3.5 hours. At Emei we transferred to a van for a one-hour drive to the city of Leshan lying at the intersection of the Min and Dadu Rivers. Our objective was a visit to the nearby Grand Buddha which, at a height of 71 metres (233 feet), is the largest Buddha in the world.

Grand Buddha

The story of the construction of the Grand Buddha is charming. At Leshan the Min River flows southerly into the Yangtze River and the Grand Buddha is carved out of a cliff facing west at the juncture with the Dadu River. The construction project was engineered by a Buddhist monk named Haitong in the belief that it would lead to a moderating of the swift currents at the confluence of the two rivers, which had caused the deaths of many fishermen. The project started in 713 AD and was completed 90 years later, well after the death of Haitong. The vast quantity of rock removed during construction, falling into the waters below, calmed the waters and solved the drowning problem.

We accessed the Grand Buddha initially by boat from Leshan and our reaction at first sight after rounding a bend in the river was of utter disbelief because of its magnitude from such close quarters. We docked near the Buddha where we were met by a van, which drove us up to a monastery. A path led to the brink of the cliff immediately above the Buddha and zig-zagged down the cliff face to its base. The lack of erosion of the Buddha is quite remarkable as it is carved out of massively bedded cream to red sandstone. The Buddha is seated with arms resting on its knees and feet resting about 7 metres above the water level at the time of our visit. I measured the length of the big toe at 8.3 metres (27 feet) and attracted two young Chinese to stand on the 1.6 metre long big toenail for a dramatic photo looking upward past Grand Buddha's head. The climb back up the stairway with its many recessed niches, some of which contain poorly- and well-preserved statues of Buddhas, took about 5 minutes. Our guide, a former teacher, mentioned that about 20,000 foreigners visit the site annually, plus 5 million Chinese – mostly from Sichuan Province. The cost for our visit, including the ferry ride was 25 yuan ($6 Canadian) each.

Emei Shan (Mt. Emei Temples)

Mt. Emei, rising to an elevation of 3099 metres, is one of China's famous sacred Buddhist mountains, with original temples dating from the first millennium. At one time there were over 135 temples on the mountain. Today there are about 20 active ones of which we visited two: Wannian (Temple of 10,000 years) at an elevation of 1020 metres and Qingyin (Pure Sound) Pavilion at 710 metres. Our ascent started from our van at about 700 metres and followed a good path of concrete with steps as required, about 1.3 metres wide, for 3 kilometres to Wannian. There are actually three temples on site, partly reconstructed and dating from 874-879 AD in this, the oldest surviving Emei monastery. The feature of the temples is a statue made in 980 AD, which was dedicated to the protector of the mountain. It is 8.5 metres high, cast in copper and bronze and weighs about 35 tons. The temples are occupied by monks who strike bronze bells as visitors bend in prayer. During our visit we noted three elderly women tending extensive gardens around the buildings. On our climb to Wannian I photographed several hardy porters carrying individuals, either tired or incapable of the hike, on simple stretchers supported by two bamboo poles on their shoulders. Qingyin is a relatively small temple and pavilion, so named to reflect the sound effects produced by a cascading creek coursing through nearby rock gorges. The setting of the temple is particularly picturesque. We accessed the temple by 3 kilometres of almost continuous steps extending downward from Wannian. From Qingyin we continued walking on a downward grade for about 2 kilometres, mostly along the bank of a local river. The path also coincides with a ditch containing water from the river for local use. An interesting feature at Wannian temple was a long row of spice stalls set up beside the final steps to the site. Items were for sale to provide income for the monks. In addition, wild monkeys live on Mt. Emei and several of these were on site, supposedly domesticated to provide income for their keepers. One of the monkeys took exception to Sisley, biting her finger. Our accommodation at Emei on the previous night was the HongZu Shan Hotel which, for a period during World War II, was headquarters for Generalissimo Chiang Kai-Shek.

Chengdu

Chengdu boasts a 2500-year history linked closely to the arts and crafts trade with a solid industrial base. It ranks as one of China's wealthiest cities on a per capita basis, with a current population of about 1.5 million people and over one million bicycles. Among its attractions are the Chengdu Zoo with its six pandas, all but one of which were dozing during our visit. However, as compensation we enjoyed an excellent view of a more active lesser panda with a rich, red-brown coat and long bushy tail, which is a member of the raccoon family, but generally less than 10 pounds in weight at maturity compared to its giant cousin at about 350 pounds. The prolific bamboo thickets of Sichuan Province on which the 1000 or so remaining pandas depend also support a bamboo weaving factory which we visited. The process is relatively simple requiring stripping and calibration of the threads for weaving purposes. The products in the sales centre were diverse and we purchased a bamboo-covered teapot (100 yuan), a 0.7 x 1.3 metre finely-painted wall hanging composed of over 2000 strips for 25 yuan – $6 Canadian and a painted fan (2 yuan).

One of China's celebrated realist poets was Du Fu (712-770 AD) who travelled throughout China and fled to Chengdu after being captured by rebels, where he built himself a humble cottage in which he lived and wrote over 240 poems on the suffering of the people over a 4-year period. This represented about one-sixth of his eventual output. The present grounds with its re-thatched and renovated cottage cover about 20 hectares and contain related memorabilia, including a bronze statue of Du Fu, Chinese and western editions of the poet's works, and replicas of paintings (originals were on tour in 1990) by Chinese artists depicting scenes based on Du Fu's poems.

We also briefly visited a Sichuan embroidery plant and the Art Institute in Chengdu where we watched workers and over 20 well-known artists creating original silk screens and paintings, all of which are signed as originals. The overall output of this labour must be prodigious. One artist that we watched, completed a 0.3 x 0.6 metre painting of a giant panda in a bamboo thicket, using only black paint, which was deftly applied with deliberate strokes. The final product, with his signature and original emblem added, was finished within 10 minutes! Before dinner we spent an hour wandering through a huge Chengdu market area. As with other markets we visited in China, we were intrigued by the wide variety of items for sale. There was a stall devoted to a variety of small snakes and eels writhing around in water-filled plastic basins; another featuring brains; a section containing a variety of baskets filled with pigeons; fresh meat of unknown origin in a rusty container behind a packed bicycle, which I photographed being weighed on the common hand-held scales used at Chinese markets. Nearby we watched with other passers-by as Stan bartered at a small stall for a beautifully embroidered opera robe which he finally bought for the equivalent of $300 Canadian. He had seen a similar item for sale in Vancouver for $1,500. Before leaving Chengdu in the evening we enjoyed a superb Sichuan dinner at a restaurant adjoining the Cultural Palace where we were conveniently screened off from a room filled with other foreigners!

Dazu Stone Sculptures

At 10:30 pm on August 22 our party of 12 boarded a first-class(?) sleeper coach at the North Railway Station for our 7.5 hour journey from Chengdu, which we understood would take us to Dazu. Our tour group had not considered our accommodation distribution and as only three compartments had been reserved, I finally suggested that we split up on the basis of four men, four women and four youths, which everyone agreed to. The compartments were unbelievably hot on boarding and it was obvious that sleeping would probably be something of an ordeal. Each compartment had a floor area of 2 x 2.3 metres and was equipped with two sets of two-tiered bunks, each about 0.7 metres wide, a straw mat on top of a sheet covering the bunk, no other sheets and two fans, one of which was not working. Fortunately, we could open the window even though this resulted in the inhalation of a certain amount of soot from the steam engine during the night. At 5:30 the next morning we were awakened by banging on our door and were told that we had five minutes to get off the train with our bags. Lack of communication between Liu and the train attendant led to the speediest exit imaginable of a party of twelve. A further surprise was that we were not at Dazu, but at a railway station at Jong Chun; it was a further 60 kilometre van drive to Dazu. While having breakfast at the Dazu Hotel a water truck drove by cleaning the streets and blaring out Christmas Carols on a loudspeaker, including Jingle Bells and Joy to the World.

The Dazu stone sculptures lie about 15 kilometres north of Dazu town at Baoding within cliffs of limestone rising on the side of a u-shaped portion of a heavily timbered river valley. The sculptures occupy an almost continuous 125 metre long section with literally hundreds of individual figures carved in scenes rising to an average height of about 7 metres, with a protective layer of protruding limestone above. As such, it has been described as grotto-art. It would appear that most of the figures were initially painted, as colours are well protected immediately below the overlying cap, and worn off to varying degrees lower down. The initial carvings and the concept is credited to Zhao Zhifeng, a Tantric Buddhist monk who started the sculptures which were produced over a 70-year period ending in 1249 AD. The carved figures range from a few centimetres to 30 metres (Reclining Buddha) in size. Virtually all are inclined progressively outward with height above the pathway, beneath the cliff-face, for ease of viewing. The range of subjects depicted in the multitude of scenes emphasizes the differences between good and evil and is perhaps unique in its adaptation of the area's natural features to illustrate these objectives. A pertinent example is a sculptured scene of a 'cowboy' with a flute next to a water buffalo with its head upturned to catch drips of water falling from a drain extending from a spring in a nearby cave. However, most of the scenes depict stages of life with an emphasis on good and evil deeds. Included are sobering sculptures on the evils of alcohol and other misdemeanors and a section showing the evolution of a family with its couple and son. On balance, there is an overwhelming focus placed on evil-doing and its results, demonic figures being particularly abundant. A feature of the 'grotto' is the so-called 'dartboard', again held by a huge demonic figure, showing the six ways to transmigrate – the latter procedure not being particularly clear. As I noted in my photo album, the stone sculptures had an obvious impact on us westerners. I have a photo of Malcolm Crane staring open-mouthed at evil-doers in one stone carving scene being swirled around in a boiling cauldron. Like the rest of us, he was obviously appalled by these hideous sights.

The Dazu stone sculptures, completed over a 70-year period in 1249 AD, were over 700 years old before access to them, other than by locals, was provided sometime before 1975. We felt particularly fortunate to be among the few foreign visitors, not only to Dazu, but to China, to witness these remarkable sculptural scenes, which have survived over so long a period. A recent threat was the fanatical Red Guards who descended on the Dazu area bent on defacing the sculptures, as had been their practice in other parts of China, but were stopped in this case, reportedly, by an urgent order from Zhou Enlai.

Dazu to Chongqing

Within Dazu and the area around it I photographed scenes which made me feel as though I had regressed into the last century, experiencing a time-warp. One was of a sparsely bearded spry old man walking towards me in sandals, bearing two heavily-loaded baskets hung from a four-foot long pole across his shoulders; another of a much younger, handsome man eyeing me somewhat speculatively as he walked by with a shallow 1.3 metre diameter wide basket slung across his shoulder. Finally, a woman and her young child spotted us from a horse-drawn cart. The boy let out a shriek of laughter, pointing directly at me. Our guide explained, "You are probably the first Big Nose (white foreigner) that he has ever seen!" Later, as we drove on a lightly-traveled asphalt road for five hours covering 185 kilometres to Chongqing, we experienced spectacular local scenery and isolated experiences, and took many photographs, reflecting rural life. Among some of my memorable photos taken on this route were a literally rice-laden road, with stacks from which rice had been beaten and separated on the side of the road to dry, over many miles. Occasionally, we would see 'blankets' of noodles hung out like washing near homes along the highway to dry. One of my most embarrassing moments during the tour to Chongqing occurred during a pre-arranged stop at a pottery factory. We were given a brief tour of the facility and shown the larger pots, about two-feet tall, all piled up in the yard near the ovens. There was a subtle difference in the pots of which I was not aware. Apparently, they were left out to dry for about three days before being burned in nearby ovens, and then stacked outside – with no appreciable difference from pre-burned pots, other than a darker colour. Curious about the weight of these items, I grasped one with the intention of hefting it to estimate its weight. Alas, the section that I lifted came away in my hand. I was mortified, and our guide said, "Don't worry about it!" However, considerably stricken, I approached the boss and insisted on paying 10 yuan, which he apparently considered fair compensation for my oversight. Later, I had my photo taken with my memento! Earlier, at the Remini Hotel, our guide had checked with their office in Chongqing concerning our arrival there. On return, he looked glum and one of us commented jokingly, "The Yangtze River has dried up and the cruise is off!" a comment which was soon to prove almost prophetic, although made in jest.

Chongqing

On arrival at the Chongqing Hotel after a full day starting at 5: 30 am, we were told to be ready for a hot-pot dinner at a restaurant on the other side of the Yangtze River, within half an hour. Everyone rebelled, and I finally suggested 7:30 pm, only a half an hour later, which appeared acceptable. En route to the restaurant we crossed the Yangtze River on a namesake concrete-bridge built in 1980 spanning over 500 metres – the longest in China. En route we passed by the 'old town', with its rickety little houses sprawled haphazardly on the side of a mountain. Chongqing was the war-time capital of the Kuomintang from 1938 onwards and refugees raised its population to over two million people during World War II when the city was subjected to continuous bombardment by the Japanese, with multitudes taking shelter in nearby caves. An unusual characteristic of Chongqing as a large Chinese city is its relative lack of bicycles because of its mountainous nature. Our charming local guide Joe, in describing this situation couldn't resist telling us a Chinese riddle, "Why is the bicycle not standing? – Because it is two-tired!" We stopped briefly at the Remini Hotel, built in 1953 which has a colourful and ornate exterior and is dominated by an enormous circular domed concert hall 70 metres high which seats 4000 people. Two large wings contain rooms costing from 5-60 yuan per night ($1-12). Again Joe quipped, "It looks like heaven, but the rooms are like hell!"

Our hot-pot dinner was a unique experience and worthy of comment. The large Lazy Susan glass contained the usual hors d'ouvres plus a centrally-located large white platter, artistically arranged using a great variety of items including boiled eggs, cantaloupes and various greens to resemble two fish and a plant. The hot pot arrived following a vast array of earlier dishes including goose, liver, chicken, rice, brain, tripe, noodle, mushrooms, beef, eel, *live* sardines, angel-hair noodles and several greens. These items were dipped into either side of the large hot pot, which was divided into two halves, one containing a light-coloured sweet liquid and the other an extremely dark one. The various items were boiled for periods of ten seconds to ten minutes, as required. Following dinner, Joe asked us to assemble in the lobby for some information. We were devastated when we were told that the Yangtze Cruise had been overbooked and that our group would probably not be accepted. Two alternatives were suggested: (1) arrive early in the morning at the dock and demand our berths by virtue of prior payment. Neither Liu nor Joe favoured this option; or (2) try to arrange an alternate booking on the following day, which would require us to get off the cruise boat a day early at Yichang for a 12-hour train ride to Wuhan in order to connect to our flight to Xian. After a meeting without our guides we opted for the alternate booking arrangement, although Rene, Samantha and I initially favoured storming the ship!

Our extra day in Chongking provided an opportunity to visit 'Red Crag' and the Artists Village. Chongking has developed into an enormous industrial city with a population of about 7 million people. As such it is a source of air pollution which obscures distant vistas within less than one kilometre. On our subsequent trip down the Yangtze River, the smog mostly derived from Chongqing but, added to by plants on the Yangtze River, did not clear for over 100 miles downstream. Red Crag, otherwise known as Chiang Kai-Shek's Criminal Acts Exhibition Hall on SACO prison lies on the outskirts of Chongquin in mountainous terrain. In 1941 the U.S. and Chiang Kai-Shek signed a secret agreement to set up the Sino-American Cooperative Organization (SACO) under which the U.S. helped to train and dispatch secret agents for the Kuomintang government. Two prisons were established to contain up to 500 communist revolutionaries, including about 40 women. The prisoners lived in close quarters in squalid conditions and were reportedly allowed out each day for only ten minutes to use toilet facilities. We were shown a room which was used as a torture chamber. The Berlitz book on China which I referenced stated that Western visitors were not usually permitted access to the room in deference to their sensibilities as they are blamed for collusion with the Nationalists in the excess of their anti-communist struggle. On November 27, 1949 the Kuomintang killed all 300 prisoners on site, except for 15 who managed to escape. The prison facilities were then burned in anticipation of the liberation of Chongqin by the communists, which occurred on November 30.

The Artists Village provides working accommodation for 17 artists and their families. At the time of our visit the ages of the artists ranged from 40-75. The work is both interpretive and realistic using watercolours, oils and woodcuts. During our visit only one artist appeared to be on site. He specialized in woodcuts – each painting he produced took about two months to complete from design to final print. Each print reportedly sold for 400-600 yuan per print. One hundred prints were produced from each woodcut. Another artist produced woodcuts as originals on pear wood tablets, each about one centimetre thick. One of these reportedly took five years to complete!

Yangtze River Cruise

The Yangtze (Chang) River, at a length of 6699 kilometres, is the third longest river in the world behind the Nile (7030 kilometres) and the Amazon (6760 kilometres). It contains over 700 tributaries, only one of which we would access in a small boat from Wushan. Our route from Chongquin to Yichang covered 640 kilometres (400 miles) and included the famed Three Gorges: Outang (8 km long), Wu (45 km) and Xiling (75 km).

After all the problems anticipated in actually completing the Yangtze River cruise, we were absolutely delighted with our accommodation on the M.S. Goddess, which we boarded on August 25. The ship was built in 1981 and was powered by 1960 horsepower twin engines capable of a maximum speed of 30 kilometres per hour. Our room on board was air-conditioned, equipped with two beds and huge windows from floor to ceiling covering the deck side of the room, plus a small bathroom.

Leaving Chongquin at 9:30 am, I photographed several other cruise boats in the smog as we passed by numerous factories along the shoreline. Rene had awakened in Chongquin with a cough and a very sore throat following an almost sleepless night. On board she visited a doctor who checked her throat and prescribed three sets of pills. These eventually helped, but she stayed on the ship at our first stop at Fengdu.

Fengdu, about 120 km east of Chongquin, with a population of 30,000 in an area of about one square kilometer, has a history dating back 1800 years to the Han Dynasty. Fables record that two men of that period, whose names were Yin and Wang and whose families were united, were mistakenly thought to be Yinwang, the King of Hell. Apparently, this was the source of legends which developed that evil ghosts inhabited the surrounding hills. Numerous temples were built on the nearby Pingdu and Ming mountains following the Tang Dynasty (618-907 AD) containing sculptures of demons and devils. These are accessed by two pathways extending southwest and northeast from the town and its ornate entrance gate. Just beyond the gateway stands an elegant old banyan tree, which the Chinese admire for its longevity. Interestingly, a well-preserved petrified stump of a tree with similar girth but unstated name or origin has been embedded alongside – perhaps to emphasize its old age. Our climb of about two kilometres and about 200 vertical metres to the temple area on Mount Ming was completed at a temperature of about 40º centigrade. Several hikers were using fans and we envied another tour group equipped with mauve-coloured umbrellas. The clusters temples on Mount Ming are particularly attractive both in shape, having ornately fluted tile roofs, setting and colour, with bright blue and white dominating. Apparently almost all of them have been rebuilt over time, particularly after the Cultural Revolution, ending in 1980. Our guide indicated that only the steps were reliably original, dating back about 500 years. In fact, several carpenter's saws on site remained little changed from those used centuries ago, as depicted by one particularly grisly display at one of the uppermost temples. These displays were exhibited in two glass-covered cases, each about 30 metres long, and were obviously intended to show the consequences of evil actions in hell. Realistic models of devils and their human victims being killed in a variety of ways, included a particularly sinful individual hung upside down by his legs being sawn through his crotch with a band saw, which appeared identical to that used in the renovation work. Even the names of the temples and one bridge on Mount Ming reflected the depressing nature of the displays, including, "Between the Living and the Dead", "Palace of the King of Hell" and "Bridge of Helplessness". However, in spite of the depressing nature of the temple's themes, the views from the pathways to the city below and the Yangtze River beyond, with its surrounding scenery, were definitely worth the hike.

Following our return to the cruise boat at Fengdu, we attended the captain's welcoming cocktail party which featured very tasty hors d'oeuvres, including meatballs, Polish sausage slices covered with cheese, French fries, cold chips, onion rings and a variety of drinks. After an acceptable but not noteworthy dinner - perhaps because of the preceding event, we attended a couple of short films on the Three Gorges. Later, we had been cruising down the Yangtze River in the dark for over an hour and from our cabin we could see the navigation spotlight system used which consisted of two lights, both very powerful with each probing the fore and aft shorelines. Occasionally, our side of the ship remained dark for a short period, suggesting that the lights were then focused on the other side.

The Three Gorges more than lived up to our expectations. The narrowest is only 100 metres wide in winter and about 140 metres wide in summer. All of them have sections of vertical cliffs rising hundreds of metres above the muddy river water with mountains above, over 1000 metres in height. Most of these sheer cliff-faces expose massive, near-horizontal beds of limestone, minor sandstone and shale. Ancient 'pathways' about 3-4 metres in height, cut by hand for several metres in depth into the cliff faces are visible for many kilometres along the river in these sections, and I noted telephone poles added in one section. Between the gorges the land slope is more moderate, with many areas of well-terraced hills near communities, extending far up the hills. Following our relatively short trip (20 minutes) through Outang Gorge we docked at Wushan, a small town on the north bank of the Yangtze River on the west side of Danning River.

Until our cruise, we had not even been aware that we would be privileged to experience a ride on a converted, partly motor-powered sampan on one of the tributaries of the Yangtze River. The experience was the highlight of the cruise. Apparently, Danning River has only recently been 'discovered' as a major attraction with its three minor gorges. This important tributary extends northerly from the Yangtze River for about 250 kilometres from its source in the Dubashan Mountain area, of which the lower 120 kilometres are *reportedly* navigable. The river, like the Yangtze, has three gorges: Dragon Gate (Longmenxia) Gorge immediately above Wushan, 3 kilometres long, followed by the Misty (Bawuxia) Gorge, just below the attractive village setting of Shuanglong and finally by Vivid Green (Dicuixia) Gorge extending for 20 kilometres with green vegetation growing down to the river, which is reportedly the most beautiful, although we only saw part of it. All of the exposed rock in these gorges was massive to well-bedded limestone.

The trip up Danning River was awesome in every respect. In the gorges we went through particularly narrow sections only about 5 metres wide with cliffs rising on either side for hundreds of metres. At one location a cave reportedly contained a coffin from ages past. Similar ones on Yangtze River had been destroyed during the Cultural Revolution. Some of the most impressive remnants of a by-gone era were visible rows of square holes about 10 centimetres per side, set about 2 metres apart and about 10 metres above the river level. About 2 metres above were a similar set only one metre apart. These provided for the supports of an ancient plank road from Danning into the interior, with a protecting layer of planks above. When war lords in the interior were threatened by others coming from the Yangtze River, the planks were simply removed.

Above Misty Gorge, the river widened and was extremely shallow, our riverboat grounding occasionally, and eventually coming to a standstill. The 'younger' passengers were asked to disembark and walk about 500 - 600 metres upstream on the gravel and rocky outcrops. The trip to Shuanglong was particularly spectacular and took about 3.5 hours. Beyond, following a light lunch, we traveled through part of the Dripping Green Gorge. In this section I assisted the two bowmen, with several others, in hauling the boat up by rope. I felt as if I had entered another time warp, as I joined in, chanting "Hai-ta", as we hauled the boat for about 10 minutes. We were rewarded by cooling off in the river up to our knees. On the way back we again stopped at Shuanglong to pick up Samantha and Sisley, both of whom had stayed behind, feeling quite sick. Just below Shuanglong, local children tried to sell tourists coloured pebbles and brass coins on the way up river. The better swimmers came back out on the return journey to splash us. A guide on our boat came up to Rene and I and said, "I hope you are not angry at the children for splashing you... they mean no disrespect... they are on summer holidays having fun," and almost as an afterthought, "they want to make you happy." Actually, any water splashed malevolently or otherwise would have been most welcome, and I'm sure that none of the passengers were disturbed.

Our remaining trips through Wu and Xiling gorges on the Yangtze River were very similar to that on the upper part of the river, with extensions of the ancient pathway, other river boats and colourful sampans with their lofty sails, several inactive factories and significant towns such as Badong, below Wu Gorge. Xiling Gorge, the lowermost of the three gorges, appears most verdant, and provides a mixture of mountain spires reminiscent of Guilin, a variety of caverns and waterfalls and a spectacular arched concrete bridge spanning a deeply entrenched river valley tributary to Xiling Gorge with small waterfalls at its mouth. At Yichang, the terminus of our cruise, I photographed a fisherman's net supported by bamboo poles, which are common throughout China. Finally, I photographed a colourful sampan above the dam and locks at Yichang. The city had a population of 420,000 in 1990. Its dam provided outlets through a system of locks for ships and also production of power. It was built between 1981 and 1989. At Yichang we visited a research station where sturgeon are studied. One, which we saw, caught below the dam, weighted 200 kilograms (440 pounds), was 16 years old and over two metres long. The dam impedes their return to spawn up-river, once every 15 years. They may live for 55-60 years. Like salmon, they travel great distances from the sea to spawn. From Yichang to East China Sea is about 1500 kilometres.

Yichang to Wuhan

From Yichang we boarded a night train with a sleeper for Wuhan similar to that from Chengdu to Dazu. Fortunately it was a much cleaner and more comfortable car and we had no problem sleeping. I was awake by 6:30 am and sat watching workers already well-established in rice paddies. Soon the sun rose dramatically as a beautiful orange-red ball, enhanced by the smog. En route I saw many farmers plowing with water buffalo. By 9:00 am, on reaching Wuhan the temperature was already 39º centigrade. As we stood waiting for our baggage we were approached by beggars, the first we had seen in China. There were several Red Army soldiers nearby and one of them kicked one of the beggars. The soldier's sandal flew off in the process and this appeared to annoy him even more, for after retrieving it, he walked quickly after the beggar and kicked him again. Although we had not seen any such intolerance of others in China, none of the nearby crowd seemed at all concerned.

We had some time to spend in Wuhan prior to our flight to Xian, scheduled at 2:50 pm. Our guides decided that we should have an early lunch at the airport – as a means of passing some time before our flight, and more importantly to get us cooled off in the air-conditioned dining room – which was much appreciated. We then drove to the Hubei Provincial Museum near East Lake, which is noted for its impressive collection of artifacts from the Zhenghouyi Tomb, unearthed in 1978. The tomb dated from the Warring States Period (433 BC) and was the final resting place of Marquis Yi, buried along with 21 female sacrificial victims, his dog and about 7000 of his favourite artifacts including a large range of well-preserved bronze vessels and musical instruments of which the most impressive is a set of intricately decorated bronze bells.

After our museum visit we drove to East Lake, claimed as the largest lake in any municipal park in China at 34 square kilometres. We sat under some shady cedar trees enjoying a cool breeze before returning to the airport for our flight, which was delayed, reportedly by mechanical problems until 5:00 pm.

Wuhan to Xian

Our flight in a twin-engine turbo-prop aircraft filled with 48 passengers was memorable, for its lack of air-conditioning and the rattling doors behind the cockpit where we sat, which required us to yell in order to be heard. We made an unscheduled stop at Xiangfan, 175 miles northwest of Wuhan and midway to Xian. We never did learn the reason for the stop, where all the passengers were required to disembark and walk across the sweltering tarmac to an equally hot waiting room where we spent the next hour. Our only entertainment was watching army personnel, three of whom went off to guard the aircraft; two others to guard the entrance and exit of the waiting room, and two remaining with us in the waiting room. About half an hour later, one of the young guards in the waiting room ran out to the aircraft and saluted the officer guarding the left front wheel, who sauntered back to the concourse. Shortly after, another guard entered the waiting room, removed his belt and before exiting handed his belt to another guard who donned it, and scurried out to the aircraft.

Finally, all the passengers re-boarded the aircraft and with the air-conditioning not working, those with fans tried to cool themselves. Half an hour later the aircraft had to climb to clear high cumulous clouds and within 10-15 minutes passengers were shivering. The latter part of our route to Xian was mountainous, with no obvious sign of habitation with the exception of rare mountain roads. Finally we landed at Xian at about 8:00 pm and to our surprise it felt beautifully cool and our spirits rose accordingly.

By 9:00 pm we had cleared the airport and arrived at the Hawaii Hotel Xian, although our baggage had not arrived with us. Dinner included excellent sweet-and-sour pork, a salty noodle referred to as "jelly fish" and a Polish pork dish with greens. By 10:30 pm our bags had still not arrived and at 11:00 pm when we were almost asleep there was a knock on our door and I put on pants expecting our bags. However, it was our Xian Guide Miss Wang who, noticing my partly bare condition, said, "Good morning", then giggled, covering her mouth, and said, "Oh no, sorry! I'm afraid I have some bad news for you!" I wondered what it could be after our four-hour-delayed arrival. Apparently our bags were lost, not just ours, but that of the entire group of twelve people. Belatedly, I realized that 11 pieces of baggage at the hotel on our arrival, assumed to be ours by our guide, belonged to another group.

After breakfast the next morning, we assembled at a van for our visit to the Terra Cotta Warrior Tomb only to learn that our baggage had not been located even after I had been told by Miss Wang only one half our earlier that it was in Wuhan and would be sent on to us at Beijing. As all the luggage was under a group ticket I felt some reassurance that it would be located. This was probably wishful thinking; our return tickets from Xian to Vancouver and our 22 rolls of exposed film were in our bags.

On driving out of Xian we passed by a horse-drawn wagon carrying a 3-metre long barrel about one-metre in diameter. Intrigued, I asked Miss Wang if she knew its contents. "Oh, not very nice!" she said, explaining that most homes were not equipped with toilet facilities, and that these were deposited into cans left outside front doors for deposit into the barrels, on a daily basis. They were then taken out to farm areas for use as fertilizer.

We were not disappointed by our long-anticipated visit to the nearby Terra Cotta Warrior Tomb at the Qin Army vaults, symbolically guarding the tomb of the first Qin-Emperor Shihuang following his death in 210 BC. His nearby tomb was yet to be excavated at the time of our visit. However, in anticipation of his death, he had conscripted hundreds of thousands of his subjects to build his tomb and construct the life-size terra cotta warriors, horses and chariots as guardians. An estimated 6000 extremely life-like 'warriors' were involved of which about 1000 had been partly exposed at the time of our visit in a 230 x 62 metre vaulted area, in eleven rows with four warriors to a row, in columns 4 metres wide. Subsequent fires during construction destroyed original colours of the terra cotta guardians. However, the scale of the operation, conceived and completed over 2000 years ago appears inconceivable, based on our knowledge of the progress of civilization and the Dark Ages in the intervening period.

On our drive back to Xian we visited the nearby Bampo village site, where excavation in the 1950s for a factory, unearthed parts of a stone-age village estimated at 6000 years old containing pottery (clay), stone and bone instruments and graves. The latter included large lidded clay pots in which young children who had died were buried, in addition to unlidded pots with air holes used for the burial of babies who had died, "to let their souls escape".

On arrival at our hotel we learned that our group's bags had been located at Canton, near Hong Kong, about 800 kilometres south of Wushan, their check-in point. After a brief wash we were driven to the Big Wild Goose Pagoda, originally located in the inner walled-city of Xian, but relocated outside in 1990. We climbed 242 steps through seven pagoda tiers to the top, about 50 metres above, for a photo. The pagoda was built in the 7th century AD to house precious Buddhist texts brought back by Xian pilgrims. Our dinner near the Bell Tower in the centre of the old city featured 16 different servings of dumplings. Our evening at Xian ended with entertainment at the Tang Dynasty Restaurant and Theatre, a lavishly decorated and maintained complex where we listened to songs and watched folk dances, with accompaniment provided by replicas of ancient musical instruments, including a 3000 year old flute.

Our experiences in China suggested that the Chinese delighted in providing either good news or bad news to their flock in the late evening. Just before settling down for our last night in Xian, Liu and Miss Wong knocked on our door to deliver each of us a 'gift' of underwear and a white T-shirt, because of our still-unavailable baggage. Rene's underwear was pink and mine orange, otherwise they were identical in shape, material and size.

Our trip to Beijing the next morning in a Russian-built Tupolev jet aircraft took only a little over one hour in hazy skies over dry mountainous terrain. It was warmer on arrival than Xian, but for once we were on schedule and quite relieved to have reached out final destination in China. At the airport we met Elaine, our local guide, and once again we had to voice our preference to be checked in at our local hotel, The Beijing Yanshan in this case, prior to starting off on a tour.

Beijing

Chinese history has been dated back 7000 years to farmers along the Yangtze River, recognized as the first to grow rice. By the end of the Warring States Period (475-221 BC), China had been unified by Qin Shi Huangdi, the first emperor, and Beijing, then known as Ji, had become the capital of the Yan Kingdom. However, it was not until the time of Genghis Khan that it became the capital of the Empire of the Great Khan.

We stayed for three days at the Beijing Yanshan Hotel in the northwest part of the city. Unlike other cities that we visited in China, which generally had one or two principal attractions, Beijing and its nearby area contains many, of which we toured the following extensively: Temple of Heaven, Forbidden City, Tiananmen Square, Summer Palace, Ming Tombs at Shisanling and Great Wall at Badaling.

Temple of Heaven

The Temple of Heaven was built in the 15th century and lies southeast of the central part of Beijing, covering 267 hectares and accessed by four entrance gates at the compass points. Its principal function was to provide a location in the very ornate Hall of Prayer for Good Harvest for the Emperor to pray once each year for a good harvest. The royal entourage took over a day to cover the four kilometre distance from the centrally located Forbidden City to the Temple of Heaven – in total silence and out of sight of all commoners. Included in the procession were elephants, chariots, troops, officials and musicians. Built in 1420, the original Temple of Heaven was completely burned in 1889; heads rolled when blame was apportioned. Nearby is the Imperial Vault of Heaven, which contained tablets of the emperor's ancestors, used in winter solstice ceremonies. It has a similar façade to the Hall of Prayer for Good Harvests, but is much smaller with a single circular roof and a one-tier terrace, compared to the three-tiered roof and three-tiered marble terrace of the larger building. It is centrally located within the highly publicized Echo Wall, a massive brick structure, over 5 metres in height and 65 metres in diameter, which has remarkable acoustics – an audible whisper reportedly travels from one end of the wall to the other. South of the Echo Wall sits the five-metre-high Round Altar constructed in 1530 and rebuilt in 1740. It is also a three-tiered marble structure, with the top tier composed of nine rings of stone, increasing outward from a centrally located stone where it is considered lucky to stand. The three tiers all contain odd numbers of stones, in increasing numbers, with the top tier representing heaven, the middle tier earth and the bottom tier man. China's long-standing penchant for odd numbers continues to the present as shown by the re-numbering of houses by some new home owners in our district in West Vancouver and elsewhere!

Forbidden City

Like so many of China's present day attractions that appear ancient, but well-preserved, Forbidden City is no exception. It was created as the Palace for emperors of the Ming and Qing dynasties, (1368-1911 AD), the last two of the great parade of dynasties in China commencing in 2200 BC. Although the original palace buildings were constructed between 1406 and 1420, in the intervening 500 years they and many newer ones with their historic contents, were to be consumed by fires, warfare and associated looting, culminating with the Kuomintang before the Communist take-over in 1949. However, 50 years later, following continuing renovations, even the shortest of visits can leave the 10,000 or more daily visitors, not just impressed but exhausted.

The Palace grounds occupy 72 hectares of parks, gardens, terraces, 800 buildings and reportedly 9,999 rooms! Completely surrounded by a water-filled moat, 50 metres wide, and a 10 metre high wall, it was off-limits to all but invited guests until after the end of imperial rule in 1911.

We entered Forbidden City from the south via the Meridian Gate, followed by the Gate of Supreme Harmony, further north, emerging into a huge courtyard capable of holding an audience of 100,000 people. Beyond, to the north and centrally located in the palace grounds are three successive halls: Hall of Supreme Harmony, Hall of Middle Harmony and Hall of Preserving Harmony. These buildings are prominently raised on a large marble terrace and were used respectively for ceremonial occasions, a transit lounge for the emperor, banquets and later for imperial examinations. Both within and flanking the great halls are numerous interesting articles, including a richly decorated immense Dragon Throne, a massive brass urn which contained water for fire-fighting, a huge bronze turtle on the terrace in front of the Hall of Supreme Harmony, which was a symbol of longevity and stability. It has a removable lid on top of its shell for use on special occasions when incense inside was lit so that smoke billowed from its mouth. Beyond the great halls to the north, and accessed by a Gate of Heavenly Purity are two palaces and one additional hall within an enclosed courtyard. The Palace of Heavenly Purity, was the residence of Ming and early Qing emperors. The Hall of Unions current attractions are over 100 clocks of all sizes obtained from all over the world by one of the last emperors. To the west and east of the above central line of buildings are many other palaces and buildings and their gardens which were used as residences, libraries, temples, and theatres. Formerly these included residential quarters for the emperor's retinue in one section and for diplomatic corps of neighbouring states and nations in another, each with their own walls. Several of these buildings currently display a vast array of museum specimens and early furniture, paintings and personal items of the empress and concubines. We exited via the northern Gate of Divine Military Genius about 2.5 hours later, not quite totally 'palaced-out' , as we were due for more on the morrow! Before then, however, we had one more visit on our schedule - Tiananmen Square.

Tiananmen Square

Immediately south of Forbidden City lies Tiananmen Square covering 40 hectares, which has held up to one million or more people at important events. Originally only one-quarter as large under the emperors, it was enlarged by Mao Zedong after the flag of a communist China was raised in the square on October 1, 1949. At the time of our visit in August 1990 only a little over one year after the massacre of pro-democratic demonstrators, more than one dozen Red Army guards were evident at various locations and Rene noted cameras, strung at intervals around the massive square on lamp standards. Mao's presence even after his death is most evident, with his huge portrait at the north end of the square on Tiananmen Gate (Gate of Heavenly Peace), built in the 15th century. On the southern side of the square stands a 36 metre high granite obelisk weighing 60 tons with carvings of significant revolutionary events. Named the Monument of the People's Heroes, it also bears calligraphic messages from both Mao and Zhou Enlai. Immediately south of the monument is the Mao Zedong Mausoleum built by mid-1977 following his death in September 1976. It was later modified as a museum in 1983 with exhibitions on the lives of Mao, Zhou Enlai and others.

Summer Palace

Before our scheduled visit to the Summer Palace, I left our hotel at about 6:50 am for a 5-kilometre run along roads leading to Kunming Lake, site of the Summer Palace, along its north shore. En route, I passed about 100 people exercising in a park in three separate groups, one of which was directed by four men wielding swords. I reached the East Palace Gate in about one-half hour before returning to the hotel for breakfast. Coincidentally, our tour starting about an hour later originated at the East Gate. The original palace, as its name implies, was used by the emperor as a summer residence and now lies within a park covering 280 hectares, dominated by Kunming Lake and including palaces, pavilions, temples and halls. The elegant buildings and furnishings in their beautifully landscaped placid setting belie the site's turbulent history. In 1860 Anglo-French troops gutted and looted it during the Opium War. Empress Dowager Cixi began rebuilding in 1888 but in 1890 foreign troops partly gutted the buildings in the Boxer Rebellion. Our first stop during our visit was the Hall of Benevolence and Longevity which houses a hardwood throne and has a courtyard with bronze animals. It was the principal building for receiving envoys and handling state affairs. Next we saw the Hall of Jade Ripples which was the emperor's private quarters and a prison after 1898, following the failure of the Wuxun Reform movement. Like most other visitors we paid an extra 3 yuan each in order to visit the Empress Dowager's quarters and the theatre area. An interesting feature of the current Summer Palace is the use of life-like mannequins and individuals in traditional dress in certain areas – particularly the Empress Cixi's quarters. I took a photo of a young girl sitting beside a mannequin of Empress Cixi inspecting her head-dress with the aid of a mirror held behind her by another mannequin. Later in the Court of Virtuous Harmony, which is lined with memorabilia of the Empress Cixi, I took a picture of Rene standing beside a young woman garbed in period dress wearing elevated shoes with a replica of bound feet showing beneath her long dress under her real shoes. The Long Corridor which extends across the north end of Kunming Lake for over 700 metres is decorated with over 1000 paintings in mythical themes hung from the ceiling, most of which probably post-date the Cultural Revolution. Beyond the west end of the long corridor a dock provides access to colourful 'dragon boats' which are available for tours on Kunming Lake. Nearby is Empress Cixi's incredible double-decker Marble Boat, beached at the edge of the lake, as it was incapable of floating because of its weight. It was built with funds she squandered from the naval budget. The Empress was an individual to be reckoned with – having her son the Emperor Guanxu placed under house arrest and finally murdered and essentially ruling on behalf of her grandson Pu Yi, the Last Emperor. One of my final photos, probably taken not only for the serenity of the scene, but also to record some more marvelous names, shows the Temple of Buddhist Virtue with the Hall of Buddhist Tenants above it, taken from the Cloud Dispelling Gate.

Ming Tombs

It is customary for tour groups to visit both the Ming Tomb and the Great Wall at Badaling on the same day because of their proximity to each other – the Ming Tomb lying 50 kilometres northwest of Beijing and Badaling an additional 20 kilometres. The Ming Dynasty existed from 1368-1644 AD and included 16 emperors, by far the most noteworthy being Yong Le, the third emperor, who raised China to its greatest power prior to his death at age 64 in 1424. He was responsible for changing China's capital from Nanking to Peking (Beijing), ordering the rehabilitation of the Grand Canal and other waterways, for foreign ventures which included far-ranging naval expeditions, and finally, the selection of the site for the Ming Tombs.

Our drive out to the Ming Tombs took considerable time as a stalled freight train had blocked the main road creating a massive traffic jam which eventually led us to circumvent the area via country roads. When we finally reached the four-lane highway beyond the traffic jam we could only travel at 40 kilometres per hour because of speed limits – a typical condition in China where many restrictions appear illogical by western standards. Our driver knew that the road was well-patrolled by police looking for speeders. This led our guide Elaine to relate some of the stories she knew about local law enforcement. Apparently police could be bribed from issuing traffic tickets by being offered packs of cigarettes – however, they had to be imported brands! She mentioned a case where a tour bus was stopped by a policeman and the driver got out to plead his case unsuccessfully. As the policeman was writing a ticket, a large German tourist got off the bus and interceded because of the delay. Realizing that his efforts were unheeded, he reached over, grabbed the ticket and tore it into small pieces. The officer gasped – not knowing what to do, and finally waved the bus on. According to Elaine, drivers refer to policemen as "white dogs" – an insult reflecting the colour of their jackets and caps.

Eventually we reached the large marble gateway, more than four centuries old, which marks the start of the seven kilometre long 'spirit way' leading to the tombs. This is the 'triumphal arch', the middle passage of which was used only once in each reign, for the delivery of the emperor's remains to his tomb. Beyond is the Red Gate which contains a giant tortoise, carved in stone, supporting the largest stele in China. The 'spirit way' is a very wide avenue lined with 36 stone statues, the initial ones animals (lion, dromedary, elephant, horse, mythical beasts) in pairs on either side of the road, standing and the next set kneeling. They are about twice life-size. The final figures are guards composed of generals, ministers, and officials, each distinguished by headgear. The avenue ends at the Dragon and Phoenix Gate. In China's past the dragon reflects 'male' and the phoenix 'female', as depicted in embroidered opera costumes and other garments.

Each burial procession from the Forbidden City took up to one month to complete, as the large retinue had to be camped and fed and speed was not a requirement. Dingling, the tomb of the Emperor Wan Li (1573-1620 AD) is the second largest tomb and the only one excavated to date. It took six years to complete the burial chambers and the outside temples using 30,000 workers. The underground passageways and caverns cover about 1200 square metres and the tombs are all faced with large granitic stone blocks. The underground tombs occupy an area of about 40 x 100 metres with tombs 7-8 metres wide, rising to arched roofs 8 metres high. They resemble ordinary underground vaults and now contain replicas of coffins, pots, etc. The only actual items of human remains in the tomb were part of the Empress' hair. According to one reference, the tomb of Yung Le contained the bodies of 16 of his concubines who were buried alive with his corpse. A museum at the site contains some of the items excavated from Wan Li's tomb, including many gold dishes, pots, models and pictures of the layout.

The Great Wall

As we drove to Badaling we could see sections of the Great Wall from our van, winding through the mountains and even from this low vantage point, the sight was awesome. Most of the 6400 kilometres of the Great Wall, recognized in 1990 as the only man-made object visible with the naked eye from from space/earth orbit, is in ruins. Certain parts, such as those at Badaling, have been rebuilt in recent times for the tourist trade, providing sections about four metres wide with short crenulated walls above the walkway area, about six metres above ground level. The 'China-side' of the wall, as it was known in earlier times, contained steps up the side of the wall at intervals to provide access for troops. Garrisons for troops occurred at much greater intervals as did watchtowers and signal-towers. From Badaling, one can currently climb either north or south to prominent lookout towers in that section of the Great Wall. The Anderson/Killin family and Rene and I selected the south route and we all climbed to the top of the main lookout tower in that area for a stunning view of the rest of the wall. Feeling in need of exercise, I went ahead and enjoyed running to the next main tower to the east, reaching it in about five minutes. No one else was in the vicinity and I could see that beyond, the wall was unrepaired with large parts reduced to ground level and steps either missing or overgrown with grass and weeds. The surrounding countryside is very mountainous and one can visualize the difficulty that marauding Mongolian hordes would have had, not only in reaching the wall, but breaching it. Our guide Elaine claimed that the Chinese never lost a battle on the wall. An estimated 180 million cubic metres of rammed earth were reportedly used to form the core of the original wall. Considerable restoration of the wall was undertaken during the Ming dynasty by facing it with bricks and stone slabs – apparently some 60 million cubic metres over a 100-year period.

The Last Supper

This pretty well ends our most memorable tour of what must be considered as the greatest nation on Earth in historical times. I have eliminated much of the irksome background which led to frequent criticism and outbursts with our guides during the tour. Our misplaced baggage arrived the day before we left Beijing to return to Vancouver.

I have added this farewell dinner experience, which even I, as a sober recorder, would be willing to question considering the bizarre events as I recollect them. We gathered at a restaurant near the Embassy area with Liu Changuing (our National guide), his director Jang from Beijing's Chinapac International Tours Services (CITS) office and the general marketing and sales manager for CITS, who was also president of CITS. The latter were present to try to make amends for the problems we encountered during our tour. The president of CITS sat at a table with the Anderson/Killan clan, while Liu and Jang sat with the rest of us. The president explained the purpose of the gathering and ended by saying he had a present for each of us to reflect the company's concern for what he referred to as the worst series of problems ever encountered by his group. He then presented each of us with a cotton embroidered table cloth, about 2.5 x 2.5 metres in area. As no one from our group volunteered any thanks or comment, I did, although as a perfunctory toast, as I did not want to give the impression that a 'present' was a satisfactory remedy for fairly gross communication and management problems. At our table, I recommended use of tested methods of baggage handling as one obvious remedy, as Chinapac appeared to be trying to rediscover the wheel in many respects. All in all, we had a pleasant, if not too communicative, dinner as Liu had to interpret for Jang who certainly did not appear to be receptive to our comments. The latter part of the evening was one of the most bizarre experiences to a formal dinner that either Rene or I had ever experienced. At the end of several toasts, the president again referred to the need to show the extent of his concern for the problems encountered. Initially, he elected to sing part of an opera as a form of restitution. Whatever the opera was, we were not familiar with it, and his rendition was not the best, to put it mildly. He then finished off by offering a 'Viking toast.' This consisted of grasping the sides of the table, then carefully lowering his head and mouth over the rim of a filled wine glass, which he grasped with his teeth. He then adroitly lifted his head, upended the glass so that its contents drained into his mouth and as suddenly lowered his head, with all the wine falling back into the glass! Needless to say, we were unprepared for this extraordinary performance and quite speechless.

The next day we started our journey back to Vancouver via Shanghai. In retrospect, our three weeks in China was by far my most memorable 'vacation' experience in our brief exposure to five or more millennia of history involving about 20 percent of our world's people.

Blood Donations – 'Friends for Life'

Following my arrival in Vancouver in 1951 I learned of the existence of the Canadian Red Cross Society (CRCS) Blood Donor Agency program, which had only been instituted in British Columbia in 1947. The blood transfusion service, as it was called in the early 1940s, evolved at a particularly critical period. It was not until 1940 that the Rh factor in blood was discovered which dramatically effected a significant increase in successful blood transfusions. In addition the medical staff of the Connaught Laboratories discovered a method of drying plasma for longer storage.

In 1942 the Society achieved an initial target of 3,000 donations per week in Eastern Canada. By January 1944, weekly donations reached 20,000 with contributions from as far west as British Columbia. During World War II about 100,000 blood donations were required to care for 17,000 casualties. By 1944 Canadian blood donations passed the one million mark and by the end of the war about 2.5 million units had been donated. (A unit is equivalent to 450 cc.)

By 1950 there were 1,385 clinics held for blood collection in Canada and 198,208 units of blood were collected, 92,893 patients receiving transfusions. By 1960 donations had increased to 641,544 units and 227,997 persons received transfusions. By 1994 there were 700,000 voluntary blood donors in Canada and the number of transfusions had only increased to 350,000.

By the mid-1960s the average heart operation required 24 units of blood. The Guinness Book of Records (1996 Edition) notes that ". . . a 50-year old hemophiliac, Warren C. Jyrich, required 2,400 donor units of blood, equivalent to 1,900 pints of blood, when undergoing open heart surgery at the Michael Reese Hospital in Chicago, Illinois in December, 1970."

I'm not sure what motivated me to attend my first blood donor clinic other than curiosity and the feeling that it could serve an important need for a small inconvenience. I had only recently returned from examining several mining properties in B.C. and I elected to make my donation on a fairly hot day in the fall. Our office was in a building on the corner of Homer and Pender Streets and the clinic was about six blocks west on Hastings Street. As I recall, there was very little involved in being registered as a donor. No information was requested on my exposure to diseases, my intake of drugs, sexual activity, nor were any tests performed other than to determine my blood type.

After making my donation I was not warned to avoid strenuous physical activity for a period of time afterwards. I had made my donation during our lunch break and had not eaten, so I jogged back to a diner across from our office on Pender Street and sat at the counter. While eating a sandwich, without any warning, I passed out – fortunately across the counter. I revived quickly to find a couple of restaurant staff steadying me at the counter. I'm sure they were quite relieved when I thanked them and left.

In the early 1950s the period required between blood donations was at least 90 days. I used to donate quite regularly; however, it was not until returning to Vancouver from our transfer to Toronto, that I reached my 50th donation in 1980 and my 75th in January, 1992. Rene and I travelled to several foreign countries following our return to Vancouver and any exposure in malaria-infected regions automatically required abstaining from blood donations for a period of one year. A similar requirement was mandatory following major surgery.

In the 46 year period of my blood donations precautionary measures to minimize risk in blood transfusions and to continuously improve the safety of the system were instituted, especially in the 1980s when HIV tainted blood was discovered to have been transfused to some patients. Since then the precautionary measures have grown at an exponential rate, leading to complaints from donors of the length of time required to complete a donation.

A donor of two decades ago would be overwhelmed by the changes. In mid-1997 the time for the actual transfusion has changed little, being in the range of 20 - 25 minutes. This allows about 5 minutes from the time of occupying a couch or bed for the transfusion to begin, followed by 10 - 15 minutes for the donation and a 5 minute rest period for some refreshments.

However, the actual preparation period, which originally consisted only of a record check to verify the donor, followed by a hemoglobin count from a small amount of blood taken from the donor's finger, which took about 3 - 5 minutes, has now increased to 20 - 25 minutes. The preparation actually includes a mini-medical. The donor is escorted to a cubicle with his paperwork in a folder, including a pamphlet on recent changes in donation requirements. The donor is expected to be familiar with this update. The donor must then answer a full page itemization of questions concerning general health, medications, previous exposure to disease-infected tropical areas, recent surgery or previous afflictions and sexual activity with partners who may have exposed the donor to HIV carriers.

Armed with a partly completed folio the donor then waits for a nurse who scrutinizes the questionnaire, asks any questions material to the donation, and follows with tests of blood pressure, pulse and temperature. Should the nurse find any "abnormality/condition" recent policy requires that the individual provide his/her physician with an External Medical Enquiry Form Letter. The letter asks the physician to provide the requisite information and return it to the Head Nurse for the donor's district. Eligibility to donate blood is then based on the physician's response to the letter.

I recall sitting drinking coffee following a donation in 1995 when a veteran donor sat down next to me and announced that he had been turned down for a donation because his blood pressure was considered too high. I commiserated, recognizing that the taking of blood pressure would not have been a requirement during most of his earlier donations. However, I accepted that the changes being implemented were not only designed to protect the CRCS and its Blood Donor Agency, but also the donors.

In the early 1980s I had a short period of heart palpitations without evident cause and Dr. Follows, my physician, referred me to a specialist who ran a series of tests including running on a treadmill at increasing speeds to fatigue level. The physician also recommended a 24 hour cardiac function stress test. I had arranged to make a blood donation on the same day and asked Dr. Follows if this would be a problem. He had no objection so I appeared, adequately garbed, at the clinic, my chest wired to a battery-operated monitor, neither of which were obvious beneath my shirt. The period of the blood donation could not be detected and the cause of my palpitation remained unresolved. However, for at least 20 years I have had cardiac arrhythmia (irregular pulse), which is most prevalent following strong exercise. During the last few years, more modern blood pressure / pulse monitoring equipment has been installed at the clinics which are capable of registering arrhythmia during the clinic meeting with the nurse.

In March, 1995 arrhythmia was detected during my visit to Lions Gate Hospital for a donation and the nurse provided me with an External Medical Enquiry Form Letter. I took it to Dr. Follows who referred me to a specialist who had me take a 24-hour cardiac function test. The specialist concluded that I had had a leaky heart valve for many years, which accounted for my irregular pulse, but promised no obvious future problems.

Prior to my visit to the specialist I was phoned for a donation by the CRCS and made an appointment for the following day. No arrhythmia was detected and between March 1995 and May 1996 I made six additional donations without any problem. Again, in May 1997 I was phoned by the Red Cross as blood shortages had intensified. Arrhythmia was detected by the Lions Gate Blood Donors Clinic and my donation refused even though I explained the same problem had been encountered a year earlier and that I had donated regularly since. By the early 1990s the allowable period between donations had been reduced to 56 days and the upper age for donations set at 71.

I realized that the form that I had taken to Dr. Follows a year earlier had never been completed and returned to the CRCS following my visit to the specialist a year earlier. I was phoned in mid-June 1997 by the Red Cross and informed that my name had been added to the "Doctor's List" and that I should have no problem in making a blood donation. I made my 100th blood donation on July 21st and I was later presented with a handsome framed diploma recording the event.

I recall the early beliefs that the donation of a unit of blood, less than 10 percent of the average (150 lb.) person's blood supply could affect athletic performance for up to one week or more as the body's blood supply was replenished. For several years I have recorded the dates of my blood donations in a training log. I have consistently avoided donating blood for up to one week prior to an athletic event (run, duathlon, triathlon) but have found that my training times have not been adversely affected after about 72 hours from a blood donation. On the following day or even up to 48 hours on there is a very definite association with abnormal fatigue and accordingly I have avoided workouts immediately after donating.

The publicity in recent years regarding hemophiliacs and other people who have developed transfusion-related infections such as human immunodeficiency virus (HIV) which leads to AIDS, and the hepatitis C virus (HCV) which can result in debilitating liver disease, resulted in public apprehension about the past, present and future safety of the Canadian blood supply.

By Order in Council P.C. 1993 - 1879, the Honourable Mr. Horace Krever was appointed Commissioner to enquire into and report on the blood system in Canada and was required to submit an interim report on the safety of the blood system, with appropriate recommendations on actions which might be taken to address any current shortcomings.

To assist him a Management Committee was established in the spring of 1994 which was composed of 10 individuals with expertise in various disciplines relevant to understanding the complexities of the blood system. The objective of the committee was to assess safety of the present blood supply and to make recommendations on ways to minimize risks and continuously improve the safety of the blood system. The study was concerned with the blood system in 1994 and <u>did not</u> include the past system, especially the events occurring during the 1980s when a quantity of tainted blood was permitted to be transfused to recipients.

The committee's analysis led to nine recommendations concerning: utilizing, governance, regulation, process, surveillance, monitoring, research and evaluation and public confidence. The committee's findings were summarized in the responses to the following questions:

Should Canadians who need blood or blood products have to worry that they are less safe in Canada than in other developed countries?

> The answer as far as the committee could assess is no. In 1994, no evidence was identified of increased risk to blood or blood product recipients in Canada compared to that in other developed countries.

Can important deficiencies be identified in the Canadian blood system?

> The answer is yes. This report provides details regarding a number of deficiencies in the blood system in 1994.

If the deficiencies are not corrected, is there significant potential risk to the safety of the Canadian blood supply in future years?

> The answer is yes. The recommendations from this study are presented as constructive steps that, when implemented, will assure continual improvement and safety of the Canadian blood system.

Dr. Krever's interim report is composed of ten chapters, his letter of transmittal, recommendations, references and annexes. His last chapter, 'Summing Up' excerpted below, provides a clear and carefully considered statement of the current situation expressed convincingly by an obvious humanitarian.

> For the foreseeable future, blood will continue to be needed and used for therapeutic purposes. Inherent in its use is a risk of harm, sometimes serious harm. Risk of harm, however, is inherent in all medical interventions in which the use of blood is appropriate, quite independently of the use of blood. In weighing the risk of using blood, consideration should also be given to the reality that in some situations the risk of not using blood is certain death.
>
> It is apparent from the report of the safety audit committee and from this interim report that the precise measurement of the various risks described cannot, on the basis of data available at the present time, be made or even reliably estimated. But a qualitative assessment is possible. When weighed against other risks associated with the many vicissitudes of health care and ill health, it would be unreasonable to characterize the risks in using blood as anything but low in those cases in which the use of blood is necessary. This is not to suggest that the risks cannot be reduced further. The recommendations in this interim report, if implemented, will produce this result.
>
> It must always be a fundamental policy of the blood system to strive to reduce risk and make the blood supply safer. Any other approach would be a betrayal of the public's trust. If the recommendations made in this report are not implemented, it will not be accurate to say that the blood supply is as safe as it should be.
>
> The opinion that the current risks to the blood supply are low, although not as low as they should be and can be made, invites comparisons of the Canadian blood supply with those of other nations. Just as it is not possible to measure precisely the safety of the Canadian blood supply, so it is impossible to be precise about the blood supply of other countries. Often data are not available, and the data that are available are not always collected in the same manner from country to country. Even if the data were available, comparison of risk at the national level would be less than helpful. The magnitude of risk is related to the incidence of infectious diseases in the blood donor population which, in turn, is related to the prevalence of infectious diseases in the general population. Transfused blood from rural Iowa is probably safer than transfused blood from Montreal or Toronto. On

the other hand, transfused blood from rural Saskatchewan is probably safer than transfused blood from New York or San Francisco.

The safety audit committee concluded that Canadians who need blood or blood products should not have to worry that they are less safe in Canada than they would be in other developed countries. This is an opinion, not based on the measurement of risk in those countries as compared with that in Canada, but based on the committee's inability to find evidence of increased risk in Canada and, more important in my view, on its members' experience with, and expert knowledge of, the blood systems of other developed countries. On the basis of information I have received from blood transfusion experts in the United States and the United Kingdom, the medical and scientific literature, and the report of the safety audit committee, I am confident in my opinion that Canada's blood supply is not less safe than that of other developed nations.

There is, however, no justification for complacency. To the question whether the tragedy of 1978 to 1985 could happen again with a different contaminating agent, I believe the answer is "Yes." How to minimize the likelihood of such a calamity is the challenge to be addressed in the further work of this Inquiry and the final report.

Among the many issues to be addressed in the continuing proceedings there is one that I believe is worth mentioning now. However low the risks to Canada's blood supply may be, it is certain that some deaths and serious disease will result from the therapeutic use of blood. It is of little consolation or even relevance to those unfortunate members of our society who suffer those results that the risks are low or that the blood supply is adjudged relatively safe. In my opinion, a system that knows that these consequences will occur and that brings them about has, at the very least, a moral obligation to give some thought to the question of appropriate relief for those affected by the inevitable events.

This report must not end without an important acknowledgement. Earlier I referred to the beneficence of the generous group of Canadian blood donors who are the heart of the blood system. All members of Canadian society, and not merely the direct beneficiaries of their generosity, owe the donors a debt that can never be repaid. They are truly life savers. It is important that I emphasize that nothing I have recommended will diminish the urgent need for donations. The history of their humanitarian action persuades me that the blood donors of Canada can be relied on to continue their selfless benefactions as long as blood is necessary for therapeutic purposes.

I look back with a deep sense of satisfaction at having been able to provide a vital benefit to others and I consider the 200 or so hours involved as one of my most rewarding investments. I am also much impressed by the careful practice and dedication of the CRCS Blood Donor Agency, notwithstanding the tragic lapse of responsibility of senior management in the 1980s.

Sporting Events

Athletics consumed a significant portion of my time in my school years as I have described in *'Upper Canada College'* and *'Uncloistered Halls'*. This was followed by a lengthy hiatus; my exploration work provided plenty of exercise, but no opportunities for sports, other than recreational skiing with my family, which began in the mid-1960s following our transfer to Toronto from Vancouver, while I was employed with Kennco. Following our return to Vancouver in 1971, this activity intensified with our family membership at Grouse Mountain and later on Cypress Ski Hill, both on the North Shore. I have both Susan and Rob to thank for the re-ignition of my interest in running beginning in the mid-1970s when I frequently ran with both of them as their running developed. As I relate later, Susan was directly responsible for my involvement in marathons, duathlons and triathlons, commencing in 1990. Rob's interest in the outdoors also led to my sharing with him and, occasionally, other friends, many memorable hikes and climbs. These included ascents of Mt. Baker in 1972, following by Mt. Rainier, Mt. Hood and Mt. Whitney, as described in *'Travel, Trekking and Climbing.'* The most memorable hikes and treks with Rene have also been described in preceding chapters, mostly, *'A Khumbu Trek.'*

My stories about Susan and Rob principally relate to athletic or outdoor interests we share. From an early age David bypassed pursuits involving slower forms of travel. Initially he focused on a mini-bike, but rapidly graduated to larger motorcycles. After nine years of accident-free riding, he hit an oil slick and, though badly injured, made his own way to a hospital. Following a preliminary examination he was sitting in the emergency waiting room, hoping to leave soon, when a doctor approached with a colleague: "Look, there he is. He has a fractured thumb, sprained wrist, double-fractured shoulder blade, shattered scapula, two broken ribs, a punctured lung and bruised kidneys. And here he is, sitting quietly reading a magazine."

At the time we believed his motorcycle days were over, but he surprised with a new Kawasaki "Ninja ZXII" which he enjoyed and admired for several years before deciding that the length of life is more precious than the pace at which it proceeds.

David graduated in electronics with honours from the British Columbia Institute of Technology in 1984, winning the Hewlett Packard Physics Award. He joined a research an development firm specializing in high-technology equipment for the mining and sawmill industries, requiring considerable innovative research. At the same time he purchased a 41-foot fibreglass power cruiser called the "Polynesian" as a live-abord, which he still enjoys. Other interests which he ahs developed include watching Formula I car racing, completing a computerized course in helcopter operating and participating in a chat group of aircraft enthusiasts and pilots.

In the following section, I recall several of my more memorable sporting events, beginning with boxing, which should probably have been described much earlier as it terminated in my late teens. My more recent duathlons and particularly triathlons, have been most fulfilling, especially through the many new friends and acquaintances that emerged through this shared interest. The opportunity to participate with the Canadian National Triathlon team in five world championships, beginning with Lausanne in Switzerland in 1998, followed by Montreal (1999), Perth in Australia (2000), Edmonton (2001), and

Cancun, Mexico (2002) fulfilled a life-long dream for a Canadian of advanced years. Opportunities and rewards in this particular sporting event continue: In 2003, I won my fifth national triathlon age-group championship in Edmonton, while qualifying for the upcoming World Triathlon Championship in Queenstown, New Zealand in December, which will be part of a vacation there with Rene. I also qualified for the 2004 World Triathlon Championship in Madeira, Portugal in May 2004.

Boxing

My boxing 'career' was inauspicious, to say the least. Until I was eleven years old I had never had an altercation with another which led to a fight. At the time I was attending Brighton, Hove and East Sussex Grammar School in England in 1939. For some reason or another, the details of which I cannot remember, I had the great misfortune to get into an argument with Bertie Banks, a classmate. At that time any such disputes which could not be settled on the spot and which were considered of sufficient significance, even among 11-year olds, were settled in a 'gentlemanly fashion' after school hours at the nearby playing field.

The rules governing combatants on such occasions were well recognized however antiquated. Each adversary was represented by his chosen second. The power and decisions of each second were recognized as final and not to be questioned. Something like a duel in much earlier times!

In my case, to cut short an oft-remembered embarrassing defeat, Bertie Banks beat the hell out of me and I was spared irreparable damage, as it appeared at the time, only by the timely intervention of my second, who probably saved my life, but alas whose name I cannot recall!

Bertie and I were to become the best of friends and I only wish I might have had the chance to meet him once again after 1940 and being evacuated to Canada.

Following a year at the Lower School at Upper Canada College in grade 8, I entered the Upper School in September 1941. At the time there was a strong emphasis placed on boxing, which carried through to graduation at grade 13. In spite of my terrible eye-sight (the faces of my opponents were featureless white blobs in which I could not distinguish any anticipatory eye movements), I was only to lose my weight class in final bouts to my very good friend Harvey J. (Mickey) McFarland, who fortuitously gained sufficient weight to permit me to win my weight class in my final three years. Among my other opponents in the Upper School were Jimmy Biddell, another good friend and Bill Hewitt, who, like his father, was to become well known for his Saturday night radio broadcasts covering National Hockey League games. I also played on the UCC 125 pound football team as a right middle lineman with Bill as our quarterback. We had a remarkably successful season in 1944, losing only two games out of 12 to opposing teams who were not restricted by a weight limit.

The Boxing Finals at UCC were terrifying affairs. Unlike the preliminaries, which attracted only a smattering of spectators, the finals were held on a Saturday night in the gym, with many adults present in formal attire. It must have been a particularly trying event for any parents who might be subjected to witnessing a son facing an opponent who obviously out-classed him. I cannot recall whether females were permitted to attend these events, my impression was only of many older men as spectators. But then my impressions remain vague in this respect as I couldn't recognize any spectators, being so short-sighted and I never had to worry about a parent being present.

In 1946, my first year of enrollment at the University of Toronto in the School of Practical Science (SPS), in Mining Geology, with a record of about 10 wins and 2 losses in pre-university boxing, I was attracted to the SPS boxing team. At the time both first and second year engineering schools were held at Ajax, a Toronto suburb within a refurbished World War II munitions plant near Whitby, Ontario, about 20 miles east of Toronto.

For me these were exciting days indeed. I was suddenly engaged working in a much-modified, far more mature milieu, and my associates, mostly much-respected war veterans, averaged 5 years older. However, as I did at UCC, I found myself cramming in every opportunity to compete in athletics. I was eventually to be a member of nine intermediate and senior University Track and Field, Harrier (cross-country) and boxing teams in my four years at U of T.

I recall Stu Gordon as one of my indefatigable heavier-weight sparring partners and a noted lightweight division boxer, even though he was probably as near-sighted as myself. In an unplanned 'raid' by second year types in our first year at Ajax, Stu ran down one of the corridors toward a threatened entrance, which he successfully guarded against all comers, much like 'Horatius on the Bridge.' En route he encountered a potential adversary at a corner of the building whom he punched out, in the heat of the moment not recognizing him as his best friend.

My training with Stu and others at Ajax for the Intramural Boxing Championships gave me considerable confidence until a week or so before the competition when I learned that there were no other opponents in my weight class except an individual named 'Hammering Hank' Henshall, who I regretted to learn was a former paratrooper with over 100 amateur fights to his credit.

Somewhat apprehensively, I went into Toronto from Ajax and witnessed my forthcoming adversary emitting horrible grunting gasps with every blow as he pummeled his 'partner' in a training bout. All too late I realized why our weight class seemed so unpopular; all the other faculties except engineering were in Toronto!

Thoroughly chastened, I spent a miserable week envisaging my ultimate demise the following Saturday night at the Intramural Boxing Finals, though the actual event was only scheduled for three three-minute rounds.

I somehow managed to survive the ordeal and, as expected, I lost in a unanimous decision. The referee, as I recall, actually stopped the slaughter to inspect Hank's gloves in the third round, I'm sure to give me a breather after some nearly knee-buckling blows from Hank and recognizing the ultimate outcome of the bout.

That night I was invited to stay in Toronto with Rod McLellan, a long-time friend from UCC, who, like myself, had lost his heavyweight bout at the Intramural Finals. We had our post-fight injuries recorded on film the next day by Rod's mother, mine being by far the worse; two black eyes, a swollen nose, cut lips and miscellaneous bruises.

A couple of days later I had one of my most memorable experiences. I was riding down to one of my lecture classes on one of the 'Green Dragons' - buses used to transport Ajax students between classes, when I overhead two individuals sitting next to where I was standing, talking about their weekend. One mentioned having attended the Intramural Boxing Championships the previous Saturday night. He said, with impressive awe, that he had never seen anyone survive such a beating - or words to that effect. His incredulous look when I said, "I'm glad you enjoyed it," left no doubt that he recognized my battered face, and left me with a real glow of satisfaction that, even though I had unquestionably lost, of which I had absolutely no doubt, with the referee's help I had done it with some aplomb! Stu Gordon, who had dispatched his adversary in a KO, was the only other SPS entry.

Senior Intercollegiate teams at U of T at that time were comprised of individuals who had won specific events, or in the case of team sports were top performers. Intermediate teams included runners-up and others. Hank Henshall went on to win his weight division every year for four years in the Senior Intercollegiate team, which held bouts against competitors from Western, McGill and Queens. As a runner-up I was in the Intermediate Intercollegiate team and I drew an opponent from Queens University. Although I thought I might have won the fight, the judges didn't see it that way and my mediocre boxing career at the university convinced me that it was a good time to 'hang up the gloves.'

World Duathlon Championships

Cathedral City, California - November 24, 1990

A casual suggestion made in late summer 1990 by my daughter Susan led to my completing my first duathlon, a 5-30-5 kilometre run-bike-run in Vancouver on September 23, dubbed the Canadian National Duathlon. At the time Susan was nearing the end of an arduous summer of cycling including a triathlon at Whistler. She had participated in the B.C. Tel Duathlon Series with her friend Maria Taylor as runner and Susan as cyclist. The relay team was named the "Nautilus." They won their age group in the series, placing first in all five of the events they entered. The Canadian National Duathlon was an affiliated race with those placing first in their respective age groups guaranteed to qualify for the inaugural World Duathlon Championships being held in the Palm Springs area on November 24, 1990.

Although I had decided to join the North Shore YMCA Marathon Clinic in training for my first marathon scheduled in May 1991 in Vancouver, I had not ridden a bike for almost 20 years, and that being a single outing after about 25 years. Accordingly, my training base for the cycling portion was totally inadequate as was my knowledge of bike maintenance. I started cycling on August 6 and had only ridden 120 kilometres prior to the event. Susan had prophetically mused, "You shouldn't have any trouble winning your age group Dad, as you'll probably be the only one entered!"

The event generated far more accolades than I deserved, as on entering Brockton Point Oval for my final running circuit, the announcer bellowed, "And here comes Dave Barr of West Vancouver, the new Canadian National Champion in the 60 - 64 age group!" My time was 2:05:18. The Nautilus team had no problem winning their event. I still treasure my 1990 Christmas present from Susan: a framed photo of the two of us in our running gear at Brockton Oval following the event, taken by Maria, which hangs in my office.

I decided to accompany Susan to Palm Springs and we left Vancouver on November 22 via American Airlines. It was a particularly convenient flight with a single stop at San Jose. Approaching Palm Springs, we gazed apprehensively at the desert below, noting the arid mountains rising to over 10,000 feet and the high summits lightly sprinkled with snow. The view from the aircraft was memorable, with a plethora of lush green golf courses near the residential areas abutting abruptly against white sand, and an abundance of lofty palm trees inter-grown among a great variety of deciduous and evergreen trees and flowering shrubs, most notably bougainvillaea in the residential sections. This prompted the chief flight attendant to welcome the passengers to Honolulu!

On our arrival near noon, the temperature was about 75 degrees Fahrenheit and reached 85 degrees Fahrenheit in the afternoon. We checked into rooms at the Doubletree, the hotel selected to host the duathlon long course and a short course, with start and finish on the hotel grounds in a somewhat isolated locale. We rented a four-door Oldsmobile from Avis, which fortuitously could handle both our bikes, one in the back seat and the other in a partly open trunk.

In the afternoon we set off in the heat to walk the running portion of the course, which took us almost two hours. We were quite intimidated by the experience; by the dramatic contrast to the cool, wet weather we had been used to in Vancouver, and by the effort required to negotiate the sandy part of the course. The long course, which we had entered, was a 10 - 62 - 10 kilometre run-bike-run event which included the infamous Desert Princess "Dirt Road from Hell" in the running portion. This covers part of a sandy dyke about 20 feet wide lying on the side of a wide dry sandy wash west of the Desert Princess Resort. The stretch covers about four kilometres of the running course and is so particularly fatiguing on the second leg of the run that it has inspired the race's alias "Run-Bike-Die."

That evening we enjoyed a very pleasant meal at a nearby Italian restaurant. After further checks of our equipment, we both decided to retire early - how early you can imagine as I slept for eleven hours! The next morning we took our bikes for the mandatory check to a nearby cycle shop. Included was a check of tire pressure, which at 50 lbs per sq. inch was less than half the recommended 110 lbs psi; an accurate indication of my cycling experience!

After breakfast, we drove the bike course which started at the Doubletree at an elevation of about 400 feet, following roads easterly to Date Palm and north across State Highway 10 to Varner Road at the foot of a rolling range of hills. The course then made a loop of about 30 miles (48 kilometres) following Varner Road west to Mountain View, then north to Dillon Road at an elevation of 900 feet and east on Dillon to its highest point at about 1340 feet, before descending to the intersection with 1000 Palms Canyon Road at 1080 feet and then south on the Canyon to Ramon Road at 280 feet. This latter section was a real breeze and restorative experience. There are only a few short steep pitches in the course, the overall gradients being moderate and the desert scenery a memorable change from vistas in Vancouver.

Following some additional sightseeing including an aborted trip to the overcrowded line-up for the Tram Ride and a short stint at poolside, we decided to commit to the carbo-dinner at the hotel. It turned out to be fairly modest. We then attended the Athletes Instruction presentation which was preceded by a half-hour talk on goal-setting by Dave Scott, the noted triathlete, who at that time was nearing the peak of his performance with three or four Hawaiian Iron Man championships to his credit.

Neither of us slept well in anticipation of the long-awaited race day and the usual apprehensive feelings. Susan had been in several very long races including the Seattle to Portland bike ride, a distance of about 200 miles, which she covered in 11.5 hours and the Ramrod - a 156-mile circuit of Mt. Rainier in which she was the first woman to finish in just over 9 hours. This was to be my longest race, and based on my training, I was shooting for about 4½ hours; my longest training run-ride-run time was 3:47 for a 10-52.4-3 kilometre distance.

The race was very well organized with police at every intersection, cones along the main streets in the city area, water and *Exceed* available at one-mile intervals on the run and about 10-mile intervals on the bike ride. Over 400 volunteers were on hand to assist over 800 runners from 21 countries. The elite athletes included Scott Tinley, Ken Souza, George Pierce, Paul Granger, Thea Sybesma (Netherlands), the Puntous sisters (Canada), and an age group range from 8 to over 70 years. The run started with an elite wave only five minutes late at 8:05 a.m. with a cool temperature in the shade. Ten minutes later the rest of us were off. I consciously tried to maintain a consistent nine-minute mile pace, but found myself loping along at a relatively effortless faster pace which I was able to continue through the "Dirt Road from Hell." At the 5-mile mark I caught up to another old fart, identified by his age mark on his calf as being 61. I greeted him with a "Hello there, son," and after a few comments I gradually pulled away. About half a mile from the transition area, he caught up with me and we ran together. He was from New Jersey and had driven with several grandchildren to California especially to participate in the duathlon. My transition time between running and commencing the bike ride was less than ten seconds as I had not graduated to the use of biking clogs from running shoes. My time for the first 10 kilometre run was 52 minutes.

I joined a flock of other cyclists and headed east toward Date Palm on recently re-surfaced 30th Street. I took the corner of Date Palm and 30th at too high a speed and crossed the sharp margin of asphalt onto sand, struggling to regain the road surface with an ominous thump on my back wheel. Everything seemed fine but a short time later, I realized that I was riding on a flat rear tire. With my heart pounding, knowing the impending time loss, I came to a stop and clambered off the bike to commence my very first rear-wheel tube change. Trying to look on the bright side, I breathed a sigh of relief for having decided to carry two spare tubes as I was less than three kilometres from the start of a 62-kilometre ride on what I knew to be a relatively rough road. I tore out the tube and had some difficulty in replacing it with a new but slightly twisted tube. As I pumped the tire up I noted with considerable alarm that it was not holding air. It took some time for me to reluctantly accept the need to remove the tube and replace it with my remaining tube. A desert is a rotten place to change tires as those tiny little nuts that hold the valves against the rim of the wheel tend to get lost! About 20 minutes later I rode on with no one in sight and with the knowledge that I would have to be very careful and lucky to avoid another flat.

Eventually I passed about eight cyclists, two of them returning early either from dejection, disqualification or sickness, or combinations thereof; two mending tires and one walking his bike. Near the end of the ride at about the 52-kilometre point, I was pumping my way up the last gradient on Varner Road between Ramon and Date Palm roads when another cyclist came up beside me. He slowed down and rode alongside for about five minutes. He was a local resident and biking enthusiast who had ridden the course hundreds of times but was not participating in the race. I realized that he was only about two metres from me and said, "I don't mean to sound unsociable, but the drafting regulations for the race require a distance of 3 metres between cyclists riding alongside." We separated the additional metre. The talk was a welcome interlude in what was otherwise a lonely ride which took me 2:48 hours instead of the 2:30 I had hoped for, the difference being the tire changes.

I spent about 20 seconds in the transition, taking off my helmet and replacing it with a recently purchased floppy white cap with a fairly long rim and a red "Rad Dad" emblem. The Rad Dad equipment is manufactured for those athletes who apparently want to do "their very best" - a naive concept as there can be very few who don't. The hats are great though, as they are completely permeable, allowing water to be poured through them, which I did at every water stop. I did not experience as much of a problem as I had anticipated in the early part of the run, although it took some time to develop a sustainable pace. Fatigue really overcame me at the start of the "Dirt Road from Hell" but I pushed on, eventually passing a total of about six runners – three of whom were walking. On reaching the first of them, I slowed down and walked with him trying to offer some encouragement and companionship. I also enjoyed the rest! After a few minutes I took off again targeting on the next distant figure and repeating the process twice more. On the final mile I maintained an even 9-minute/mile pace and raised my arms in 'victory' when Susan welcomed me near the finish line for a photo. I also showed an appreciative smile for the official photographer beneath the Finish-Line banner recording a time of 4:40:51 hours.

I was delighted to learn that Susan had no problems and had finished in fourth place in her age group with a time of 3:35 hours. That evening at the awards ceremony we were both to receive medals awarded eight deep for the individual events and five deep for the relays. I found that I had placed sixth in my 60 - 64 age group out of seven competing - all the rest being Americans. In spite of my tire problems, I was delighted with the results.

Susan arranged to sit with several other Canadian athletes who she knew at the awards ceremony and she saved me a seat beside her in the front row, where we took photos of each other receiving our awards and of several of the elite group including the men's overall winner Ken Souza (2:35:43); second - George Pierce (2:37:22); fifth - Scott Tinley (2:39:16); ninth - Paul Granger (2:41:44); the first woman - Thea Sybesma (2:58:13); second - Donna Peters (2:59:35); third Sylviane Puntous (3:00:55) and fourth - Patricia Puntous (3:00:59).

The only competitor in men's 70 plus age group, and the oldest, was Norton Davey, a tall American who looked more like 55 and whose time was 4:34:51. The youngest overall competitor was an 8-year old boy who completed the course in 5:45, the final finisher.

In the men's 25 - 29 age group, the first twenty finishers all broke the course record, the top finisher recording 2:47:07. David Rudd from Vancouver, in the 55 - 59 age group, was the winner in 3:11:07. He only competed in a few more events, suffering from back problems. The winner in my 60 - 64 age group was Dick Robinson in 3:22:14.

The ADT London Marathon – April 12, 1992

On May 13, 1991 I described "My First (and Last?) Marathon" following completion of the Vancouver Marathon on May 5, 1991. Later in the year Rene and I started to plan for a vacation in England in April, which is somewhat earlier that we would normally plan a visit because of the weather.

On December 25 I air-mailed a request for an application form to the London Marathon, receiving a prompt response on January 7 which indicated that the application, complete with a registration fee of fifty pounds, had to be received in Britain before December 31, 1991 in order to be included in the ballot in early January, 1992. I air-mailed my application off the same day with my fingers crossed.

The first London Marathon was conceived and organized in 1981 by Chris Brasher, currently Chairman of the Board of London Marathon Limited, and winner of the 3,000-metre steeplechase at the 1956 Olympic Games. He is probably best remembered as the individual who paced Roger Bannister through the first two and a half laps of the world's first sub-four-minute mile in 1954. Brasher's inspiration occurred in 1979 after having watched the New York City Marathon in October. Following a year of painstaking planning, the first London Marathon attracted 7,747 runners, increasing dramatically to 18,059 in 1982. Designed as a fun-run, the event has nevertheless included an elite field of world-class marathoners. Course records are held by Steve Jones of Great Britain, 2:08:16 in 1985 and Ingrid Kristiansen of Norway, 2:21:06 in 1985.

The London Marathon is so popular that, for many for years, it has been necessary to regulate the number of runners by means of a ballot. The principal limitations are the narrow roads on part of the route, the staggered start required and the need to control the number of finishers accessing Westminster Bridge at the finish line to provide a relatively constant profile. Although over 80,000 runners apply annually, about 35,000 were approved in 1992 of which about 27,000 started and 23,656 finished. (The last noted in The Independent published on April 13 with the times of all competitors as "... and finally B. Baldridge 9:01:33.") The difference between the number approved and the number competing is partly due to a loophole which stipulates that those applicants accepted who cannot appear at race-time because of injury or sickness are guaranteed an entry the following year. As the results of the ballot are not known, in some cases as in 1992, until mid-February, many aspirants plan to compete in the marathon the year following acceptance, feigning sickness or injury, to avoid committing to a three to four month training program before they know they're eligible. I actually committed to a training program in late December, starting from a solid distance base at the 10-mile level, but was not to learn of my confirmation in the race until Rob phoned me in Maui from Vancouver in mid-February. Needless to say, I was elated by his recorded announcement which included a taped excerpt from "Chariots of Fire" plus a dubbed congratulatory message he concocted in a very convincing British accent. Because of the uncertainty of my application being approved, we delayed confirming our holiday plans for Britain until we returned from Maui in late February.

Following my marathon training with the North Shore YMCA in 1990-91 in preparation for the Vancouver Marathon, I decided to follow the less-demanding Galloway Training schedule with its bi-weekly long runs, in which two additional miles are added every two weeks, followed by an easy weekly run of one-half the previous long run distance. In addition a minimum of two weekday runs, each about four miles long, are also required. I found the schedule ideal, experiencing none of the soreness which had plagued my training in 1990-91 in the latter stages following the weekly long runs on Sundays. My longest training run was about 23-24 miles three weeks before the London Marathon.

We left Vancouver for London on an Air Canada Boeing 767 at 7:00 p.m. on April 7, arriving ten hours later following a delay in Edmonton, and emerging somewhat bleary-eyed to a sunny but cool afternoon at Heathrow. We enjoyed the airbus ride into Marble Arch, having a great view of the surrounding area from front seats on the upper deck. Following check-in at the centrally located Regency Palace Hotel in Piccadilly Square, we snoozed for several hours before dinner and were asleep again by 10:30 p.m. I had an easy 8-mile run through St. James and Hyde parks the next morning, but was to feel the effects of jet lag until late Friday.

On Thursday afternoon we walked over the to the Exhibition/Registration area for the marathon at Jubilee Gardens on the 'South Bank' of the Thames, immediately south of Hungerford Bridge. There can be little doubt that the London Marathon is the best organized major marathon in the world. As Chris Brasher claims, "Having been to the other major marathons, I haven't seen one that gives better service to its runners than London. There's nothing that matches London."

Foreign participants such as myself, including representatives from over 45 countries, must bring their approved selection form with its designated number to the "Trouble Desk." How the name was selected I never learned. In signing for my race number (15,168), the official compared my signature with that on my application form! We spent about an hour wandering through the adjoining Exhibition Tent with its diverse assortment of runners' paraphernalia. Accumulated items are conveniently deposited into a large, sturdy plastic bag marked "ADT London Marathon Kit Bag," complete with sturdy drawstrings, issued to participants at registration. It serves a dual purpose, as marathoners can deposit any unrequired garments or articles in the Kit Bag on race day at the starting area in Greenwich for pick-up at the finishing area. Identification is provided by a large gummed label bearing the runner's number, also issued at the registration area.

At the Registration area I also purchased a copy of the ADT London Marathon Official Magazine which contains the names of all pre-registered runners in their age groups in addition to much useful information on the event and several articles of related interest.

From the magazine I was able to identify Jenny Allen, aged 80, the oldest female participant, whom we met in the Exhibition area and whom I photographed with her companion. It may not be her last marathon as she commented that she loved them *"because they're so much fun!"* (Jenny finished 23,142nd in a very creditable time of 5:59:16.)

Marathon Day dawned dull, cold and windy. After a very modest breakfast of orange juice, rolls, jam, and coffee, I left the hotel with my kit bag, fanny-pack containing a few emergency supplies, a 24-exposure 35mm. disposable camera weighing about six ounces with 400 ASA film, and assorted garments. I walked to the nearby subway station at Piccadilly and boarded a train for Charing Cross. I caught the 7:28 a.m. train at Charing Cross along with a horde of other runners bound for Blackheath. The organizers had arranged with British Rail to allow all official marathon entrants free travel from several stations, including Charing Cross, to either Greenwich or Blackheath, being the nearest stations to assembly areas at three separate locations which merged with the main race route at points one mile and 3.5 miles from the starting areas.

The crowd travelling with me to Blackheath were fairly subdued at this time of the morning. The vistas of the surrounding countryside were not particularly inspiring as the line passes through an area of dull industrial buildings, row-housing with only occasional green areas and rare flowering trees. The bleakness of the day and the grimy train windows were probably contributing factors as well.

On emerging from the train at Blackheath Station, I joined the multitudes of runners making their way slowly uphill through this small centre toward the nearby heath with its prominent church and extensive greenbelt, now cordoned off to provide the assembly area for some 15,000 runners. Within the assembly area were all the required support facilities, including change tents, refreshments (tea, coffee, Spa water bottles, Gatorade), portable toilets, medical-aid personnel and vehicles, plus about 45 closely-parked vans to receive the kit bags which were to be transported to the finish area near Westminster Bridge.

Shortly after 9:00 a.m. four parachutists dropped from an aircraft high above the Assembly area, two of them supporting large flags. They slowly descended amidst emerging smoke plumes gracefully landing in the central portion of the assembly area. All four were to participate as runners in the marathon.

At about 9:10 a.m. a loud blast from a Royal Artillery gun signalled the start of the Elite Women and the AAA of England championships competitors at St. John's Park on the edge of Blackheath. Terry Waite, the noted British journalist who was held hostage for over three years by terrorists in Beirut lives in Blackheath and was the official starter for the race. The wheelchair competitors, male and female, started five to six minutes later and included Daniel Wesley from British Columbia, who had competed in the two previous marathons and who was to win his first London Marathon in the record time of 1:51:52, ahead of four others, all of whom finished within a record-breaking time of 1:52:48 in a highly contested race.

At 9:30 a.m. the main mass of the marathoners departed, stretched out from the starting-line according to a marked-off location corresponding to their race numbers. My group, numbering from 15,000 to 15,999 did not move for over five minutes, and I did not reach the starting-line until almost ten minutes after the official start. The runners in the middle of the pack formed a fairly solid mass which did not start to split up for several miles from the starting line. I was later to hear a commentator state that it was almost physically impossible for the so-called 'fun-runners' to average more than 9-9½ minute miles during the first two miles of the run. Throughout much of the marathon this congested condition prevailed and I estimated that those runners seeking to improve their positions probably added on a significant distance while swerving around other runners -- perhaps as much as one mile or more over the length of the marathon.

The route of the London Marathon has remained essentially unchanged since its inception in 1981. From the three starting points on Blackheath, nearby St. John's Park, and Greenwich Park, designated the Blue, Green, and Red starts, respectively, the route follows several roads trending easterly. The Blue and Green branches merge near Shooters Hill Road, about one mile from the start and follow Charlton Park Lane and Grand Depot Road for about two miles to merge with the Red start branch following Charlton Way, Charlton Road, Hillreach, and Artillery Place to merge at about three miles from the start with the rest of the runners at John Wilson Street. The route turns westerly following Woolwich Road and Trafalgar Road near the South Bank of the Thames to Greenwich Village at the 6-mile point. The runners detour under the prow of the historic Clipper Ship 'Cutty Sark' at 6½ miles and continue around a major meander of the Thames following Creek Road, Evelyn Street, West Ferry, Red, Salter, Brunel, and Jamaica roads to the south side of Tower Bridge at 12 miles.

Throughout this portion of the marathon the streets were solidly lined with spectators and several bands urging the runners onwards. Children were given vantage points on the curbs and held their hands out, along with adults, for a passing pat or slap. I must have slapped almost 1,000 hands of various sizes and colours, as it seemed such a deeply satisfying reciprocal gesture of friendship and mutual participation in the event. Several of the exchanges included religiously inspired vocal wishes for my welfare, hopefully not prompted by my appearance! In the latter part of the run, from about 19 miles onward, the hands of children and adults alike offered candies which were welcomed by many runners, including myself.

The London Marathon includes a high percentage of runners participating at least partly to raise money for charitable organizations through pounds-per-mile donations pledged by families or friends. Several of these appear garbed in costumes, some of which are so burdensome that the runners understandably fail to reach the finish line. I photographed two runners very realistically attired as rhinoceri. A televised video of the marathon, which I later saw, provided an excellent view of these two creatures having a battle as they ran by. Other runners appeared as convicts, a gorilla, a Gatorade bottle, and a clown. I photographed and spoke with the latter, Paul Townsend of Rothampton, a London suburb, who was appearing in his fourth London Marathon and second dress-up and who expected to complete the run in about 4½ hours. We had a brief exchange when I passed him at the 4-mile mark.

Several of the runners were actively soliciting donations during the run and I noted several generous contributions being placed into pails provided for this purpose.

At the 8-mile mark I saw a pair of runners passing by, one of whom had a brightly printed pattern on the back of his T-shirt above the notation 'Colombia.' I ran alongside and asked if he was from Colombia. He responded in a very Spanish accent, "Yes, do you know Colombia?" I said "I was born in Cali." His running partner poked his head around that of his companion and said, "I live in Cali!" We shook hands, like long lost friends, spoke for a while about the marathon, after which I bid them both farewell, as it was evident that they were running at a much faster pace than I could maintain.

Considering the constricted nature of several of the narrower portions of the route, where only one lane was open for the runners, the flow was maintained with relatively minor disruption. Where slowdowns occurred, several moving 'dams' were the cause, these being created by groups of runners occupying almost the entire width of the course. One of these held a very large flag horizontally; another very entertaining group included about a dozen Italian runners who burst into operatic songs periodically, much to the delight of runners and spectators alike. By contrast, the very imposing Essex police contingent replete with bobbies' hats and truncheons ran two-abreast and were not an impeding obstacle. I mentioned the constrictions encountered during the run as one of several topics I covered with recommendations in a letter to Alan Storey, the Race Director. He responded very promptly with the following comments:

As you may be aware we limit the entry to ensure that not more than 25,000 runners finish the event. Experience has shown that this is the maximum number that the course can cope with, without log-jams occurring. We do not allow wheeled contraptions, animals or anything that threatens the safety and comfort of other competitors. We did not sanction the flag you mention, but it is not too easy to remove from the course once the event is under way. We do try to control the event as effectively as possible without causing too many disturbances. One of the problems is that we do not grant team entries, the group you mention had entered as individuals, and therefore gave us no warning of their intentions.

From Tower Bridge, with its mass of spectators, the route turns abruptly eastward following Cable Street and Commercial Road toward the ominous Isle of Dogs, Europe's largest building site, dominated by Canary Wharf Tower and its surrounding complex. This stretch of the route, which extends from Tower Bridge at 12½ miles to the eastern side of Canary Wharf Tower, back west to the Tower, and then south for a 3-mile loop before resuming to Tower Bridge at 22 miles, is almost devoid of spectators, with the exception of the Tower Bridge area and part of the south loop which extends through a residential area. As we turned southward from Canary Wharf Tower we were greeted by the sight of runners three miles ahead, turning westward on the other side of a large wire fence separating the two roads, en route back to Tower Bridge. A veteran runner alongside turned and commented, "Now there's a depressing sight for you!" In my letter to Alan Storey I commented that ideally the southern loop should be added on to a more historic portion of London, perhaps Hyde Park west of Buckingham Palace, within the park boundary to reduce traffic problems, noting that this had probably been considered in the past. In his letter he said:

We would dearly love to use Hyde Park - it would be ideal as a finishing area -but the disruption it would cause to traffic (public transport in particular) prevents us.

Nearing Tower Bridge on my return from the Isle of Dogs, I spotted the 20-mile marker, a psychological barrier because of its association with the dreaded "wall." I realized that I felt fine, rationalizing that I only had 10K or a loop of Stanley Park left before the finish-line. I stopped to take a quick photo and resumed running, stepping up my pace. Shortly after, a runner came up beside me and fell into my stride. We soon started a discussion and I learned that he was Dr. Don Kaiserman, a neurologist from Los Angeles, 55 years old, running in his 14th marathon, and his first outside North America. He mentioned that he had seen me run past him and decided to tag along. We were back among the crowds of spectators and re-inspired by their encouraging yells. A dramatic change of scenery and mood occurs west of Tower Bridge with its many vistas of historic landmarks. We supported each other, chatting much of the time and with our stepped up pace to 7½-8 minutes per mile, Don estimated that we probably passed about 1,000 runners in the next four miles.

Beyond Tower Bridge the route follows the coarse cobblestone pavement alongside the Thames and the Tower of London which was partly covered by a mat for the benefit of the wheelchair competitors. The route continues along Upper Thames Street and Victoria Embankment passing historic London, Southwark, Blackfriars, and Waterloo bridges, before turning away from the river via Northumberland Avenue to Trafalgar Square, and entering The Mall beneath Admiralty Arch. The wide scenic Mall leads directly to the Queen Victoria Memorial in front of Buckingham Palace, the turning point for the final approach to Westminster Bridge via Brigade Walk and the finish-line on the east side of Westminster Bridge.

On exiting Tower Bridge, I raced ahead of Don to take his photo amidst other runners with the full extent of Tower Bridge in the background. Part of my euphoric state in that final phase of the race was caused by the approaching and attainable finish-line and partly by the continuous cheering of the spectators as we ran along, a remarkable portion of which appeared to be directed to Don by name. I happily shared in this ovation; however, it was not until several days later when my photo of Don near Tower Bridge was printed that I realized that "Cheer for Don!" was boldly inscribed on the front of his T-shirt. Coincidentally the photo taken at the 20-mile marker shows Don's back, front and centre.

I had missed seeing Rene, my cousin Mary Stump and her husband David whom I had expected to catch a glimpse of either at Tower Bridge or on the Mall near Buckingham Palace. We never did see each other, although they were on the Mall. Eventually we met at the Jubilee Garden Assembly area where the marathon organizers had anticipated the need to arrange for meeting areas and had appropriately placed markers from A to Z on trees for identification purposes.

My last photo of this memorable run shows the finish line at Westminster Bridge. My official time of 4:25:46 is shared by four other finishers, reflecting the well-controlled congestion from start to finish. Allowing for the time required to reach the starting-line and to photograph the event, this is quite similar to my 4:12:15 closing for the 1991 Vancouver Marathon. A minor difference was my placing 16,858th out of 23,656, compared with 726th out of 900!

I have mentioned the encouragement provided to the runners by the spectators along the route, who were estimated to number over one million. A most notable part of the entire organization, as in most events, is the role played by more than 1,000 volunteers, including those tending a variety of facilities in the assembly and finish areas and throughout the route -- particularly at the many water and Gatorade stops. I will never forget those St. John Ambulance personnel standing aligned in their immaculate uniforms at various water stops, patiently holding out hands covered with rubber gloves bearing globs of Vaseline for those runners suffering from rubbed areas.

I doubt that I will ever again experience such an exhilarating marathon; however, one never knows!

Vancouver Triathlon – July 7, 1997

The Cyclepath Vancouver Triathlon was a companion event carried out at the same location and a month later by the same sponsoring groups as the Locarno Beach Sprint Triathlon. The principal difference was the length of the course, double that of its predecessor, a full Olympic distance; a 1.5 km swim followed by a 40 km cycle and a 10 km run.

The principal differences in the course included a double loop of the previous swim course – this time defined by 6-foot high pyramidal red floats easily visible from shore – responding to complaints following the previous event. The water temperature was about 65º F. The cycle route added an extra leg up the infamous hill to UBC plus additional legs down Chancellor Blvd. to the vicinity of Blanca Road and back – about 4 km each to make up the 40 km distance. The run included an extra 5 km loop into neighbouring

Jericho Park with an intricate layout on gravel- and bark-covered trails, junctures of which were marked with limy arrows or attended by volunteers.

On leaving the Locarno Beach area a month earlier I had chatted with Rob Daniel and John Fettis of North Vancouver, long-time triathlon/duathlon friends. Rob had asked what would be my next triathlon event and I had said, "I'm finally convinced that I couldn't handle anything longer than these short courses, so this will probably be it for the year." Memory can be extremely short, for a week later I had ordered a new "Orca" long-sleeve, long-leg wet-suit from the B.C. dealer in Penticton which arrived a couple of days later. I couldn't believe the difference in comfort the suit provided in cold water after test swims in our unheated pool and in the ocean. I increased my training times to levels approaching the two Olympic distance Vancouver triathlons I had entered in 1991 and 1992 – my only previous full-length events. I pre-registered for the Cyclepath Vancouver triathlon on June 19th.

It rained much of July 6 and the prediction for the following day was for continuing unsettled weather. However, apart from a sprinkle of rain enroute to Locarno Beach and some fog, the conditions couldn't have been much better, with the sun breaking through cloud cover during the latter part of the event.

There were 137 'competitors' entered in the triathlon, including 90 males, 35 females and four 3-person 'mixed' relay teams. There were also about 30 individuals competing in a related duathlon event.

Unlike my swim a month earlier I was not nearly as cold on exiting the water and even managed to clip my helmet on without trouble. However my feet didn't thaw out until the end of the second of three cycle loops. The derailleur of my bike malfunctioned twice and I had to stop both times to re-attach the chain, thoroughly greasing up my hands as I had forgotten to put on my bike gloves at the transition. The grease came off quite easily when I wiped my hands on my swimming trunks. My split time for the swim was 41:41 minutes and with my bike time and a relatively slow transition following the swim my time was 2:22:41 hours starting the run. I had set a goal of about 3:20 hours as a best possible time but by the time I started the run my only objective was to finish the event.

About 1 km into the run I was passed by a young woman who I later learned was Linda Chramosta of Maple Ridge, competing in the 14 - 19 year age group with her two brothers. Denny and Jim Chramosta are about 17 years old and probably destined for great futures as triathletes. The two placed 18 and 19 overall in the Locarno Beach triathlon, only 5 minutes behind Graeme Martindale of Vancouver, the winner in 1:57:20 minutes. Martindale was also the winner of the Vancouver triathlon in 2:02:02 with Denny Chramosta 14th overall in 2:17:40.

As I came to an apparently unmarked intersection without a volunteer to point the way I caught a glimpse of Linda rounding a bend on the trail ahead. Checking the ground I could see an almost obliterated arrow pointing to the other pathway. I sprinted as best I could after her and hollered "come back!" Fortunately she heard and turned back. I checked her once more a little later after she passed me again, wondering whether she had missed a turnoff.

The run was the most tiring I can recall in recent years; however, I was surprised to see that my 10 km time for the Sun Fun Run in 1995 was even slower. I was welcomed at the finish line by Dave Harrison who had finished first again in the 60-plus age group. I was more than delighted to finish, let alone to finish second once more.

The awards for the two Cyclepath triathlon and duathlon events were mostly donated by sponsors and support groups such as Power Bar Foods (Power Bars, Power Gel, Bathing Caps, T-shirts, water bottles), Sea to Sky Products Inc. (Body Ayde drink); but included some quite elegant vests and diverse tickets and coupons for products or events. By far the most awesome of these were the cases containing 12-2 litre bottles of Body Ayde – advertised as a highly nutritional drink. They were piled three tiers high and formed a most imposing barricade in front of the awards desk. The first, second and third finishers in each age group were announced and requested to come forward to receive their awards. These they chose among plastic bags containing a variety of the above items, a large water bottle plus 4 power bars and a case of Body Ayde. The winner had first choice, the second place finisher had second choice. Invariably the third place finisher had to lug a heavy case of Body Ayde away.

John Fettis and Rob Daniel both came away with cases of Body Ayde and we chuckled at the rather novel transition in award presentations from 'gold, silver and bronze' medals that had been the tradition for so long and the 'eenie, meenie, minie' deliberations of the recipients.

The Vancouver triathlon took its toll, as is so often the case in athletic events. Five starters failed to finish, including four that completed the swim but not the cycling event. One young woman was hypothermic following the swim and Jim Hyslop, an experienced and prominent triathlete, had the misfortune of a flat tire immediately after leaving the transition area, which he replaced, followed by a blow-out on the hill up to U.B.C. which put him out of the race. An ambulance was called for one competitor and, although a stretcher was taken to the First Aid tent, the individual apparently recovered sufficiently that hospital treatment wasn't required.

Penticton Peach Classic – July 19, 1998

Ever since my first year in triathlons I had heard consistent reports about the challenging course held annually in Penticton in July to mark the annual B.C. Provincial Championship. It was described as challenging because of the hills encountered in both the cycle and run sections and the typically draining summer heat of the Okanagan. I had dubbed it a 'maybe' in my 1998 schedule. It remained in this category until I had an opportunity to cycle all but an 8 km stretch of the cycling portion of the course on a moderately cool day in June while returning from the Alberta Provincial Championships at Allan Lake near Edmonton. Although the hilly course had a couple of fairly lengthy steep pitches, I reasoned that it was no worse than many of the routes on the North Shore and mailed in my registration form after reaching Vancouver.

Although Rene would have accompanied me to Penticton I talked her out of it as there would not have been any particular joy for her in the heat. She cheerfully insisted on my driving in her Cressida with its most welcome air-conditioning and the added attraction of a disc-player. I left Vancouver under cloudy skies and a temperature of 15°C at 8:15 am on July 7 and by the time I reached Hope and the start of the Coquihalla Highway heading to Merritt and Lake Okanagan, the skies were clear.

I reached Penticton in 4¾ hours. Viewing the city and the south end of Okanagan Lake from Highway 97 as I approached, the entire south end of the lake was shimmering with what appeared to be closely spaced lights. As I drove along Lakeshore Drive toward the check-in area in Gyro Park, I realized that the spectacle was caused by sunlight reflecting from the front windows of an unbroken series of parked vehicles facing the crowded beaches.

The temperature rose to 35°C by mid-afternoon. After completing my bike check and obtaining my registration package at Gyro Park, I quickly donned my swimming trunks in the car parked within a block of the beach and swam out to a series of buoys marking the edge of the public swimming area. I later learned that the lake temperature was 20°C

(72°F) and that wetsuits would be permitted for the triathlon. According to the International Triathlon Union (ITU), regulation wetsuits are prohibited only if the water temperature exceeds 22°C.

After registering at the Sandman Inn, I drove over the run route which ran from the transition area opposite Gyro Park west along Lakeshore Drive for about 2 km before starting up the hill at West Bench, rising about 100 metres in 1.2 km before reaching West Bench Drive. The course turns north along West Bench Drive over an undulating route to the turnaround at the 5 km point leading back over the route to the finish line. I envisaged this route as extremely challenging on a hot day at the end of an equally challenging bike course.

I also drove out to find the end portion of the bike route – locating it at a place locally called Lava Rocks – a popular viewing spot across Okanagan Lake. I inwardly groaned at the additional hills in the 5 km stretch to the north of my earlier bike ride on the east side of Okanagan Lake.

After a fairly restless night at the Sandman Inn I had breakfast at a nearby restaurant – the Sandman's restaurant being packed with two bus tours from the U.S. even at 6:15 am. After checking out I drove off, parking our car within two blocks of Gyro Park and the transition area in a parking lot across Lakeshore Road in the Rotary Park area. I joined the horde converging on the check-in area and after parking my bike and gear at a designated slot in the transition area, I got in line for the usual body mark-up with my number and receipt of the numbered velcro wrist band required for the swim split time.

After arranging and checking and re-checking my gear in the transition area I joined the 450 triathletes in the starting zone area for the swim, which lay on the beach immediately to the east of a walking pier jutting out into Okanagan Lake. Just about everyone I spoke with had a comment about the rough water conditions caused by a strong wind blowing from the north creating fairly large white-cap waves. The first swimming 'wave' consisted of all male athletes under 40 years of age, in maroon caps. Due to start at 8:00 am sharp, the swim did not commence until about 8:15 am because of a late decision to move the huge red buoys outlining the margin of the course. About 150 males in the plus-39 age group followed the first group about 5 minutes later, and were followed by the 150 female triathletes in the event about 3 minutes later.

Unlike a swim in relatively calm water, this swim was daunting and the first distant buoy seemed like an unachievable target. The conditions were discouraging enough that about five of the swimmers elected to get pulled out before completing the swim, and 14 additional competitors did not continue the event after completing the swim.

The 'race manual' for the Penticton Classic Triathlon noted that, "As well as the lifeguards you see on surface vessels we have scuba divers following the pack and sweeping the course. If you see bubbles, don't panic, it is not Ogopogo." Even with this warning I was startled in rounding the last buoy to see a goggled visage looking up at me from the somewhat murky depths! The section of the swim between the first and last buoy was almost as difficult to negotiate for many like me, who had to turn their heads into the wind to get a gulp of air. I took one large mouthful of water into my lungs, and spluttered for half a minute while continuing. My friend Mike Koseruba in the 50 - 54 age group wasn't able to continue for a couple of minutes after a painful water intake.

The strong north wind was a mixed blessing. It made all the difference in the succeeding cycle leg of the event as it had a cooling effect and assisted in the hills on returning from Lava Rocks, with an overall drop in elevation of over 100 metres in the final two kilometres of the undulating course through scenic vineyards. Apparently three cyclists were unable to complete the event because of mechanical problems and one was hospitalized after a serious crash.

I knew that I was going to have problems in the final running section – virtually everyone except the elite and younger triathletes were in the same boat. The ITU rules provided prior to the race had the following initial comment which I had not observed before and which certainly did little to lift my spirits prior to the event: "No form of locomotion other than running, walking or crawling is allowed." I had walked briefly in portions of other running events, but the swim and cycle sections had been more challenging than any other triathlon that I could remember, and my running time was to be the poorest I can recall for a 10 km run. I was not alone; a comparison I made between results for other competitors of varying age groups between the 1998 Penticton event and the 1997 Kelowna Apple Triathlon indicated an average of about 15 percent slower in the 1998 event – attributable to the relatively flat Kelowna running and cycling courses and the better swimming conditions. A slightly lower ratio applied for the cycling event with a surprising but consistent improvement in swim times averaging about 5 percent, which suggests a difference in the course lengths.

The most appreciated support provided in the run event was at the aid stations at 2.0 km, 3.0 km, 3.5 km, 4.5 km and 5.0 km prior to the turn-around and of course at the same points on the way back. Included were water sprays at two of the aid stations which I particularly relished for a complete body soaking – in spite of the soggy shoes that resulted.

I could hear Steve King's incomparable voice booming out over a loudspeaker at the finishing line as he welcomed in the runners, while I plodded home about 0.5 km from the end. I was given a memorable welcome by Steve and the crowd as the winner of the 70+ age group in the B.C. Championships, and more personally from several friends. Mike Koruba who was at the finish line with his wife insisted on having her take a photo of the two of us which should be a reminder of my toughest triathlon to date.

The event was superbly organized – a tribute to Investors Group, the principal sponsor, and the many years that the Penticton Classic has attracted dedicated and enthusiastic volunteers. By the end of the excellent buffet and awards ceremony at about 2:30, copies of a 14-page booklet of detailed results with each athlete's split times, overall place in category and total event, were made available to the attendees.

I was deeply moved to be the winner of my 70+ age group with an overall time of 3:22.06 hours and to receive an impressive free-standing plaque as a memento of the event. Fred Cox of Armstrong, whom I had raced against in the past, was second at 3:53.14. The winning time for males was 1:54.25 by Stefan Jakobsen (20 - 24 age group) of Sidney, B.C. and for females by another world-class competitor, Carol Montgomery (30 - 34 age group) of North Vancouver in a record breaking meet time of 2:04.25.

Squamish Triathlon – August 16, 1998

The inaugural Squamish Triathlon was an unusual event in two ways. First, it was dedicated to the memory of Bob McIntosh, a member of Triathlon B.C. for many years who had raced and represented his age group many times both locally and at various national and world championship events. A respected lawyer and a resident of Squamish, Bob was beaten to death by a 'person or persons unknown' on New Years Eve, 1997 when he and a companion checked on a rowdy party of about 150 young people in progress at a neighbour's home in the absence of the parents.

The event fulfills Bob's dream of establishing a triathlon in his hometown. It is also a fund-raising vehicle for the Robert W. McIntosh Scholarship Fund – awarded annually to a Squamish high school graduate who has demonstrated outstanding achievements and contributions in athletics, scholarship and citizenship.

Secondly, unlike normal triathlons organized around a common transition area, the organizers elected to design a more challenging point-to-point course, originating in Alice Lake Park, about 4 km northeast of the community of Brackendale on the east side of Squamish River Valley and ending at an oval running track in the Brackendale Secondary School grounds.

Because of restricted parking facilities at Alice Lake, most athletes and volunteers were required to park vehicles at Brackendale School and leave their running gear at a transition zone equipped with numbered bike racks. Competitors then loaded their bikes on to vans and boarded buses with their swimming and cycling gear for the ride to Alice Lake. On arrival, they collected their bikes and went to the transition area to prepare for the swim/bike events. Swimming cap distribution, wrist timing bracelet attachment and body marking were carried out nearby.

The swimming course at Alice Lake was well marked out with a series of buoys and large floats around an L-shaped course. Wet suits were allowed as the water temperature was about 70°F (19°C). I later learned from Mike Koseruba – one of my many triathlete acquaintances, that he had gone to Alice Lake to look over the course the evening before, after competing in a triathlon at Whistler. On arrival he was told that it was doubtful that the swim would be held at Alice Lake as a 22-year old man had drowned there that day and his body had not been located. The unfortunate individual's body was located that evening and alternative plans were not required.

The swimming portion of the triathlon was completed without too many problems. Greg Crompton, one of the organizers of the Squamish Triathlon with Runners World in Vancouver and a triathlete suffered a bad cut on his head when he slammed into a wooden raft marking one of the turning points of the swim course, but completed the event.

The cycle course was mostly flat and fast with a very welcome drop of about 350 metres from Alice Lake into the Squamish River Valley. Several speed bumps at the entrance to Alice Lake and one sharp turn near the bottom of the hill had been pointed out as possible problem areas. A crash of two cyclists at the sharp turn ended their event with minor injuries. The course ran north up the Squamish River road – a secondary feature mostly asphalted but containing three sections covering about 2 km of recently applied seal coat with a very rough pebbly surface and loosened gravel fragments.

The turning point was at the 18 km mark and I watched eagerly for a sign of the lead cyclist on the return route down the valley to Brackendale. He eventually emerged followed by two other cyclists about a minute apart and then a group of about 10 - 12 in a closely packed peloton[29]. So much for the dire warnings against drafting, a practice generally unacceptable and subject to penalty, other than in elite races.

En route down the valley I encountered three poor souls frantically trying to deal with bike problems – mostly flat tires. As I neared Brackendale I realized that my swim/bike time was going to be below two hours – a time which I had rarely achieved. I managed to find my bike rack without much trouble and was soon ready for the run.

Having only seen the run route on a sketch map, I was aware that there were many turns along asphalted roads leading southwesterly to the east side of Squamish River where the route turned north. The next section followed an old dirt and gravel road with a very

---

[29] peloton: The main group of riders in a cycling race, generally regarded as of elite calibre.

rough irregular surface for about 4 km which required considerable attention to one's footing. The course finally turned east to Government Road, a most welcome asphalted surface, leading back toward Brackendale School on an overall downward slope. Steve King's voice was audible over the loudspeaker for some distance from the final oval track and provided a stimulus for the final stretch. My time of 2:52.04 was a personal best, exceeding the next best – the Vancouver International Triathlon on August 9, 1992 – by 2 minutes, 20 seconds.

Although I was competing in a 70+ age group category, when the results for the 60+ group were announced I was very surprised to learn that I was first, followed by Hilmar Lewandowski (2:53.58), David Harrison (2:57.46) and Barrie Street (2:59.30), all in the 60 - 64 age category, whom I had competed against, always from behind, for several years.

Even though Olympic distance triathlons all claim lengths of 1.5 - 40.0 - 10 km for the swim-bike-run sections, not only do the distances vary, but hills and particularly hot, muggy weather conditions can radically affect performance. In addition, everyone has their 'good' and 'bad' days. David Harrison had been suffering from a knee problem for much of the 1998 season and had told me at the start of the swim at Alice Lake that he was not sure whether he would be able to complete the run section. A veteran triathlete and duathlete, David has competed in three World Triathlon Championships and is well aware of the need to provide time for recovery from injuries.

Canadian National Triathlon Championship

Winnipeg, Manitoba, August 8, 1998

Although I had qualified for the National Triathlon Team at both the Alberta and British Columbia Provincial Triathlon Championships in 1998, I had never competed at a triathlon event at the national level and elected to do so in my 71st year (purist dating!).

I travelled to Winnipeg via Air Canada on a fully occupied Airbus A320 jet which landed us on schedule at about 5:15 pm on August 6. My travel arrangements were simplified by my having agreed with Susan to split the cost of a Trico Sports Iron Case designed for safe bike transport on aircraft. The case, when full, measures about 45 x 27 x 12 inches and is composed of two heavy-duty plastic shells, which are interlocked by numerous outside straps. The bike components, with wheels, seat, handle-bars and pedals removed, are sandwiched between three rubberized pads. There is plenty of room to accommodate most triathlon equipment, e.g. helmet, shoes, swimming gear, etc. My additional gear was ensconced in a small carry-on bag with a large shoulder strap. On arrival in Winnipeg, the bike case, weighing about 65 pounds, was disgorged at the baggage terminal along with standard baggage although it had required special baggage treatment on loading in Vancouver.

A very genial cab driver cheerfully helped me stow the bike case on the back seat of his cab and delivered us at the Lombard Hotel at the famous intersection of 'Portage and Main' – the coldest corner in Winnipeg in winter – but quite the opposite at this time, at 35ºC, with little to no wind. The Lombard was most convenient as it had been selected as the registration and bike check site and the relatively elegant room at $74/night was a bargain compared to downtown Vancouver.

On arrival, following check-in, I spent a hectic 30 - 40 minutes unpacking and re-assembling my bike – a task made possible only by the patient instructions provided by both Susan and Rob prior to leaving Vancouver plus a dry-run on my own. I had to have it checked in for transportation to Birds Lake Park, the triathlon site, by 8:00 pm.

My arrival at the Lombard also coincided with that of about 20 young (18 – 25-year-old) members of the Alberta Provincial Triathlon team, all bedecked with smart sleeveless team jackets and similar Iron cases for their bikes. I had met several of them at the Provincial Triathlon Championship in Alberta. I also met several other triathletes with whom I had competed who were much nearer my age group, including Lloyd Johnson (50 - 54) and Paul Poffenroth (60 - 64); also K.C. Emerson (women's 35 - 39) from Vancouver, who had come 4th in her age group at the World's Triathlon Championship in Perth, Australia in 1997. In addition, I chatted briefly with well-travelled Steven King who had just driven in from his home in Penticton after participating in a 24-hour road relay on the previous day.

Arrangements were made for interested triathletes to be driven by bus to Birds Lake Park, leaving the Lombard the next day at 10:00 am and returning by 2:00 pm, thus providing time for a test run of the bike circuit, a check on the run circuit and a brief swim in the recently completed artificial lake.

The triathlon facilities at Bird Hill Provincial Park are probably second to none. Featuring hills and ridges of glacial origin, the park covers 35 sq. km, and lies about 24 km northeast of Winnipeg. It contains an artificial lake, oak and aspen forests among cleared land, deer and wild turkeys, water fowl and songbirds.

Winnipeg hosted the 1967 Pan American Games and roads in the park were asphalted to serve spectators expected to watch equestrian events. When Winnipeg again won the bid to host the Pan American games in 1999, the Manitoba Triathlon association with the Manitoba Dept. of Natural Resources successfully lobbied the provincial government to renovate the site to accommodate the triathlon event. The lake was dredged, graded and provided with fresh aerated well water. All perimeter roads around the park received two layers of new asphalt and a 5 km bicycle trail around the lake was re-asphalted producing a very smooth surface. The entire renovations, completed this summer, cost over $3 million.

The organization for the 1998 National Triathlete Championships was awesome with over 400 volunteers, 15 motorbike officials throughout the course, extra-wide bike racks and a spacious distribution of support facilities throughout the transition area.

I boarded a bus at the Lombard with about 20 other athletes at 5:30 am on August 8 for the 45 minute ride to the park. I had plenty of time to re-check my bike which had been transported with others to the transition area the evening of my arrival. In my practice circuit of the 3-lap course the day before, I had started off from the transition area to find myself labouring while pedaling. The transportation the evening before had partly dislodged the rear wheel which was rubbing on the frame. There were no further problems.

The age group men received preferential treatment for an unstated reason as we all started first, promptly at 8:00 am from a line drawn between two towers on the 2-lap course of the lake. The announcer stated that anyone touching the line prior to the starting siren would be disqualified! No wet-suits were allowed as the water temperature was about 25ºC – well in excess of the ITU limit of 22ºC for wet-suit use. Apart from the usual mauling experienced by closely packed swimmers, the swim went off with only one serious problem and was well-patrolled by kayakers plus a cycle-powered flat-bottomed craft. Carol Montgomery of Vancouver, one of the world's top elite triathletes and an obvious favourite, withdrew 100 metres into the swim when she was kicked in the head – a not uncommon event in the mass starts.

There was a 200 m run from the lake exit along an asphalt pathway skirted by grass, into the bike transition area. The bike lap of 13.3 km was designed to circle the park along several access roads, which were completely devoid of ruts, cracks or debris because of the recent asphalting. The course was incredibly flat, no grades exceeding 4-5%. This was a surprising contrast to the profile of the course in the athlete's information manual which, through extreme exaggeration, looked like a cross-section of the Alps. Following each lap the course was briefly re-routed through the transition area for the benefit of spectators who were provided with a long section of tiered stands. I saw only one unfortunate young woman replacing a flat tire during my ride.

When re-entering the transition zone at the end of my bike ride and dismounting, I was followed to my bike rack portion by a photographer armed with a heavy video-camera. I chatted with him while making as hasty a change as possible into my running gear which took at least 1 - 2 minutes. It was quite an experience having someone standing about a metre away, not saying anything but focusing on my cycling equipment as it was being removed along with other items, followed by running gear being donned. He didn't have any trouble keeping up with me for a while as I ran out of the transition area!

Although the run course was partly protected from the hot sun by trees, most of it was in the open and many of the more fatigued age group triathletes suffered accordingly – including myself. I later learned that the heat was intense enough to force a number of the competitors to stop short of completing the event.

Steve King, the announcer, mentioned me at least three times as I passed through the transition area, insisting that the crowd recognize the oldest competitor. It was a little embarrassing, but inspirational, as I didn't dare look as if I wasn't having a delightful time! The final cheer from the crowd as I approached the finish line was a most welcome one and I was quite overwhelmed by the number of people that offered congratulations after the race. My time was 3:15.00, which was quite satisfying and, of course, I was very relieved not to have suffered from my recently injured left knee.

As I was taking my bike and gear out of the transition zone, Steve King announced the arrival of Jean-Sebasti Rioux, a 17-year old Junior Elite from Quebec, about 90 seconds behind Matias Optiz of Ontario, another Junior Elite who had topped his field with a 1:58.27 clocking. The cheers from the crowd were abruptly silenced when he started to uncontrollably flail his arms, head and body as he raced for the finish line. The crowd shrieked and groaned as he fell on the asphalt across the finish line. Fortunately his injuries were minor and he was sitting up in the medical tent being tended by an array of individuals as I passed on my way to the parking lot.

At the awards ceremony that evening – held at the Rendezvous Club, about a 15 minute walk from the Lombard, Steve King, as MC, reversed the normal awards procedure by announcing the oldest age group first, which comprised only myself. He also commented on my time, making the rather unprecedented observation that it was faster than both the gold and silver medalists in the 65-9 category, who were Julien Hutchinson from Ontario and Mikey Stokotelny from Manitoba, who had both competed for Canada at previous World Triathlon Championships.

World Triathlon Championships

Lausanne, Switzerland – August 30, 1998

I was so anxious to be a member of the Canadian National Triathlon Team that I actually qualified with first-place age group finishes at three separate events - Alberta Provincial Championships, near Edmonton on June 21; B.C. Provincial Championships at the gruelling Penticton event on July 19; and finally the Canadian Championships near Winnipeg on August 8. Apart from wanting to improve my conditioning from the start of the triathlon season in March, participating in three different events at new locations was appealing. It also meant exposure to many new individuals sharing a common interest.

Although I had learned that the World Triathlon Championships for 1998 would be held in Lausanne, Switzerland, I was initially deterred from any serious thoughts of being a participant. The first reference I saw to the triathlon course at Lausanne was in the January 1998 issue of TriBC Word magazine. A short article by K.C. Emerson stated that, ". . .if you ever wanted to pretend you were taking part in the Tour de France, this looks like the year you want to qualify." The 1998 ITU Triathlon World Championships brochure provided by Triathlon Canada described the 40 km bike course consisting of four 10 km loops in the city: "Tough continuous climbs, jolting cobbles, steep descents and sharp turns make for a physically challenging and technical course." That really sounded inviting to a 70-year old!

Eventually a detailed profile of the 10 km cycle course was provided. It showed a gradual climb, interrupted by short descents or flat areas, rising for 165 m in 4.5 km from sea level to the summit of the hill dominating Lausanne on which is perched the 13th century cathedral. There is a 17% slope near the summit which reportedly is about 150 m in length. An acquaintance at the Winnipeg Triathlon had told me it was 500 m long which sounded impossible, and mercifully it was.

Initially I actually planned to remove my old bike pedals from my ancient but faithful Claude Butler model (1960s version) originally purchased for Rob. This would permit me to easily get off the bike and push it up the hill, rather than deal with cleated bike shoes which could prohibit running and could be difficult to dislodge from the standard pedals. As I gained confidence during successive triathlon and training rides, I dismissed this as an option.

When I obtained the bike course profile for Lausanne, I decided to try to emulate it in a practice course in West Vancouver. I got quite close by using an altimeter and my car's odometer. The course I selected climbed 140 m in 2.5 km compared to Lausanne's 165 m in 4.5 km. However, the steepest grade was probably no more than 10%. In addition, it dropped back to the base level at the starting point in 5 km compared with the 5.5 km at Lausanne. I managed two rides of 3 laps and one ride of 4 laps. They weren't easy. Additional training included three 40 km[1] rides from home, two of them with Susan who provided me with some most useful and much-appreciated advice on dealing with hill climbs and descents. She also offered encouragement to the old man as we parted on our last ride on August 25, as she rode on to cycle back across the Lion's Gate Bridge to her home near UBC.

August 26

I had hoped to have a triathlon-related seat companion on Air Canada's B767M flight to Toronto - the first leg of my trip on August 26 to Lausanne. The empty window seat beside me remained that way until the aircraft started to fill up with unconfirmed passengers. No athletic-looking individuals appeared to be in my area and the aircraft was almost completely filled. As the plane rolled out of the boarding area, a Chinese woman was ushered into the window seat from the back of the plane to complete the loading and as she didn't speak any English, our vocal chords remained at rest during the flight.

Following a scheduled two-hour stop in Toronto, our Air Canada Flight 878 for Zurich was delayed for an hour and we took off for the 7½-hour flight at about 8:30 p.m. With the time change of 9 hours between Vancouver and Zurich, it was already after 2:00 a.m.

---

[1] All to the top of Cypress Ski Hill road (2200' vert.)

Zurich time. I had recently read advice on reducing jet lag by trying to match the amount of daylight with that at the destination. I planned to make an effort for early shut-eye and some sleep with an eyeshade and a sleeping pill, but that would have to await 'dinner.' As in the previous flight to Toronto, the aircraft was full and my travelling companion this time was a German lady returning home with a tour group. Her English was non-existent and my German not much better, so once again our communication was restricted. Although I took half a sleeping pill, an eye-shade, a pillow and made use of innovative stiff adjustable 'flaps' at the extremities of the head-rest, I didn't sleep - but I rested relatively peacefully - my ears plugged with Kleenex!

August 27

Although it was cloudy, our passage on reaching the west coast of Europe was graphically plotted on a monitor on all screens. I had just under an hour to make my way through the air terminal at Zurich from the Air Canada section to the appropriate departure gate for Swiss Air's flight to Geneva. I kept my fingers crossed, hoping that our bike case would make the connection, but to no avail. Uniform cloud cover at about 3,000 metres blanketed most of the ground during the short flight, but the crest of the Alps to the south was evident throughout the flight with occasional glimpses of the neatly arranged rural areas of Switzerland below.

A Swissair attendant in the Lost Baggage Office was most helpful in locating the bike case in Zurich and within five minutes produced a detailed hard copy from her computer of its description and routing. She gave me a five franc voucher for light refreshments in the terminal building and instructions on how to locate the nearby railway depot for my train to Lausanne so that I could arrange for a ticket. As requested, I returned about an hour later at 1:00 p.m. and watched as the last special baggage was picked up by a group of British triathletes. "That's all" insisted the baggage man, certifying that the flight from Zurich had arrived. Back I went to the Lost Baggage Office, to waiti for an attendant. As I was not going to be served for about 10 minutes, I went back for a last look at the special baggage outlet, and was much relieved to see my bike case making an exit. I loaded it on to a buggy and half ran through the long terminal building to the escalator leading down to the railway platforms - locating the waiting train scheduled to leave at 1:30 p.m. precisely. The British triathletes were already on board, having also selected 2nd Class and one of them helped me lug the bike case onto the train.

The ride along the north shore of Lake Leman (Geneva) provided glimpses of the road that Rene and I had travelled by a U-Drive rented at Geneva in September, 1989 when we drove past Lausanne en route to the Eagles Nest in Germany and on to Vienna. I managed to locate the only taxi waiting outside the railway station and arrived at the Hotel Novotel about 10 km northwest of Lausanne about twenty minutes later. It had been selected as one of two hotels to accommodate the Canadian National Team. I had plenty of time to locate a couple of the team committee and was kindly provided with a ride to the Olympic Museum for a team photo at 4:30 p.m.

Lausanne attracts a significant number of visitors annually with its scenic and cultural attractions. In addition, it has developed as an international sports capital, building on the establishment of Lausanne as the official headquarters of the International Olympic Committee in 1915 and more recently as the site for the Olympic Museum, inaugurated by the IOC in 1993. The Olympic flame, one of the symbols of the Olympic Games, flutters in a large urn near the front of the museum entrance and this was selected as the site for our team photo. Preparations for the photo consumed the better part of a half-hour as the photographer and various committee members tackled the problem of coping with almost one hundred team members, a goodly number of whom kept straggling in over a 20-minute period following the supposed deadline.

I talked to several people whom I had originally met both at the Edmonton and Winnipeg triathlons. Included was Paul Poffenroth, a Civil Engineer from Calgary (60 - 64 age group) and his wife Holly, daughter Cheryl Lowery and friend Sue Scott. They had rented a car at Zurich about a week earlier and were planning some additional touring of Switzerland following the triathlon before returning to Edmonton. After the photo was taken, I walked with them over to the square near Chateau d'Ouchy on the waterfront which had been selected as the assembly point for the Opening Ceremony, scheduled to start at 6:00 pm. Eventually most of the 2500 age group and 300 elite triathletes from 64 countries were arranged for a parade in alphabetical order, each country led by small local boys carrying appropriate placards. The Canadian team, numbering 103, were by far the largest on a per capita basis, although fifth overall, and the noisiest. The parade was led into a large open area at Place de la Navigation on the waterfront for welcoming presentations by a number of dignitaries. Entertainment was provided by an energetic local band, a group of six Alpenhorn blowers and about a dozen individuals carrying enormous cow bells which emitted mellow sounds as they were banged against the knees of these sturdy young men.

The opening ceremony culminated in a further walk to the west along Avenue de Rhodanie where an enormous tent had been erected for a pasta party at Place Bellevire. This was the only item which could have been improved by a more efficient provision of pasta and cold drinks. The mass of humanity, quite famished by 8:00 pm, came to an abrupt halt as an immovable column about six to eight deep. Numerous individuals began circumventing the column which could have been foreseen with such a competitive crowd. Our small party decided to seek nourishment elsewhere and we eventually located the American Pasta restaurant in nearby Ouchy with a selection of about 100 varieties. I collapsed into bed at about 10:30 pm after over 30 hours without sleep.

<u>August 28</u>

The Novotel provided an excellent buffet breakfast included in the cost of the room. Somewhat apprehensively I left the hotel on my bike to ride over to Coubertin Stadium at the site of the triathlon competition to join a cycle tour of one of the 10 km laps of the 40 km course. Most of the rest of the Canadian team had ridden the course the previous day and had travelled as a group from the hotel to the stadium. Even so, they got lost. Armed with a map provided to me by Canadian team committee member Pam Fralick, on which she had marked the correct route, I had no trouble in reaching the stadium well before the proposed departure time of the tour. By the time we left there were 500 or more triathletes strung out on the course. I was vastly relieved to find that, as expected, the overall gradient to the top of the hill was far less than my practice route in West Vancouver. And, as suggested by several of the Canadians who had ridden the course, the dreaded '17% hill' was negotiated without any real trouble, thanks to my added lower gears. Having said that, it was to be a different story on the 3rd and 4th laps on race day! In addition the cobbles were very flat and presumably well worn over the years and would certainly not deter cyclists negotiating them at a reasonably fast rate, providing they remained dry.

Returning to the stadium area I drove down to check the swimming course which was well marked with many yellow buoys disappearing into the distance. I decided not to check out the running course as it was also well marked out on the information map and quite flat. After riding back to Hotel Novotel I had a light lunch of raclette and salad and returned to my room for a rest before completing an easy 30-minute run along asphalt paths criss-crossing and surrounding a large area of corn fields on the summits of the local hills. The 'big news' on the notice board at the lobby area that evening was that Edmonton had been awarded the 2001 World Triathlon Championships. Next year is Munich, Germany, and surprisingly Perth will again host the Championship in 2000 – repeating 1997's on a slightly modified course.

Kathy wood of Alberta, one of the Canadian triathletes, broke her left arm during a cycling course tour the day I arrived. She crashed after hitting a cat crossing the road. A note on the notice board from her husband Mark, thanking the team for a box of chocolates and flowers, indicated that her elbow had been pinned with two screws and wire that morning. A 100 dollar reward was offered for the "tail of an orange cat with broken ribs."

August 29

Thanks to a sleeping pill – taken in two halves – I slept from about 10:30 pm to an unbelievable 9:30 am the following morning. I checked and re-checked my gear in preparation for a group cycling departure at 1:00 pm to the stadium area, as we were required to check in our bikes and leave them overnight in the transition area. In preparation, I realized that unless I took along a pair of running shoes I would be stuck with cleated bike shoes for the rest of the day as I did not plan to return until late evening after watching the men's elite race scheduled to start at 5:30 pm. I walked around the hotel and notified a number of the Canadian team to pass the word around, as most had not considered the potential problem.

We left on schedule, everyone pretty well clad in the team uniform and I got a couple of photos of the group and Holly and Paul in front of the hotel. Security at the bike check-in was strict and a yellow form had to be completed and initialled by an attendant indicating all the details of the equipment being stored overnight on well-numbered racks in the transition area. As Holly had driven down in their car, we all clambered in and drove up to a point on the bike course and parked near a prominent sharp turn for viewing purposes. I took several photos of the elite male cyclists, marvelling at the speed they could generate on both the uphill and downhill sections, while maintaining control. Such was not to be the case for the age-groups on the next day.

We had dinner nearby at the outside of La Rhapsody – an excellent restaurant serving unbelievably large portions of varied pasta dishes. Most of us only finished about half the amount we were served in large tureens and we advised a few Canadian arrivals to share a serving. I had a welcome opportunity to pay the bill under much genuine protest – but insisted on returning the Poffenroth's hospitality.

August 30

We were all up to leave with Holly driving by 5:40 am as Cheryl and Sue's events were due to start between 7:00 - 7:10 am, while Paul and I in the plus 40 male age group were set to start the swim at 8:10 am. We spent the waiting period under one of the many tarps set up in military fashion in a field near the transition area. About 400 triathletes were included in the group – all in white swimming caps. We created quite a splash as the gun suddenly went off without much warning. The water temperature was reportedly about 71°F (19°C) and pleasantly cool in spite of the activity. After the usual early pandemonium I got into an acceptable rhythmic crawl amid a crowd of swimmers trying to stay near the line of yellow marker buoys strung out on our right. Deviation to the wrong side was grounds for disqualification, and I understand that at least one suffered the consequences, although there were numerous swimmers on the wrong side in the early part of the swim.

Near the mid-way mark of the course I noticed a swimmer clad in a blue-coloured rear portion of a wet-suit slowly edging past me on my right. I remembered that Paul wore such a suit and decided to try and follow him. I was successful and actually managed to draft him, being careful to avoid touching his toes. Near the final turn to shore, I couldn't maintain the pace without over-stressing so I cut back. He exited the ramp about 40 seconds ahead of me and was still in the transition area when I arrived. He was obviously surprised to see me – although I was aware that swimming was his poorest split, whereas it had become my best. We both had a swim split in the 32 - 33 minute range but he was to make up over 37 minutes on my final time in completing the triathlon in 2:51:56 as Canada's top finisher in 19th place in the 60 - 64 age group (age 61).

The cycling section of the triathlon was going to be far more stressful in the initial lap than I could have foreseen. Although I was somewhat winded leaving a relatively steep section of the route at the start, I managed to gain the lower part of the final hill in relatively good condition and time. Having been exposed to the narrow road width – more like a lane in the steep section – I stayed over on the extreme right hand side as close as I dared to free-standing steel fence sections which provided a safe, but narrow walkway for pedestrians. Concentrating on the route I was suddenly broadsided by a young woman cyclist who knocked me into the fence which tumbled over – she landed partly on top of me. I rapidly checked my bike and a couple of spectators helped right the fence. It was, of course, impossible to board the bike on that 17% grade, and both the woman and I pushed our bikes 50 metres up to the top of the hill in our bike shoes before re-boarding and starting back down the steep hill. I completed that section rather tentatively as there were many cyclists on the route, several braking sharply and others flying by. The descent back to the start of the next 10 km lap was a welcome diversion from the climb and, of course, provided time for some recovery.

As we turned into the stadium area for the next lap, I unwittingly followed a woman cyclist who turned into the finish lane instead of turning into the lane marked "laps 2 - 4." The next I knew the crowd at the disembarking point of the transition area were hollering, "Yea Canada!" – not recognizing a thorn amid a bunch of young roses. I quickly stopped and an official most sympathetically said, "It's fine, just go back (about 100 m) carefully on the far right of the road and turn into the other lane". I probably lost no more than one minute correcting this oversight – having completely forgotten that the female triathletes had started an hour earlier than us.

Mercifully the rest of the cycle route held no further surprises – except one which could have been somewhat disastrous. During the third lap I noticed that my right shoe was slipping in the pedal attachment. I completed the cycle section without stopping to check the problem, enjoying the speedy descents of over four kilometres to a relatively flat section before starting the next lap with its gradual climb. Following the race, when I retrieved my bike and cycling gear I found that the cleat of my shoe had lost two of the three screws which secured it to the sole. Within another kilometre or two I would not have had any way of securing my wobbly shoe to the pedal!

I was very tired both physically and emotionally from the bike ride, having completed it in 1:43.27 with its 660 metres of overall vertical rise (2000 feet) and somewhat painful distractions (sore buttock from my fall). However, my time compared most favourably with flatter courses I had recently completed in about 90 minutes. In addition, we later learned from three bike odometers that the route actually measured about 41.5 kilometres, which would account for an additional four minutes in my time.

When I started the run, I was exposed to the full sun with little respite and I simply didn't have enough energy left for a good run. I was not alone as many of the other competitors felt the same. It was probably the most painful run I have had without resorting to partial walking. My running rate varied very little until the very last 500 metres. Throughout much of the cycling event we were supported by encouraging yells from the crowd. However, there was almost continuous encouragement throughout much of the running event, not just from Canadians but from spectators of many other nationalities. In addition I received several pats of encouragement from passing runners – again not only Canadians. It was a most emotional experience and it certainly was a stimulus. As I ran down the final 100 metre stretch to the finish line, past the spectators' stands, a Canadian on the sideline thrust out one of the small Canadian flags that we had used at the Opening Ceremony. I grabbed it and waved to the crowd, more as a joke than a real gesture of national pride. They applauded in appreciation to a very tired individual.

I was delighted with my overall time of 3:29.58, as I had set a goal of 3:30 hours. The time was good enough for 6th place overall in my 70 - 74 age category.

The first and second places were taken by Ken Nash and Robert Eazor of the USA in 3:07.28 and 3:08.25 respectively. Both are outstanding age group triathletes with many previous world championship appearances and titles to their credit. Egon Demmler of Germany was third in 3:15.24 followed by Rudi Schuster of the USA in 3:23.53 and Edward Hoy of Great Britain in 3:25.40. Vaughn Kimbrough of the USA, who was competing in his 6th world triathlon championship was 7th in 3:36.41. Walt Palmer, also of the USA, was disqualified. Age takes its toll. The only competitor in the men's 75 - 79 age group was 76-year-old Bill Schweitzer of the USA, the well-known age-group triathlete, who completed the event in 3:30.37.

Blake Jasper won a bronze medal for Canada in the 20 - 24 age group, finishing in 2:04.47, and was the only age group or elite Canadian male to mount the podium. By contrast, the Canadian female contingent had a field day, winning two gold: Marci Aitken (25-29), Leith Drury (55-59) and two silver: Nancy Burden (35-39), Edie Fisher (40-44).

Paul, his daughter Cheryl, her friend Sue, and I were all pleased with our results, feeling that we couldn't have done any better, given the conditions and competitors. We learned that there had been many crashes, several disqualifications and a number who did not finish. Two men and one woman in the Junior Elite Class whom we had watched in the running event, finished despite severe cuts and abrasions from bike crashes.

The Age Group Triathlon was followed by a Junior Elite male and female event and finally by a Women's Elite event, late in the afternoon. We watched parts of these triathlons, while awaiting permission to enter the transition area to retrieve our bikes and equipment from the swim and cycle sessions. At about 4:30 pm Cheryl, Sue and I got wearily on our bikes for the cycle back to the hotel. Paul had injured one of his knees so he later drove back to the hotel with Holly.

The five of us left the hotel about 7:00 pm for the Awards Ceremony and banquet – again held in that large tent, but this time far better organized with set tables and a remarkably diverse array of tasty edibles available from booths distributed around the walls. Included were pasta, raclette, salads, pastries, seafoods and even a few oysters on the half-shell. Each table also had bottles of red and white wine and soda water.

We had selected a table in the centre of the floor, half-way from other end of the tent, which accommodated a stage. When dining was in its latter stages, an individual suddenly arrived with a garment bag next to our table. He pulled out a variety of pieces of international triathlon clothing which he lay on the floor for trading purposes. Within minutes, Holly exchanged one of the team's coveted hats for a Mexican T-shirt for Paul – a process I photographed as I thought it rather novel. This was far from the case, for participants at previous World Triathlons were well prepared for a trading bee. It grew as the evening wore on, continuing unabated during the awards ceremony. We left at about 10:30 but the show didn't end until after 1:00 am.

I used up my little remaining energy to dismantle and pack up my bike for the trip home, starting with a 6:30 am wake-up call. I finally slept but not before realizing how lucky I had been, for competing in a World Triathlon Championship for Canada was a dream I never seriously considered achievable. As I have written elsewhere, Susan and I competed as Canadian representatives in the World Duathlon Championship in Palm Springs in California in 1990. However, there was really little comparison between the two events either in numbers competing, countries represented, organization or the associated hoopla. Having said, that, there was a rather remarkable coincidence, for there I also placed sixth in the 60 - 64 age group as the only Canadian participant, with *all* the others being from the USA.

1999 International Triathlon Union (ITU) World Championship Montreal, September 11 - 12

I arrived at Dorval airport at about 5:00 pm on Thursday, September 9, on an Air Canada flight with four other triathletes, including Norm Thibault of Victoria who managed to get bumped to a business-class seat. We were met by an ITU official who had us fill out a form for our return bus trip to the airport. He then guided us out to the small waiting van which the driver deftly packed with our bike cases and luggage, leaving six of us in the front two seats.

Norm Thibault and I were both deposited at the Chateau Royale Hotel – featuring only suites which were ideal for many of the triathletes sharing costs and for family groups. The hotel location is an active one, lying as it does within a block of Crescent Street in the downtown section, filled with indoor/outdoor restaurants, bars and dance halls.

I met Fred Cox, my triathlete friend from Armstrong, BC in the hotel lobby the following morning and we biked down to Windsor Station where we were shuttled by bus to Ile Notre Dame in the nearby St. Lawrence River, the site of the triathlon event. The island was created in the early 1970s for the 1976 Montreal Olympics from material dredged from the St. Lawrence River and piled partly against a wall of the Seaway Canal. The island is accessed from Montreal via the Concord Bridge to Ile Sainte-Hélène (site of Expo 67), or by the Metro subway and a short bridge to the island. Ile Notre Dame is best known for Circuit Gilles Villeneuve, the 4.4 km racecar track which was the site of the triathlon cycle stage. The circuit lies adjacent to the 2 km x 200 metre wide Olympic Basin, the site for the swimming stage, also used for other aquatic events. The running circuit looped around the Olympic Basin from the transition zone at one end, to finish in front of nearby grandstands.

After obtaining my registration package, Fred and I completed two laps of the cycle circuit before returning by shuttle bus with our bikes to Montreal. That afternoon we returned to Windsor Station with Barry Patterson, a phenomenal triathlete from Vernon, BC, in the 55 – 59 age group, to again catch a shuttle bus, in order to deposit our bikes in the transition area, as required before 7:00 pm. At Windsor Station we joined a small group of triathletes, several of whom had been waiting for over 30 minutes for a shuttle bus. After another 15 minutes, one of our group checked the Metro at Windsor Station, only to find that no bikes were allowed during the rush hour – starting at 3:00 pm. About 10 of us finally boarded our bikes and managed to find our way out to Ile Notre Dame. There we joined about 500 others in a long line entering the transition area. So lengthy was the line-up, officials made no attempt to complete a bike check – a first for most of us.

In the transition area, we met Ydo Buencher, a friend of Barry's, who introduced us to Roman Jezek of Beaconsfield, Quebec, at 85 the oldest competitor and, like James Ward of the U.S. in the 80 – 84 age group, sole competitors in their respective age groups. Ydo then drove us back to our hotel. That evening Fred and I joined my friends Paul and Holly Poffenroth and their daughter Cheryl Lowery and her friend Susan Paul at Windsor Station for a carbo-load pasta meal open to all the registered triathletes and guests.

The age-group event, with 389 women starters and 651 males from 36 countries, originally scheduled to start at 1:00 pm, was moved ahead to the morning at 10:30 am, with 50-minute intervals provided between succeeding waves. That was an improvement for the older triathletes, although afternoon heat was not to be a severe problem for most of the competitors.

The triathlon event started early in the day with a citizens' wave, i.e. the annual Montreal Triathlon with 322 competitors. It was followed by the women's 40-plus age group at 10:30 am sharp with the men's 50-plus age group, the largest wave at 130 individuals at 11:20 am. Six succeeding waves were generally composed of increasingly speedy (younger) competitors, the objective being to have all competitors finished by about 6:30 pm.

Wet suits were not allowed as the water temperature was just above 22º centigrade (76ºF), the ITU limit. Our 50-plus age group were all clustered in a relatively confined starting area and I wisely placed myself well back in the group. As it was, there were the usual flailing arm and leg kicks, which cannot be avoided for the first minute or so. The route was very well marked by a series of large red floats strung together by a line extending for 750 metres to a turning point in the Olympic Basin. I had one of my best swims of the year and caught up on quite a few swimmers in the latter part of the course, emerging with a split time of 35:54 minutes, which was fourth fastest in my 70 – 74 age group composed of 10 competitors. There was a 100-metre run at the exit to my equipment rack in the transition area and I only took about one minute to don my bike gear and run an additional 100 metres up a short incline to the start line.

The 40-km cycle course involved 9 laps of the 4.4 km race track – a relatively flat course with only one short incline over a bridge and several fairly sharp bends, one on entering and exiting and another at one of the corners bounded by a horde of enthusiastic spectators. My real concern, shared by a multitude of others, was remaining sufficiently focused to keep accurate track of the number of circuits. At my cycling speed I estimated it would take me about 10 minutes per circuit. An added circuit would be calamitous for the best triathletes, as it would eliminate them from a podium finish. A missed circuit could lead to disqualification. As is increasingly common practice, all competitors had been issued a timing chip secured to their ankles at check-in on race day. These were activated at appropriate points in the triathlon to provide split and final times and were collected following completion of the event. They were also programmed to record laps completed on the bike circuit. The competitors check-in package included a piece of waxed paper containing 8 circular orange tabs, which could be pulled off only with relatively long finger nails. This enabled the competitors to keep track of their laps. I had opted to stick short lengths of paper half-way under the tabs to assist in pulling them off, and it was to work very well. I ripped the tabs off successively as I rounded the sharp bend near the exit area. Several times I heard my name yelled encouragingly in this area, and as I completed the bend for my final lap, I yelled back "That's my last one!" as I threw the tab into the air with considerable relief.

During the bike ride I passed only a few cyclists and was passed by a multitude of others. I provided unnecessary encouragement on one of my first laps as I passed Gillian Palejko of North Vancouver in the 60–64 age group on one of her final laps. She was to finish the triathlon in 3:03:29 hours, about 11 minutes ahead of me, to win a silver medal, after only her third year in triathlon events.

On re-entering the transition area, I took an extra minute before starting the 10-km run to tape up my left knee, which had plagued me for most of the year. The tape shifted my patella to a less painful position and was secured by a newly purchased tensor. Edie Fisher, in the women's 40–44 age group, from Kenora, Ontario, whom I had met at Lausanne, was in the transition area at the same time. Spotting me apparently loafing, she yelled out, "Come on Dave – get moving!" Edie was also to win a silver medal as at Lausanne.

My running time was to be the fastest and least painful of the five triathlons I completed in 1999 at the Olympic 10-km distance. I had driven myself as hard as possible during the swim and cycle sections, with the intent of finishing the same way in the run. However, I recall being quite shocked when I passed the 2-km marker alongside the Olympic Basin, only to reach a 7-km marker about 200 metres ahead – a case of reverse déjà vu. During the run, which was essentially a flat, winding course around the Olympic Basin, I was passed only by one of my age group – most others being ahead. Equivalent age group competitors could be recognized by the age category mark on their calves. I was also passed by a stream of fleet-footed females in the 15 – 39 age group – an uplifting experience! I received encouragement from many of them and responded appropriately, as I admired their seemingly effortless stride.

The finish line was a particularly welcome sight and I was pleased with my time of 3:14.51 hours – my best of the year and 15 minutes faster than the hilly Lausanne cycle course in 1998. There were 10 competitors in my age group, including six from the U.S., two from Canada and one each from Germany and Australia. The winner was Ken Nash from the U.S. with a time of 2:41.10, which compared with his winning time in the same age group at Lausanne in 1998 of 3:07.29. Terry Kenney of the U.S. placed second in 2:48.11 and Gerhard Krauss placed third in 2:54.16. I placed 8th, ahead of Harold Beal Jr. of the U.S. in 3:37.03 and Fred Cox in 4:14.25.

Fred had a very rough time as he had not practised without a wet suit and this affected his swimming time severely. Another friend, Tim Hartley, in the 25 – 29 age category finished in 2:12.32 but unfortunately added an extra lap in the cycle event, which would have increased his overall time by about 7:25 minutes.

I was extremely tired after the event and as Fred had not checked into the transition area when I went to pick up my bike and other gear, I assumed that he would still be with his son and daughter-in-law who had come to see him from Toronto. I returned to Windsor Station by the Metro and met Fred later to attend the awards dinner held at Windsor Station that evening.

The Canadian team, the largest at the event with over 200 triathletes, had their best-ever world performance. Age group competitors won four gold, six silver and five bronze medals. As in 1998 at Lausanne, where women won two gold and two silver medals to a single bronze won by a man – they were dominant in Montreal winning eleven of the fifteen well-earned medals.

Obviously, as a senior age-grouper in triathlons, I have cogitated on performance levels of world-class triathletes. This led me to prepare and submit an article published in the January 1999 issue of Triathlon British Columbia's 'The Word' magazine. My follow-up article published in the Nov./Dec. 1999 issue on 'Winning Triathlon Performance' compared world-class triathlete's performance levels between hilly (Lausanne) and flat (Montreal) cycle courses, the running courses both being relatively flat. For those interested the latter article is reproduced here.

Triathlon BC Newsletter (The Word) Vo. 99, Issue 10:

The January 1999 issue of The Word examined the effect of age on winning triathlon performances by men and women at the Canadian Nationals in Winnipeg and the worlds at Lausanne in 1998. This article provides more comparable data from the 1998 Worlds at Lausanne and the 1999 Worlds in Montreal – also benefitting from comparisons of performances by many of the same competitors at both events.

The principal difference between the two courses was the hilly Lausanne cycle course with over 600 metres of vertical change in 41.5 kilometres from Lake Geneva to the top of the city in four technically challenging laps. By contrast, Montreal offered the speed associated with the Gilles Villeneuve racetrack in nine laps totalling 40 kilometres. Both 10 kilometre runs were flat, and wet suits were permitted in the cool waters of Lake Geneva, but not in the Olympic Basic on Ile Notre Dame in Montreal.

The accompanying graph, comparing time/age relationships in current winning triathlon performances at world age group levels, provides some interesting observations.

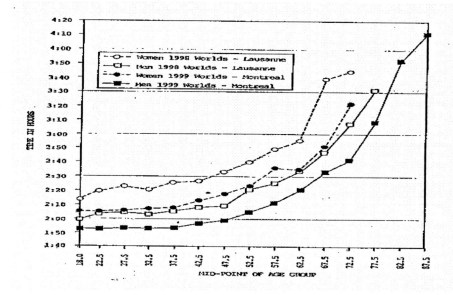

There is a remarkably uniform top performance level for both men and women from 16 to 40 years of age in the flat Montreal course. For men there is an overall variation of only 54 seconds in winning times in the 24-year period. For women, the total is 6 minutes or an average annual increase of about 15 seconds. This top performance level of apparent perpetual youth gradually begins to take its toll. For both men and women it shows an increase of about one minute per year to age 60 and, from very limited data, by about 4 minutes per year for women and 2 minutes per year for men to age 70. Results based on single performances by men in the two 80-plus age groups indicate a further increase to about 6 minutes per year at these levels. The oldest male and female competitors, both singles in their age groups, were Canadians – Roman Jezek of Beaconsfield, Quebec in the men's 80-90 age group winning his third world title and Phyllis Goodlad of Oromocto, New Brunswick, in the women's 70-74 category.

The flatter Montreal cycle course resulted in 10-20% faster cycle times for both men and women up to age 70 (allowing for the 1.5 extra kilometres at Lausanne). This increased to 25% for men up to age 80, with no comparable data available for women.

Canadians won four gold, six silver and five bronze medals in age-group competition in Montreal, the country's best ever performance at a Worlds triathlon event. It compares with two gold, two silver and one bronze in 1999 at Lausanne. This is a striking difference in one year and can partly be related to home turf as shown by the following breakdowns for the four principal competing countries.

	1998 Lausanne		1999 Montreal	
No. of age group competitors	1139		1044	
Men	767		655	
Women	372		389	
No. of countries represented	48		36	
Principal countries; % of total competitors (medals)				
–USA	6.1	(15)	22.4	(29)
–Great Britain	15.2	(8)	15.2	(6)
–Australia	11.2	(10)	13.1	(8)
–Canada	7.8	(5)	18.8	(15)
			69.5	

2003 ITU Triathlon World Championships
Queenstown, New Zealand, December 6-7, 2003

I did not expect to record this event because the draft of this book was in the final editing stage. It turned out to be the most memorable triathlon, which attracted over 1500 triathletes from 52 nations, supported by over 500 volunteers.

We had planned the trip almost a year earlier and received a most unexpected and welcome present from Susan, who provided us with Executive Class passage to Aukland via Los Angeles and return, on her Aeroplan account. We left Vancouver at 5 pm on November 26 and connected with a Boeing 747-400 flight on Air New Zealand which left Los Angeles for Aukland at 9:30 pm. The accommodation and service were fabulous and we actually slept from about midnight until breakfast, and we arrived at Aukland 13 hours later at 7:30 am New Zealand time, having crossed the international date line to arrive on November 28!

We stayed in a spacious suite on the ground floor of the Aurum Hotel, a 5-minute walk from the city center. The hotel was selected by Triathlon Canada as the preferred accommodation for its team, so we met many members, several of whom I had seen at national events in Canada and at previous Worlds.

Although Queenstown was the name of the location for the Triathlon World Championships, the actual event, with minor differences in lengths and routes for the elite, age group and senior (sprint distance) courses, started at Lake Hayes, one of New Zealand's most picturesque lakes, 12.5 kilometres north-east of Queenstown. The course had two transition zones; one at Lake Hayes and another 7.8 kilometres north of the lake at Millbrook Resort, the site of the 10-kilometre run course, composed of four laps around a fairly open golf course with two short hills per lap.

I had been aware for many months prior to the event that the age group bike course included an 800-metre-long hill with an 8-9 percent grade which had to be negotiated three times. I agonized about this hill for months before our arrival, because it was definitely at the limit of my ability, or so I believed. I got a little comfort from cycling up the steepest portion of the hill in West Vancouver on the approach to the Cypress Ski Resort. It was actually about two kilometers long, and I only completed it once because it was a bitterly cold day, about two weeks before our departure to New Zealand. Eventually, after arriving in New Zealand, I realized I could always, as a last resort, get off the bike and walk up the hill.

During the week leading up to our event, I had only a few workouts, including two swimming sessions at the Queenstown High School pool where Triathlon Canada had reserved time for its triathletes. One day, I cycled from Queenstown to lake Hayes, then up to Arrowtown above the 800 metre hill and down to Millbrook Resort, following the cycle route for the event. From there I completed one of the two 16-kilometre laps before cycling back to Queenstown.I also ran on a path alongside the main road from Queenstown to Frankton, 6 kilometres east of Queenstown, and back, rather than trying to complete a run at Millbrook Resort, which was really too difficult to access.

One morning prior to the event, Rene and I walked about one kilometer to the base of the Skyline gondola which rises at an angle of about 50 degrees to a height of about 600 metres above beautiful Wakatipu Lake in Queenstown – itself at an altitude of 310 metres. The view from the top was magnificent. Triathlon Canada also arranged for a team photo and luncheon at this site several days later, which shows all available team members in a crescent-shaped cluster wuth Queenstown below and the surrounding mountains as a backdrop.

At the lunch we sat at a table which included my old friend Mikey Stokotelny (70-4 age group) from Manitoba and Jamie Wilson (35-9) from The Pas in Manitoba, whom I had not met before. Jamie had traveled to New Zealand with his wife, one young child, and his brother-in-law, and had rented a van for accommodation. He very kindly volunteered to drive me out to lake Hayes on December 5 with my bicycle and helmet which had to be left in the transition area the day before the event. We also had to leave our running gear at the Millbrook Resort transition area.

For me, the most memorable part of December 5 was the convenience of a relaxing drive to the area that we would be challenged by on the following day. While at lake Hayes, Jamie said he wanted to complete the 1.5 kilometre swim route. He asked me if I'd like to accompany him, but I said, "Oh no, I haven't brought a swimming suit; I hadn't planned to swim." He said, "I have a spare pair of trunks that you are welcome to." I put them on and accompanied him down the steep slope to the lake below the transition area. He was in his wet suit. There were several people gathered at the lakeside, and one woman said, "Now there is a real man," pointing to me. She was presumably aware of the water temperature of about 15°C, and the fact that every other swimmer was clad in a wet suit. I hastened to respond, "I'm just checking out the water temperature." I dove in and found after a few strokes that I could actually stand it, and eventually swam out to the 400 metre mark before returning to shore, thus completing the equivalent of 800 metres of the 1500 metre course. This, together with my one lap of cycling up the hill toward Arrowtown, gave me considerable encouragement for the following day's event. After leaving lake Hayes we drove to Millbrook Resort and dropped our running gear at the transition zone which was fortunately a complete duplicate of the one at Lake Hayes. This made finding one's spot among 1500 others much simpler.

When we returned to our room we found a "facs" from Susan slipped under the door. It read, "As my friend John Hughes would say, "HFDF" (have fun don't fall). It basically sums it up, doesn't it? I'll be thinking of you tomorrow, Love Susan."

I left Queenstown by bus with other competitors at 8:00 am on December 6. It had rained during the night and the transition area at Lake Hayes was quite soggy. After placing the clothing for the bike ride in a separate bag and stashing the clothing I was wearing in another marked with my competition number 1972, I donned my swimming gear. After a cup of coffee I eventually joined friends Mike Stokotelny and Mike Ellis (65-9) near the start of the swim wave for the men's 60-plus age group scheduled to start at 10:40 am. The water was rough, with whitecaps and a stiff on-shore wind. When the starting whistle blew we took off straight into the wind in water reported to be 16°C. Halfway across the first 600 metre leg of the course I noticed a swimmer with his arm in the air; almost immediately a boat came and picked him up. It seemed to take forever to reach the buoy marking the turn at 600 metres. The next leg of 300 metres and the final 600 metres back to shore also seemed to be into the wind and I got several mouthfuls of water. I staggered out of the lake and had started running up the steep path to the transition area when I felt a sharp cramp in my right thigh. I walked the rest of the way up the hill and ran slowly through the transition area, hoping to alleviate some of the pain from the cramp.

Transition took me more than five minutes. My bike ride went well though the winds and the 800-metre-long hill were taxing. I waved to Rene who was standing with others near Arrowtown, and later learned that she had begun wondering where I was. It was a relief to finally reach the transition area at Millbrook. My run was significantly slower than normal as my legs were tired and sore from the cramp. However, it was an excellent course, almost totally on pavement and paths with crowds of people yelling encouragement most of the way around the four laps. At one point a group of American supporters had a garden hose and I welcomed a dousing at close quarters on each lap; it really picked me up. On the final lap I caught up with Lorene Hatelt in the 'Athlete With a Disability' (AWAD) class for Canada, who had recently injured a leg and was running painfully. I stopped, gave her a hug and ran with her for a minute. She has been a real inspiration for many of her teammates. I will never forget the roar of the crowd as I finally turned off the lap circuit onto the 300-metre finishing lane. I have also never had a tougher triathlon or felt more elated at the finish line. This was a feeling expressed by most of the other triathletes. Much slower times were posted than normal.

Until the final results were announced, I was never quite sure whether there were only two of us in the 75-9 class. There had been three in the class at Cancun, the winner being Charlie French of the U.S. He was also the winner at Queenstown by a significant margin, and I was second. When all the results were announced, I learned that the Canadian team, totalling 152 members, had won one silver and three bronze medals, compared to the host team's 21 medals and Australia's 17. My article, "Winning Triathlon Performances," published in the November/December 1999 issue of the Triathlon BC newsletter, points out the advantages of home turf on overall performance; Canada won four gold, six silver and five bronze medals at the 1999 World Triathlon Championships in Montreal.

Ramblings of a 75-year-old Triathlete

Triathlons and duathlons did not exist in my younger days. Triathlons originated in California in the early 70s. The first recognized event that I have been able to reference occurred in 1974 when two members of the San Diego Track Club, Don Shanahan and

Jack Johnstone, added a bike ride to John Pain's Birthday Biathlon, conceived in 1972 which consisted of a 6.2 mile run and a half-mile swim (Scott Tinley reference).

The sport developed rapidly from its modest, unofficial beginnings with the first Iron Man Triathlon (2.4 mile swim, 112 mile bike ride and 26 mile run) being held in Hawaii in 1977. Twelve men finished.

My interest in triathlons also evolved rapidly. In the late 1970s and through the mid to late 1980s, I had run erratically with Susan and Rob and also had done much hiking and climbing with Rob. I had also completed a number of local runs, starting in 1982, with Susan such as the Terry Fox Run, Vancouver Seawall and Khatsalano 15 km Road Race.

My training for marathons, which started in August, 1990 with the North Shore Marathon Clinic, was motivated by Susan's marathon activities and a long standing wish to complete one. Almost simultaneously, Susan suggested that I should consider training for duathlons, another of her competitive activities. I did my first bike ride of 20 km on August 6 which produced two flat tires – the first less than 100 m from my parked car on Marine Drive in West Vancouver.

Seven weeks later, with a total base of only 100 km of biking on four different days separated by our 3-week visit to China, I placed first in the National Duathlon Championship in Vancouver for the 60 - 64 age group. It was a 5-30-5 km, run-bike-run event. Susan also won her age group as a cyclist in the relay event. At the time she was still recovering from a broken femur sustained in the Seattle Half-Marathon. One of my many treasured memories was the trip to Palm Springs in November when both of us competed in the Desert Princess World Duathlon Championship, which I have described elsewhere.

I graduated to my first triathlon in much the same way, after completing the Vancouver International Marathon in May, 1991 – my first marathon, and starting to train for runs and duathlon events that summer. In June, I learned that the Vancouver International Triathlon, featuring a number of world-class triathletes, was scheduled on August 11, 1991. The cross training required for a triathlon immediately appealed to me as a relief from the many hours of running that I had devoted to the marathon. I started to train for the event in mid-June and although I had several out-of-town business trips in Canada and the U.S. and a 10-day vacation with Rene in northern Ontario which prevented any cycling, there were plenty of opportunities for swims and runs. Prior to the event, which I have also recorded elsewhere, I had a training base of 22-410-130 km in swimming, cycling and running – quite an improvement over my first duathlon.

Dealing with injuries has always been a fact of life for most athletes. Much has been written since I started competing in my teens about preventive measures to reduce injuries – particularly stretching exercises before and after events, plus 'listening to your body.' Fortunate indeed is the athlete in the late stages of a competitive life, even in the amateur field, who has had no serious injuries or significant recovery problems from surgery to deal with.

My situation was no different. In 1991 while training for the Vancouver International Marathon, I had sore knees for two months, which affected my training, but fortunately cleared up with the use of knee tensors and walks supplanting scheduled training runs. I suffered a torn knee ligament in January, 1992 while on a training run in the Ambleside Park area in West Vancouver which had cleared up by the end of April in time for the North Shore Spring Triathlon in late May. In mid-June I broke my left clavicle and tore associated ligaments when I took a header from my bike on the Upper Levels Highway in West Vancouver. That ended my duathlon-triathlon season but I was repaired enough to compete in the B.C. Senior Games in early September.

After starting training for the 1994 season events and completing several runs in April, a medical check-up in mid-May led to the detection of prostate cancer. I completed two triathlons in late May before surgery on June 9 and the long recovery period afterward. I ran my first 10 km since April 10 in the Richmond Flatlands race on October 16, placing third in 54:00 minutes. Recovery from my third hernia operation on November 8 coupled with excessively sore knees in training sessions in the 1995 season led to arthroscopic surgery on my left knee on October 18, 1995. I only completed one triathlon and four competitive runs in 1996 before fully recovering for injury-free, bountiful 1997-8 seasons. This only culminated in late July after the Penticton Peachland Classic triathlon, my seventh since March, 1998, when I once again suffered a painful left knee several days after the event.

I tried a very tentative run but had to quit my increasingly limping plod after 3 km. I decided to avoid any further aggravation of the injury by not running, and found that it didn't affect my biking or swimming. To be on the safe side, I had my knee checked by Dr. Follows who immediately referred me to Dr. Jack Taunton of the UBC Sports Clinic. I was examined the next day and it was concluded that I had strained my left tibular ligament and muscles below the patella. I was placed on an accelerated walk/run plus physiotherapy program with the objective of competing in my next scheduled (and paid-up) triathlon – the National Triathlon Championship in Winnipeg on August 8. The physiotherapy treatment and increased exercises worked wonders and by August 5, I had completed a pain-free 8 km run.

In the preceding section I have summarized recollections of only a few of the 62 duathlons and triathlons in which I have participated between 1990–2003. Most importantly for older competitors in particular is to recognize the need to increase training mileage only moderately and to allow enough time for injuries to heal. In October, 1998 I tore my right achilles tendon in a training run at the 14 km mark after too sudden a mileage increase and it took me over 4 months to recover.

I have mentioned my knee problems, which commenced as early as 1991. Since then, these have recurred periodically, but I have been able to cope with an acceptable level of discomfort by a combination of periodic rest periods with walk/run sessions, an excellent tensor bandage, regular ice-packs (20 minutes) on knees following runs, and frequent rubs with Myoflex, an anti-arthritic unguent rubbed over the knee caps, and finally, quadriceps enhancing exercises with weights. Why don't I quit? The answer is simple. As a 75-year-old, the pleasure of association I enjoy with other triathletes, albeit mostly far younger, far outweighs the discomfort level. Furthermore, as a Canadian, the opportunity to compete athletically, albeit as an age-grouper, at the world-class level, is an honour I never anticipated in my earlier years.

In Retrospect

On my 18th birthday in 1946, I was still at Upper Canada College in my final year. Viewing it as a somewhat auspicious event, even though I celebrated it alone without fanfare, I recall looking into the distant future and making three wishes. They were not necessarily made in order of prominence, but in reality, of course, they had to occur in a certain order.

First was to run a 4-minute mile. Hey, I was only 18 and I had run one in 4:50! This long-heralded goal, sought by others far closer to the mark than I, was not to be achieved until a wet and windy evening at the Iffley Road track in Oxford on May 6, 1954 when Roger Bannister ran the mile in 3:59.4 minutes. My closest was to be 4:29.0 in 1950, two weeks after breaking the indoor mile record at the University of Toronto with a time of 4:29.2 minutes.

Second was to be lucky enough to locate a life mate with whom to share the future. As I write, approaching our 50th wedding anniversary on December 29, 2003, I have been granted this – my most important wish – beyond my greatest expectations. I cannot conceive of a life without Rene and our children.

Third was to be able to look back, late in life and to be able to say that if I had life to live over again I wouldn't have wanted it changed too much. In retrospect, there are several events which I, of course, wish had never occurred, some within my control, others not. I have described the tragic DuPont helicopter accident which claimed five lives in 1980. It was the lowest point of my life. The safety committee I founded as a result and chaired for 20 years has undoubtedly contributed to a significant reduction in accidents and fatalities in the mineral exploration sector, but it cannot compensate for that tragic loss.

Then again, there are others where opportunities taken have yielded great satisfaction. The opportunity to start my career with an aggressive and successful organization as a junior geologist led to the perhaps unique experience of carrying out mineral exploration and/or mining property examinations in every province and territory in Canada, except P.E.I. Furthermore, I had the opportunity to observe mineral exploration and mining properties in Mexico, Central America and parts of Africa. I am only too aware of how fortunate I have been when I think of the hundreds of millions of souls around the world who live only to survive.

Finally, I know of only two major events which have affected my whole life. I had no control over the first, which was the decision by my parents which led to my conception and the opportunity for life. God bless them, for I would have really missed being here to give them my belated thanks!

My meeting with Irene McLellan on January 30, 1953 was truly love-at-first-sight. It led to the births of Susan Irene, November 29, 1954; Robin McLellan, January 14, 1957 and David Walter Stuart on August 31, 1959, the four blessed, inter-related parts of the second major event.

We don't have any control over being blessed with life. Great expectations come initially from others, rather than ourselves. Yet, if we owe any debt in life it's usually to others long since gone and to those who have lived life with us. My experience has been no different. Many individuals have influenced my life and career. Among the latter, I benefited from the advice of the father of Bert Maxwell, my best friend at university, a mining executive who envisaged a great future for Canadian geologists in the mining industry. My career was also strongly influenced and supported by the late John Sullivan, whom I succeeded on his retirement as president of Kennco Explorations Canada Limited. The many others are reflected in the preceding pages.

And, of course, I have a close family relationship which continues to enrich our lives after almost fifty years. This has had by far the greatest impact on my life.

Acknowledgements

Many individuals and organizations, apart from my family, influenced my life, and I have attempted to recognize their contributions within the narrative. However, several who contributed to the preparation of my story warrant particular emphasis.

Chrissy Bux, of Accutype Business Services in West Vancouver, prepared most of the original draft over a period of about ten years. Michael Barnes, a long-time family friend and accomplished author of many books concerning Ontario's towns, mines, characters and organizations, commented on my original draft and was instrumental in connecting me with John Stevens of Toronto, who painstakingly edited the final draft and provided many useful suggestions.

Harbour Publishing Co. Ltd. of Madeira Park, BC, has kindly permitted the reprinting of several paragraphs from a chapter in "Spilsbury's Coast," a west coast classic by Jim Spilsbury, which appears in "The Sherwood Mine Saga."

I am also indebted to my friend Brian Abraham of Vancouver, a geologist and lawyer who provided legal guidance.

Selected Bibliography

A Guide to Trekking in Nepal, Second Edition 1982, by Stephen Bezruchka, Second Edition, 1982.

Beyond Everest – Quest for the Seven Summits by Patrick Morrow, Camden House, 1986.

Chamberlain J.A. (1967), Sulfides in the Muskox Intrusion, Canadian Journal of Earth Sciences, Vol. 4, No. 1

China – a travel survival kit, 2nd Edition, Lonely Planet Publications, May 1988. Berlitz China, 1986/1987 Edition.

Commission of Inquiry on the Blood System in Canada; Interim Report, 1995 and Interim Report Annexes.

Dahl, C.L. (1969) Geology of the Copperbelt, Zambia (Draft), Kennecott Explorations, Inc., Exploration Services, Geologic Research Division, March, 1969.

DuPont of Canada Exploration Limited, Corporate Profile, January 1984.

Fifty Years in the Canadian Red Cross, P.H. Gordon, undated (last date referenced - 1967).

Garson, M.S. and W.C. Smith (1958) Chilwa Island, Nyasaland Geological Survey, Memoir 1.

Geological Aspects of Ore Reserve Estimates, Pine Point District Zinc-Lead Deposits, N.W.T., A.W. Randall, D.A. Barr and G.H. Giroux, presented at SME-AIME Fall Meeting, Albuquerque, New Mexico, October 16-18, 1985.

Gunning, H.C. (1962) Exploration in Rhodesia, Talk to Mineral Exploration group, February 21, 1962. C.F. Memorandum by D.A. Barr, February 27, 1962.

Heinrich, E.W. (1966) The Geology of the Carbonatites, Rand McNally, Vol. 48.

Insight Guides, Nepal, Second Edition, APA Productions (HK) Ltd., 1985.

Kennco Explorations (Canada) Limited: 'Visit to Mines and Areas of Geologic Interest, Republics of Zambia and South Africa, April 15M^{3}30, 1969, by D.A. Barr'.

King, B.C. and D.S. Sutherland (1960) Alkaline rocks of eastern and southern Africa, Part 1, Distribution, ages and structures, Science Progress, Vol. 48.

Lombard, A.F. (1962) Geology and Exploration of the Copper Deposit in Carbonatite at Loolekop, Palabora Complex, Palabora Mining Company Report.

'Lonely Planet Travel Survival Kit'. Islands of Australia's Great Barrier Reef', by Tony Wheeler, 1990.

Mendelsohn, F. (1961) The Geology of the Northern Rhodesian Copperbelt, Editor, Macdonald & Co., London.

Minerals Yearbook (1966), U.S. Bureau of Mines.

'Mississippi Valley-Type Lead-Zinc Deposits'. G.M. Anderson and R.W. Macqueen in Ore Deposit Models, edited by R.G. Roberts and P.A. Sheahan. Geoscience Canada Reprint Series 3, 1988.

Palabora, Engineering Mining Journal, November, 1967.

Pelletier, R.A. (1964) Mineral Resources of South-Central Africa, Oxford University Press, London.

'Pine Point, A Northern Legacy of Success'. D.L. Johnston, CIM 1988 Annual General Meeting, May 11, 1988.

'Pine Point Memories – The Earth's Physical Resources'. Cominco Ltd. Video, Vancouver Corporate / Legal Library.

Premier (Transvaal) Diamond Mining Co. Ltd., Annual Report, 1967.

Premier (Transvaal) Diamond Mining Co. Ltd., Information Booklet.

Report on Lacana Evaluation, by D.A. Barr, December 1, 1977 (Distributed March 31, 1978).

Report of the Standing Industry Advisory Group on Blood Issues; A Report to the British Columbia Ministry of Health; Final Report, May 23, 1996.

Smith, C.H. (1962), Notes on the Muskox Intrusion, Coppermine River Area, District of MacKenzie, G.S.C. Paper 61-25.

'The Cominco Story, Part 7: Brains, Guts and Luck'. Orbit, Cominco's Quarterly Magazine, Spring 1990.

'The Guinness Book of Records'. 1996 Edition.

The Vancouver Sun 'Asia Pacific Report', June 1, 1990.

To all Men, The Story of the Canadian Red Cross, by McKenzie Porter, 1960.

Van Rensburg, W.C.J. (1965) Copper Mineralization in the Carbonate Members and Phoscorite, Phalaborwa, South Africa, Ph.D. Thesis, University of Wisconsin, 1965.

Western Deep Levels Limited, 11th Annual Report, 1968.

Western Deep Levels general information booklet, undated.

World Mining (1968), Why Rustenburg means Platinum, August, 1968.

Index of Proper Names

"Harry," 26-7
"John and Jane," 218-20
Aarea, Tony, 76
Abercrombie, Lt. W.R., 297
Abraham, Brian, 462
Adam, Gordon, 31
Aho, Aaro, 74, 263, 265-6
Ainsworth, Ben, 308
Aird, Charlie, 76
Aitken, Marci, 450
Alexander, Albert, 48
Alldrick, Dani, 329, 332, 335, 337
Allen, Jenny, 433
Allen, Lt. Henry T., 297
Alvarez, Antonio, 196-9
Alvarez, George, 350
Alvidiez, Octavio, 197
Anderson, Jack, 368
Anderson, Joan & John, 398
Anderson, John, 49-53, 55-6, 73-4, 79-85, 111, 113, 125-6, 129, **236**, **246**, 263, 272, 276, 280, 370
Anderson, Nancy, 52
Anderson, Samantha, 398, 412
Angelo, Mark, 287
Angold, Dr., 275
Angus, 20
Anti-Rose, Michael, 5
Antunez, 196
Apperloo, Gerrit, 320
Armour, John, 19, 32, 36
Atwater, M.M., 123-4
Aunt Isa, 9
Austin, Alan (Mac), 26, 28-9
Austin, Allan, 26
Austin, Bill, 26
Austin, Chuck, 26
Austin, Elizabeth (Solsberg), 32
Austin, Jack, 26
Austin, Jim, 26
Austin, John, 26, 28-9, 32-3
Austin, Mac, 32-3

Austin, Mac, 33
Austin, Mr., 26-7
Austin, Richard, 26, 33
Babcock, Russ, 136
Badley, Adam, 20
Baker, Art, 186, 188, 191, 195-7, 199, 215, 281
Baldridge, B., 431
Banas, Ed, 94, 97, 100
Bangs, Richard, 302
Banks, Bertie, 426
Bannister, Roger, 38, 57, 431, 460
Barakso, John, 111, 125
Barnes, Ivy, 92
Barnes, Michael, 462
Barnett, Denis, 218-20
Barr, David, 214, **235-9**, **246-54**
Barr, David Anderson, 5, 24
Barr, David Walter Stuart, 39, 92, **240**, 425, 460
Barr, James, 5
Barr, Nancy, 54
Barr, Rene, 11, 15, 19, 38, 50, 58, 66, 70, 92, 105-6, 108, 110, 117, 119, 126, 133-5, 205, **237**, **240**, **250**, **254**, 264, 272, 282, 307, 378-91, 395, 398, 402-3, 410, 414, 417, 419, 431, 436, 438, 455, 457-8, 460
Barr, Robin McLellan (Rob), 145, 155, 214, **240**, **250**, 302, 307, 378, 386, 392, 395, 425, 442, 458, 460
Barr, Susan Irene, 54, **240**, **252**, 302, 379, 385, 392, 425, 428-31, 442, 445, 451, 455, 457, 460
Barron, Mr., 10
Barrow, Barbara, 15
Barr (Willis), Edyth Maude ("Nina"), 5, 6, 8, 10
Bartlett, Fara, 14
Bauer, Jacques, 169, 174
Baycroft, Charlie, 279-80
Beal, Harold Jr., 453
Beatty, Harry, 14

Bedford, Doug, 41–2
Bell, Dr. Robert, 221
Bell, Malcolm, 203
Belmas, Fred, 111
Bendicksen, Cariboo John, 48, 51
Bennett, W.A.C., 133
Berglund, Axel, 63, 65, 67–8, 71
Betts, Vern, 320–2, 327, 332, 334
Beveridge, Bill, 209
Beveridge, Bonnie, 209
Bibb, Leon, 307–8
Biddell, Jim, 20, 36
Biddell, Jimmy, 16, 426
Biggar, Jim, 23–5, 31
Billy, Allen, 372–6
Bird, Alfred, 339
Bisset, Kathy, 320
Black, Dr. Phil, 143
Black, Robert, 37
Blackader, Ealt, 301
Blann, Brian, 379
Blann, Chris, 379
Bloom, Harold, 469
Blusson, Ross, 79, 89, 91–2, 101, 113, 125, **246**, 467
Blusson, Stuart, 232
Borgheutz, C.J., 161
Bottomer, Lindsay, 325–8
Bourquin, Jim, 326, 328, 334–5
Bowles, Dick and Jessie, 152
Bradfield, Bob, **236**
Bradshaw, Peter L., 231
Brady, Blake, 94
Bragg, Don, 72, 79, 308
Brasher, Chris, 431–2
Bremner, Pete, 20, **235**
Bridcut, Stan, 298
Britton, Jim, 333
Britton, Tom, 219–20
Brock, John S., 263, 265, 311
Brodell, Hugo, 76
Brown, Atholl Sutherland, 282
Bruland, Tor, 308

Brummer, Joe, 96, 137, 262
Brundtland, Gro Harlem, 289
Bucholz, William, 93, 98
Bucholz, William
Buencher, Ydo, 452
Bull, Hon. Mr. Justice, 144, 154
Burden, Nancy, 450
Burgess, Charles H., 128
Burgess, Harry, 189
Burns, Ann, 308
Burns, Robert, 308
Burns, Ross, 308
Burrows, John W. "Bunt," 26, 28–9, 33
Bush, Jack, **246**
Butchett, Cliff, 163
Butler, Ed, 184–5
Callbreath, Ed, 106, 108
Callison, Pat, 273
Cameron, Bob, 275
Campbell, Dick, 72, 74
Campbell, Gordon, 327, 336
Campbell, Kayce, 55
Candido, Larry M., 144–7, 149, 152, 154
Cannon, Walter, 63, 68, 71
Careless, Ric, 303–4, 307
Cargill, George, 115–6
Carlson, Charles, 14
Carlson, Dr. Gerald ("Gerry"), 87, 231, 308
Carpenter, Ted, 94
Carpenter, W.S., 210
Carson, Mr., 23
Carstensen, Gary, 113–4, 368–9
Carter, Nick, 346, 350, 352–4
Cashore, Hon. John, 291, 313
Cathro, Bob, 212, 281
Cecire, Dominic, 94
Chandler, Terry, 299
Chaplin, Bob, 262–8
Chappell, Doug, 113, 261, 363, 365
Charteris, Stan, 299
Chiang Kai-Shek, Generalissimo, 406
Chisholm, Dunc, 114

Chislett, Albert, 183–4
Chramosta, Denny & Jim, 437
Chramosta, Linda, 437
Christian, Wilf, 65, 72, 79, 85, 88, **236**
Clark, Bill, 307
Clark, Jack, 42, 46
Clark, Lewis J., 91
Clark, Rusty, 41
Clarke, Robert Cecil, 206–7, 210
Classey, Buzz, 30
Clay, Pat, 113, 119
Clutten, John, 162
Coates, Donald W. "Doc," 43, 116–7, 255–62
Coates, Mr., Q.C., 350, 354
Coburn, Don, 19
Cole, Dick "Cozy," 26
Coles, John, 160
Cook, Capt. James, 339
Cook, E.H., 201
Cook, Pat, 65, 72
Cooke, Frank, 52, 262, 266–9
Cooper, Andy, 45
Cooper, John, 14
Copeland, Jake, 72
Copeland, Mrs., 72
Cox, Fred, 440, 451–3
Crane, Malcolm, 398, 400, 407
Cripps, Milo, 14
Cripps, Sir Stafford, 14
Crompton, Fred, 37–8
Crowhurst, Jack, 59
Cruise, Bill, 277
Cunningham, Jacquie, 324
Cunningham, Mitch, 324
Cuttrell, Becky and Herbert, 6, 12
Cuttrell, Jerry, 12
Cuttrell, Patricia, 12
Dallaire, Réjean, 185
300 Dalton, Jack, 300
Dalziel, George, 52, 66–7, 71, 276
Daniel, Rob, 437–8
Darke, Ken, 65

Davey, Norton, 431
Davidson, "Skook," 48, 51–2, 63, 73, 268–9
Davidson, Don, 217–8, 220
Davies, Gordon, 85, 88, 212–3, **236**
Davies, Gordon, Jr., 79
Davis, R.E. Gordon, 94, 96–7, 113, 124, 134, 137, 263–5, 467
Davis, Hon. Jack, 347
Dawson, Ed, 162
Dawson, Grace, 60
Dawson, Wendell, 58–62, 366
Day, Norman, 323
de Wet, David, 176
Dean, Bill, 56, 280
Dean, Sandy, 55–7, **236, 238**
Delane, Gerry, 308
DeLeen, John, 276
Demmler, Egon, 450
Denver, John, 307
Devenny (Hammond), Sybil, 50, 52, 280
Devon, Howie, 45
Dewaal, Peter, 164
Dewar, Doug, 105–6
Dickson, Gordon, 82
Diefenbaker, John, 151
Dirom, Gavin, 289, 326, 330, 334, 336–7
Donaldson, Herb, 115
Douglass, William C., 296
Doull, George, 37–8, **235**
Downing, Bruce, 307–8
Downing, Bruce
Drinkwater, Joe, 339
Drown, Tom, 191, 213
Drury, Bill, 79
Drury, Leith, 450
du Pont de Nemours, Eleuthère Irénée, 186
Dumont, Bill, 307–8
Dunlop, Jack, 308
Dunlop, Wendy, 308
Dunn, Lyle, 76
Dupas, Jacques, 206
Duval, David, 202

Duval, W.E., 40
Eazor, Robert, 450
Eby, Steven, 371, 373–7
Eccles, Louise, 78, 193, 210
Edmonds, Carl, **249**
Edwards, Hon. Anne, 291, 313
Edwards, Hon. Anne
Edwards, Phil, 37
Elaine, 414, 418–9
Ellis, Mike, 457, **253**
Elwyn, Bruce, 110
Emdick, Dr. Andrew, 301
Emerson, K.C., 443, 445
Emslie, Ron, 185
Enermark, Gordon, 332–3
Evans, Harvey, 114
Fahrig, Robert B., 43–4, 255–9
Faille, Albert, 69, 466
Farquharson, Graham, 195
Farrow, Moira, 52
Ferguson, Richie, 38
Ferris, A. Boyd, Q.C., 147–8
Ferris, Glenda, 320, 326, 334–5
Fettis, John, 437–8
Field, Cyrus W., 96–7, 125
Fields, Mark, 132, 169, 231
Finlayson, Eric, 169, 184
Finley, Jim, 183
Fipke, Charles (Chuck), 168, 232
Fisher, Edie, 450, 453
Fitzgerald, James, 16
Fiva, Gunn, 103
Fleming, Hal, 114, 157
Fletcher, Dave, 65, 68
Follows, Dr., 385, 421–2, 459
Formholz, Amy, 60
Formholz, Bill, 59
Fortescue, John, 96, 137, 469
Fournier, Mike, 337
Fowler, Howard, 58, 366–368
Fox, Peter, **249**
Fox, Rosemary, 288, 320
Fralick, Pam, 447

Fraser, Dan & "Ma," 49
Fraser, Hon. Russ, 347
Freberg, Ray, 133, 135,
Freeman, Ed, 65, 68
Freeman, Pete, 162
Freeman, Theresa (Terry), 162
Freer, Willard, 51, 73–4
Freer, William, 48
French, Charlie, 457
Friend, Johnny, 66
Gableman, Colin, 313
Gabrielse, Hugh, 52
Gale, Robert, 264
Gale, Stuart, 327, 332
Gardiner, Shari, 327
Gaunt, Jim, 374
Gawinski, Tom, 38
Gayer, Robert B., 343–4
Gibson, Grant, 98
Gifford, Bob, 55–6
Gill, John, 79
Gillies, Fred, 273
Gillies, Marion, 273
Gilroy, Bob, 272, 363–5
Glanville, Ross, 353
Glave, Edward, 300
Goddard, Harley, 79, 89, 91–2, 94, 467
Goldie, Ray, 180
Golem, Elmer,45–6
Golem, Terry, 46
Goodlad, Phyllis, 454
Gordon, Stu, 427
Gorgenyi, Bela, 79, 87, 94, 96, 126–7, **246**, 467
Gormi, Stan, 276
Gower, Jack, 117, 145, 157
Grady, Gordon, 274
Granger, Paul, 429, 431
Green, Dunc, 37–8
Greenaway, Gwen, 54
Greenaway, John, 49, 53–4, 65, 67–8
Groat, Buster, 48, 63–5, 67
Gross, W.H. (Bill), 186, 196

Guay, Albert, 128
Guiler, Bill, 136
Gullison, Ray, 79
Gunn, Christopher B., 191, 195, 201–2, 205–6, 210,
Gunn, Ian and Erica, 209
Gunn, Nina, 209
Hachey, Oz, 276
Hale, Denis, 79
Hall, Simian, 45
Hamer, Edmund, 21
Hamilton, Cynthia, 205
Hammond (Deveney), Sybil, 50, 52, 280
Hammond, Paul, 49–52, 280
Hanman, Ted, 347
Hanman, Ted, 350, 356–7
Hansuld, John, 229
Harbotle, Bud, 82
Harcourt, Premier Mike, 313
Hardin, Michael, 288
Harp, D.G., 333
Harris, Jack, 41–2
Harris, John, 21
Harrison, "Wild Bill," 79, 134
Harrison, David, 437, 442
Harron, Gerald, 191, 195, 199, 227–8,
Hart, Lillian, 344
Harter, Alvin, 200
Hartley, Tim, 453
Hatelt, Lorene, 457
Havard, Eric, 327
Hawkes, Herbert, 469
Hawkins, Brent, **250**, 379
Haynes, B.J., 43
Heard, R. Terry, 346, 351–4
Heddle, Dunc, 129
Henry, Eric, 395, 397
Henry, Lyla, 395–6
Henshall, "Hammering Hank," 427–8
Hepler, Meade, 298
Herchmer, Doug, 285–8
Hermitson, Mr., 259
Hermon, Dick, 192

Hernandez, Mr., 198
Hess, Merv, 111
Hetherington, Robert, 39
Hewitt, Bill, 20, 426
Hewitt, Foster, 20
Hewitt, John, 72
Hickel, Governor, 306
Hill, William, 195, 198
Hillhouse, Neil, 20
Hoffman, Slim, 45
Hoffman, Stan, 308
Holland, Terry, 345–6
Holmgren, Lisa, 60
Hope, Diana, 73
Hope, Glen, 73
Hope, Glenda, 73
Hope, Mrs., 73
Hope, Ralph, 73
House, William, 55
Houston, Bill, 42
Howard, David, 155
Hoy, Edward, 450
Huestis, Spud, 262, 266–8
Hughes, E.R., 128
Hughes, Hughie, 114, 270–1
Hughes, John, 457
Hugo, 87
Hume, Stephen, 306–8
Hunderi, Jake, 65, 67–8, 72
Hutchinson, Julien, 444
Hutchison, Bob, 94
Hyslop, Jim, 438
Israelson, Arnie, 301
Jackes, Don, 111–2, 114
Jacks, Mrs., 40
Jakobsen, Stefan, 440
Jalbert, Rudy, 79, 97
Jang, 419
Jasper, Blake, 450
Jaworski, Bert, 321
Jaworski, Joe, 398
Jeffery, Dr. W.G., 127
Jenkins, Dave, 308

Jezek, Roman, 452, 454
Johansson, Ingemar, 95
Johnson, Arthur W., 147, 154
Johnson, Lloyd, 443
Johnson, Mr., 149, 152
Johnston, Art, 114
Johnston, D.L., 222
Johnston, David, 288–9
Johnston, Sheila, 6–7
Johnstone, Bruce, 204
Johnstone, Jack, 204, 458
Jones, Ed, **236**
Jones, Keith, 191, 193
Jones, Larry, 337
Jones, Steve, 431
Jyrich, Warren, 420
Kaiserman, Dr. Don, 435–6
Kamm, Carl C., 469
Kasmer, Ed, 65, 72
Keen, Jim, 110
Keenen, Joe, 53
Kennedy, Robert Jr., 294
Kenney, Terry, 453
Kerr, Forrest A., 91
Ketchum, Philip, 26
Kidd, Dr. Des, 60–1
Kierans, Martin, 193, 213
Kiernan, Hon. W.K., 128
Kilby, Ward, 293
Killam, Cheryl, 398, 401
Killam, Laura, 398
Killam, Sisley, 398, 406, 412
Kimbrough, Vaughn, 450
Kimura, Ed, 155
King, Mr., 203
King, Steven, 440, 442–4
Kitagawa, Kit, 37
Kongode, Michael, 162
Koona Alk Sal, 300
Korenic, John, 191
Koseruba, Mike, 439, 440–1
Kowalchuk, John, 191, 321
Krauss, Gerhard, 453

Kremenchuks, Tony, **248**
Krever, Hon. Mr. Horace, 422–3
Kristiansen, Ingrid, 431
Kulan, Al, 263–6
Kvale, Einar, 60–1
Laanela, Hugo, 79, 87
Lackie, Bob, 102
Lampard, Ron, 350, 353
Landy, John, 38, 57
Lanella, Hugo, 126
Langford (Lindley), Helen, 53, 205
Langford, Walton, 205
Lank, Herbert, 193–4
Lannin, Leo, 264
Law, Terry, 213
Lawrence, Ed, 94, 101
Lawson, Nancy, 31
Lawton, Roland, 162
Leake, Dr. Charles, 209
Lechner, George, 79
Ledingham, Brian, 366
Leedy, Mr., 44
Lefebure, Dave, 329
Legatt, Sandy, 162
Legge, Bill, 55–6, 58
Legge, John, 366–7
Legge, Roland E. (Roly), 366–8
Leggett, S.R., 143, 147, 150–3, 155
Lehmann, Dr. Wolfgang, 201
Leonard, Al, 372
Leuty, Bill, 20
Lewandowski, Hilmar, 442
Lewis, Jim, 336–7
Liford, Phil and Joyce
Light, Denis, 114
Linardos, Jane, 384, 390
Linardos, Jim, 384
Lindahl, Gunar, 6
Linder, Harold, 133–4
Lindley, Case K., 66
Lindsay, Thayer, 32
Lister, Diane, 308
Lister, Gordon, 157

Little, Margaret, 273
Liu Changuing, 419
Livingstone, Gary, 328
Lockwood, Charlie, 42
Lonergan, Al, 205, 365–7
Longe, Robert, 308
Louie, Chief Louis, 323
Loutchan, Joe, 272
Lowery, Cheryl, 447–8, 450, 452
Lowes, Barry, **235**
Lubbock, Eric "Yah," 24–6, 30
Lund, Jim, 66–7
Lund, Rita, 66–7
MacDermot, Terry, 25
MacDonald, Gary, 347, 356
Mackenzie, Dave, 232
Mackenzie, Warren, 42
Mackillop, Al, 72, 83
Macklem, H., 43
Macklem, Michael, 31
MacLean, Alison, 209
MacLean, Andrea and Kirsten, 209
MacLean, Keith Alexander (Sandy), 193, 195, 202, 205–6, 210, 215
Maddison, Mr. Justice Harry C.B., 265
Maguire, Vic, 69
Mahoney, Ted, 307–8
Mair, Ken, 261
Malan, Dr. B.J., 163
Malaspina, 339
Maley, Len, 65
Mallet, Freddie, 21
Mallot, Mary Lou, 325
Maltz, Maxwell, 261
Manhard, Roy, 148
Mao Zedong, 416
Markham, R., 263
Marshall, Dave, 116
Marshall, Jack, 65, 72
Martin, Denis, 353
Martin, Joe, 41
Martindale, Graeme, 437
Mason, Peter, 224

Mathews, Lt.–Col., 123
Matic, Frank, 125–6
Matic, Otto, 126
Matney, Harvey, 203
Mawer, Bruce, 365
Maxwell, Walter Bernard (Bert), 31–2, 35, 460
May, Corporal, 220
McAuslan, David, 134, 180, 182–4, 191, 214, **247**, **250**, 379, 392, 469
McAusland, Jim, 79, 94, 96–7, 467
McCatty, Winston, 21–2
McCreary, Gordon, 191, 213
McCutcheon, Archie, 200
McDermot, Mr., 13
McDonagh, Mike, 79, 86, 88, 127
McDonald, Gary, 353
McDonald, Rod ("Bud"), 116–7
McDougall, B.W.W., 352
McDougall, Jim, 298–9
McEwen, Donald, 229, 231
McEwen, Robert, 231–3
McFarland, Harvey J. (Mickey), 16, 426
McFarlane, Mr. Justice, 154
McGivern, Hugh J., Q.C., 147–9, 152
McIntosh, Robert W., 440
McIntyre, Mr. Justice, 154
McKamey, Ray, 73
McKay, Tom, 348
McKenzie, L.M. "Butch," 22, 25
McKinnon, Don, **246**, 370–1
McKnight, Bruce, **249**, 312, 337
McLaren, Graeme, 325, 332–3, 338, 352
McLaren, Kenneth, 78
McLean, Norm, 334
McLellan, Ann, 53
McLellan, Bob, 52, 378
McLellan (Barr), Margaret Irene (Rene), 52–54
McLellan, Rod, 427
Mclennan, Rod, 36
McLeod, Bill, 114, 270–1
McLeod, Dave, 55
McLeod, Frank, 63

McLeod, Willie, 63
McMullen, George, 37, **235**
McNeil, Clifford ("Cap"), 148
McNeil, Don, 148
McOnie, Al, 183
McQuillan, Tom, 60-1, 203-5, 367
Medland, Brian, 371, 374
Medland, Julie, 371, 374
Melissen, Mel, 219
Menard, Martin, 130
Merensky, Dr. Hans, 164, 176
Merrill, Larry, 40, 42
Messenger, Robin, 169
Messmer, Hon. Ivan, 347
Midge, 6
Mikes, John, 302, 379
Mikes, Johnny, 302, 307
Milbourne, Gordon, 64, 280
Millar, Len, 72
Miller, Dan, 313, 333
Miller, Dave, 134
Miller, Fritz, 78
Miller, Hon. Dan, 291, 325-6, 336
Milliken, Frank R., 128-9
Minaar, Col, 175
Mitchell, "Mitch," 45
Mitchell, Lynn, 384, 389, 391
Moffat, Denis, 350
Molyneux, Tom, 158-9, 163-4, 174-5
Montgomerie, Ann, 5
Montgomery, Carol, 440, 443
Moon, Lowell B., 128, 138, 158, 188-90
Moore, Neely, 129
Moore, Peter, 114
Moreno, Jarier, 196
Morrison, Murray, 93, 98
Morrow, Patrick, 385
Mortifee, Ann, 307-8
Mouat, Robert, 350, 353-5, 357
Muir, Bill, 213
Muirhead, Mary, 5
Muirhead, William, 5
Mullqueen, Duke, 53-4

Munday, Bert, 55
Munday, Don, 55
Murkowski, Frank, 111
Murphy, Jim, 102, 114, 125, 467
Murphy, Marlin, 321-2
Murray, Peter, 164
Muscroft, Joe & Mary, 54
Myllyla, Eino, 123
Nagle, Ed, 221-2
Nash, Ken, 450, 453
Nathan, Mr., 165
Nelson, Ned, 40
Neufeld, Richard, 336
Nevin, Ken, 37, 53
Ney, Charles, 55-6, 62, 79-85, 140-1, 145, 147, **239**, **247**, 272, 276, 282, 362, 369-70
Niblock, Pete, **235**
Nielsen, Bud, 215
Nikolai, Chief, 297
Nimsick, Hon. Leo, 281
Nino, 6
Noel, Gerry, 53, 63-4, 71, 280, 466
Norman, George, 38
Norquist, Bill, 353
North, Mr., 163, 175-6
Northcote, Ken, 129
Nuppunen, John, 125-6, 130, **246**
Nussbaumer, Margaret, 209
Nussbaumer, Peter, 209
Nussbaumer, Ruth, 195, 206, 209-10
O'Shaughnessy, M., 129
Ogilvie, Cam, 76
Ogryzlo, Larry, 129
Okabe, Curly, **236**
Olivier, Wilhelm, 163
Opperman, K., 162
Optiz, Matias, 444
Ostashek, John, 306
Ostergard, Peter, 333
Owen, Stephen, 291, 312-3, 315
Pain, John, 458
Pakula, Dan, 323

Palejko, Gillian, 452
Palmer, Baird, 79, 88
Palmer, Walt, 450
Paquet, Maggie, 310-1, 323
Parker, Len, 55
Parkinson, Denny & Beth, 397
Parnell, J.M., 150, 152
Parnell, Jack, 141-2, 147
Patmore, Dr. William Henry, 136, 143-4, 146-51, 153
Patterson, Barry, 451
Patterson, Bob, 79, 85
Patterson, Floyd, 95
Patterson, Jack, 294, 309
Patterson, John W., 258
Patterson, Raymond, 69-70
Paul, Susan, 452
Pawikowski, J., 114
Paxton, Jim, 213
Peach, Barney, 46, 279
Pearce, Bill, 346-7, 350, 353-4, 356-7
Pearce, William, 338
Pearson, "Rocky," 111
Peaver, Jim, 361
Pegg, Dan, 308
Pemba, 384, 390-1
Perkins, Doug, 308
Persson, Eric, **236**
Peters, Donna, 431
Peters, Jim, 38, 57-8
Peterson, Doris ("Susie"), 80, 273
Peterson, Ed, 296
Peterson, Herman, 79-80, 82-3, 85, 89, 100-1, **239**, 270, 272-8, 365
Peterson, Mr., 162
Petrunia, "Buttons," 45
Pfeiffer, Harold, 466
Phee, Earl, 38
Philip, Gord "Slim," 45
Philipp, Al, 228
Phillips, Geoff, 320
Phillips, Hec, 21, 37-8
Pieper, Paul, 58-9, 61-2

Pierce, George, 429, 431
Pieterse, J.J., 175
Poce, Paul, 38
Poffenroth, Holly, 447-8, 450, 452
Poffenroth, Paul, 443, 447-50, 452
Polkosnik, Frank, 120
Pollock, Gerry, 183
Poole, Don, 101-2, 114, 467
Pope, Mabel, 9-10
Porter, John, 72-4
Portz, Max, 272, 363-5
Potts, Maj. Gen. Arthur, 22
Preston, Dave, 37
Pringle, Geoff, 31
Prinsloo, Joachim, 166
Puntous, Patricia, 429, 431
Puntous, Sylviane, 429, 431
Puusaari, Annikki, 191, 209
Quigley, Steve, 327
Rache, Ed, 128-9
Rae, Bob, 275
Ramsey, Walter, 103
Randall, Alf, 224-6
Randall, Marie, 224
Ratel, John and Maris, 209
Rayner, Gerry, 94, 96, 99-101, 113, 115-6, 125, 360-2, 467
Read, Gordon, 374
Reed, Pop, 45
Rennie, 20
Rennie, Cliff, 308
Retty, J.A., 157
Reynolds, Pat, 268
Rice, David L., 145, 147, 154
Rich, Jack, 65, 70
Richardson, Alan, 27
Richardson, Bob, 225
Riordan, Dr., 274
Rioux, Jean-Sebasti, 444
Ritchie, Al, Jr., 106, 108-9
Ritchie, Al, Sr., 106, 110
Roadhouse, Don, 127
Roberts, Wayne, 311

Robertson, Gordon, 221
Robinson, Dick, 431
Robson, Angus, 384
Robson, Setsuko, 384, 390
Rolfe, Mr., 147
Rolls, John, 265
Roots, E.F. (Fred), 63–4
Roscoe, Bob, 213, 216–7, **248**
Ross, Browning, 38
Ross, Don, 59
Rowlands, Peter, 288, 310, 323
Rudd, David, 431
Rudiger, Gerry, 45
Russell, Bob, 79
Rutherford, Jim, 68, 71–2
Ryan, Bruce, 184
Ryback-Hardy, Vic, 155
Sadler, Dr. J.A., 164
Salfinger, Rod, 332
Salmond, Cliff, 21
Samchuk, George, 45
Sampras, Pete, 129
Sampson, Bill, 286, 288, 320, 323
Sandburg, Gene, 45
Sanft, Godfrey Frederick, 136–7, 139, 141–4, 146–8, 153
Sargent, Dr. Hartley, 128, 342–3, 351
Schedler, George, 44
Scheepers, H.J., 164–5
Schmidt, Oscar, 65–6, 69–70
Schmidt, Rolf, 333
Schoeman, Dr. J.J., 164
Schroeter, Tom, 282, 337
Schuster, Rudi, 450
Schweitzer, Bill, 450
Schwindt, 349
Scott, Dave, 429
Scott, Jean, 58
Scott, Jim, 56, 58, 68, 71, 73
Scott, Sue, 447–8, 450
Sebastian, Rick, 183
Seel, George, 137, 139, 141–2
Sevensma, Peter, 129

Sewell, E.J.B., 165
Shanahan, Don, 457
Shaw, Ian Ross, 195, 206, 210
Shaw, Rosemarie, 209
Shaw, Tom, 195, 206, 209
Shearer, I.K., 9
Shearer, Mr., 30
Sherman, Paddy, 55, 57
Sherwood, Wally, 339, 342, 344
Shouk, 300
Simmons, George, 273
Sinclair, Alastair, 231, 345–6
Singh, Arun, 381
Skoda, Ed, 354
Smillie, Al, 65, 67–8, 70
Smith, "Preacher," 102–3
Smith, "Smitty," 45
Smith, Colin, 337
Smith, Donald Alexander (Lord Strathcona), 339
Smith, Jock, 169
Smith, Marshall, 78, 191, 193, 199, 210
Smith, Paul, 111
Smitheringale, Bill, 55, 58–9, 61–2, 205–6, 210, **238**
Smyth, Ron, 305
Smyth, Sam K., 469
Snyder, Hugh, 223–4
Soehl, Tom, 320, 322, 325, 329, 332–4
Solsberg, Elizabeth (née Austin), 32
Soregaroli, Art, 223–4, 282
Soutar, Glen, 114
Souther, Dr. Jack, 87–8, 316
Souza, Ken, 429
Souza, Ken, 431
Spencer, Bruce, 223–4, 227
Spencer, John, 18
Spencer, Victor, 18
Spilsbury, Jim, 344
Springer, Karl, 60–1
Steenkamp, Nick, 164
Steiner, Mr., 149
Stephen, Alan, 13, 23

Stephenson, John, 192
Stevens, John, 462, 466
Stevenson, Bob, 108, 139, 213–4, 231, 285
Stevenson, Scotty, 42
Stewart, Bill, 20
Stewart, George, 140, 142, 147–8
Stewart, Jim, 41–2
Stokes, Ron, 281
Stokotelny, Mike, **253**, 444, 456–7
Stone, Curtis, 37
Storey, Alan, 435
Strachan, Hon. Bruce, 341
Stradiotti, Alda, 105–6
Stradiotti, Napoleon, 105–6
Stradiotti, Ricardo, 105–6
Stratmoen, Louis, 130
Street, Barrie, 442
Stump, Mary, 436
Sulkowski, Ed, 162
Sullivan, John, 63, 68, 71, 116, 124, 128, 157–8, 461
Sumanik, Ken, 294, 337, 374
Sybesma, Thea, 429, 431
Tashoots, Chief Yvonne, 320
Tattersall, Mrs. Merna, 344
Taunton, Dr. Jack, 459
Taylor, Maria, 428
Tegart, Peter, 220
Tener, Mr., 341, 345
Tennant, A., 169
Terry, David, **249**
Thibault, Norm, 451
Thompson, Bob, 262
Thompson, Derek, 333–4
Thompson, Ed, 231
Thompson, J.W., 73
Tilson, Herb, 37
Tingley, Clarence, 276
Tinley, Scott, 429, 431, 458
Todd, David, 20
Todd, Ian, 372–3
Tough, Sherman, 40, 42

Townsen, Alf, 42
Townsend, Paul, 434
Tracey, Percival White, 166
Turbitt, Barney, 296
Turner, Alan, 169
Turner, Dick, 94
Tustin, John, 140, 142–3, 145–8
Ure, Bob, 42
Vallee, Marcel, 231, 345
van Blommestein, Peter, 165
van Kralinger, Win, 164–5
Van Lionden, Mr., 169
Van Tassell, Dutch, 232–3
Vance, Bill, 160, 162
Vander Zalm, Hon. William, 89, 107, 284
Veasey, Dr. H.B., 122
Verbiski, Chris, 183–4
Versosa, Raul, 321
von Gruenewaldt, Gero, 163–4
von Gruenewaldt, Judy, 164
Wager, Franz, 169
Wagner, Fred, 181
Waite, Terry, 433
Walker, Alistair, 162
Walker, Greg, 114, 271
Wallace, Bob, 176–8
Wallace, Emerson, 368
Wang, Miss, 413–4
Ward, James, 452
Ward, Syd, 275
Warren, Dr. Harry, 188
Waters, Dr. Arnold E., 159, 165
Waters, Dr., Jr., 169
Watson, 20
Watson, Frank, 65
Weatherhead, Bruce "Casanova," 45
Webb, Henry, 180–1
Webster, Geddes, 299–300, 302
Webster, George, **235**
Webster, Tom, 180, 182–3
Weese, Stan, 398, 400, 407
Weir, Charlie, 94–7, 100–1, 467
Werner, Joe, 55

West, Charlie, 114
Westervelt, Ed, 231
Westervelt, Ralph, 345–6
Wheeler, Mr., 19
White, Arthur, 232
White, Barbara, 398
White, Bill, 262
White, Howard, 344
White, Len, 76
White, Paul, 264
Whitlock, Jack, **250**, 379, 392
Whitman, Gordon, 206
Wickmeyer, Jim, 384
Wiegand, W.B., 23
Wiessner, Fritz, 55
Wigglesworth, Mr., 89
Wilkinson, Bill, 298
Williams, Doug, 43–4, 255
Williams, E.L., 193
Williams, Hon. Chief Justice, 189
Williams, Keith, 94, 102, 114
Williamson, Ray, 72
Williamson, Ron, 45
Willis, Harry, 6
Willis, Laurie, 46
Wilson, Gordon, 37
Wilson, Jamie, 456
Wiseman, Eric, 16–7, 23
Wittman, Gord, 281
Wodjak, Paul, 316, 335
Wolfe, Bill, 132
Wood, Brian, 322
Wood, J.E.R., 148
Wood, Kathy, 448
Wood, Mark, 448
Woodcock, Dick, 78, 86–7, 92, 103, 127, 358
Woods, J.D., 24–5
Wortman, Dave, 353, 356
Wright (Nolet), Irene, 280
Wright, Fluor Daniel, 202
Wright, S.W., 35
Wright, Shirley, 53
Wyatt, Joan, 53–4
Yeager, Dave, 325
Young, Dean C.R., 34
Young, Reg, 261
Yourt, G.R., 40
Zeemel, Albert, 42
Zenger, Ned, 111
Zhou Enlai, 416
Zumba, 384, 388
Zweck, Elizabeth, 288

End Notes

[i] Although laborious, the procedure for map preparation is relatively simple. Adjacent photos on each flight line within the area, called stereo pairs, are used for reference. As the centre of each photo lies vertically below the camera on the aircraft, it is plottable without correction. The radial line plot is completed on film overlays on the stereo pairs oriented according to their common photo centres, which appear on both photos. Common features of interest on the stereo pairs, e.g. river intersections, mountain tops, geologic formations, are then plotted by resection from the common centres on the stereo pairs, i.e. the intersection of lines from the centres of the photos to the feature.

[ii] Available geologic information within the western part of the proposed exploration area and that obtained during the 1954 exploration program under the direction of Gerry indicated that copper-lead-zinc mineral occurrences in the region were related to limestone-chert sequences in Paleozoic strata marginal to stocks and batholiths of granitic composition, which were themselves not particularly favourable foci for sulfide mineral concentrations.

The study permitted the recognition of granitic areas through characteristic features, namely higher summits, precipitous nature of high relief, serrated ridges, dendritic pattern of gully and freshet heads, radial drainage and uniformity of texture patterns. This was subjected to reconnaissance observations in the early part of the program which I completed as helicopter time became available. I was also able to provide data on trends of the sedimentary facies for preliminary guidance of our prospectors.

[iii] My report continued:

"Prospectors have been landed individually at several locations where ridges are clear of snow and travel, although limited, is possible for several miles. Several small lenses of pyrrhotite and chalcopyrite were discovered in limestone and porphyry adjacent to a granite contact at a point 12 miles northerly from the north end of Hyland Lake. The mineralization is encouraging but not important except as an indication of possible extensions.

"Aerial reconnaissance to date indicates that a much larger proportion of the area surrounding Hyland Lake is underlain by granitic rocks than was previously believed. Much of the Hyland Lake segment will therefore be eliminated as a favourable prospecting area."

[iv] Only small scattered mineral occurrences consisting of pyrrhotite, pyrite, arsenopyrite and chalcopyrite with minor magnetite were discovered, none of which warranted staking.

[v] Although located in a similar geologic environment to the Axel showing, the Buster showing contained predominantly lead-zinc-copper mineralization, again localized in argillite and limestone near a granitic contact

[vi] "Albert Faille was a legendary trapper in the Nahanni area. In the 1920s, he heard the legend of the McLeod brothers' lost gold mine in the South Nahanni River valley. Every spring he would load up all his supplies and follow the Liard River from the Mackenzie to Nahanni Butte and enter the South Nahanni. He would beach below Virginia Falls, completely dismantle his boat and portage all his gear and every plank of his boat to the river above the falls, a heroic job in itself, as the falls have a 316 foot drop. Faille worked as engineer on the *M.V. JP Murphy* from 1944 until 1950. Then he went back to his prospecting on the Nahanni, returning every year until 1967. He never found that rich gold strike but he found many friends for he was a very generous and thoughtful fellow and everyone who knew him thought the world of him." From *The Man Who makes Heads With His Hands: The Art & Life of Harold Pfeiffer, Sculptor* by John A. Stevens and Harold Pfeiffer, 1997 General Store Publishing House.

[vii] Gerry Noel succinctly described the setting of the mineral occurrences in his 1954 report as follows:

> "At location (3), pyrrhotite, pyrite, and chalcopyrite have massively replaced limestone beds over a thickness of 120 feet and along a strike length of 1000 feet. In section, the replacement is lenticular and the thick central portion would only have a length of 300-500 feet. The mineralization is exposed in the crest of a fold structure involving the well-bedded limestone sequence and can be traced down-dip to the southwest for about 100 feet to where it passes under drift and overlying barren beds. The individual limestone beds have been selectively replaced with the most massive replacement in the upper thirty feet. The characteristic lime silicate alteration is mainly developed below, and in the lower part of, the sulfide body. A representative sample of the massive sulfides, which are remarkably uniform, assayed 0.60% copper, and trace gold".

[viii] The Fitzob showing occurs in a similar geologic setting to the Axel deposit, 12 miles to the northwest although the intensity of alteration and mineralization is much less. A thick sequence of thin-bedded chert and silicified limestone is anticlinally folded over a central granitic core and mineralized over a rusty area of about 1000 by 2500 feet. Mineralization consists of disseminations and fracture replacements of pyrrhotite, pyrite, chalcopyrite, arsenopyrite and tetrahedrite over a stratigraphic thickness of 300 feet. The showings occur at an elevation of about 6500 feet on a steep mountain slope.

[ix] Encouragement to pursue this work was based on reports of favourable mineralization occurring in the McEvoy Lake area within the region and a previous recommendation for prospecting along the northeast margin of the Anvil batholith made by Gerry Noel. It was also hoped that work in Paleozoic rocks at some distance from large batholiths, such as those encountered to the southeast in 1955 would result in the discovery of large tonnage bedded replacement deposits rather than contact metamorphic types which tended to be smaller and more erratically mineralized.

[x] During 1960 in portions of July to September, our crew, consisting of Gerry Rayner, Ross Blusson, Gord Davis, Harley Goddard, Bela Gorgenyi, Jim McAusland and myself, completed 29 square miles of reconnaissance geological mapping, one square mile of detailed mapping, 4982 lineal feet of sampling, principally on six deposits, 2.5 line miles of dip-needle surveys and detailed stream sediment sampling at 47 sites in addition to the claim staking. Apart from Charlie Weir, we were supported by Jim Murphy and finally Don Poole as helicopter pilots. The cost of this program was an unbelievable $20,000!

[xi] These have included stream sediment, soil and rock geochemical methods, detailed mapping to define petrologic and alteration trends and a variety of geophysical methods, principally magnetics and induced polarization surveys. Anomalies have been explored by trenching prior to embarking on the more costly drilling programs, be they diamond, percussion or rotary, to obtain acceptable subsurface indications of extent and grade of deposits, prior to considering underground exploration.

[xii] 1961 Galore Creek Estimate Report:

Deposit	Average Grade %	Mineralized Area sq. ft.	Remarks
HB	1.0	200,000	Mineralization in three Areas separated by drift covered or low-grade sections.
Junction	1.5	150,000	Includes extension to DDH No. 1
Butte	2.0	45,000	
Dendritic Creek Breccia	2.0	15,000	
N. Junction	1.0	25,000	
South 110 Creek	1.0	50,000	
Saddle	1.0	20,000	
South Butte	2.0	15,000	
Total	1.0 - 2.0	520,000	

Estimates show that the mineralized deposits indicated above constitute two percent of the outcrop area of the Galore Creek Complex, which in itself represents ten percent of the ten square mile area believed to be underlain by potentially favourable Complex units.

[xiii] 1964 Galore Creek Estimate Report:

"The 1964 report focused on the potential of the Central Zone, rather than the total including the many satellite deposits. It noted the following conclusions:

Ore (tons)	Avg. Grade % Cu	Cutoff % Cu	Waste Rock (tons)	Till (tons)	Stripping Ratio Rock	Rock & Till
95,000,000	1.32	0.6	349,560,000	80,600,000	3.7 : 1	4.5 : 1
69,500,000	1.42	0.6	232,650,000	59,300,000	3.3 : 1	4.2 : 1
61,500,000	1.54	0.7	240,700,000	59,300,000	3.9 : 1	4.9 : 1

The following reserves are indicated and inferred from diamond drilling in higher grade near-surface portions of the Central Zone, and are recommended for early production purposes.

Ore (tons)	Avg. Grade % Cu	Cutoff % Cu	Waste Rock (tons)	Till (tons)	Stripping Ratio Rock	Rock&Till
6,000,000	1.75	1.0	13,750,000	5,700,000	2.3 : 1	3.3 : 1
9,600,000	1.98	1.0	36,500,000	16,000,000	3.8 : 1	5.5 : 1

A total of 61,500,000 tons of 1.54% copper have been indicated and inferred by drilling to an average elevation of 1600 feet (average depth of 700 feet below surface) within a strike length of 6500 feet. Drill intercepts of similar grade have been obtained at elevations varying between 1000-1500 feet. The possibility for the existence of additional tonnages of similar grade at depth are excellent. The northern extension of the Central Zone warrants considerable detailed drilling as a result of the excellent results obtained in DDH's 120 and 137. An additional 60 million tons of plus 1.0% copper would appear a realistic objective as possible tonnage to an elevation of 1000 feet within the Central Zone.

Additional tonnages indicated or inferred with 1.2% and 1.5% cutoffs on other deposits are as follows:

1.2% cutoff	Tons	Grade	Remarks
North Junction	9,337,000	1.82	To 500' depth
Junction	2,000,000	1.50	" " "
Southwest	700,000	2.11	" " "

1.5% cutoff	Tons	Grade	Remarks
North Junction	7,723,000	1.97	To 500' depth
Junction	2,000,000	1.50	" " "
Southwest	700,000	2.11	" " "

Several deposits, notably the West Fork Glacier, Butte, Saddle, South 110 Creek and West Rim, have been investigated and contain drill hole intercepts or surface assays of interest. Additional drilling will be required on these deposits before reserve estimates can be classed as indicated.

The distribution of deposits is such that the largest, the Central Zone, is situated in the lowest portion of the basin which exposes units of the Complex. Other deposits are situated at higher elevations, the Butte, Junction, North Junction and West Rim lying at elevations between 3000 - 4000 feet, or 2000 - 3000 feet above the lowest mineralized drill intercepts on the Central Zone. There is no evidence which would indicate that the mineralized structures containing the deposits would not persist to equivalent elevations at which mineralization has been encountered in the Central Zone

[xiv] All creeks draining this 750-square mile plutonic complex were sampled by a team lead by John Fortescue at half-mile intervals. The minus 80 mesh fraction was analyzed at a Vancouver laboratory for copper and zinc using methods developed by Harold Bloom and Herbert Hawkes.

[xv] The work done demonstrated that copper and molybdenum mineralization is associated with a circular stock of quartz monzonite porphyry of Eocene age about 2,000 feet in diameter, which intrudes volcanic and minor sedimentary rocks of Middle Jurassic age. Most of the important mineralization occurs within altered rocks surrounding the stock and extending for as much as 600 feet from its periphery.

The dominant type of mineralization is chalcopyrite and molybdenite; the latter showing a preference for the immediate vicinity of the stock and falling in grade from an average of about 0.10% molybdenite to 0.03% molybdenite within 600 feet of the contact. The lead-zinc-silver veins staked by the Lead Empire Syndicate in 1948 occur about 2,000 feet easterly from the eastern contact of the stock.

In 1967 Sam K. Smyth of Kennco had completed preliminary reserve estimates on the Berg Deposit showing a mineable open-pit reserve of 74 million tons grading 0.48% copper and 0.065% molybdenite using a 0.4% copper cut-off and 45 degree pit slopes for a stripping ratio of 1.1 tons of waster per ton of ore. Following additional diamond drilling in the northeast part of the Berg deposit, Carl C. Kamm of Kennco calculated a strikingly similar reserve in 1971 totalling 75 million tons grading 0.51% copper, 0.061% molybdenite using a 0.3% copper cut-off and 45 degree pit slopes for a stripping ratio of 1.0 tons of waste per ton of ore.

[xvi] Major deformation has affected the ore-bearing formations and ore deposits commonly occur on the limbs of folded structures, marginal to granitic domes reflecting irregularities in the basement. The ore-bearing formations in Zambia comprise a sequence of argillites or shales, quartzites and impure dolomites with thicknesses ranging from zero to about 70 metres.

[xvii] By 1970 it had been explored by diamond drilling and underground developments with about 100 million tons grading 0.4% copper, 0.2% nickel and recoverable precious metal credits indicated in a tabular zone with a crescentic cross-section, up to 30 metres thick, 500 metres wide and traced for a length of about 3200 metres, still open on strike at depth.

[xviii] Earlier work by the Geological Survey of Canada, substantiated by Kennco's observations, indicated that the intrusion was a layered complex. "An inward dipping group of olivine gabbros at the margin of the intrusion is succeeded inwards by a sequence of relatively flat-lying feldspar-rich cumulates including leucogabbro, leucotroctolite and anorthosite. The overall geometry of the intrusion is considered to be funnel-shaped. Late block faulting has affected the Harp Lake mass with maximum displacements of 700 metres."

[xix] Dave McAuslan's write-up in the 1971 Annual Report concluded:

"The best indication of grade obtained in the 1971 program consists of a layer of disseminated pyrrhotite-chalcopyrite mineralization which assayed 0.6% copper, 0.08% nickel, 0.03% cobalt across a thickness of four feet, adjacent to a zone 21 feet thick grading 0.29% Cu, 0.05% Ni. All surface showings in areas investigated show evidence of surface leaching so that a reliable indication of grade is not possible from surface sampling alone. Selected samples assayed for Pt and Pd returned low values, the best being 130 ppb Pt, 70 ppb Pd. Magnetite-rich outcrop and magnetite-ilmenite float were analyzed for TiO_2 and V_2O_5, the best results being 16.5% TiO_2 and 0.57% V_2O_5 on separate samples."

Printed in the United States
57915LVS00002B/3-6